S0-CBI-350

Methods in Enzymology

Volume 359

NITRIC OXIDE

Part D

Nitric Oxide Detection, Mitochondria and Cell Functions, and Peroxynitrite Reactions

METHODS IN ENZYMOLOGY

EDITORS-IN-CHIEF

John N. Abelson   Melvin I. Simon

DIVISION OF BIOLOGY
CALIFORNIA INSTITUTE OF TECHNOLOGY
PASADENA, CALIFORNIA

FOUNDING EDITORS

Sidney P. Colowick and Nathan O. Kaplan

Methods in Enzymology

Volume 359

Nitric Oxide

Part D

Nitric Oxide Detection, Mitochondria and Cell Functions, and Peroxynitrite Reactions

EDITED BY

Enrique Cadenas

SCHOOL OF PHARMACY
UNIVERSITY OF SOUTHERN CALIFORNIA
LOS ANGELES, CALIFORNIA

Lester Packer

SCHOOL OF PHARMACY
UNIVERSITY OF SOUTHERN CALIFORNIA
LOS ANGELES, CALIFORNIA

ACADEMIC PRESS

An imprint of Elsevier Science

Amsterdam Boston London New York Oxford Paris
San Diego San Francisco Singapore Sydney Tokyo

This book is printed on acid-free paper. ∞

Academic Press
An imprint of Elsevier Science.
525 B Street, Suite 1900, San Diego, California 92101-4495, USA
http://www.academicpress.com

Academic Press
84 Theobalds Road, London WC1X 8RR, UK
http://www.academicpress.com

International Standard Book Number: 0-12-182262-1

PRINTED IN THE UNITED STATES OF AMERICA
02 03 04 05 06 07 MM 9 8 7 6 5 4 3 2 1

Table of Contents

Section I. Detection of Nitric Oxide

Section II. Nitrosothiols and Nitric Oxide in Cell Signaling

Section III. Nitric Oxide and Mitochondrial Functions

Section IV. Peroxynitrite

Section V. Nitric Oxide Synthases

Contributors to Volume 359

Article numbers are in parentheses following the names of contributors.
Affiliations listed are current.

BARRY W. ALLEN (11), *Department of Anesthesiology, Duke Center for Hyperbaric Medicine and Environmental Physiology, Duke University Medical Center, Durham, North Carolina 27710*

RICHARD R. ALMON (40), *Department of Biological Sciences, State University of New York, Buffalo, New York 14260*

MARÍA NOEL ALVAREZ (32), *Department of Biochemistry, Universidad de la República, 11800 Montevideo, Uruguay*

SILVIA ALVAREZ (30), *Laboratory of Free Radical Biology, School of Pharmacy and Biochemistry, University of Buenos Aires, Buenos Aires, Argentina*

GIUSEPPE ARIENTI (21), *Department of Internal Medicine, Università di Perugia, 06122 Perugia, Italy*

SILVIA LORES ARNAIZ (30), *Laboratory of Free Radical Biology, School of Pharmacy and Biochemistry, University of Buenos Aires, Buenos Aires, Argentina*

GILBERT G. A. BALAVOINE (33), *Laboratory of Coordination Chemistry, CNRS, F-31077 Toulouse Cedex, France*

MICHAEL BALAZY (35), *Department of Pharmacology, New York Medical College, Valhalla, New York 10595*

RUI M. BARBOSA (10), *Laboratory of Instrumental Analysis, Center for Neurosciences, University of Coimbra, 3000 Coimbra, Portugal*

CARLOS BATTHYÁNY (18), *Department of Biochemistry, Universidad de la República, 11800 Montevideo, Uruguay*

ERIC BLASKO (39), *Department of Cardiovascular Research, Berlex Biosciences, Richmond, California 94804*

LISARDO BOSCÁ (42), *Institute of Biochemistry, Universidad Complutense, 28040 Madrid, Spain*

HORACIO BOTTI (18), *Department of Biochemistry, Universidad de la República, 11800 Montevideo, Uruguay*

ALBERTO BOVERIS (30), *Laboratory of Free Radical Biology, School of Pharmacy and Biochemistry, University of Buenos Aires, Buenos Aires, Argentina*

ALEJANDRO D. BOVERIS (30), *Laboratory of Free Radical Biology, School of Pharmacy and Biochemistry, University of Buenos Aires, Buenos Aires, Argentina*

MARIE-CHRISTINE BROILLET (12), *Institute of Pharmacology and Toxicology, University of Lausanne, CH-1005 Lausanne, Switzerland*

PAUL S. BROOKES (28), *Department of Pathology, University of Alabama, Birmingham, Alabama 35294*

UWE B. BRUCKNER (7), *Division of Surgical Research, Department of Surgery, University Medical School, University of Ulm, 89073 Ulm, Germany*

GARRY R. BUETTNER (1), *Free Radical and Radiation Biology/ESR Facility, University of Iowa, Iowa City, Iowa 52242*

JUANITA BUSTAMANTE (30), *Laboratory of Free Radical Biology, School of Pharmacy and Biochemistry, University of Buenos Aires, Buenos Aires, Argentina*

MARÍA CECILIA CARRERAS (37), *Laboratory of Oxygen Metabolism and Department of Clinical Biochemistry, University Hospital, University of Buenos Aires, 1120 Buenos Aires, Argentina*

JEAN-YVES CHATTON (12), *Institute of Physiology, University of Lausanne, CH-1005 Lausanne, Switzerland*

YI CHEN (41), *Department of Physiology, School of Medicine, Tokai University, Isehara 259-1193, Kanagawa, Japan*

DEBORAH CITRIN (8), *Tumor Biology Section, Radiation Biology Branch, National Institutes of Health/National Cancer Institute, Bethesda, Maryland 20892*

CAROL A. COLTON (8), *Division of Neurology, Duke University Medical Center, Durham, North Carolina 27710*

DANIELA CONVERSO (37), *Laboratory of Oxygen Metabolism, University Hospital, University of Buenos Aires, 1120 Buenos Aires, Argentina*

LOUIS A. COURY, JR. (11), *Analytical Development, Eli Lilly and Company, Tippecanoe Laboratories, Lafayette, Indiana 47909*[*]

ANDREAS DAIBER (34), *Department of Cardiology, University Hospital Hamburg-Eppendorf, 20246, Hamburg, Germany*

VICTOR M. DARLEY-USMAR (28), *Department of Pathology, University of Alabama, Birmingham, Alabama 35294*

TARA P. DASGUPTA (20), *Department of Chemistry, University of the West Indies, Mona Campus, Kingston 7, Jamaica*

ANDRÉ DEJAM (14), *Department of Medicine, Heinrich-Heine-Universität, D-40225 Düsseldorf, Germany*

ANA DENICOLA (18), *Department of Biochemistry, Universidad de la República, 11800 Montevideo, Uruguay*

DEBRA C. DUBOIS (40), *Department of Biological Sciences, State University of New York, Buffalo, New York 14260*

MICHAEL GRAHAM ESPEY (8), *Tumor Biology Section, Radiation Biology Branch, National Institutes of Health/National Cancer Institute, Bethesda, Maryland 20892*

MARTIN FEELISCH (8), *Department of Molecular and Cellular Physiology, Louisiana State University Medical Center, Shreveport, Louisiana 71130*

JOÃO FRADE (10), *Center for Neurosciences, University of Coimbra, 3000 Coimbra, Portugal*

EMILIA M. GATTO (37), *Abnormal Movement Section, University Hospital, University of Buenos Aires, 1120 Buenos Aires, Argentina*

YURII V. GELETII (33), *Institute of Problems of Chemical Physics RAS, Chernogolovka, Moscow Region, 14243 Russia*

PEDRAM GHAFOURIFAR (31), *Department of Pharmacology and Therapeutics, Louisiana State University Health Sciences Center, Shreveport, Louisiana 71130*

MATTHEW B. GRISHAM (8), *Department of Molecular and Cellular Physiology, Louisiana State University Medical Center, Shreveport, Louisiana 71130*

MASANORI HATASHITA (26), *Departments of Experimental Radiology and Health Physics and Otorhinolaryngology, Fukui Medical University, Matsuoka, Fukui 910-1193, Japan*

SACHIKO HAYASHI (26), *Departments of Experimental Radiology and Health Physics and Otorhinolaryngology, Fukui Medical University, Matsuoka, Fukui 910-1193, Japan*

[*]When this work was done Dr. Coury was at the Department of Chemistry, Duke University Medical Center, Durham, North Carolina 27710.

SANDRA J. HEWETT (8), *Department of Neuroscience, University of Connecticut Health Center, Farmington, Connecticut 06030*

SONSOLES HORTELANO (42), *Institute of Biochemistry, Universidad Complutense, 28040 Madrid, Spain*

JOHN S. HOTHERSALL (22), *Institute of Urology and Nephrology, University College London, London W1P 7EJ, United Kingdom*

ANNONG HUANG (19), *Evans Memorial Department of Medicine and Whitaker Cardiovascular Institute, Boston University School of Medicine, Boston, Massachusetts 02118*

ALBERT HUISMAN (3), *Department of Clinical Chemistry, University Medical Center, Utrecht, The Netherlands*

TAKANOBU ISHIDA (15), *Department of Chemistry, State University of New York, Stony Brook, New York 11794*

KAZUO ISHIWATA (41), *Department of Physiology, School of Medicine, Tokai University, Isehara 259-1193, Kanagawa, Japan*

PHILIP E. JAMES (5), *Department of Cardiology, Wales Heart Research Institute, University of Wales College of Medicine, Cardiff CF14 4XN, Wales, United Kingdom*

ZHAO-HUI JIN (26), *Departments of Experimental Radiology and Health Physics and Otorhinolaryngology, Fukui Medical University, Matsuoka, Fukui 910-1193, Japan*

DAVID JOURD'HEUIL (8), *Center for Cardiovascular Sciences, Albany Medical College, Albany, New York 12208*

EIICHI KANO (26), *Departments of Experimental Radiology and Health Physics, Fukui Medical University, Matsuoka, Fukui 910-1193, Japan*

KATALIN KAUSER (39), *Department of Cardiovascular Research, Berlex Biosciences, Richmond, California 94804*

JOHN F. KEANEY, JR. (19), *Evans Memorial Department of Medicine and Whitaker Cardiovascular Institute, Boston University School of Medicine, Boston, Massachusetts 02118*

MALTE KELM (14), *Department of Medicine, Heinrich-Heine-Universität, D-40225 Düsseldorf, Germany*

STEFAN KERBER (14), *Department of Medicine, Heinrich-Heine-Universität, D-40225 Düsseldorf, Germany*

PETER KLATT (23), *Department of Immunology and Oncology, Centro Nacional de Biotecnología, Campus of Cantoblanco, 28049 Madrid, Spain*

PETRA KLEINBONGARD (14), *Department of Medicine, Heinrich-Heine-Universität, D-40225 Düsseldorf, Germany*

ANDREI L. KLESCHYOV (4), *Division of Cardiology, University Hospital Hamburg-Eppendorf, 20246 Hamburg, Germany*

ANDREI M. KOMAROV (2, 6), *Department of Physiology and Experimental Medicine, George Washington University Medical Center, Washington, DC 20037*

SANTIAGO LAMAS (23, 25), *Department of Structure and Function of Proteins, Centro de Investigaciones Biológicas, Instituto Reina Sofía de Investigaciones Nefrológicas, E-28006 Madrid, Spain*

JOÃO LARANJINHA (10), *Laboratory of Biochemistry, Center for Neurosciences, University of Coimbra, 3000 Coimbra, Portugal*

ANA LEDO (10), *Center for Neurosciences, University of Coimbra, 3000 Coimbra, Portugal*

DIANA A. LEPORE (16), *Pituitary Research Unit, Murdoch Children's Research Institute, Royal Children's Hospital, Parkville, Victoria, Australia 3052*

CHUN-QI LI (29), *Biological Engineering Division, Massachusetts Institute of Technology, Cambridge, Massachusetts 02139*

ALI R. MANI (24), *Center for Hepatology, Department of Medicine, Royal Free and University College Medical School, University College London, London NW3 2PF, United Kingdom*

SEAN M. MARTIN (1), *Free Radical and Radiation Biology/ESR Facility, University of Iowa, Iowa City, Iowa 52242*

GABRIJELA MATANOVIC (13), *Biomedical Mass Spectrometry Facility, University of New South Wales, Sydney NSW 2052, Australia*

HIDEKI MATSUMOTO (26), *Departments of Experimental Radiology and Health Physics, Fukui Medical University, Matsuoka, Fukui 910-1193, Japan*

MITSUFUMI MAYUMI (15), *Department of Pediatrics, Fukui Medical University, Fukui 910-1193, Japan*

MARIANA MELANI (37), *Laboratory of Oxygen Metabolism, University Hospital, University of Buenos Aires, 1120 Buenos Aires, Argentina*

JANET MEURER (39), *Department of Molecular Pharmacology, Berlex Biosciences, Richmond, California 94804*

KATRINA M. MIRANDA (8), *Tumor Biology Section, Radiation Biology Branch, National Institutes of Health/National Cancer Institute, Bethesda, Maryland 20892*

FUMIO MIZUTANI (9), *Biosensing Technology Group, National Institute of Advanced Industrial Science and Technology, Tsukuba, Ibaraki 305-8566, Japan*

KEVIN P. MOORE (24), *Center for Hepatology, Department of Medicine, Royal Free and University College Medical School, University College London, London NW3 2PF, United Kingdom*

THOMAS MÜNZEL (4), *Division of Cardiology, University Hospital Hamburg-Eppendorf, 20246 Hamburg, Germany*

HIROE NAKAZAWA (41), *Department of Physiology, School of Medicine, Tokai University, Isehara 259-1193, Kanagawa, Japan*

ANA NAVARRO (30), *Department of Biochemistry and Molecular Biology, University of Cádiz, 11003 Cádiz, Spain*

YEE KONG NG (38), *Department of Anatomy, National University of Singapore, Kent Ridge, Singapore*

XI-LIN NIU (41), *Department of Medicine, Duke University Medical Center, Durham, North Carolina 27705*

ALBERTO A. NORONHA-DUTRA (22), *Institute of Urology and Nephrology, University College London, London W1P 7EJ, United Kingdom*

ANDREAS K. NUSSLER (7), *Department of Surgery, Humboldt University of Berlin, Campus Virchow, Campus Charité, 13353 Berlin, Germany*

RURIKO OBAMA (41), *Department of Neurology, School of Medicine, Tokai University, Isehara 259-1193, Kanagawa, Japan*

TAKEO OHNISHI (26), *Department of Biology, Nara Medical University, Kashihara, Nara 634-8521, Japan*

TOSHIO OHTSUBO (26), *Departments of Experimental Radiology and Health Physics and Otorhinolaryngology, Fukui Medical University, Matsuoko, Fukui 910-1193, Japan*

ANN ORME (39), *Gene Therapy Research, Berlex Biosciences, Richmond, California 94804*

CARLO A. PALMERINI (21), *Department of Biochemical Science and Molecular Biotechnology, Università di Perugia, 06122 Perugia, Italy*

ROBERTO PALOMBARI (21), *Department of Chemistry, Università di Perugia, 06122 Perugia, Italy*

NAZARENO PAOLOCCI (8), *Division of Cardiology, Departments of Medicine and Biomedical Engineering, Johns Hopkins Medical Institutions, Baltimore, Maryland 21287*

RAKESH P. PATEL (28), *Department of Pathology, University of Alabama, Birmingham, Alabama 35294*

GONZALO PELUFFO (27), *Department of Biochemistry, Universidad de la República, 11800 Montevideo, Uruguay*

LUCÍA PIACENZA (27), *Department of Biochemistry, Universidad de la República, 11800 Montevideo, Uruguay*

CLAUDE A. PIANTADOSI (11), *Departments of Medicine and Anesthesiology, Duke Center for Hyperbaric and Environmental Physiology, Duke University Medical Center, Durham, North Carolina 27710*

ESTELA PINEDA-MOLINA (25), *Department of Structure and Function of Proteins, Centro de Investigaciones Biológicas, Instituto Reina Sofía de Investigaciones Nefrológicas, E-28006 Madrid, Spain*

JUAN JOSÉ PODEROSO (37), *Laboratory of Oxygen Metabolism, University Hospital, University of Buenos Aires, 1120 Buenos Aires, Argentina*

PETER RADERMACHER (7), *Division of Pathophysiology and Process Development, Department of Anesthesiology, University Medical School, University of Ulm, 89073 Ulm, Germany*

RAFAEL RADI (18, 27, 32), *Department of Biochemistry, Universidad de la República, 11800 Montevideo, Uruguay*

TIENUSH RASSAF (14), *Department of Medicine, Heinrich-Heine-Universität, D-40225 Düsseldorf, Germany*

ANDREAS REIF (2), *Department of Psychiatry and Psychotherapy, University of Würzburg, D-97080 Würzburg, Germany*

NATALIA RIOBÓ (37), *Department of Cell and Developmental Biology, University of Pennsylvania, Philadelphia, Pennsylvania 19104*

HOMERO RUBBO (18), *Department of Biochemistry, Universidad de la República, 11800 Montevideo, Uruguay*

HARALD H. H. W. SCHMIDT (2), *Justus-Liebig University, Rudolf Buchheim Institute for Pharmacology, D-35392 Giessen, Germany*

SRUTI SHIVA (28), *Department of Pathology, University of Alabama, Birmingham, Alabama 35294*

YUTAKA SHOYAMA (41), *Department of Physiology, School of Medicine, Tokai University, Isehara 259-1193, Kanagawa, Japan*

JAMES N. SMITH (20), *University of Technology, Kingston 6, Jamaica*

GEORGE A. SMYTHE (13), *Biomedical Mass Spectrometry Facility, University of New South Wales, Sydney NSW 2052, Australia*

HAROLD M. SWARTZ (5), *EPR Center, Department of Radiology, Dartmouth Medical School, Hanover, New Hampshire 03755*

STEVEN R. TANNENBAUM (29), *Biological Engineering Division and Department of Chemistry, Massachusetts Institute of Technology, Cambridge, Massachusetts 02139*

DOUGLAS D. THOMAS (8), *Tumor Biology Section, Radiation Biology Branch, National Institutes of Health/National Cancer Institute, Bethesda, Maryland 20892*

SHANE R. THOMAS (19), *Evans Memorial Department of Medicine and Whitaker Cardiovascular Institute, Boston University School of Medicine, Boston, Massachusetts 02118*

ANDRÉS TROSTCHANSKY (18), *Department of Biochemistry, Universidad de la República, 11800 Montevideo, Uruguay*

LAURA J. TRUDEL (29), *Biological Engineering Division, Massachusetts Institute of Technology, Cambridge, Massachusetts 02139*

MADIA TRUJILLO (32), *Department of Biochemistry, Universidad de la República, 11800 Montevideo, Uruguay*

HIROKAZU TSUKAHARA (15), *Department of Pediatrics, Fukui Medical University, Fukui 910-1193, Japan*

VOLKER ULLRICH (34), *Department of Biology, Universität Konstanz, 78457 Konstanz, Germany*

LAURA VALDEZ (30), *Laboratory of Free Radical Biology, School of Pharmacy and Biochemistry, 1113 Buenos Aires, Argentina*

ALBERT VAN DER VLIET (36), *Department of Pathology, University of Vermont, Burlington, Vermont 05405*

ERNST E. VAN FAASSEN (3), *Debye Institute, Section of Interface Physics, Utrecht University, Utrecht, The Netherlands*

ANATOLY F. VANIN (3), *Institute of Chemical Physics, Russian Academy of Sciences, 119991 Moscow, Russia*

SUJATHA VENKATARAMAN (1), *Free Radical and Radiation Biology/ESR Facility, University of Iowa, Iowa City, Iowa 52242*

JOSEPH A. VITA (17), *Section of Cardiology, Boston Medical Center, Boston, Massachusetts 02118*

JOSEF VOGT (7), *Division of Pathophysiology and Process Development, Department of Anesthesiology, University Medical School, University of Ulm, 89073 Ulm, Germany*

DAVID A. WINK (8), *Tumor Biology Section, Radiation Biology Branch, National Institutes of Health/National Cancer Institute, Bethesda, Maryland 20892*

GERALD N. WOGAN (29), *Biological Engineering Division and Department of Chemistry, Massachusetts Institute of Technology, Cambridge, Massachusetts 02139*

PATRICK S.-Y. WONG (36), *Department of Internal Medicine, University of California, Davis, California 95616*

PETER T.-H. WONG (38), *Department of Pharmacology, National University of Singapore, Kent Ridge, Singapore*

TERESA L. WRIGHT (29), *Department of Pharmacology, Surgery, and Toxicology, Charles River Laboratories Discover and Development Services, Worcester, Massachusetts 01608*

MEI XU (38), *Department of Biology, Georgia State University, Atlanta, Georgia 30302*

MIRIAM ZEINI (42), *Institute of Biochemistry, Universidad Complutense, 28040 Madrid, Spain*

Preface

The discovery that nitrogen monoxide or nitric oxide (NO) is a free radical formed in a variety of cell types by nitric oxide synthase and is involved in a wide array of physiological and pathophysiological phenomena has ignited enormous interest in the scientific community. One of the unique features of nitric oxide is its function as an intercellular messenger and, in this capacity, its involvement in the modulation of cell signaling and mitochondrial respiration. Nitric oxide metabolism and the interactions of this molecule with multiple cellular targets are currently areas of intensive research and have important pharmacological implications for health and disease.

Accurately assessing the generation, action, and regulation of nitric oxide in biological systems has required the development of new analytical methods at the molecular, cellular, tissue, and organismal levels. This was the impetus for *Methods in Enzymology* Volumes 268, 269, and 301, Nitric Oxide Parts A, B, and C, respectively. Only a few years later, this new Volume 359 reflects the amazing development of new and important tools for the assessment of nitric oxide action. Nitric Oxide, Part D contains five major sections: Detection of Nitric Oxide, Nitrosothiols and Nitric Oxide in Cell Signaling, Nitric Oxide and Mitochondrial Functions, Peroxynitrite, and Nitric Oxide Synthases.

In bringing this volume to fruition, credit must be given to the experts in various specialized fields of nitric oxide research who have contributed outstanding chapters to these sections on nitric oxide methodology. To these colleagues, we extend our sincere thanks and most grateful appreciation.

ENRIQUE CADENAS
LESTER PACKER

METHODS IN ENZYMOLOGY

VOLUME I. Preparation and Assay of Enzymes
Edited by SIDNEY P. COLOWICK AND NATHAN O. KAPLAN

VOLUME II. Preparation and Assay of Enzymes
Edited by SIDNEY P. COLOWICK AND NATHAN O. KAPLAN

VOLUME III. Preparation and Assay of Substrates
Edited by SIDNEY P. COLOWICK AND NATHAN O. KAPLAN

VOLUME IV. Special Techniques for the Enzymologist
Edited by SIDNEY P. COLOWICK AND NATHAN O. KAPLAN

VOLUME V. Preparation and Assay of Enzymes
Edited by SIDNEY P. COLOWICK AND NATHAN O. KAPLAN

VOLUME VI. Preparation and Assay of Enzymes (*Continued*)
Preparation and Assay of Substrates
Special Techniques
Edited by SIDNEY P. COLOWICK AND NATHAN O. KAPLAN

VOLUME VII. Cumulative Subject Index
Edited by SIDNEY P. COLOWICK AND NATHAN O. KAPLAN

VOLUME VIII. Complex Carbohydrates
Edited by ELIZABETH F. NEUFELD AND VICTOR GINSBURG

VOLUME IX. Carbohydrate Metabolism
Edited by WILLIS A. WOOD

VOLUME X. Oxidation and Phosphorylation
Edited by RONALD W. ESTABROOK AND MAYNARD E. PULLMAN

VOLUME XI. Enzyme Structure
Edited by C. H. W. HIRS

VOLUME XII. Nucleic Acids (Parts A and B)
Edited by LAWRENCE GROSSMAN AND KIVIE MOLDAVE

VOLUME XIII. Citric Acid Cycle
Edited by J. M. LOWENSTEIN

VOLUME XIV. Lipids
Edited by J. M. LOWENSTEIN

VOLUME XV. Steroids and Terpenoids
Edited by RAYMOND B. CLAYTON

VOLUME XVI. Fast Reactions
Edited by KENNETH KUSTIN

VOLUME XXXVI. Hormone Action (Part A: Steroid Hormones)
Edited by BERT W. O'MALLEY AND JOEL G. HARDMAN

VOLUME XXXVII. Hormone Action (Part B: Peptide Hormones)
Edited by BERT W. O'MALLEY AND JOEL G. HARDMAN

VOLUME XXXVIII. Hormone Action (Part C: Cyclic Nucleotides)
Edited by JOEL G. HARDMAN AND BERT W. O'MALLEY

VOLUME XXXIX. Hormone Action (Part D: Isolated Cells, Tissues, and Organ Systems)
Edited by JOEL G. HARDMAN AND BERT W. O'MALLEY

VOLUME XL. Hormone Action (Part E: Nuclear Structure and Function)
Edited by BERT W. O'MALLEY AND JOEL G. HARDMAN

VOLUME XLI. Carbohydrate Metabolism (Part B)
Edited by W. A. WOOD

VOLUME XLII. Carbohydrate Metabolism (Part C)
Edited by W. A. WOOD

VOLUME XLIII. Antibiotics
Edited by JOHN H. HASH

VOLUME XLIV. Immobilized Enzymes
Edited by KLAUS MOSBACH

VOLUME XLV. Proteolytic Enzymes (Part B)
Edited by LASZLO LORAND

VOLUME XLVI. Affinity Labeling
Edited by WILLIAM B. JAKOBY AND MEIR WILCHEK

VOLUME XLVII. Enzyme Structure (Part E)
Edited by C. H. W. HIRS AND SERGE N. TIMASHEFF

VOLUME XLVIII. Enzyme Structure (Part F)
Edited by C. H. W. HIRS AND SERGE N. TIMASHEFF

VOLUME XLIX. Enzyme Structure (Part G)
Edited by C. H. W. HIRS AND SERGE N. TIMASHEFF

VOLUME L. Complex Carbohydrates (Part C)
Edited by VICTOR GINSBURG

VOLUME LI. Purine and Pyrimidine Nucleotide Metabolism
Edited by PATRICIA A. HOFFEE AND MARY ELLEN JONES

VOLUME LII. Biomembranes (Part C: Biological Oxidations)
Edited by SIDNEY FLEISCHER AND LESTER PACKER

VOLUME LIII. Biomembranes (Part D: Biological Oxidations)
Edited by SIDNEY FLEISCHER AND LESTER PACKER

VOLUME LIV. Biomembranes (Part E: Biological Oxidations)
Edited by SIDNEY FLEISCHER AND LESTER PACKER

VOLUME LV. Biomembranes (Part F: Bioenergetics)
Edited by SIDNEY FLEISCHER AND LESTER PACKER

VOLUME LVI. Biomembranes (Part G: Bioenergetics)
Edited by SIDNEY FLEISCHER AND LESTER PACKER

VOLUME LVII. Bioluminescence and Chemiluminescence
Edited by MARLENE A. DELUCA

VOLUME LVIII. Cell Culture
Edited by WILLIAM B. JAKOBY AND IRA PASTAN

VOLUME LIX. Nucleic Acids and Protein Synthesis (Part G)
Edited by KIVIE MOLDAVE AND LAWRENCE GROSSMAN

VOLUME LX. Nucleic Acids and Protein Synthesis (Part H)
Edited by KIVIE MOLDAVE AND LAWRENCE GROSSMAN

VOLUME 61. Enzyme Structure (Part H)
Edited by C. H. W. HIRS AND SERGE N. TIMASHEFF

VOLUME 62. Vitamins and Coenzymes (Part D)
Edited by DONALD B. MCCORMICK AND LEMUEL D. WRIGHT

VOLUME 63. Enzyme Kinetics and Mechanism (Part A: Initial Rate and Inhibitor Methods)
Edited by DANIEL L. PURICH

VOLUME 64. Enzyme Kinetics and Mechanism (Part B: Isotopic Probes and Complex Enzyme Systems)
Edited by DANIEL L. PURICH

VOLUME 65. Nucleic Acids (Part I)
Edited by LAWRENCE GROSSMAN AND KIVIE MOLDAVE

VOLUME 66. Vitamins and Coenzymes (Part E)
Edited by DONALD B. MCCORMICK AND LEMUEL D. WRIGHT

VOLUME 67. Vitamins and Coenzymes (Part F)
Edited by DONALD B. MCCORMICK AND LEMUEL D. WRIGHT

VOLUME 68. Recombinant DNA
Edited by RAY WU

VOLUME 69. Photosynthesis and Nitrogen Fixation (Part C)
Edited by ANTHONY SAN PIETRO

VOLUME 70. Immunochemical Techniques (Part A)
Edited by HELEN VAN VUNAKIS AND JOHN J. LANGONE

VOLUME 71. Lipids (Part C)
Edited by JOHN M. LOWENSTEIN

VOLUME 72. Lipids (Part D)
Edited by JOHN M. LOWENSTEIN

VOLUME 73. Immunochemical Techniques (Part B)
Edited by JOHN J. LANGONE AND HELEN VAN VUNAKIS

VOLUME 74. Immunochemical Techniques (Part C)
Edited by JOHN J. LANGONE AND HELEN VAN VUNAKIS

VOLUME 75. Cumulative Subject Index Volumes XXXI, XXXII, XXXIV–LX
Edited by EDWARD A. DENNIS AND MARTHA G. DENNIS

VOLUME 76. Hemoglobins
Edited by ERALDO ANTONINI, LUIGI ROSSI-BERNARDI, AND EMILIA CHIANCONE

VOLUME 77. Detoxication and Drug Metabolism
Edited by WILLIAM B. JAKOBY

VOLUME 78. Interferons (Part A)
Edited by SIDNEY PESTKA

VOLUME 79. Interferons (Part B)
Edited by SIDNEY PESTKA

VOLUME 80. Proteolytic Enzymes (Part C)
Edited by LASZLO LORAND

VOLUME 81. Biomembranes (Part H: Visual Pigments and Purple Membranes, I)
Edited by LESTER PACKER

VOLUME 82. Structural and Contractile Proteins (Part A: Extracellular Matrix)
Edited by LEON W. CUNNINGHAM AND DIXIE W. FREDERIKSEN

VOLUME 83. Complex Carbohydrates (Part D)
Edited by VICTOR GINSBURG

VOLUME 84. Immunochemical Techniques (Part D: Selected Immunoassays)
Edited by JOHN J. LANGONE AND HELEN VAN VUNAKIS

VOLUME 85. Structural and Contractile Proteins (Part B: The Contractile Apparatus and the Cytoskeleton)
Edited by DIXIE W. FREDERIKSEN AND LEON W. CUNNINGHAM

VOLUME 86. Prostaglandins and Arachidonate Metabolites
Edited by WILLIAM E. M. LANDS AND WILLIAM L. SMITH

VOLUME 87. Enzyme Kinetics and Mechanism (Part C: Intermediates, Stereochemistry, and Rate Studies)
Edited by DANIEL L. PURICH

VOLUME 88. Biomembranes (Part I: Visual Pigments and Purple Membranes, II)
Edited by LESTER PACKER

VOLUME 89. Carbohydrate Metabolism (Part D)
Edited by WILLIS A. WOOD

VOLUME 90. Carbohydrate Metabolism (Part E)
Edited by WILLIS A. WOOD

VOLUME 91. Enzyme Structure (Part I)
Edited by C. H. W. HIRS AND SERGE N. TIMASHEFF

VOLUME 92. Immunochemical Techniques (Part E: Monoclonal Antibodies and General Immunoassay Methods)
Edited by JOHN J. LANGONE AND HELEN VAN VUNAKIS

VOLUME 93. Immunochemical Techniques (Part F: Conventional Antibodies, Fc Receptors, and Cytotoxicity)
Edited by JOHN J. LANGONE AND HELEN VAN VUNAKIS

VOLUME 94. Polyamines
Edited by HERBERT TABOR AND CELIA WHITE TABOR

VOLUME 95. Cumulative Subject Index Volumes 61–74, 76–80
Edited by EDWARD A. DENNIS AND MARTHA G. DENNIS

VOLUME 96. Biomembranes [Part J: Membrane Biogenesis: Assembly and Targeting (General Methods; Eukaryotes)]
Edited by SIDNEY FLEISCHER AND BECCA FLEISCHER

VOLUME 97. Biomembranes [Part K: Membrane Biogenesis: Assembly and Targeting (Prokaryotes, Mitochondria, and Chloroplasts)]
Edited by SIDNEY FLEISCHER AND BECCA FLEISCHER

VOLUME 98. Biomembranes (Part L: Membrane Biogenesis: Processing and Recycling)
Edited by SIDNEY FLEISCHER AND BECCA FLEISCHER

VOLUME 99. Hormone Action (Part F: Protein Kinases)
Edited by JACKIE D. CORBIN AND JOEL G. HARDMAN

VOLUME 100. Recombinant DNA (Part B)
Edited by RAY WU, LAWRENCE GROSSMAN, AND KIVIE MOLDAVE

VOLUME 101. Recombinant DNA (Part C)
Edited by RAY WU, LAWRENCE GROSSMAN, AND KIVIE MOLDAVE

VOLUME 102. Hormone Action (Part G: Calmodulin and Calcium-Binding Proteins)
Edited by ANTHONY R. MEANS AND BERT W. O'MALLEY

VOLUME 103. Hormone Action (Part H: Neuroendocrine Peptides)
Edited by P. MICHAEL CONN

VOLUME 104. Enzyme Purification and Related Techniques (Part C)
Edited by WILLIAM B. JAKOBY

VOLUME 105. Oxygen Radicals in Biological Systems
Edited by LESTER PACKER

VOLUME 106. Posttranslational Modifications (Part A)
Edited by FINN WOLD AND KIVIE MOLDAVE

VOLUME 107. Posttranslational Modifications (Part B)
Edited by FINN WOLD AND KIVIE MOLDAVE

VOLUME 143. Sulfur and Sulfur Amino Acids
Edited by WILLIAM B. JAKOBY AND OWEN GRIFFITH

VOLUME 144. Structural and Contractile Proteins (Part D: Extracellular Matrix)
Edited by LEON W. CUNNINGHAM

VOLUME 145. Structural and Contractile Proteins (Part E: Extracellular Matrix)
Edited by LEON W. CUNNINGHAM

VOLUME 146. Peptide Growth Factors (Part A)
Edited by DAVID BARNES AND DAVID A. SIRBASKU

VOLUME 147. Peptide Growth Factors (Part B)
Edited by DAVID BARNES AND DAVID A. SIRBASKU

VOLUME 148. Plant Cell Membranes
Edited by LESTER PACKER AND ROLAND DOUCE

VOLUME 149. Drug and Enzyme Targeting (Part B)
Edited by RALPH GREEN AND KENNETH J. WIDDER

VOLUME 150. Immunochemical Techniques (Part K: *In Vitro* Models of B and T Cell Functions and Lymphoid Cell Receptors)
Edited by GIOVANNI DI SABATO

VOLUME 151. Molecular Genetics of Mammalian Cells
Edited by MICHAEL M. GOTTESMAN

VOLUME 152. Guide to Molecular Cloning Techniques
Edited by SHELBY L. BERGER AND ALAN R. KIMMEL

VOLUME 153. Recombinant DNA (Part D)
Edited by RAY WU AND LAWRENCE GROSSMAN

VOLUME 154. Recombinant DNA (Part E)
Edited by RAY WU AND LAWRENCE GROSSMAN

VOLUME 155. Recombinant DNA (Part F)
Edited by RAY WU

VOLUME 156. Biomembranes (Part P: ATP-Driven Pumps and Related Transport: The Na, K-Pump)
Edited by SIDNEY FLEISCHER AND BECCA FLEISCHER

VOLUME 157. Biomembranes (Part Q: ATP-Driven Pumps and Related Transport: Calcium, Proton, and Potassium Pumps)
Edited by SIDNEY FLEISCHER AND BECCA FLEISCHER

VOLUME 158. Metalloproteins (Part A)
Edited by JAMES F. RIORDAN AND BERT L. VALLEE

VOLUME 159. Initiation and Termination of Cyclic Nucleotide Action
Edited by JACKIE D. CORBIN AND ROGER A. JOHNSON

VOLUME 160. Biomass (Part A: Cellulose and Hemicellulose)
Edited by WILLIS A. WOOD AND SCOTT T. KELLOGG

VOLUME 161. Biomass (Part B: Lignin, Pectin, and Chitin)
Edited by WILLIS A. WOOD AND SCOTT T. KELLOGG

VOLUME 162. Immunochemical Techniques (Part L: Chemotaxis and Inflammation)
Edited by GIOVANNI DI SABATO

VOLUME 163. Immunochemical Techniques (Part M: Chemotaxis and Inflammation)
Edited by GIOVANNI DI SABATO

VOLUME 164. Ribosomes
Edited by HARRY F. NOLLER, JR., AND KIVIE MOLDAVE

VOLUME 165. Microbial Toxins: Tools for Enzymology
Edited by SIDNEY HARSHMAN

VOLUME 166. Branched-Chain Amino Acids
Edited by ROBERT HARRIS AND JOHN R. SOKATCH

VOLUME 167. Cyanobacteria
Edited by LESTER PACKER AND ALEXANDER N. GLAZER

VOLUME 168. Hormone Action (Part K: Neuroendocrine Peptides)
Edited by P. MICHAEL CONN

VOLUME 169. Platelets: Receptors, Adhesion, Secretion (Part A)
Edited by JACEK HAWIGER

VOLUME 170. Nucleosomes
Edited by PAUL M. WASSARMAN AND ROGER D. KORNBERG

VOLUME 171. Biomembranes (Part R: Transport Theory: Cells and Model Membranes)
Edited by SIDNEY FLEISCHER AND BECCA FLEISCHER

VOLUME 172. Biomembranes (Part S: Transport: Membrane Isolation and Characterization)
Edited by SIDNEY FLEISCHER AND BECCA FLEISCHER

VOLUME 173. Biomembranes [Part T: Cellular and Subcellular Transport: Eukaryotic (Nonepithelial) Cells]
Edited by SIDNEY FLEISCHER AND BECCA FLEISCHER

VOLUME 174. Biomembranes [Part U: Cellular and Subcellular Transport: Eukaryotic (Nonepithelial) Cells]
Edited by SIDNEY FLEISCHER AND BECCA FLEISCHER

VOLUME 175. Cumulative Subject Index Volumes 135–139, 141–167

VOLUME 176. Nuclear Magnetic Resonance (Part A: Spectral Techniques and Dynamics)
Edited by NORMAN J. OPPENHEIMER AND THOMAS L. JAMES

VOLUME 177. Nuclear Magnetic Resonance (Part B: Structure and Mechanism)
Edited by NORMAN J. OPPENHEIMER AND THOMAS L. JAMES

VOLUME 178. Antibodies, Antigens, and Molecular Mimicry
Edited by JOHN J. LANGONE

VOLUME 179. Complex Carbohydrates (Part F)
Edited by VICTOR GINSBURG

VOLUME 180. RNA Processing (Part A: General Methods)
Edited by JAMES E. DAHLBERG AND JOHN N. ABELSON

VOLUME 181. RNA Processing (Part B: Specific Methods)
Edited by JAMES E. DAHLBERG AND JOHN N. ABELSON

VOLUME 182. Guide to Protein Purification
Edited by MURRAY P. DEUTSCHER

VOLUME 183. Molecular Evolution: Computer Analysis of Protein and Nucleic Acid Sequences
Edited by RUSSELL F. DOOLITTLE

VOLUME 184. Avidin-Biotin Technology
Edited by MEIR WILCHEK AND EDWARD A. BAYER

VOLUME 185. Gene Expression Technology
Edited by DAVID V. GOEDDEL

VOLUME 186. Oxygen Radicals in Biological Systems (Part B: Oxygen Radicals and Antioxidants)
Edited by LESTER PACKER AND ALEXANDER N. GLAZER

VOLUME 187. Arachidonate Related Lipid Mediators
Edited by ROBERT C. MURPHY AND FRANK A. FITZPATRICK

VOLUME 188. Hydrocarbons and Methylotrophy
Edited by MARY E. LIDSTROM

VOLUME 189. Retinoids (Part A: Molecular and Metabolic Aspects)
Edited by LESTER PACKER

VOLUME 190. Retinoids (Part B: Cell Differentiation and Clinical Applications)
Edited by LESTER PACKER

VOLUME 191. Biomembranes (Part V: Cellular and Subcellular Transport: Epithelial Cells)
Edited by SIDNEY FLEISCHER AND BECCA FLEISCHER

VOLUME 192. Biomembranes (Part W: Cellular and Subcellular Transport: Epithelial Cells)
Edited by SIDNEY FLEISCHER AND BECCA FLEISCHER

VOLUME 193. Mass Spectrometry
Edited by JAMES A. MCCLOSKEY

VOLUME 194. Guide to Yeast Genetics and Molecular Biology
Edited by CHRISTINE GUTHRIE AND GERALD R. FINK

VOLUME 195. Adenylyl Cyclase, G Proteins, and Guanylyl Cyclase
Edited by ROGER A. JOHNSON AND JACKIE D. CORBIN

VOLUME 213. Carotenoids (Part A: Chemistry, Separation, Quantitation, and Antioxidation)
Edited by LESTER PACKER

VOLUME 214. Carotenoids (Part B: Metabolism, Genetics, and Biosynthesis)
Edited by LESTER PACKER

VOLUME 215. Platelets: Receptors, Adhesion, Secretion (Part B)
Edited by JACEK J. HAWIGER

VOLUME 216. Recombinant DNA (Part G)
Edited by RAY WU

VOLUME 217. Recombinant DNA (Part H)
Edited by RAY WU

VOLUME 218. Recombinant DNA (Part I)
Edited by RAY WU

VOLUME 219. Reconstitution of Intracellular Transport
Edited by JAMES E. ROTHMAN

VOLUME 220. Membrane Fusion Techniques (Part A)
Edited by NEJAT DÜZGUÜNES

VOLUME 221. Membrane Fusion Techniques (Part B)
Edited by NEJAT DÜZGÜNES

VOLUME 222. Proteolytic Enzymes in Coagulation, Fibrinolysis, and Complement Activation (Part A: Mammalian Blood Coagulation Factors and Inhibitors)
Edited by LASZLO LORAND AND KENNETH G. MANN

VOLUME 223. Proteolytic Enzymes in Coagulation, Fibrinolysis, and Complement Activation (Part B: Complement Activation, Fibrinolysis, and Nonmammalian Blood Coagulation Factors)
Edited by LASZLO LORAND AND KENNETH G. MANN

VOLUME 224. Molecular Evolution: Producing the Biochemical Data
Edited by ELIZABETH ANNE ZIMMER, THOMAS J. WHITE, REBECCA L. CANN, AND ALLAN C. WILSON

VOLUME 225. Guide to Techniques in Mouse Development
Edited by PAUL M. WASSARMAN AND MELVIN L. DEPAMPHILIS

VOLUME 226. Metallobiochemistry (Part C: Spectroscopic and Physical Methods for Probing Metal Ion Environments in Metalloenzymes and Metalloproteins)
Edited by JAMES F. RIORDAN AND BERT L. VALLEE

VOLUME 227. Metallobiochemistry (Part D: Physical and Spectroscopic Methods for Probing Metal Ion Environments in Metalloproteins)
Edited by JAMES F. RIORDAN AND BERT L. VALLEE

VOLUME 228. Aqueous Two-Phase Systems
Edited by HARRY WALTER AND GÖTE JOHANSSON

VOLUME 229. Cumulative Subject Index Volumes 195–198, 200–227

VOLUME 230. Guide to Techniques in Glycobiology
Edited by WILLIAM J. LENNARZ AND GERALD W. HART

VOLUME 231. Hemoglobins (Part B: Biochemical and Analytical Methods)
Edited by JOHANNES EVERSE, KIM D. VANDEGRIFF, AND ROBERT M. WINSLOW

VOLUME 232. Hemoglobins (Part C: Biophysical Methods)
Edited by JOHANNES EVERSE, KIM D. VANDEGRIFF, AND ROBERT M. WINSLOW

VOLUME 233. Oxygen Radicals in Biological Systems (Part C)
Edited by LESTER PACKER

VOLUME 234. Oxygen Radicals in Biological Systems (Part D)
Edited by LESTER PACKER

VOLUME 235. Bacterial Pathogenesis (Part A: Identification and Regulation of Virulence Factors)
Edited by VIRGINIA L. CLARK AND PATRIK M. BAVOIL

VOLUME 236. Bacterial Pathogenesis (Part B: Integration of Pathogenic Bacteria with Host Cells)
Edited by VIRGINIA L. CLARK AND PATRIK M. BAVOIL

VOLUME 237. Heterotrimeric G Proteins
Edited by RAVI IYENGAR

VOLUME 238. Heterotrimeric G-Protein Effectors
Edited by RAVI IYENGAR

VOLUME 239. Nuclear Magnetic Resonance (Part C)
Edited by THOMAS L. JAMES AND NORMAN J. OPPENHEIMER

VOLUME 240. Numerical Computer Methods (Part B)
Edited by MICHAEL L. JOHNSON AND LUDWIG BRAND

VOLUME 241. Retroviral Proteases
Edited by LAWRENCE C. KUO AND JULES A. SHAFER

VOLUME 242. Neoglycoconjugates (Part A)
Edited by Y. C. LEE AND REIKO T. LEE

VOLUME 243. Inorganic Microbial Sulfur Metabolism
Edited by HARRY D. PECK, JR., AND JEAN LEGALL

VOLUME 244. Proteolytic Enzymes: Serine and Cysteine Peptidases
Edited by ALAN J. BARRETT

VOLUME 245. Extracellular Matrix Components
Edited by E. RUOSLAHTI AND E. ENGVALL

VOLUME 246. Biochemical Spectroscopy
Edited by KENNETH SAUER

VOLUME 247. Neoglycoconjugates (Part B: Biomedical Applications)
Edited by Y. C. LEE AND REIKO T. LEE

VOLUME 266. Computer Methods for Macromolecular Sequence Analysis
Edited by RUSSELL F. DOOLITTLE

VOLUME 267. Combinatorial Chemistry
Edited by JOHN N. ABELSON

VOLUME 268. Nitric Oxide (Part A: Sources and Detection of NO; NO Synthase)
Edited by LESTER PACKER

VOLUME 269. Nitric Oxide (Part B: Physiological and Pathological Processes)
Edited by LESTER PACKER

VOLUME 270. High Resolution Separation and Analysis of Biological Macromolecules (Part A: Fundamentals)
Edited by BARRY L. KARGER AND WILLIAM S. HANCOCK

VOLUME 271. High Resolution Separation and Analysis of Biological Macromolecules (Part B: Applications)
Edited by BARRY L. KARGER AND WILLIAM S. HANCOCK

VOLUME 272. Cytochrome P450 (Part B)
Edited by ERIC F. JOHNSON AND MICHAEL R. WATERMAN

VOLUME 273. RNA Polymerase and Associated Factors (Part A)
Edited by SANKAR ADHYA

VOLUME 274. RNA Polymerase and Associated Factors (Part B)
Edited by SANKAR ADHYA

VOLUME 275. Viral Polymerases and Related Proteins
Edited by LAWRENCE C. KUO, DAVID B. OLSEN, AND STEVEN S. CARROLL

VOLUME 276. Macromolecular Crystallography (Part A)
Edited by CHARLES W. CARTER, JR., AND ROBERT M. SWEET

VOLUME 277. Macromolecular Crystallography (Part B)
Edited by CHARLES W. CARTER, JR., AND ROBERT M. SWEET

VOLUME 278. Fluorescence Spectroscopy
Edited by LUDWIG BRAND AND MICHAEL L. JOHNSON

VOLUME 279. Vitamins and Coenzymes (Part I)
Edited by DONALD B. MCCORMICK, JOHN W. SUTTIE, AND CONRAD WAGNER

VOLUME 280. Vitamins and Coenzymes (Part J)
Edited by DONALD B. MCCORMICK, JOHN W. SUTTIE, AND CONRAD WAGNER

VOLUME 281. Vitamins and Coenzymes (Part K)
Edited by DONALD B. MCCORMICK, JOHN W. SUTTIE, AND CONRAD WAGNER

VOLUME 282. Vitamins and Coenzymes (Part L)
Edited by DONALD B. MCCORMICK, JOHN W. SUTTIE, AND CONRAD WAGNER

VOLUME 283. Cell Cycle Control
Edited by WILLIAM G. DUNPHY

VOLUME 284. Lipases (Part A: Biotechnology)
Edited by BYRON RUBIN AND EDWARD A. DENNIS

VOLUME 285. Cumulative Subject Index Volumes 263, 264, 266–284, 286–289

VOLUME 286. Lipases (Part B: Enzyme Characterization and Utilization)
Edited by BYRON RUBIN AND EDWARD A. DENNIS

VOLUME 287. Chemokines
Edited by RICHARD HORUK

VOLUME 288. Chemokine Receptors
Edited by RICHARD HORUK

VOLUME 289. Solid Phase Peptide Synthesis
Edited by GREGG B. FIELDS

VOLUME 290. Molecular Chaperones
Edited by GEORGE H. LORIMER AND THOMAS BALDWIN

VOLUME 291. Caged Compounds
Edited by GERARD MARRIOTT

VOLUME 292. ABC Transporters: Biochemical, Cellular, and Molecular Aspects
Edited by SURESH V. AMBUDKAR AND MICHAEL M. GOTTESMAN

VOLUME 293. Ion Channels (Part B)
Edited by P. MICHAEL CONN

VOLUME 294. Ion Channels (Part C)
Edited by P. MICHAEL CONN

VOLUME 295. Energetics of Biological Macromolecules (Part B)
Edited by GARY K. ACKERS AND MICHAEL L. JOHNSON

VOLUME 296. Neurotransmitter Transporters
Edited by SUSAN G. AMARA

VOLUME 297. Photosynthesis: Molecular Biology of Energy Capture
Edited by LEE MCINTOSH

VOLUME 298. Molecular Motors and the Cytoskeleton (Part B)
Edited by RICHARD B. VALLEE

VOLUME 299. Oxidants and Antioxidants (Part A)
Edited by LESTER PACKER

VOLUME 300. Oxidants and Antioxidants (Part B)
Edited by LESTER PACKER

VOLUME 301. Nitric Oxide: Biological and Antioxidant Activities (Part C)
Edited by LESTER PACKER

VOLUME 302. Green Fluorescent Protein
Edited by P. MICHAEL CONN

VOLUME 303. cDNA Preparation and Display
Edited by SHERMAN M. WEISSMAN

VOLUME 304. Chromatin
Edited by PAUL M. WASSARMAN AND ALAN P. WOLFFE

VOLUME 358. Bacterial Pathogenesis (Part C: Identification, Regulation, and Function of Virulence Factors)
Edited by VIRGINIA L. CLARK AND PATRIK M. BAVOIL

VOLUME 359. Nitric Oxide (Part D: Nitric Oxide Detection, Mitochondria and Cell Functions, and Peroxynitrite Reactions)
Edited by ENRIQUE CADENAS AND LESTER PACKER

VOLUME 360. Biophotonics (Part A) (in preparation)
Edited by GERARD MARRIOTT AND IAN PARKER

VOLUME 361. Biophotonics (Part B) (in preparation)
Edited by GERARD MARRIOTT AND IAN PARKER

Section I

Detection of Nitric Oxide

[1] Electron Paramagnetic Resonance for Quantitation of Nitric Oxide in Aqueous Solutions

By SUJATHA VENKATARAMAN, SEAN M. MARTIN, and GARRY R. BUETTNER

Introduction

Nitric oxide ($^{\bullet}NO$) has many complex and diverse biological functions.[1–3] For example, it functions as an endothelium-derived relaxing factor (EDRF), a vascular antioxidant, and as a messenger molecule in the cardiovascular and immune systems. Nitric oxide also plays important roles in the biochemical aspects of a plethora of disorders. A molecule of such significance needs reliable methods for its detection and quantitation so that detailed mechanistic studies are possible.

Because $^{\bullet}NO$ is a reactive free radical and oxidizes rapidly to nitrite and nitrate, the quantitation of $^{\bullet}NO$ in an aqueous solution is challenging. There are a number of methods for the detection of nitric oxide,[4–8] but due to lack of specificity and sensitivity, only a few are able to provide actual absolute quantitation of $^{\bullet}NO$.[9–11] The choice of the method depends on the objective of the analysis: whether it is qualitative or quantitative.

Of the methods proposed so far, electron paramagnetic resonance (EPR) appears to be very useful and reliable. EPR has been used extensively to unravel the role of $^{\bullet}NO$ in biological systems.[12,13] Despite $^{\bullet}NO$ being a paramagnetic compound, it cannot be detected directly using EPR. However, diamagnetic compounds that form stable paramagnetic adducts with $^{\bullet}NO$ can be used to detect $^{\bullet}NO$ by EPR. The high affinity of $^{\bullet}NO$ for certain classes of Fe^{2+} complexes plays an

[1] S. Moncada and A. Higgs, *N. Engl. J. Med.* **329,** 2002 (1993).
[2] S. Moncada, R. M. J. Palmer, and E. A. Higgs, *Pharmacol. Rev.* **43,** 109 (1991).
[3] Q. Feng and T. Hedner, *Clin. Physiol.* **10,** 407 (1990).
[4] T. Malinski and L. Czuchajowski, *in* "Methods in Nitric Oxide Research" (M. Feelisch and J. S. Stamler, eds.), p. 319. Wiley, New York, 1996.
[5] J. F. Leikert, T. R. Rathel, C. Muller, A. M. Vollmar, and V. M. Dirsch, *FEBS Lett.* **506,** 131 (2001).
[6] D. M. Hall and G. R. Buettner, *Methods Enzymol.* **268,** 188 (1996).
[7] A. F. Vanin, *Methods Enzymol.* **301,** 269 (1999).
[8] B. Kalyanaraman, *Methods Enzymol.* **268,** 168 (1996).
[9] R. W. Nims, J. C. Cook, M. C. Krishna, D. Christodoulou, C. M. B. Poore, A. M. Miles, M. B. Grisham, and D. A. Wink, *Methods Enzymol.* **268,** 93 (1996).
[10] S. Venkataraman, S. M. Martin, F. Q. Schafer, and G. R. Buettner, *Free Radic. Biol. Med.* **29,** 580 (2000).
[11] K. Tsuchiya, M. Takasugi, K. Minakuchi, and K. Fukuzawa, *Free Radic. Biol. Med.* **21,** 733 (1996).
[12] Y. Henry, C. Ducrocq, J.-C. Drapier, D. Servent, C. Pellat, and A. Guissani, *Eur. Biophys. J.* **20,** 1 (1991).
[13] R. F. Lin, T.-S. Lin, R. G. Tilton, and A. H. Cross, *J. Exp. Med.* **178,** 643 (1993).

0076-6879/02 $35.00

important role in biological processes. The most frequently used traps for ${}^{\bullet}$NO also take advantage of this property. For example, ${}^{\bullet}$NO reacts with Fe^{2+} hemoglobin and other ferrous hemes, as well as Fe^{2+}(dithiocarbamate)$_2$ complexes forming stable paramagnetic species. These complexes have been widely used in the study of the chemical and biological aspects of ${}^{\bullet}$NO. The use of iron complexes to detect ${}^{\bullet}$NO in animal studies was first reported by Vanin *et al.*[14]

A new series of paramagnetic compounds, nitronyl nitroxides (NNR), has been used to detect ${}^{\bullet}$NO. Although the chemistry of these compounds with ${}^{\bullet}$NO is quite different from that of iron complexes, these compounds have good specificity and can be used to quantitate ${}^{\bullet}$NO in aqueous solutions using EPR. Extensive studies have been reported on the use of nitronyl nitroxides as a probe for vasodilation and as a ${}^{\bullet}$NO antagonist in perfused organs.[15,16] Nitronyl nitroxides have also been used in other areas of biology to study the antimicrobial action of ${}^{\bullet}$NO.[17]

This article presents methods using $Fe^{2+}(MGD)_2$ (MGD, *N*-methyl-D-glucamine dithiocarbamate) and nitronyl nitroxides to quantitate ${}^{\bullet}$NO in aqueous solution by EPR. Our goal is to provide methods that will direct the chemistry involved to obtain reliable and accurate results.

${}^{\bullet}$NO Chemistry and Detection Strategy

It is always important to understand the chemistry behind methods of quantitation. Nitric oxide has a relatively short half-life,[18] it reacts spontaneously with dioxygen in the gas phase with a stoichiometry of 2 : 1 [Eq. (1)] and 4 : 1 in water [Eq. (2)].[10,19]

$$2\,{}^{\bullet}NO + O_2 \rightarrow 2\,{}^{\bullet}NO_2 \qquad \text{(gas phase)} \qquad (1)$$

$$4\,{}^{\bullet}NO + O_2 + 2H_2O \rightarrow 4NO_2^- + 4H^+ \qquad \text{(aqueous phase)} \qquad (2)$$

These reactions are the basis for several methods of detection and quantitation of ${}^{\bullet}$NO, including the estimation of [${}^{\bullet}$NO] using an oxygen monitor,[10] spectrophotometric methods such as UV-Vis,[9] or fluorescence[20] analysis of nitrite using various substrates.

[14] A. F. Vanin, P. I. Mordvintcev, and A. L. Kleschyov, *Stud. Biophys.* **102,** 135 (1984).

[15] A. Konorev, M. M. Tarpey, J. Joseph, J. E. Baker, and B. Kalyanaraman, *Free Radic. Biol. Med.* **18,** 169 (1995).

[16] E. A. Konorev, M. M. Tarpey, J. Joseph, J. E. Baker, and B. Kalyanaraman, *J. Pharmacol. Exp. Ther.* **274,** 200 (1995).

[17] K. Yoshida, T. Akaiki, T. Doi, K. Sato, S. Uiri, M. Suga, M. Ando, and H. Maeda, *Infect. Immunol.* **61,** 3552 (1993).

[18] L. J. Ignarro, *Annu. Rev. Pharmacol. Toxicol.* **30,** 535 (1990).

[19] P. C. Ford, D. A. Wink, and D. M. Stanbury, *FEBS Lett.* **326,** 1 (1993).

[20] A. M. Miles, D. A. Wink, J. C. Cook, and M. B. Grisham, *Methods Enzymol.* **268,** 105 (1996).

Nitrite is also measured based on the Griess reaction.[21] In this diazotization method, nitrite is reacted with Griess reagent to form a colored azo product that is measured spectrophotometrically.[22,23]

A chemiluminescent method of detection of gas-phase $^{\bullet}NO$ uses its reaction with ozone. In the gas phase, O_3 reacts with $^{\bullet}NO$ according to Eqs. (3) and (4):

$$^{\bullet}NO + O_3 \rightarrow {}^{\bullet}NO_2^* + O_2 \quad (3)$$

$$^{\bullet}NO_2^* \rightarrow {}^{\bullet}NO_2 + h\nu \quad (4)$$

where $^{\bullet}NO_2^*$ denotes the $^{\bullet}NO_2$ molecule in the excited state and $h\nu$ represents an emitted photon. Chemiluminescence resulting from these reactions provides a basis for the Sievers NOA detection system for $^{\bullet}NO$.[24] This method is ideal for anaerobic $^{\bullet}NO$ solutions and is highly sensitive, but can be time-consuming and expensive for occasional quantitative analysis.

Nitric oxide reacts spontaneously with heme proteins. This forms the basis of the spectrophotometric method that measures the amount of metHb formed from the reaction of $^{\bullet}NO$ with oxyhemoglobin.[25,26] Also, $^{\bullet}NO$ reacts with deoxyhemoglobin and forms a stable nitrosylated adduct, which is measured using EPR.[27,28]

Nitric oxide is also measured by other techniques, such as electrochemical detection[4] and other spectroscopic methods based on its physical and chemical properties.

EPR spectroscopy continues to play a significant role in the evolution of our understanding of $^{\bullet}NO$ biology. EPR faithfully reproduces the spectral features of the indicator molecule, revealing the identity of the $^{\bullet}NO$ complexes formed, a result not achieved by other methods. Because $^{\bullet}NO$ is a paramagnetic compound, it should be detectable by EPR. However, the quantum mechanical properties of diatomic free radicals (fast spin-orbit relaxation resulting in very, very broad lines) render these radicals essentially EPR silent at room temperature. Therefore, a spin trap is needed that has high affinity for $^{\bullet}NO$ and also forms a stable paramagnetic complex. Classic spin traps such as 5,5-dimethyl-1-pyrroline N-oxide (DMPO) do not trap $^{\bullet}NO$, as $^{\bullet}NO$ has little tendency to form the covalent bond with nonradical molecules required by the typical spin-trapping reaction. However, $^{\bullet}NO$ rapidly forms relatively stable ligand bonds, forming nitrosylates, with certain

[21] J. P. Griess, *Phil. Trans. R. Soc. (Lond.)* **154,** 667 (1864).
[22] S. Archer, *FASEB. J.* **7,** 349 (1993).
[23] J. M. Hevel and M. A. Marletta, *Methods Enzymol.* **233,** 250 (1994).
[24] A. Fontijn, A. J. Sabadell, and R. J. Ronco, *Anal. Chem.* **42,** 575 (1970).
[25] M. E. Murphy and E. Noack, *Methods Enzymol.* **233,** 240 (1994).
[26] M. Feelisch, D. Kubitzek, and J. Werringloer, *in* "Methods in Nitric Oxide Research" (M. Feelisch and J. S. Stamler, eds.), p. 455. Wiley, New York, 1996.
[27] A. Wennmalm, B. Lanne, and A.-S. Petersson, *Anal. Biochem.* **187,** 359 (1990).
[28] Q. Wang, J. Jacobs, J. DeLeo, H. Kruszyna, R. Smith, and D. Wilcox, *Life Sci.* **49,** PL55 (1991).

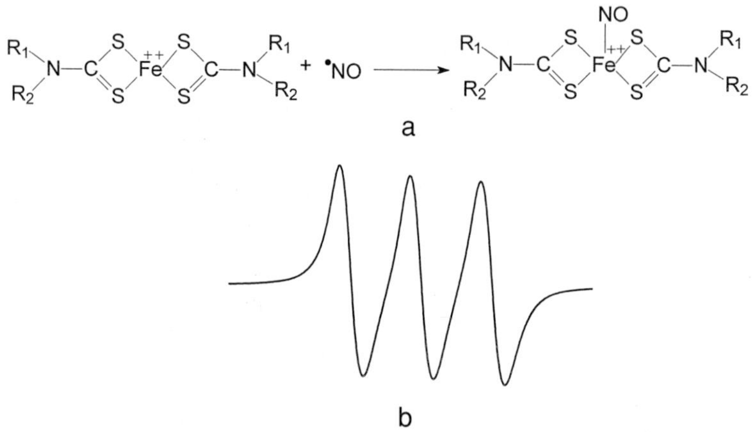

FIG. 1. (a) The trapping reaction of $^{\bullet}NO$ by $Fe^{2+}(MGD)_2$ [$R_1 = CH_3$ and $R_2 = CH_2(CHOH)_4$-CH_2OH] or $Fe^{2+}(DETC)_2$, [$R_1 = R_2 = C_2H_5$]. (b) Representative room temperature EPR spectrum for NO-$Fe^{2+}(MGD)_2$ ($a^N = 12.9$ G; $g_{iso} = 2.04$) or for NO-$Fe^{2+}(DETC)_2$ ($a^N = 12.7$ G; $g_{iso} = 2.04$).

iron complexes. Stable iron(II) complexes of $^{\bullet}NO$ yield characteristic EPR spectra. Without $^{\bullet}NO$ as a ligand, these ferrous complexes are usually EPR inactive. Example complexes are the ferrous dithiocarbamate products NO-$Fe^{2+}(DETC)_2$ and NO-$Fe^{2+}(MGD)_2$, which give EPR spectra that can be used to identify and quantitate $^{\bullet}NO$.

Nitronyl nitroxides are also used to quantitate $^{\bullet}NO$ with EPR. These are organic compounds that have both nitrone, $\gtrdot\overset{+}{N}-O^{-}$ (a functional group corresponding to a spin trap), and nitroxide, $>N-O^{\bullet}$ (a functional group corresponding to a spin label or spin adduct) moieties. Both the NNR and the product formed after its reaction with $^{\bullet}NO$ are stable nitroxides that have distinguishable EPR spectral features.

This article presents the chemistry behind $^{\bullet}NO$ and its reactions with ferrous dithiocarbamate complexes, $Fe^{2+}(DETC)_2$ and $Fe^{2+}(MGD)_2$ (Fig. 1), and nitronyl nitroxides (Fig. 2) used to detect $^{\bullet}NO$. Representative EPR spectra and their parameters are presented. Also, important rate constants related to $^{\bullet}NO$ quantitation are summarized in Table I.

Materials

The nitric oxide donor, DEANO [2-(*N,N*-diethylamino)diazenolate-2-oxide, sodium salt], and PTIO (2-phenyl-4,4,5,5-tetramethylimidazoline-l-oxyl-3-oxide) are from Alexis Corporation (San Diego, CA). The spin label 3-CP (3-carboxy proxyl) is from Aldrich (Milwaukee, WI). Sodium nitrite, ferrous sulfate, glacial acetic acid, and potassium iodide are from Fisher Scientific (Pittsburgh, PA). The chelating resin is obtained from Sigma Chemical Co. (St. Louis, MO). All

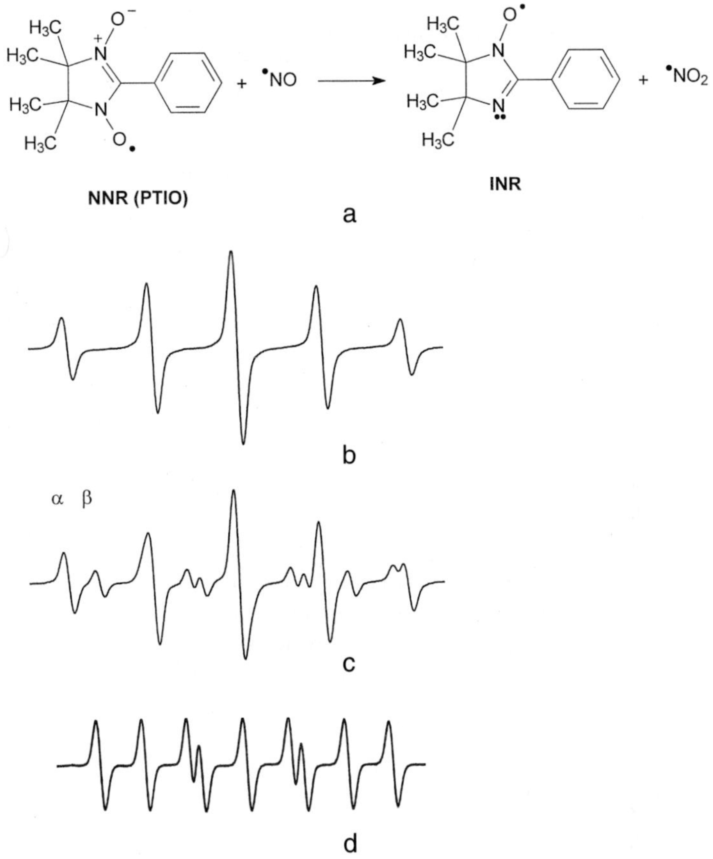

FIG. 2. (a) Reaction of $^{\bullet}$NO with nitronyl nitroxide (NNR), (b) the ESR spectrum of NNR alone ($a^N = 8.1$ G), (c) mixture of NNR and imino nitroxide (INR) with α and β indicating the low field lines of NNR and INR, respectively, and (d) INR alone ($a_1^N = 4.4$ G and $a_2^N = 9.9$ G).

chemicals and reagents used are of analytical grade. High-purity gaseous nitric oxide is obtained from AGA Specialty Gas (Maumee, OH). Argon and nitrogen are from Air Products (Allentown, PA).

Methods

Phosphate-buffered saline (PBS) (pH 7.4) is prepared and stirred very gently overnight in the presence of the chelating resin to remove adventitious catalytic metals. The ascorbate test is used to verify their removal.[29]

[29] G. R. Buettner, *J. Biochem. Biophys. Methods* **16,** 27 (1988).

TABLE I
REACTIONS AND RATE CONSTANTS OF $^{\bullet}NO$ IN AQUEOUS SOLUTION[a]

Reaction	Rate constant	Ref.[b]
$^{\bullet}NO + {}^{\bullet}NO_2 \rightarrow N_2O_3$	$k = 1.1 \times 10^9\ M^{-1}\ s^{-1}$	(a)
$^{\bullet}NO_2 + {}^{\bullet}NO_2 \rightarrow N_2O_4$	$k = 4.5 \times 10^8\ M^{-1}\ s^{-1}$	(a)
$N_2O_4 \rightarrow {}^{\bullet}NO_2 + {}^{\bullet}NO_2$	$k = 6.9 \times 10^3\ s^{-1}$	(b)
$N_2O_3 \rightarrow {}^{\bullet}NO + {}^{\bullet}NO_2$	$k = 8.0 \times 10^4\ s^{-1}$	(b)
$N_2O_3 + H_2O \rightarrow 2HNO_2$	$k = 5.3 \times 10^2\ s^{-1}$	(b)
$N_2O_4 + H_2O \rightarrow HNO_2 + H^+ + NO_3-$	$k = 1.0 \times 10^3\ s^{-1}$	(c)
$4^{\bullet}NO + O_2 + 2H_2O \rightarrow 4HNO_2$	$k = 1.58 \times 10^6\ M^{-2}\ s^{-1}$	(d)
$Fe^{2+}(MGD)_2 + O_2 \rightarrow Fe^{3+}(MGD)_2 + O_2^{\bullet -}$	$k = 5 \times 10^5\ M^{-1}\ s^{-1}$	(e)
$Fe^{2+}(MGD)_2 + {}^{\bullet}NO \rightarrow NO\text{-}Fe^{2+}(MGD)_2$	$k = 1.21 \times 10^6\ M^{-1}\ s^{-1}$	(f)
$Fe^{2+}(DETC)_2 + {}^{\bullet}NO \rightarrow NO\text{-}Fe^{2+}(DETC)_2$	$k \approx 10^8\ M^{-1}\ s^{-1}$	(g)
$Fe^{2+}(DTCS)_2 + {}^{\bullet}NO \rightarrow NO\text{-}Fe^{2+}(DTCS)_2$	$k = 1.71 \times 10^6\ M^{-1}\ s^{-1}$	(f)
$NNR + {}^{\bullet}NO \rightarrow INR + {}^{\bullet}NO_2$	$k \approx 10^4\ M^{-1}\ s^{-1}$	(h)
$^{\bullet}NO + O_2^{\bullet -} \rightarrow O = NOO^-$	$k \approx 6.7 \times 10^9\ M^{-1}\ s^{-1}$	(i)

[a] MGD, *N*-Methyl-D-glucamine dithiocarbamate; DETC, diethyl dithiocarbamate; DTCS, *N*-(dithiocarboxy)sarcosine; NNR, nitronyl nitroxide; INR, imino nitroxide.

[b] Key to references: (a) M. P. Doyle and J. W. Hoeksta, *J. Inorg. Biochem* **14,** 351 (1981). (b) M. Grätzel, J. Henglein, J. Lillie, and G. Beck, *Ber. Bunseges. Phys. Chem.* **73,** 646 (1969). (c) P. C. Ford, D. A. Wink, and D. M. Standbury, *FEBS Lett.* **326,** 1 (1993). (d) D. A. Wink, J. F. Darbyshire, R. W. Nims, J. E. Saavedra, and P. C. Ford, *Chem. Res. Toxicol.* **6,** 23 (1993). (e) K. Tshuchiya, J. J. Jiang, M. Yoshizumi, T. Tamaki, H. Houchi, K. Minakuchi, K. Fukuzawa, and R. P. Mason, *Free Radic. Biol. Med.* **27,** 367 (1999). (f) S. Pou, P. Tsai, S. Porusuphatana, H. J. Halpern, G. V. R. Chandramouli, E. D. Barth, and G. M. Rosen, *Biochim. Biophys. Acta* **1427,** 216 (1999). (g) A. F. Vanin, *Methods Enzymol.* **301,** 269 (1999). (h) T. Akaike, M. Yoshida, Y. Miyamato, K. Sato, M. Kohno, K. Sasamoto, K. Miyazaki, S. Ueda, and H. Maeda, *Biochemistry* **32,** 827 (1993); Y. Y. Woldman, V. V. Khramtsov, I. A. Grigor'ev, I. A. Kiriljuk, and D. I. Utepbergenov, *Biochem. Biophys. Res. Commun.* **202,** 195 (1994). (i) R. E. Huie and S. Padmaja, *Free Radic. Res. Commun.* **18,** 195 (1993).

Nitric Oxide Stock Solutions

Nitric oxide gas is either obtained from a nitric oxide gas tank or prepared from an acidified sodium nitrite solution.[30,10] Because the nitric oxide from either source can be contaminated with other oxides of nitrogen, it is purified by passing it through NaOH (4 *M*) and then deionized (DI) water. The purified $^{\bullet}NO$ gas is then bubbled through a gas sampling bottle containing degassed DI water and stored. DEANO is used as a source of $^{\bullet}NO$ for some experiments. DEANO has a half-life of about 2 min in PBS, pH 7.4, at 37°, ideally, releasing two molecules of $^{\bullet}NO$

[30] G. Brauer (ed.), "Handbook of Preparative Inorganic Chemistry," 2nd Ed., p. 486. Academic Press, New York, 1963.

and one molecule of diethylamine[31] (under our conditions 1.5 molecules of ${}^{\bullet}NO$ per one molecule of diethylamine).

Aliquots of the ${}^{\bullet}NO$ stock solutions are delivered to the various assay systems using argon-flushed, gas-tight microliter syringes (Hamilton Co., Reno, NV). Prior to filling the syringes with the ${}^{\bullet}NO$ stock solution, an equal volume of argon is injected into the bottle containing ${}^{\bullet}NO$ solution to minimize contamination with oxygen. It should be kept in mind that the container to which the ${}^{\bullet}NO$ stock solution is added should have a small surface-to-volume ratio so that the amount of ${}^{\bullet}NO$ escaping to the headspace is minimized.

Preparation of NO-Fe^{2+}(Dithiocarbamate)$_2$ Complexes

Diethyldithiocarbamate (DETC) is water soluble, but when complexed with Fe^{2+}, the $Fe^{2+}(DETC)_2$ formed is insoluble. To overcome the solubility problem of $Fe^{2+}(DETC)_2$, Mordvintcev *et al.*[32] incubated the complex with yeast membranes, thereby making a usable suspension. For the same reason, Tsuchiya *et al.*[33] used porcine serum albumin (PSA) to carry $Fe^{2+}(DETC)_2$ into aqueous solutions. Once in solution, oxygen should be minimized to prevent any side reactions with ${}^{\bullet}NO$. ${}^{\bullet}NO$ can then be added to these solutions. It is worth mentioning that if the physical arrangement of the EPR cavity allows the flat cell to be placed horizontally, potential sedimentation problems can be avoided.[34]

MGD, a derivative of DETC, is synthesized as described previously.[35] All solutions are prepared with argon-purged DI water. Stock solutions of $Fe^{2+}(MGD)_2$ are prepared by dissolving MGD sodium salt and ferrous sulfate in DI water, with a molar ratio of 5 : 1, respectively. The NO-$Fe^{2+}(MGD)_2$ samples are prepared by first pipetting different amounts of $Fe^{2+}(MGD)_2$ stock solution (50–400 μl) into a 10-ml test tube kept under argon. To this solution, the desired volume of ${}^{\bullet}NO$ stock solution (to achieve a specific final [${}^{\bullet}NO$]) is added using gas-tight syringes, resulting in the formation of NO-$Fe^{2+}(MGD)_2$. The concentration of $Fe^{2+}(MGD)_2$ is always in three- to fivefold excess of the [${}^{\bullet}NO$] to ensure that all the ${}^{\bullet}NO$ is trapped. The NO-$Fe^{2+}(MGD)_2$ samples, once formed, are transferred quickly to an argon-filled flat cell to avoid oxidation of the complex, and spectra are recorded within 5–10 min.

3-Carboxy proxyl is used as an additional standard to verify the concentrations of NO-$Fe^{2+}(MGD)_2$ standards,[10] after correcting for the difference in their g values using the factor g(3-CP)/g (NO-$Fe^{2+}(MGD)_2$) = 2.006/2.04.[36,37]

[31] L. K. Keefer, R. W. Nims, K. M. Davies, and D. A. Wink, *Methods Enzymol.* **268,** 281 (1996).

[32] P. Mordvintcev, A. Mulsch, R. Busse, and A. Vanin, *Anal. Biochem.* **199,** 142 (1991).

[33] K. Tsuchiya, M. Takasugi, K. Minakuchi, and K. Fukuzawa, *Free Radic. Biol. Med.* **21,** 733 (1996).

[34] Y. Kotake, L. A. Reinke, T. Tanigawa, and H. Koshida, *Free Radic. Biol. Med.* **17,** 215 (1995).

[35] L. A. Shinobu, S. G. Jones, and M. M. Jones, *Acta Pharmacol. Toxicol.* **54,** 189 (1984).

[36] J. A. Weil, J. R. Bolton, and J. E. Wertz, "Electron Paramagnetic Resonance: Elementary Theory, and Practical Applications," p. 498. Wiley-Interscience, New York, 1994.

[37] S. S. Eaton and G. R. Eaton, *Bull. Magn. Reson.* **1,** 130 (1980).

Preparation of INR Compound

In experiments in which PTIO is used to quantitate $^{\bullet}$NO, DEANO is used as the source of $^{\bullet}$NO. The concentration of PTIO used is always at least three times of that needed to capture all the $^{\bullet}$NO. A known concentration of PTIO is aliquoted into a 2-ml gas-tight, screw-capped vial containing PBS and is deoxygenated with N_2. A known amount of deoxygenated DEANO solution is injected into the vial to release $^{\bullet}$NO. The sample is then incubated for 20 min at 37°. This solution is transferred to the flat cell and EPR spectra are recorded. Spectra are recorded over time to ensure the completion of the reaction and to verify the stability of the product.

Measurements

Optical Aspects

All UV-Vis spectrometric measurements are done at room temperature using an HP 8453 diode-array spectrophotometer (Hewlett-Packard, Wilmington, DE.). The PTIO solutions are standardized using $\varepsilon_{560} = 1020 \pm 50\, M^{-1}\text{cm}^{-1}$ in water. The optical parameters of other compounds used in this study are 3-CP ($\varepsilon_{234} = 2370\, M^{-1}\text{cm}^{-1}$)[10] in water and DEANO ($\varepsilon_{250} = 6500\, M^{-1}\text{cm}^{-1}$)[31] in dilute alkali.

Aqueous nitric oxide solutions are standardized using a Sievers 280 nitric oxide analyzer (NOA) as described.[10]

EPR Spectral Aspects

The first derivative EPR spectra are obtained using a Bruker EMX spectrometer (X-band) at room temperature. Typical instrument settings are 9.75 GHz microwave frequency; 100 kHz modulation frequency; 10 mW, a nonsaturating microwave power,[38] 1 G modulation amplitude; 3418 G center field for NO-$Fe^{2+}(MGD)_2$, 3475 G for 3-CP; 100 G/84 sec scan rate; and 82 ms time constant. Settings for PTIO-$^{\bullet}$NO experiments are 0.5 G modulation amplitude; 3480 G center field; 50 G/168 sec scan rate; and a time constant of 164 ms.

Computer Aspects

The computer simulation of EPR spectra is done using the program PEST Winsim and LMB-EPR.[39,40] Deconvolution and double integration of peaks are performed using WINEPR (Bruker) software.

[38] G. R. Buettner and K. P. Kiminyo, *J. Biochem. Biophys. Methods* **24,** 147 (1992).
[39] epr.niehs.nih.gov
[40] D. R. Duling, *J. Magn Reson. B* **104,** 105 (1994).

Results

Comparison of Various Available Methods

Quantitation $^{\bullet}NO$ with $Fe^{2+}(DETC)_2$. Mordvitcev *et al.*[32] were among the first to use iron dithiocarbamate complexes to quantitate $^{\bullet}NO$ by EPR. They used DETC with ferrous iron to trap $^{\bullet}NO$. A detailed method to synthesize $Fe^{2+}(DETC)_2$ complex with $^{\bullet}NO$ is given by Vanin *et al.*[41] The complex NO-$Fe^{2+}(DETC)_2$ gives a three-line EPR spectrum at room temperature ($g_{iso} = 2.04$; $a^N = 12.7$ G) (Fig. 1).

The actual quantitation of $^{\bullet}NO$, as NO-$Fe^{2+}(DETC)_2$, is accomplished using $Fe^{2+}(NO)(S_2O_3)_2$ as an EPR standard.[32] Although $Fe^{2+}(DETC)_2$ has had many successes as a tool to study nitric oxide chemistry and biology, its solubility in aqueous solution poses problems. This issue has limited the use of DETC and spurred the development and use of $Fe^{2+}(MGD)_2$ as a tool.

Quantiation of $^{\bullet}NO$ with $Fe^{2+}(MGD)_2$. MGD is a derivative of DETC introduced by Komarov *et al.*[42] for the detection of $^{\bullet}NO$ to overcome the solubility problems with $Fe^{2+}(DETC)_2$. It is readily soluble in water and forms a water-soluble NO-$Fe^{2+}(MGD)_2$ complex. It has an EPR spectrum very similar to that of NO-$Fe^{2+}(DETC)_2$, a three-line spectrum with $g_{iso} = 2.04$, $a^N = 12.9$ G in aqueous solution (Fig. 1).

Quantitation of $^{\bullet}NO$ using $Fe^{2+}(MGD)_2$ is done at room temperature under anaerobic conditions. The concentration of $^{\bullet}NO$ is determined by adding various amounts of $^{\bullet}NO$ stock solutions to solutions of $Fe^{2+}(MGD)_2$ in DI water and their EPR spectra are recorded. The concentration can be determined using peak intensity or by double integration. The concentration of this nitrosyl complex, NO-Fe^{2+}(MGD), is obtained by double integration of the three lines and is calibrated against 3-CP standards. The concentration of $^{\bullet}NO$ solutions determined by EPR measurements correlates well (slope = 1.06) with those determined from the NOA (Fig. 3). The double integrated areas of EPR spectra for various concentrations of $^{\bullet}NO$ vary linearly with [$^{\bullet}NO$] (Fig. 4). Thus, standard curves such as this can be used to extrapolate the concentration of the unknown $^{\bullet}NO$ solutions as NO-$Fe^{2+}(MGD)_2$.

It has been shown that nitrite in the presence of $Fe^{2+}(MGD)_2$ could be a source of $^{\bullet}NO$, which in turn would form NO-$Fe^{2+}(MGD)_2$.[43–45] Therefore, it is important to determine whether the presence of nitrite, as a contaminant, interferes with the estimation of $^{\bullet}NO$. The $Fe^{2+}(MGD)_2 - NaNO_2$ solutions produce no detectable EPR signals with our instrument settings and experimental protocols until

[41] A. F. Vanin, P. I. Mordvintcev, and A. L. Kleshcev, *Stud. Biophys.* **102,** 135 (1984).

[42] A. Komarov, D. Mattson, and M. M. Jones, *Biochem. Biophys. Res. Commun.* **195,** 1191 (1993).

[43] A. Samouilov, P. Kuppusamy, and J. L. Zweier, *Arch. Biochem. Biophys.* **357,** 1 (1998).

[44] K. Tsuchiya, J. Jiang, M. Yoshizumi, T. Tamaki, H. Houchi, K. Minakuchi, K. Fukuzawa, and R. P. Mason, *Free Radic. Biol. Med.* **27,** 347 (1999).

[45] K. Hiramoto, S. Tomiyama, and K. Kikugawa, *Free Radic. Res.* **27,** 505 (1997).

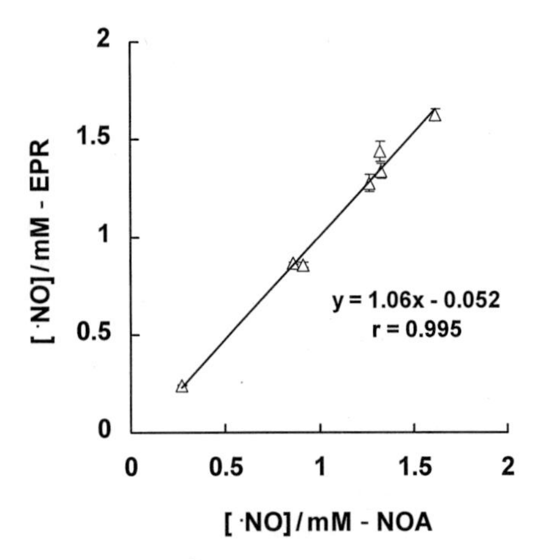

FIG. 3. Standardization of the EPR $Fe^{2+}(MGD)_2$ spin-trapping technique for the quantitation of $^{\bullet}NO$ with a NOA [from S. Venkataraman, S. M. Martin, F. Q. Schafer, and G. R. Buettner, *Free Radic. Biol. Med.* **29,** 580 (2000)]. Each data point represents the mean of three independent measurements. Error bars represent standard error.

$[NaNO_2] > 500\ \mu M$. EPR spectra corresponding to the NO-$Fe^{2+}(MGD)_2$ complex are detected only when the concentration of $NaNO_2$ is 1 m*M* or greater in the time frame of these measurements. However, with time the signal intensity increases. The signal intensity after 30 min corresponds to 110 μM of NO-$Fe^{2+}(MGD)_2$. This observation is consistent with the conversion of nitrite to $^{\bullet}NO$ in the presence

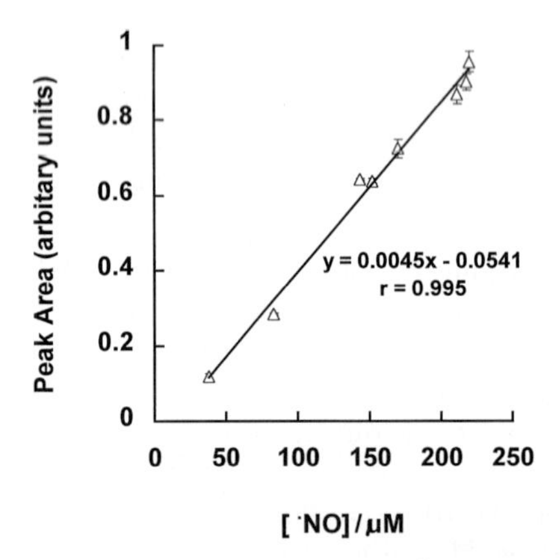

FIG. 4. Double integrated EPR area measurements of NO-$Fe^{2+}(MGD)_2$ spectra. [$^{\bullet}NO$] varied from 38 to 220 μM as determined by chemiluminescence measurement from a nitric oxide analyzer. The $Fe^{2+}(MGD)_2$ concentration was three times that of the respective $^{\bullet}NO$ concentration. Each data point represents $n = 3$. Error bars represent standard error.

of $Fe^{2+}(MGD)_2$.[46] In another control experiment, the $^{\bullet}NO$ stock solution is purged with argon for 30 min to ensure that the solution is free of $^{\bullet}NO$ and contains only nitrite, present as a contaminant. When this solution is added to the $Fe^{2+}(MGD)_2$ solution, no EPR signals are observed, even though the NOA reveals the presence of nitrite to be 150–250 μM. However, when the modulation amplitude and the time constant of the EPR spectrometer are increased, 1 to 3 G and 82 to 327 ms, respectively, the three-line NO-$Fe^{2+}(MGD)_2$ signal can be seen from solutions containing nitrite as low as 100 μM. Thus, the potential interference from nitrite must be considered. Use of reducing agents that will convert NO_2^- back to $^{\bullet}NO$ result in overestimating the amount of free $^{\bullet}NO$ in aqueous solution. Care should be taken to use these traps in anaerobic environments to avoid oxidation of Fe^{2+} to Fe^{3+}, resulting in loss of trap as well as subsequent production of superoxide (Table I). However, with appropriate protocols, $Fe^{2+}(MGD)_2$ can be used to quantitate $^{\bullet}NO$ stock solutions in the range of 500 nM to 1.9 mM (by peak height measurements or double integration).

Nitronyl Nitroxides (NNR) as Nitric Oxide Probe. Although EPR quantitation of $^{\bullet}NO$ by various Fe^{2+} complexes can be sensitive and reliable, their use is limited by the possible overestimation of $^{\bullet}NO$ in the presence of nitrite. An alternate approach is to use nitronyl nitroxides to quantitate $^{\bullet}NO$. Nitronyl nitroxides were first synthesized by Ulmann and co-workers[47] and are now available commercially. Since their discovery, this class of stable radicals has attracted considerable interest as probes for $^{\bullet}NO$. An example structure is provided in Fig. 2. These compounds are paramagnetic and give a characteristic five-line EPR spectrum. Nitronyl nitroxides react and form an unstable adduct with $^{\bullet}NO$ that rearranges to give an imino nitroxide (INR) and $^{\bullet}NO_2$.[48] Both nitronyl nitroxides and imino nitroxides are stable and have distinctly different EPR spectra, which forms the basis for their quantitation of $^{\bullet}NO$. The advantages of using these compounds are that they have a high efficiency of trapping $^{\bullet}NO$, most are crystalline and water soluble, and they do not react with nitrite or $^{\bullet}NO_2$.

The important aspect of using nitronyl nitroxide to quantitate $^{\bullet}NO$ is to find the exact stoichiometry of the reaction between NNR and $^{\bullet}NO$. Akaike and Maeda[49] first reported that the reaction between NNR and $^{\bullet}NO$ is in the ratio of 1 : 1. Because one of the products, $^{\bullet}NO_2$, has the propensity to react with $^{\bullet}NO$ spontaneously (Table I), the ratio of $^{\bullet}NO$ reacting with NNR was actually found to be close to 2 : 1[50]; 1 : 1 is observed only at low concentrations of $^{\bullet}NO$,[51] where the amount of

[46] K. Tsuchiya, M. Yoshizumi, H. Houchi, and R. P. Mason, *J. Biol. Chem.* **275,** 1551 (2000).

[47] E. F. Ullman, J. H. Osiecki, D. G. B. Boocock, and R. Darcy, *J. Am. Chem. Soc.* **94,** 7049 (1972).

[48] J. Joseph, B. Kalyanaraman, and J. S. Hyde, *Biochim. Biophys. Res. Commun.* **192,** 926 (1993).

[49] T. Akaike and H. Maeda, *Methods Enzymol.* **268,** 211 (1996).

[50] N. Hogg, R. J. Singh, J. Joseph, F. Neese, and B. Kalyanaraman, *Free Radic. Res.* **22,** 47 (1995).

[51] Y. Y. Woldman, V. V. Khramtsov, I. A. Grigor'ev, I. A. Kiriljuk, and D. I. Utepbergenov, *Biochem. Biophys. Res. Commun.* **202,** 195 (1994).

$^{\bullet}$NO reacting with $^{\bullet}NO_2$ is negligible. These seemingly disparate observations on the stoichiometry of the reaction between $^{\bullet}$NO and NNR leave the quantitation of $^{\bullet}$NO in question. We examined this stoichiometry so that $^{\bullet}$NO could be quantitated with confidence.

We found that as the concentration of $^{\bullet}$NO increases, the ratio between $^{\bullet}$NO and NNR gradually changes from 1 : 1 to 2 : 1. We present here a step-by-step method to determine $^{\bullet}$NO concentrations using NNR.

The representative nitronyl nitroxide (NNR) used for this study is PTIO. The concentration of PTIO is determined spectrophotometrically ($\varepsilon_{560} = 1020 \pm 50\ M^{-}cm^{-1}$). Different amounts of DEANO at pH 7.4 are used as the source of $^{\bullet}$NO. The concentration of PTIO used is at least three times that of the "total" concentration of $^{\bullet}$NO. The EPR sample is prepared as described in the methods section.

Because the unpaired electron of PTIO interacts with two equivalent nitrogens, the EPR spectrum of PTIO consists of five lines with intensities 1 : 2 : 3 : 2 : 1.[52,53] The EPR spectrum of the product (INR) consists of nine lines with intensities 1 : 1 : 1 : 1 : 1 : 1 : 1 : 1 : 1, which is distinctly different from that of PTIO.

When the concentration of PTIO is in excess of $^{\bullet}$NO, then the resulting EPR spectrum is due to the spectrum of the unreacted NNR and the spectrum of the product, INR (Fig. 2). The low field line of both species (NNR and INR) is taken as the representative peak for the individual compounds as they have the least overlap. The low field lines, marked α (NNR) and β (INR) in Fig. 2(c), correspond to one-ninth of the sum of total intensities of all peaks for each species. In general, an easy way to estimate the concentrations is by measuring the EPR peak heights. However, in this case, NNR and INR have different line widths and line shapes so direct comparison of peak height measurements will not yield correct information on relative concentrations. Double integration of the individual peaks would usually allow determination of the concentration of these species. However, the two peaks overlap; this overlap in conjunction with their different line shapes provides a challenge. For exact quantitation the peaks have to be deconvoluted.

For this purpose, EPR spectra recorded at different concentrations of $^{\bullet}$NO are simulated. The simulation of spectra gives the concentration of the two species by area calculation. The spectral parameters used to simulate the two species are for PTIO ($a_1^N = a_2^N =$ 8.2 G) and INR ($a_1^N = 4.4$ G and $a_2^N = 9.9$ G). The line shape and line width and g variation are optimized, and the spectra are simulated with a correlation coefficient of 0.999. The ratio of the two species, [$^{\bullet}$NO]/[NNR], changes as a result of the addition of different amounts of $^{\bullet}$NO, showing varying stoichiometry (Fig. 5) obtained from simulation of spectra and double integration of low field lines. The stoichiometry of the reaction of $^{\bullet}$NO with NNR is close to

[52] T. Akaike, M. Yoshida, Y. Miyamato, K. Sato, M. Kohno, K. Sasamoto, K. Miyazaki, S. Ueda, and H. Maeda, *Biochemistry* **32,** 827 (1993).

[53] J. S. Nadeau and D. G. B. Boocock, *Anal. Chem.* **49,** 1672 (1977).

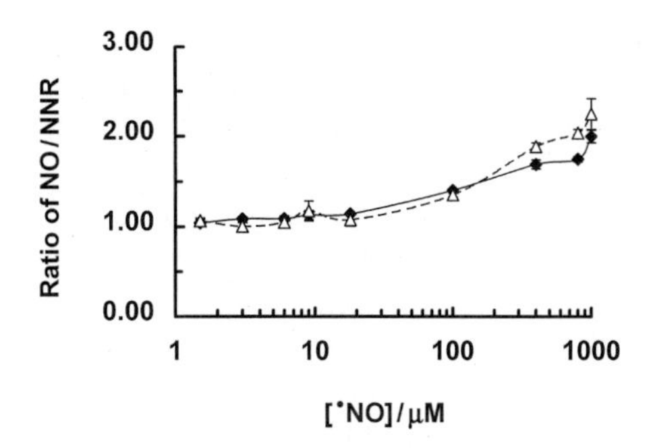

FIG. 5. The stoichiometry of $^{\bullet}$NO to NNR as determined by (Δ) simulation of experimental EPR spectra and (◆) double integration of deconvoluted low-field EPR lines of imino and nitronyl nitroxides. EPR spectra were obtained by adding different concentrations of DEANO to an anaerobic PBS (pH 7.4) solution containing nitronyl nitroxide [PTIO] and incubating for 20 min at 37°. The [PTIO] was always 3× [$^{\bullet}$NO] based on maximal release from DEANO. EPR spectra were simulated with a correlation coefficient of 0.999. Values represent mean of $n = 3$ experiments. Error bars represent standard error.

1 : 1 at low concentrations, [$^{\bullet}$NO] < 20 μM and [PTIO] $= 60$ μM, but thereafter the ratio increases gradually with the increase in [$^{\bullet}$NO] and approaches 2 : 1 at high $^{\bullet}$NO levels.

It is clear that one can use the ratio of 1 : 1 at low [$^{\bullet}$NO], $\leq$20 μM, and 2 : 1 at high [$^{\bullet}$NO], $\geq$100 μM, but the range in between is a problem. To quantitate $^{\bullet}$NO in a systematic manner and to avoid errors due to the apparent change in stoichiometry, the following method is recommended for quantitating standard solutions of $^{\bullet}$NO using PTIO. Because the concentration of $^{\bullet}$NO in saturated aqueous solution cannot exceed $\approx$2 mM (room temperature and 1 atm), we first assume an $^{\bullet}$NO concentration of approximately 2 mM. Second, into a PTIO solution of 60 μM (or greater), introduce a volume of $^{\bullet}$NO solution to achieve a dilution of about 100-fold. Even if the concentration of unknown is only 100 μM, this dilution will result in 1 μM $^{\bullet}$NO; EPR is sensitive enough to measure this amount of INR easily. These conditions should result in a reaction ratio close to 1 : 1 between $^{\bullet}$NO and NNR. Then by double integrating the deconvoluted low field line of PTIO and comparing the area with the standard area obtained from known concentrations of PTIO, the amount of $^{\bullet}$NO reacted with PTIO can be estimated. Another way of calculating the [$^{\bullet}$NO] is by double integrating the area under the low field line of INR and standardizing it with the area of known concentrations of a stable standard nitroxide radical such as 4-phenyl-2,2,5,5-tetramethylimidazoline-1-oxyl,[49] which is available commercially. With this dilution, if no INR is detected with EPR, the [$^{\bullet}$NO] is too low and the amount added will need to be increased. However, this first dilution step will lead to a point where a decision can be made on how to achieve a more favorable dilution. The important aspect is to direct the chemistry so that the stoichiometry of the reaction of $^{\bullet}$NO with NNR is 1 : 1.

To check whether $^{\bullet}NO_2$ competes with NNR compounds, in a separate experiment we introduced gaseous $^{\bullet}NO_2$ to an NNR solution and the EPR spectrum was recorded. As reported earlier,[48] no significant reaction of $^{\bullet}NO_2$ with PTIO was found. There also appears to be no interference from the presence of nitrite for the quantitation of $^{\bullet}NO$ in aqueous solution using PTIO.

Taken together, the use of $Fe^{2+}(MGD)_2$ or PTIO (NNR) combined with EPR can be used for the quantitation of $^{\bullet}NO$. The choice of method for quantitation will depend on the nature of the system being studied.

Other Potential Traps for $^{\bullet}$*NO.* Another method developed to quantitate $^{\bullet}NO$ is based on the use of cheletropic spin traps. A compound such as *o*-quinodimethane, otherwise called nitric oxide cheletropic trap (NOCT), reacts with $^{\bullet}NO$ and forms stable nitroxide radicals. These traps are formed by the photolysis of a parent ketone of the type 1,1,3,3-tetramethylind-2-one.[54] These cheletropic traps are water soluble, thermally stable at physiological temperature, and can be used to monitor $^{\bullet}NO$ with EPR techniques. The spectrum of the $^{\bullet}NO$ adduct consists of three lines with $a^N = 13–15$ G and a g value of 2.005, similar to other nitroxides.[8] These cheletropic traps react with both $^{\bullet}NO$ and $^{\bullet}NO_2$, but the EPR spectrum of both have different a^N values that make it easy to differentiate between the two. The intermediate, which traps the $^{\bullet}NO$ radical, is short-lived and needs to be produced *in situ* by photolysis for each experiment. These compounds are not susceptible to reactions with O_2, $O_2^{\bullet-}$, NO_2^-, or NO_3^-.[55] Although NOCT can be an ideal trap for the quantitation of $^{\bullet}NO$, this method can be time-consuming and the exact amount of intermediate formed is difficult to quantitate.

Another potential probe for the detection of nitric oxide in aqueous solutions is the *aci* form of nitromethane, which reacts with $^{\bullet}NO$ in alkaline solutions (pH > 12) and gives a stable adduct. The EPR signal of the resulting spin adduct gives characteristic hyperfine splitting, which facilitates the detection of $^{\bullet}NO$.[56,57]

Use of classic spin traps, nitrone and nitroso compounds, for the detection of $^{\bullet}NO$ as reported earlier[58,59] are not suitable for quantitation of $^{\bullet}NO$ simply because they do not form stable adducts.[60]

[54] H.-G. Korth, R. Sustman, P. Lommes, T. Paul, A. Ernst, H. de Groot, L. Hughes, and K. U. Ingold, *J. Am. Chem. Soc.* **116,** 2767 (1994).

[55] H.-G. Korth and H. Weber, *in* "Methods in Nitric Oxide Research" (M. Feelisch and J. S. Stamler, eds.), p. 383. Wiley, New York, 1996.

[56] K. J. Reszka, C. F. Chignell, and P. Bilski, *J. Am. Chem. Soc.* **116,** 4119 (1994).

[57] K. J. Reszka, P. Bilski, and C. F. Chignell, *J. Am. Chem. Soc.* **118,** 8719 (1996).

[58] C. M. Arroyo and M. Kohno, *Free Radic. Res. Commun.* **14,** 145 (1991).

[59] L. Pronai, K. Ichimori, H. Nozaki, H. Nakazawa, H. Okino, A. J. Carmichael, and C. M. Arroyo, *Eur. J. Biochem.* **202,** 923 (1991).

[60] S. Pou, L. Keaton, W. Suricharmorn, P. Frigillana, and G. M. Rosen, *Biochim. Biophys. Acta* **1201,** 118 (1994).

Discussion

The EPR methods presented here to quantitate or detect $^{\bullet}NO$ are good and can be made ideal if they are tailored to avoid shortcomings. The issues to address to achieve accurate quantitation of nitric oxide are highlighted here.

1. Designing the protocol to ensure that all $^{\bullet}NO$ reacts with the trap (or probe) under study: This is done by keeping the $^{\bullet}NO$ trap (or probe) in excess ($\geq 3\times$) of $^{\bullet}NO$ to increase the probability of $^{\bullet}NO$ reacting with the trap. Also, the headspace volume of the air-tight container should be minimized to reduce the partitioning of $^{\bullet}NO$ from the aqueous phase. Nitric oxide quantitation is ideally accomplished using anaerobic solutions as $^{\bullet}NO$ reacts spontaneously with oxygen. When $Fe^{2+}(MGD)_2$ is used as a trap, the conversion of Fe^{2+} to Fe^{3+} will be minimized under anaerobic conditions. As seen in Table I, $Fe^{2+}(MGD)_2$ autoxidizes to produce a molecule of superoxide ($O_2^{\bullet -}$). This $O_2^{\bullet -}$ will interfere with $^{\bullet}NO$ measurement due to very fast reactions to produce peroxynitrite. Reducing agents such as $Na_2S_2O_4$ can be used to convert Fe^{3+} back to Fe^{2+}.

2. Ensuring stability of the product. To ensure product stability ($^{\bullet}NO$ + trap), EPR spectra should be collected over a period of time. This will also provide assurance that the reaction of $^{\bullet}NO$ is complete.

3. Avoiding chemistry that may alter the apparent $^{\bullet}NO$ concentration. $Fe^{2+}(MGD)_2$ will reduce nitrite to $^{\bullet}NO$. Unreacted $Fe^{2+}(MGD)_2$ will react with this $^{\bullet}NO$, resulting in overestimation of the true concentration of nitric oxide. However, by setting up the EPR experiment appropriately, potential interference from nitrite can be avoided. Nitrite does not interfere with the nitronyl nitroxide assay. However, $^{\bullet}NO_2$, one of the products of $^{\bullet}NO$ reacting with NNR, can compete for $^{\bullet}NO$. If experiments are carried out at low concentrations of $^{\bullet}NO$ and excess NNR is used, this problem is avoided and the 1 : 1 stoichiometry holds. Also, when using the $^{\bullet}NO$ solution, the slow addition of $^{\bullet}NO$ and vigorous stirring are preferred to minimize locally high concentrations of $^{\bullet}NO$.

4. Producing a standard curve from EPR measurements. In cases where $Fe^{2+}(MGD)_2$ is used, 3-CP or EPR active iron complexes that have comparable EPR spectral/line parameters can be used as a standard to calculate the concentrations of $^{\bullet}NO$. Similarly, when using nitronyl nitroxides to quantitate $^{\bullet}NO$, stable nitroxide radicals that can be accurately quantified spectrophotometrically are used for EPR standards.

Cautionary Notes

Room temperature for EPR studies of $Fe^{2+}(DETC)_2$ is preferred over low temperature because additional signals from $Cu^{2+}(DETC)_2$, from the yeast, could complicate the spectrum at low temperature. Considering the complexity due to the precipitation of NO-$Fe^{2+}(DETC)_2$ from aqueous solutions, other spin traps may be preferred.

Because $Fe^{2+}(MGD)_2$ is toxic above a certain concentration,[61] care needs to be taken in use of this spin trap for *in vivo* and *in vitro* cell studies. Also, care should be taken to avoid or understand possible interference from nitrite.

Use of NNR as a spin trap for $^{\bullet}NO$ in biology is limited by possible interference from $O_2^{\bullet -}$ in the assay.[62] Also, NNR and INR compounds are susceptible to various reducing agents, such as thiols and ascorbate.

Acknowledgment

This work was supported by NIH Grants CA66801, CA81090, and CA84462.

[61] J. L. Zweier, P. Wang, and P. Kuppusamy, *J. Biol. Chem.* **270,** 304 (1995).

[62] R. F. Haselof, S. Zollner, I. A. Kirilyuk, I. A. Grigor'ev, R. Reszka, R. Bernhardt, K. Mertsch, B. Roloff, and I. E. Blasig, *Free. Radic. Res.* **26,** 7 (1997).

[2] *In Vitro* Detection of Nitric Oxide and Nitroxyl by Electron Paramagnetic Resonance

By ANDREI M. KOMAROV, ANDREAS REIF, and HARALD H. H. W. SCHMIDT

Introduction

Nitric oxide (NO) and nitroxyl (NO^-) are closely related reactive nitrogen species, but their chemistry is distinctly different. NO detection in the aqueous phase with the use of electron paramagnetic resonance (EPR) spectroscopy requires NO trapping by iron complexes forming paramagnetic mononitrosyl- or dinitrosyl-iron derivatives.[1] Likewise, nitroxyl reacts with metals to give nitrosyl complexes. Trapping of nitroxyl with ferrihemoproteins and the dithiocarbamate–iron complex can therefore be used for EPR detection of NO^-.[2,3] This article describes the methodology and several caveats of EPR detection of both NO and NO^-, with a special emphasis on their reaction with iron and dithiocarbamate–iron complexes.

[1] A. M. Komarov, *Methods Enzymol.* **359,** [6], 2002 (this volume).

[2] D. A. Bazylinski and T. C. Hollocher, *J. Am. Chem. Soc.* **107,** 1982 (1985).

[3] A. M. Komarov, D. A. Wink, M. Feelisch, and H. H. H. W. Schmidt, *Free Radic. Biol. Med.* **28,** 739 (2000).

0076-6879/02 $35.00

NO, Nitroxyl, and Iron-Nitrosyl Complexes

The electron distribution in the orbitals of NO and O_2 molecules in their ground state is similar, except that the $2p\pi^*$ antibonding orbital of NO contains one unpaired electron (thus making nitric oxide a free radical), whereas oxygen molecules contain two unpaired electrons. Formally, the nitroxyl anion is a reduction product of NO ($NO + 1e^- \rightarrow NO^-$). Nitroxyl is isoelectronic with molecular oxygen and contains two unpaired electrons on separate orbitals. As a result, NO^- exists in a triplet state, when both spins have parallel orientation, or in a singlet state, when the electrons have opposite spins.[4] On binding of NO to ferrous heme, the Fe-N$=$O complex acquires a "bent" shape, which gives the nitrosyl group a nitroxyl character. It is therefore not surprising that nitrosyl complexes of hemoproteins such as cytochromes and hemoglobin may serve as a source of nitroxyl.[5,6]

Typically, NO^- reacts with ferrihemoproteins, forming a ferrous nitrosyl complex, which is detectable by EPR. For example, nitroxyl reacts with methemoglobin, yielding the corresponding nitrosyl complex in which the central iron atom is reduced (Table I).[2–8] Generally, NO has a higher affinity to reduced iron than to ferric iron complexes,[9,10] nevertheless, it reacts with both methemoglobin and metmyoglobin. Methemoglobin reacts with NO as follows[7]:

$$Hb^{3+} + NO \rightarrow Hb^{3+}NO\ (Hb^{2+}NO^{+}) \quad (1)$$

$$Hb^{2+}NO^{+} + OH^{-} \rightarrow Hb^{2+} + NO_2{}^{-} + H^{+} \quad (2)$$

$$Hb^{2+} + NO \rightarrow Hb^{2+}NO \quad (3)$$

On reaction with NO, methemoglobin forms a ferrous iron–nitrosyl complex, which is detectable by EPR.[7] In contrast, NO binding to metmyoglobin yields a ferric nitrosyl complex, which is not detectable by EPR (Table I).[5] In some cases, such as the ferric form of cytochrome *c*, heme irons are also reduced by nitroxyl, without formation of a nitrosyl–iron complex.[11]

The iron(II) dithiocarbamate complex reacts only with NO, whereas the ferric form of the complex reacts with both NO and NO^-. Under aerobic conditions,

[4] M. N. Hughes, *Biochim. Biophys. Acta* **1411,** 263 (1999).

[5] M. A. Sharpe and C. E. Cooper, *Biochem. J.* **332,** 9 (1998).

[6] A. J. Gow and J. S. Stamler, *Nature* **391,** 169 (1998).

[7] V. G. Kharitonov, J. Bonaventura, and V. S. Sharma, *in* "Methods in Nitric Oxide Research" (M. Feelisch and J. S. Stamler, eds.), p. 38. Wiley, Chichester, 1996.

[8] M. E. Murphy and H. Sies, *Proc. Natl. Acad. Sci. U.S.A.* **88,** 10860 (1991).

[9] A. F. Vanin, X. Liu, A. Samoilov, R. A. Stukan, and J. L. Zweier, *Biochim. Biophys. Acta* **1474,** 365 (2000).

[10] D. A. Wink, J. S. Beckman, and P. Ford, *in* "Methods in Nitric Oxide Research" (M. Feelisch and J. S. Stamler, eds.), p. 29. Wiley, Chichester, 1996.

[11] M. P. Doyle, S. N. Mahapatro, R. D. Broene, and J. K. Guy, *J. Am. Chem. Soc.* **110,** 593 (1988).

TABLE I
EPR DETECTION OF IRON–NITROSYL COMPLEXES FORMED BY NO AND NITROXYL

Trapping agent	Reactive nitrogen species	Trapping efficiency
Methemoglobin[a,b]	NO and NO^-	50% (NO); not determined (NO^-)
Metmyoglobin[a–d]	NO^-	80% (NO^-)
Dithiocarbamate–iron(III)[e,f]	NO and NO^-	50–90% (NO); 99% (NO^-)

[a] D. A. Bazylinski and T. C. Hollocher, *J. Am. Chem. Soc.* **107,** 1982 (1985).
[b] V. G. Kharitonov, J. Bonaventura, and V. S. Sharma, *in* "Methods in Nitric Oxide Research" (M. Feelisch and J. S. Stamler, eds.), p. 38. Wiley, Chichester, 1996.
[c] M. A. Sharpe and C. E. Cooper, *Biochem. J.* **332,** 9 (1998). NO reacts with Mb^{3+} to produce the EPR-silent $Mb^{3+}NO$ complex. However, NO^- and Mb^{3+} yield $Mb^{2+}NO$ with EPR signal at $g = 2$.
[d] M. E. Murphy and H. Sies, *Proc. Natl. Acad. Sci. U.S.A.* **88,** 10860 (1991).
[e] A. M. Komarov, D. A. Wink, M. Feelisch, and H. H. H. W. Schmidt, *Free Radic. Biol. Med.* **28,** 739 (2000).
[f] A. F. Vanin, X. Liu, A. Samoilov, R. A. Stukan, and J. L. Zweier, *Biochim. Biophys. Acta* **1474,** 365 (2000).

iron(II) in this complex is oxidized easily, making the trap nonselective for the different redox species of NO.[3]

The chemistry of nitroxyl is complicated by the generation of H_2O_2 and hydroxyl radical ($OH^•$) in the presence of oxygen.[12] Accordingly, we could detect hydroxyl radical formation from the widely used nitroxyl donor Angeli's salt (AS) using EPR and the spin trap 5,5-dimethyl-1-pyrroline *N*-oxide (DMPO; 100 m*M*).[13] These reactive species may, in part, be responsible for the oxidative properties of nitroxyl. Most importantly, H_2O_2 oxidizes ferrous iron complexes readily (Fig. 1). Nitric oxide synthases in the presence of flavins can serve as a source for H_2O_2 as well.[13a] In addition, mixtures of NO and NO^- can become a source of hydroxyl radical.[13b] This should be kept in mind when EPR and iron complexes are utilized for the differentiation between NO and nitroxyl.

The chemistry of NO^- in aerobic conditions may be further complicated by its reaction with molecular oxygen, yielding peroxynitrite (Fig. 1). However, only the triplet state of the molecule reacts with O_2 to give peroxynitrite, whereas Angeli's salt is known to give singlet NO^-.[14]

[12] H. Ohshima, I. Gilibert, and F. Bianchini, *Free Radic. Biol. Med.* **26,** 1305 (1999).
[13] A. M. Komarov, unpublished data (2000).
[13a] A. Reif, Z. Shutenko, M. Feelisch, and H. H. H. W. Schmidt, unpublished data (2001).
[13b] J. M. Fukuto, J. Y. Cho, and C. H. Switzer, *in* "Nitric Oxide Biology and Pathobiology" (L. J. Ignarro, ed.), p. 23. Academic Press, San Diego, 2000.
[14] M. N. Hughes and R. Cammack, *Methods Enzymol.* **301,** 279 (1999).

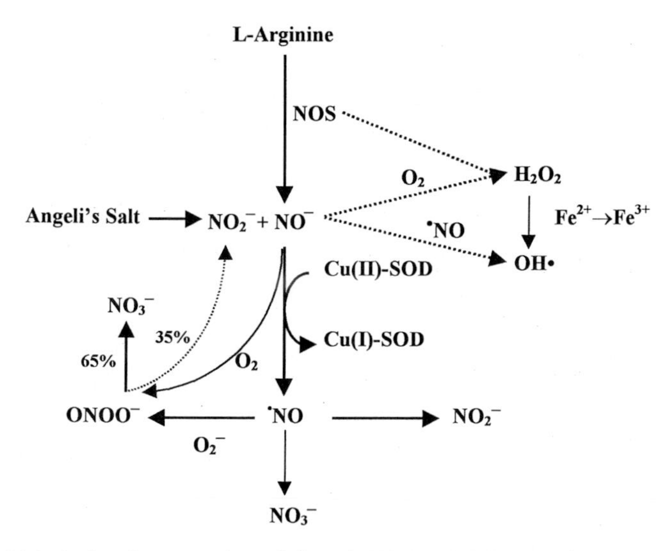

FIG. 1. Biologically relevant reactions of nitroxyl (NO^-) and nitric oxide ($^•NO$). Nitroxyl can be generated by Angeli's salt (AS) or enzymatically from L-arginine by nitric oxide synthase (NOS), especially when H_4Bip levels are subsaturating; it is further converted to NO by superoxide dismutase (SOD). SOD in this case is a stoichiometric and not catalytic reactant. Most importantly, the singlet state of nitroxyl reacts with O_2 to give H_2O_2, which can release hydroxyl radical ($OH^•$) on decomposition. Alternatively, NOS or the presence of flavins can serve as a source for H_2O_2, and mixtures of NO and NO^- can generate the hydroxyl radical as well. Only the triplet state of the NO^- reacts with O_2 to give peroxynitrite. AS is known to give singlet NO^-; triplet NO^- is generated by AS via photolysis. Another source of peroxynitrite is the reaction of NO and superoxide anion. Both peroxynitrite and NO yield nitrite and nitrate as stable products. In aqueous, neutral solutions, peroxynitrite decomposes to NO_2^- and NO_3^- with a ratio of approximately 1 : 2 in the favor of nitrate [S. Pfeiffer, A. C. F. Gorren, K. Schmidt, E. R. Werner, B. Hansert, D. S. Bohle, and B. Mayer, *J. Biol. Chem.* **272,** 3465 (1997)]. Solid and broken lines indicate major and minor reaction pathways, respectively.

Experimental Procedures

Trapping of NO and Nitroxyl

Superoxide dismutase (SOD) from bovine erythrocytes, myoglobin from horse heart, and dithionite can be obtained from Sigma (St. Louis, MO). The NO donor *S*-nitroso-*N*-acetyl penicillamine (SNAP) is available from Cayman Chemical Company (Ann Arbor, MI) and should be dissolved in ethanol. The SNAP concentration is determined spectrophotometrically using an extinction coefficient $\varepsilon_{330} = 711\ M^{-1}\ cm^{-1}$.[15] Dilutions of the nitroxyl donor Angeli's salt should be made in

[15] W. R. Mathews and S. W. Kerr, *J. Pharmacol. Exp. Toxicol.* **267,** 1529 (1993).

10 mM of NaOH immediately before use to avoid degradation. Its concentration is determined using an extinction coefficient of $\varepsilon_{250} = 8000\ M^{-1}\ \text{cm}^{-1}$.[3] Angeli's salt to date is the best known donor compound for nitroxyl and is widely used as a source of nitroxyl *in vitro*. However, the chemistry of AS is rather problematic, as is the quality and availability from commercial sources. On dissolution and even in solid form, if not stored starkly desiccated, AS decomposes spontaneously, yielding equal amounts of nitroxyl and nitrite.[16,17] Validation of the identity and purity of AS is therefore crucial in experiments using this compound and should be carried out as an internal control in all studies. It is also important to rule out that AS effects are not due to nitrite release.

MGD can be purchased from Calbiochem (San Diego, CA). A fresh stock solution of MGD (100 mM) is prepared in double-distilled water and is kept on ice until use. The stock solution of $FeSO_4$ (10 mM) is prepared in double-distilled water and is kept under pure argon. All experiments are performed in 100 mM N-(2-hydroxyethyl)piperazine-N'-(2-ethanesulfonic acid) buffer (pH 7.5) containing 1 mM of $CaCl_2$ and 1 mM of magnesium acetate. Aliquots of the MGD and Fe stock solutions are mixed to yield final concentrations of 5 mM of MGD and 1 mM of $FeSO_4$ (lower final concentrations of MGD and Fe can be used, but it is important to maintain a 5 : 1 MGD to Fe ratio).[1] Immediately thereafter, the resulting MGD–Fe complex is incubated with SNAP, Angeli's salt, or sodium nitrite in the absence or presence of SOD (3000 U/ml) for 1 hr at room temperature. SOD converts nitroxyl to NO, which can be elegantly used to check the specificity of the trapping agent toward nitroxyl. EPR spectra are recorded at room temperature with an X-band EPR spectrometer using 50-μl capillaries.

Enzymatic Reactions

Nitric oxide synthases (iNOS and nNOS) can be obtained from Calbiochem or Sigma (St. Louis, MO); cofactors L-arginine and N-nitro-L-arginine methyl ester (L-NAME) are available from Sigma (St. Louis, MO). $^{15}N_2$-guanidino-L-arginine can be purchased from Cambridge Isotope Laboratories (Woburn, MA). Typically, several micrograms per milliliter of enzyme are required for this assay. Reaction conditions for NO or nitroxyl detection from NOS are listed in Table II.[18–22]

[16] R. Z. Pino and M. Feelisch, *Biochem. Biophys. Res. Commun.* **201,** 54 (1994).

[17] D. A. Wink and M. Feelisch, *in* "Methods in Nitric Oxide Research" (M. Feelisch and J. S. Stamler, eds.), p. 403. Wiley, Chichester, 1996.

[18] H. H. H. W. Schmidt, H. Hofmann, U. Schindler, Z. S. Shutenko, D. D. Cunningham, and M. Feelisch, *Proc. Natl. Acad. Sci. U.S.A.* **93,** 14492 (1996).

[19] Y. Xia and J. L. Zweier, *Proc. Natl. Acad. Sci. U.S.A.* **94,** 12705 (1997).

[20] Y. Xia, A. J. Cardounel, A. F. Vanin, and J. L. Zweier, *Free Radic. Biol. Med.* **29,** 793 (2000).

[21] A. J. Hobbs, J. M. Fukuto, and L. J. Ignarro, *Proc. Natl. Acad. Sci. U.S.A.* **91,** 10992 (1994).

[22] H. Yoneyama, H. Kosaka, T. Ohnishi, T. Kawazoe, K. Mizoguchi, and Y. Ichikawa, *Eur. J. Biochem.* **266,** 771 (1999).

TABLE II

EXPERIMENTAL CONDITIONS FOR NO/NO^- DETECTION FROM NITRIC OXIDE SYNTHASES[a]

Type of NOS	Reaction conditions
nNOS (NOS I)[b]	50 n*M* CaM, 10 μM H_4Bip, ≥3 μM Ca^{2+}, 10–100 μM Arg, 0.1–1 m*M* NADPH, 5 μM FAD, 5 μM FMN, 5000 U/ml SOD[c]
nNOS (NOS I)[d]	10 μg/ml CaM, 10 μM H_4Bip, 0.5 m*M* Ca^{2+}, 1 m*M* Arg, 1 m*M* NADPH[e]
nNOS (NOS I)[d]	10 μg/ml CaM, 10 μM H_4Bip, 0.5 m*M* Ca^{2+}, 0.5 m*M* Arg, 1 m*M* NADPH, 10 μg/ml BSA[f]
nNOS (NOS I)[b]	2.5 μg/ml CaM, 5 μM H_4Bip, 1 m*M* Ca^{2+}, 0.5 m*M* Arg, 1 m*M* NADPH, 50 μM FAD[g]
iNOS (NOS II)[b]	5 μM H_4Bip, 0.5 m*M* Arg, 1 m*M* NADPH, 50 μM FAD, 10,000 U/ml SOD[g]
nNOS (NOS I)[h]	100 μg/ml CaM, 10 μM H_4Bip, 0.2 m*M* Ca^{2+}, 1 m*M* Arg, 1 m*M* NADPH, 1 mg/ml BSA, 40 U/ml SOD, 100 U/ml catalase[i]

[a] CaM, calmodulin; H_4Bip, tetrahydrobiopterin; Arg, L-arginine; BSA, bovine serum albumin.
[b] SOD-dependent formation of NO from NOS.
[c] H. H. H. W. Schmidt, H. Hofmann, U. Schindler, Z. S. Shutenko, D. D. Cunningham, and M. Feelisch, *Proc. Natl. Acad. Sci. U.S.A.* **93,** 14492 (1996).
[d] NOS-dependent formation of MGD–Fe–NO complex.
[e] Y. Xia and J. L. Zweier, *Proc. Natl. Acad. Sci. U.S.A.* **94,** 12705 (1997).
[f] Y. Xia, A. J. Cardounel, A. F. Vanin, and J. L. Zweier, *Free Radic. Biol. Med.* **29,** 793 (2000).
[g] A. J. Hobbs, J. M. Fukuto, and L. J. Ignarro, *Proc. Natl. Acad. Sci. U.S.A.* **91,** 10992 (1994).
[h] Inhibition of NOS by MGD–Fe complex.
[i] H. Yoneyama, H. Kosaka, T. Ohnishi, T. Kawazoe, K. Mizoguchi, and Y. Ichikawa, *Eur. J. Biochem.* **226,** 771 (1999).

Examples of Nitric Oxide and Nitroxyl Detection by EPR

Nitroxyl Generation by Angeli's Salt

On mixing of MGD-Fe with AS, the characteristic triplet signal of the MGD–Fe–NO complex can be detected ($a^N = 12.5$ G and $g_{iso} = 2.04$) (Fig. 2), which is increasing with time. After 1 hr of incubation at room temperature, the yield of the NO complex is complete.[3] AS has a half-life of about 17 min at 25° and 2.8 min at 37°.[4] The SNAP-dependent EPR signal has similar kinetics and therefore SNAP can be used as a standard in this experiment (Figs. 2a and 2b). The conversion of nitroxyl to NO is not necessary for the generation of the MGD–Fe–NO complex. This can be demonstrated by the addition of SOD (3000 U/ml) to the reaction mixture (Fig. 2c): the ratio of the EPR signal amplitude in the AS sample with SOD versus the corresponding sample without SOD is 1.1 ± 0.1 ($n = 3$).[3] The lack of effect of SOD on MGD–Fe–NO complex formation by AS may be explained in part by the inactivation of SOD in the presence of dithiocarbamate. Nevertheless, we have demonstrated previously that 5 m*M* of MGD reduced SOD activity by only 25% even when the initial SOD activity was 20-fold less than in the present

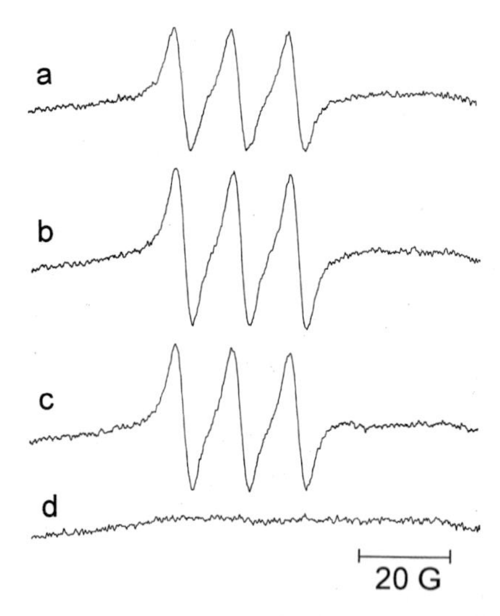

FIG. 2. MGD–Fe EPR signals with NO and nitroxyl. EPR signal of the MGD–Fe–NO complex formed under aerobic conditions (5 m*M* of MGD and 1 m*M* of $FeSO_4$, final concentration) after a 1-hr incubation at room temperature with 10 μM SNAP (spectrum a); 12.6 μM of AS (spectrum b); 12.6 μM of AS plus 3000 U/ml of SOD (spectrum c); and 10 μM of $NaNO_2^-$ (spectrum d). All experiments were performed in 100 m*M* of *N*-(2-hydroxyethyl)piperazine-*N*′-(2-ethanesulfonic acid), 1 m*M* $CaCl_2$, 1 m*M* of magnesium acetate, pH 7.5. The ratio of EPR signal amplitude in the AS sample with SOD and the sample without SOD (AS + SOD/AS) was 1.1 ± 0.1 ($n = 3$), and the yield of the MGD–Fe–NO complex after a 1-hr incubation with AS was 99 ± 3% ($n = 3$) based on the SNAP standard. Data are presented as mean ± standard error. EPR conditions are as follows: microwave frequency, 9.72 GHz; microwave power, 100 mW; modulation frequency, 100 kHz; modulation amplitude, 2.5 G; center field, 3415 G; scan rate, 30 G/min; and time constant, 0.5 sec. Reprinted from A. M. Komarov, D. A. Wink, M. Feelisch, and H. H. H. W. Schmidt, *Free Radic. Biol. Med.* **28,** 739 (2000), copyright © 2000, with permission from Elsevier Science.

experiment.[23] The reactivity of MGD–Fe with NO is further demonstrated by using SNAP to generate a NO complex. The possibility that MGD–Fe–NO is generated from AS-derived nitrite can be ruled out, as nitrite does not produce the MGD–Fe–NO complex under identical conditions (Fig. 2d).

EPR Detection of Reaction Products from Nitric Oxide Synthase

Reaction conditions for the enzymatic conversion of L-arginine to NO or nitroxyl are listed in Table II. The primary reaction product of NOS is a subject of debate and might be regulated by the availability of H_4Bip.[23a] To further

[23] C.-S. Lai and A. M. Komarov, *in* "Bioradicals Detected by ESR Spectroscopy" (H. Ohya-Nishiguchi and L. Packer, eds.), p. 163. Birkhauser Verlag, Basel, 1995.

[23a] S. Adak, Q. Wang, and D. J. Stuehr, *J. Biol. Chem.* **43,** 33554 (2000).

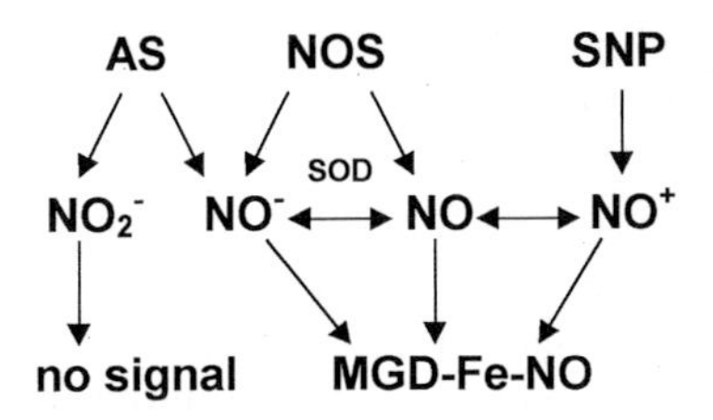

FIG. 3. Diagram of AS decomposition to nitrite or nitroxyl yielding MGD–Fe–NO EPR signals similar to NO and independently of SOD. NOS generates either NO^- or NO (in the presence of SOD). Sodium nitroprusside (SNP) is the source of nitrosonium ion (NO^+), which is detectable by EPR using the MGD–Fe trap [A. M. Komarov, D. Mattson, M. M. Jones, P. K. Singh, and C.-S. Lai, *Biochem. Biophys. Res. Commun.* **195,** 1191 (1994)]. Thus, the MGD–Fe trap and EPR allow direct detection of AS-derived NO^-, SNP-derived NO^+ and NO, and does not discriminate between NOS-derived NO and NO^-. Reprinted from A. M. Komarov, D. A. Wink, M. Feelisch, and H. H. H. W. Schmidt, *Free Radic. Biol. Med.* **28,** 739 (2000), copyright © 2000, with permission from Elsevier Science.

complicate matters, the enzyme is also capable of producing H_2O_2, even when H_4Bip levels are saturating,[13a] which might interfere with EPR measurements (see earlier discussion). In many cases, attempts to detect NO in the absence of SOD directly from purified enzyme were unsuccessful.[18,21]

Several EPR experiments argue for direct NO production from NOS: the incubation of MGD–Fe trap with nitric oxide synthase and cofactors yields the characteristic MGD–Fe–NO complex,[19,20] which can be prevented by NOS inhibitors.[19] Furthermore, the MGD–Fe–^{15}NO complex is formed if ^{15}N-labeled L-arginine is used as a substrate.[19] The interpretation of these experiments, however, requires some caution, as both NO and nitroxyl may contribute an "-NO group" to the MGD–Fe–NO complex (Fig. 3).[3,24]

Ideally, the MGD–Fe(II) trap should yield an EPR-visible NO complex only when reacting with NO, but not with nitroxyl to obtain reliable results. However, under aerobic conditions, MGD–Fe(II) is partially oxidized to MGD–Fe(III), which is capable of reacting with NO^-, yielding an EPR-active MGD–Fe–NO complex as well.[3] Lipophilic dithiocarbamate–iron complexes may be more stable to oxidation in aerobic conditions, but insolubility in water is the obvious disadvantage of these trapping agents. Another possibility of avoiding iron oxidation in the MGD–Fe complex is the addition of 10 m*M* ascorbate as an antioxidative agent,[20] which, however, will cause secondary NO release from nitrite present in NOS samples.[25] Thus the use of ascorbate cannot be recommended for the detection of the primary NOS reaction product. The rate constant of NO scavenging by hydrophilic iron(III) and iron(II) complexes is in the magnitude

[24] A. M. Komarov, D. Mattson, M. M. Jones, P. K. Singh, and C.-S. Lai, *Biochem. Biophys. Res. Commun.* **195,** 1191 (1994).

[25] D. Cornforth, *in* "Nitric Oxide: Principles and Actions" (J. Lancaster, ed.), p. 259. Academic Press, San Diego, 1996.

of 10^8 M^{-1} s^{-1} (the upper limit).[26,27] Two mechanisms have been proposed for this reaction.[9,26] According to the first hypothesis, the MGD–Fe(III)–NO complex is reduced to MGD–Fe(II)–NO in the presence of reductants such as ascorbate or another MGD–Fe complex (according to this mechanism, NO trapping efficiency is close to 100%).[9,26] The alternative mechanism is analogous to the reaction of NO with Fe(III) hemoproteins [Eqs. (1)–(3)]: MGD–Fe(III)–NO is converted to the MGD–Fe(II)–NO^+ complex, which is then hydrolyzed to nitrite and MGD–Fe(II), with the latter reacting with a second molecule of NO to yield MGD–Fe(II)–NO (NO trapping efficiency is thus 50%).[9] The net result of this intricate redox chemistry of the MGD–Fe complex in aerobic conditions is that the trap is not selective for NO versus NO^-.

Furthermore, SOD does not affect NOS-dependent formation of the MGD –Fe–NO complex.[20] This can be explained easily by the fact that the trapping agent is not selective for either NO or nitroxyl, highlighting the importance of proper control experiments. Alternatively, the presence of ferric iron in the MGD–Fe complex is detectable by absorption spectroscopy at 340, 385, and 520 nm (extinction coefficients 20,000, 15,000, and 3000 M^{-1}, respectively) or by EPR spectroscopy of frozen aqueous solutions at $g = 4.3$ (77K; characteristic EPR signal of ferric iron),[9] which can be used to exclude MGD–Fe(III) formation.

When NOS-dependent MGD–Fe–NO complex formation and L-citrulline formation are compared, the later is always found to be higher.[19] This may be due to the inhibition of purified NOS on exposure to the dithiocarbamate–iron complex (for MGD–Fe complex $IC_{50} = 25.1 \pm 2.9\ \mu M$).[22] It should be noted that in living cells, the hydrophilic dithiocarbamate–iron complex and NOS enzyme are present in separate compartments, which minimizes the possibility of direct NOS inhibition.

EPR Detection of Nitrosylmyoglobin Generation

Metmyoglobin can be used as a trapping agent for nitroxyl.[2,8] Nitrosylmyoglobin (MbNO) is prepared by mixing a 100 μM solution of Mb^{3+} with 100 μM Angeli's salt.[11,28] The rate of MbNO generation in this reaction is 6.4×10^5 M^{-1} s^{-1}.[29] Alternatively, 90 μM of Mb^{3+} is incubated with 150 μM of NO or a NO donor in the presence of 1 mM dithionite in anaerobic conditions.[5] Most importantly, under aerobic conditions, NO and Mb^{2+} do not form MbNO but rather oxyMb is formed, which reacts further with NO to generate Mb^{3+} and nitrate. EPR measurement of the MbNO complex is performed at 77K in a quartz finger

[26] S. Fujii and T. Yoshimura, *Coord. Chem. Rev.* **198,** 89 (2000).

[27] S. V. Paschenko, V. V. Khramtsov, M. P. Skatchkov, V. F. Plysnin, and E. Bassenge, *Biochem. Biophys. Res. Commun.* **225,** 577 (1996).

[28] E. A. Konorev, J. Joseph, and B. Kalyanaraman, *FEBS Lett.* **378,** 111 (1996).

[29] J. Goretski and T. C. Hollocher, *J. Biol. Chem.* **263,** 2316 (1988).

Dewar (Wilmad, Buena, NJ) using the following spectrometer conditions: microwave frequency, 9.26 GHz; microwave power, 45 mW; modulation frequency, 100 KHz; modulation amplitude, 3.2 G, center field, 3214 G; scan rate, 100 G/min; time constant 64 ms.[5,28] MbNO is detected by EPR as a broad absorption peak at $g = 2.08$.[28] The trapping efficiency of nitroxyl by Mb^{3+} has been reported to be between 16 and 80%.[2,30] Subsequently, molecular oxygen slowly displaces the NO group from MbNO, yielding Mb^{3+} and nitrate.[28] NO binding to Mb^{3+} only causes a decrease of the high-spin $g = 6$ signal of Mb^{3+}, indicating formation of an EPR-silent Mb^{3+}–NO complex.[5]

Conclusion

EPR detection of iron–nitrosyl complexes could be used for the quantification of NO and nitroxyl generation *in vitro*. However, in the case of ferric compounds, formation of the ferrous nitrosyl complex not always indicates nitroxyl production. Both NO and nitroxyl generate EPR-visible iron–nitrosyl complexes from methemoglobin as well as the ferric iron dithiocarbamate complex. Furthermore, the ease of aerobic oxidation for iron dithiocarbamate traps makes them nonspecific for the different redox species of NO.[3] However, metmyoglobin forms an EPR-visible nitrosyl–iron complex with nitroxyl and an EPR-silent Mb^{3+}–NO complex with NO and is therefore usable for discriminating between nitroxyl and NO.

[30] J. J. Poderoso, M. C. Carreras, F. Schöpfer, C. L. Lisdero, N. A. Riobo, C. Giulivi, A. D. Boveris, A. Boveris, and E. Cadenas, *Free Radic. Biol. Med.* **26,** 925 (1999).

[3] Iron Dithiocarbamate as Spin Trap for Nitric Oxide Detection: Pitfalls and Successes

By ANATOLY F. VANIN, ALBERT HUISMAN, and ERNST E. VAN FAASSEN

Introduction

Various derivatives of dithiocarbamate ligands are known[1–5] to enhance the affinity of Fe^{2+} tremendously for the volatile radical nitric oxide (NO). This property of iron–dithiocarbamate complexes was used successfully for scavenging

[1] L. Cambi and A. Cagnasso, *Atti Accad. Naz. Lincei Cl. Sci. Fis. Mat. Natur. Rend.* **254,** 809 (1931).
[2] J. Gibson, *Nature* **196,** 64 (1961).
[3] B. A. Goodman, J. B. Raynor, and M. C. R. Symons, *J. Chem. Soc.* (A) 2572 (1969).
[4] H. Büttner and R. D. Feltham, *Inorg. Chem.* **11,** 971 (1972).
[5] O. A. Ileperuma and R. D. Feltham, *Inorg. Chem.* **16,** 1876 (1977).

0076-6879/02 $35.00

and detection of NO in animal organisms for the first time at the Semenov Institute in Russia.[6–8] Binding NO with hydrophobic Fe^{2+}–diethyldithiocarbamate (Fe^{2+}-DETC) complexes resulted in the formation of paramagnetic mononitrosyl iron complexes with DETC (MNIC-DETC) in tissues, which can be detected and evaluated by electron paramagnetic resonance (EPR) spectroscopy at 77K and ambient temperature.[6–8] Subsequently, researchers from the United States proposed using water-soluble Fe^{2+} complexes with *N*-methyl-*D*-glucamine dithiocarbamate (MGD) for the detection and quantification of NO in living systems.[9,10]

This technique is strongly reminiscent of the spin trapping method in which diamagnetic molecules (the spin traps) are added to detect short-living paramagnetic radical species. The reaction between the trap and the radical produces a long-living paramagnetic adduct, which is subsequently detected by EPR. Therefore, the straightforward binding of NO to the Fe^{2+}–dithiocarbamate complex is often referred to as a trapping reaction of NO. This trapping reaction serves as the main (but not the only) pathway for MNIC formation. This pathway is referred to as pathway I (Scheme 1):

$$Fe^{2+}\text{–(dithiocarbamate)}_3 + NO \longrightarrow NO\text{–}Fe^{2+}\text{–(dithiocarbamate)}_2 + \text{dithiocarbamate} \qquad \text{(Scheme 1)}$$

Following these early developments, iron–dithiocarbamates have been used in numerous studies concerning NO in biological systems. The physicochemical properties of the Fe–dithiocarbamate traps and MNIC adducts have been determined. Various derivatives of dithiocarbamate were synthesized. The peculiarities of MNIC formation and methodological aspects were investigated. Most applications focus on the detection of endogenous NO production in animal tissue and cell culture. The results of trapping experiments have been compiled in a number of excellent reviews and monographs,[11–15] and are not repeated here. The purpose of this article is to identify and discuss the possible pitfalls and artifacts that may

[6] A. F. Vanin, P. I. Mordvintcev, and A. L. Kleschyov, *Stud. Biophys.* **102,** 135 (1984).

[7] V. J. Varich, A. F. Vanin, and L. M. Ovsyannikova, *Biofizika* (*Rus*) **32,** 1064 (1987).

[8] L. N. Kubrina, W. S. Caldwell, P. I. Mordvintcev, I. V. Malenkova, and A. F. Vanin, *Biochim. Biophys. Acta* **1099,** 233 (1992).

[9] A. M. Komarov, D. Mattson, M. M. Jones, P. K. Singh, and C.-S. Lai, *Biochem. Biophys. Res. Commun.* **195,** 1191 (1993).

[10] A. M. Komarov and C.-S. Lai, *Biochim. Biophys. Acta* **1272,** 29 (1995).

[11] Y. A. Henry, A. Guissani, and D. Ducastel, "Nitric Oxide Research from Chemistry to Biology: EPR Spectroscopy of Nitrosylated Compounds," p. 293. RG Landes, Austin, TX, 1996.

[12] B. Kalyanaraman, *Methods Enzymol.* **268,** 168 (1996).

[13] D. J. Singel and J. R. Lancaster, *in* "Methods in NO Research" (M. Feelish and J. S. Stamler, eds.), p. 341. Wiley, New York, 1996.

[14] A. F. Vanin and A. L. Kleschyov, *in* "Nitric Oxide in Transplant Rejection and Antitumor Defense" (S. Lukiewicz and J. L. Zweier, eds.), p. 49. Kluwer Academic, Norwell, MA, 1998.

[15] A. F. Vanin, *Methods Enzymol.* **301,** 269 (1998).

be encountered when using Fe–dithiocarbamate traps. The interpretation of such trapping experiments is particularly complicated by the redox activity of both Fe–dithiocarbamate complexes and their nitroso derivatives. This redox activity affects the yield of MNIC adducts and is particularly troublesome in the complicated biochemistry found in biosystems. However, we do not feel that these pitfalls limit the usefulness of Fe–dithiocarbamate complexes for the detection of NO. Rather, these complications should be considered as "points for growth," ensuring the further development and refinement of this method. This article demonstrates that redox chemistry may even be turned into a definite advantage in certain cases, providing new information as well as enhanced sensitivity for the detection of NO.

This article is organized into four major sections. The two sections following this introduction are devoted to the redox activities of the Fe–dithiocarbamate traps and of the MNIC adducts, respectively. The final section reports several illustrative trapping experiments.

Redox Activity of Iron–Dithiocarbamates

Pathways for Adduct Formation

The straightforward binding of NO to the Fe^{2+}–dithiocarbamate complex was described previously. It is the most important reaction mechanism for adduct formation and is referred to as pathway I. It has become clear that redox chemistry provides alternative pathways from the NO trap to the paramagnetic MNIC adduct.

In aqueous solution, Fe^{2+}–dithiocarbamate complexes are oxidized easily by dissolved oxygen to form Fe^{3+}–dithiocarbamate complexes. This reaction is clearly visible to the naked eye as the color of the solution turns from colorless to deep orange or black.[16,17] The rate constant of the reaction for water-soluble Fe^{2+}–MGD complexes is determined[18] as $5 \times 10^5 \, M^{-1} \, s^{-1}$ at ambient temperature. The reaction is accompanied by O_2 consumption from the solution. In contrast, no oxygen is consumed by solutions of Fe^{3+}–$(MGD)_3$ complexes.[17] The Fe^{3+} ion at the center of this complex is paramagnetic and usually forms a mixture of high-spin ($S = 5/2$) and low-spin ($S = 1/2$) states.[19] EPR absorption of the complex does not interfere with EPR detection of the MNIC adducts.[17]

However, Fe^{3+}–(dithiocarbamate)$_3$ complexes may introduce an artifact by consuming NO molecules. This reaction results in the formation of diamagnetic iron nitrosyl derivatives, respectively.[16–19] The latter are rather unstable and may

[16] K. Tsuchiya, M. Yoshizumi, H. Houchi, and R. P. Mason, *J. Biol. Chem.* **275,** 1551 (2000).

[17] A. F. Vanin, X. Liu, A. Samouilov, R. A. Stukan, and J. L. Zweier, *Biochim. Biophys. Acta* **1474,** 365 (2000).

[18] K. Tsuchiya, M. Takasugi, K. Minakuchi, and K. Fukuzawa, *Free Radic. Biol. Med.* **21,** 733 (1996).

[19] C. McGrath, C. O'Connor, C. Sangregorio, J. Seddon, E. Sinn, F. Sowrey, and N. Young, *Inorg. Chem. Comm.* **2,** 536 (1999).

transform spontaneously into paramagnetic MNIC–dithiocarbamates.[16–18,20] This sequence of events (oxidation to Fe^{3+}, NO binding, decay into MNIC) is referred to as pathway II. This pathway was proposed[17,18] in analogy to the reductive nitrosylation of ferrihemoproteins proposed by Hoshino and colleagues.[21] According to that mechanism, the NO molecule in the Fe^{3+}–NO–(dithiocarbamate)$_2$ complex reduces iron to the Fe^{2+} state, followed by NO^+ release from the complex. The NO^+ subsequently hydrolyzes to the nitrite ion.[17] Note that the release of NO^+ leaves the complex as a Fe^{2+}–(dithiocarbamate)$_2$ with high affinity for binding NO. The result is a paramagnetic MNIC–dithiocarbamate complex. The final step of pathway II has the form shown in Scheme 2:

$$\begin{aligned} Fe^{3+}\text{–(dithiocarbamate)}_3 + 2NO \longrightarrow\ & NO\text{–}Fe^{2+}\text{–(dithiocarbamate)}_2 \\ & + \text{dithiocarbamate} + NO_2{}^- \qquad \text{(Scheme 2)} \end{aligned}$$

This final step can be accelerated sharply by the addition of various reducing agents as the latter transform the diamagnetic NO–Fe^{3+}–(dithiocarbamate)$_2$ complexes to paramagnetic MNIC.[17]

MNIC formation via pathway II greatly complicates quantitative analysis of data from NO trapping experiments. The problem centers on the number of NO molecules needed to produce a single MNIC adduct. According to pathway I, a single NO molecule is sufficient. According to pathway II, two NO molecules are necessary. The first NO is released in the form of a nitrite ion, and the second is needed for the paramagnetic MNIC. Therefore, the adduct yield via pathway II is only half the NO yield in cells and tissues. In actual experiments, pathway I would still be open and bring the total adduct yield per NO to a number somewhere between a half and one. Accurate assessment of total NO production would only be possible by blocking pathway II, e.g., by choosing deoxygenated or reductive conditions that maintain Fe–dithiocarbamate complexes in the Fe^{2+} state. This is not feasible in biological systems.

To complicate matters even more, a third pathway (III) has been described.[22,23] It assumes that NO modulates only the redox potential when ligating to the metal center. It is hypothesized that the shift in redox potential allows NO–Fe^{3+}–(dithiocarbamate)$_3$ complexes to be reduced by nonnitrosylated Fe^{3+}–dithiocarbamate complexes. This reaction is accompanied by the release of dithiocarbamate disulfide. The reaction is referred to as pathway III (Scheme 3):

$$\begin{aligned} 2Fe^{3+}\text{–(dithiocarbamate)}_3 + NO \longrightarrow\ & NO\text{–}Fe^{2+}\text{–(dithiocarbamate)}_2 \\ + Fe^{3+}\text{–(dithiocarbamate)}_3 + 1/2\ & \text{dithiocarbamate disulfide} \qquad \text{(Scheme 3)} \end{aligned}$$

[20] S. Fujii, T. Yoshimura, and H. Kamada, *Chem. Lett.* 785 (1996).

[21] M. Hoshino, M. Maeda, R. Konishi, H. Seki, and P. C. Ford, *J. Am. Chem. Soc.* **118,** 5702 (1996).

[22] S. Fujii, K. Kobayashi, S. Tagawa, and T. Yoshimura, *J. Chem. Soc. Dalton Trans.* 3310 (2000).

[23] S. Fujii and T. Yoshimura, *Coord. Chem. Rev.* **198,** 88 (2000).

In this third pathway, NO molecules would not act as agents that reduce Fe^{3+} directly. They remain bound to the iron center throughout and can be detected with EPR spectroscopy after the diamagnetic Fe^{3+}–NO–(dithiocarbamate)$_3$ is reduced to paramagnetic MNIC adducts by Fe^{3+}–dithiocarbamate complexes.

If confirmed, pathway III would enhance the reliability of NO detection greatly using Fe–dithiocarbamates even under oxidizing conditions in which the complexes accumulate in cells and tissues in the Fe^{3+} state. The second-order rate constant of the reaction NO binding to Fe^{3+}–dithiocarbamate complexes is reported as $4.8 \times 10^8\ M^{-1}\ s^{-1}$ (calculated[22] at room temperature for Fe^{3+} complexes with *N*-dithiocarboxysarcosine, DTCS). A comparable value of $1.1 \times 10^8\ M^{-1}\ s^{-1}$ has been reported[24] for Fe^{2+}–L-proline dithiocarbamate. Both numbers are large enough to ensure efficient NO trapping *in vivo* because they exceed the binding constant $3.7 \times 10^7\ M^{-1}\ s^{-1}$ reported for oxyhemoglobin.[25,*] It implies that modest quantities of hemoglobin will not prevent NO trapping by Fe^{3+}–(dithiocarbamate)$_3$ via pathway III. Subsequent rapid reduction of NO–Fe^{3+}–(dithiocarbamate)$_3$ complexes by Fe^{3+}–(dithiocarbamate)$_3$ to paramagnetic MNIC adducts with the second-order rate constant $2.5 \times 10^7\ M^{-1}\ s^{-1}$ would provide a reliable quantification of *in vivo* NO production with Fe–dithiocarbamates.

In principle, the reduction of Fe^{3+} to Fe^{2+} by NO molecules can be reversed by the presence of oxygen. The outcome of these antagonistic reactions ultimately depends on the reaction rates and on the relative concentrations of the participating molecules. A high reaction rate of $1.1 \times 10^8\ M^{-1}\ s^{-1}$ was reported[24] between NO and Fe^{2+}–L-proline dithiocarbamate. This compares favorably with a much lower rate reported[26] for the reaction between oxygen and Fe^{2+}–dithiocarbamate ($5 \times 10^5\ M^{-1}\ s^{-1}$). For that reason, we expect that the reduction step will only be reversed by oxygen if NO levels are very low. We note the low stationary concentrations of oxygen in animal tissues *in vivo* (20–30 μM) as compared with those of NO. The latter are estimated to range from zero up to a maximum of

*We note that Pou *et al.*[32] quote much lower values for the rate constant for the spin trapping of NO with Fe^{2+}–(DTCS)$_2$ or Fe^{2+}–(MGD)$_2$: 1.7×10^6 or $1.2 \times 10^6\ M^{-1}\ s^{-1}$, respectively. The authors determined their numbers from competitive NO trapping between oxyhemoglobin and Fe–dithiocarbamates. We consider the values of Fujii *et al.*[22] and Pashenko *et al.*[24] more reliable because we fear that the oxyhemoglobin competition assay may be compromised by interference between the oxyhemoglobin and the dithiocarbamate traps. The addition of oxyhemoglobin to anaerobic Fe^{2+}–dithiocarbamate solutions could result in superoxide accumulation due to the reaction of oxygen from oxyhemoglobin with Fe^{2+}–dithiocarbamates. This would artificially diminish the amount of EPR-detected MNIC adducts and thereby decrease the calculated values of the rate constants for NO capture by Fe–dithiocarbamate traps.

[24] S. V. Pashenko, V. V. Khramtsov, M. P. Skatchkov, V. F. Plyusnin, and E. Bassenge, *Biochem. Biophys. Res. Commun.* **225,** 577 (1996).

[25] M. P. Doyle and J. W. Hoekstra, *J. Inorg. Biochem.* **14,** 351 (1981).

[26] K. Tsuchiya, J. J. Jiang, M. Yoshizumi, T. Tamaki, H. Houchi, K. Minakuchi, K. Fukuzawa, and R. P. Mason, *Free Radic. Biol. Med.* **27,** 347 (1999).

ca. 100 μM. Therefore, even when oxidized into a predominantly Fe^{3+} state, Fe–dithiocarbamates retain their capacity to bind NO and can subsequently be reduced to paramagnetic MNIC adducts.

This last point is important to keep in mind: even when Fe–dithiocarbamates accumulate in living systems, mainly in the oxidized Fe^{3+} state, the formation of paramagnetic MNIC–dithiocarbamate complexes is not prevented.

Nonenzymatic Sources of NO

In addition to oxygen, other oxidants could also be potential agents for transforming the iron atom from the divalent to the trivalent state. Nitrite ions in particular are also capable of oxidizing Fe^{2+}–dithiocarbamates to the Fe^{3+} state. The following reaction mechanism has been proposed[16] (Scheme 4):

$$2(MGD)_2\text{–}Fe^{2+} + NO_2^- + MGD + H^+ \longrightarrow (MGD)_3\text{–}Fe^{3+} + NO\text{–}Fe^{2+}\text{–}(MGD)_2 + OH^- \quad \text{(Scheme 4)}$$

However, this reaction has the very slow rate of 4.8 $M^{-2}\,s^{-1}$. This low value makes the reaction irrelevant under conditions found in animal tissue because it cannot compete with the presence of nitric oxide and oxygen at physiological levels. *In vitro* experiments[16] demonstrated that only 5% of nitrite is transformed into NO within 30 min when 1 mM nitrite is incubated in the presence of 5 mM Fe^{2+}–$(MGD)_2$ at pH 7.3 under anaerobic conditions. Similar results have been obtained (see Ref. 27). These data support the following important assertion: even when nitrite ions accumulate in tissues, they cannot function as NO donors at normal *in vivo* conditions.

The situation is less clear for tissue under very pathological conditions, as may arise in ischemic tissues, for example. Here it has been shown[28] that NO can be generated by the direct reduction of nitrite ions under acidotic and highly reduced conditions. This alternative source of NO should be kept in mind when performing experiments on tissue. In practical applications, the enzymatic NO production should always be assessed by duplicating experiments in the presence of inhibitors of nitric oxide synthase (NOS) enzyme.

We have seen that the presence of oxygen initiates the oxidation of Fe^{2+}–dithiocarbamates to the trivalent state. Further complications arise from side products in the form of reactive oxygen species such as O_2^-, $^{\bullet}OH$, and H_2O_2. These have been implicated in a wide range of adverse effects in living organisms. In particular, the $^{\bullet}OH$ radicals can decompose hydroxyurea.[29] This would open another

[27] S. Venkataraman, S. M. Martin, F. Q. Schafer, and G. R. Buetner, *Free Radic. Biol. Med.* **29,** 580 (2000).

[28] J. L. Zweier, A. Samouilov, and P. Kuppusamy, *Nature Med.* **1,** 804 (1995).

[29] K. Tsuchiya, J. Jiang, M. Yoshizumi, T. Tamaki, H. Houchi, K. Minakuchi, K. Fukuzawa, and R. Mason, *Free Radic. Biol. Med.* **27,** 347 (1999).

nonenzymatic pathway for the generation of NO. NO release from the decomposition of other nitrogen-containing compounds cannot be excluded. Therefore, selective or nonselective NOS inhibitors should always be used to separate enzymatic from nonenzymatic NO production.

S-Nitrosothiols, found abundantly in animal tissues, can also act as nonenzymatic sources of NO. We found that these compounds release NO when exposed to Fe–dithiocarbamate complexes (A. F. Vanin, in preparation). In this case, the NO release is an artifact from the trapping technique itself. Our experiments showed that incubation of 1 m*M* *S*-nitrosoglutathione (GS-NO) with 5 m*M* Fe^{2+}–$(MGD)_3$ induces complete decomposition of GS-NO in less than 5 min. The released NO is efficiently trapped by the remaining Fe^{2+}–$(MGD)_2$, as shown by the formation of paramagnetic MNIC-MGD adducts to a final concentration of ca. 1 m*M*.

Further complications arise from the possible interference of Fe–dithiocarbamate traps with the enzymatic production of NO itself. In mammals, NO is synthesized by dimers of the enzyme nitric oxide synthase. Experiments[30] have shown that the addition of Fe^{2+}–$(MGD)_2$ traps has a marked inhibitory effect on the NO synthesis by isolated neuronal NOS (nNOS) preparations. It was proposed that this trap catalyzes oxidation of the reductase domain of nNOS by oxygen. This impedes the electron flow from NADPH to the heme domain of the enzyme and inhibits the enzyme. The inhibition is dose dependent, with an IC_{50} value of 25 μM. The same phenomenon was observed with purified inducible NOS (iNOS). The presence of Fe^{2+}–$(MGD)_2$ clearly inhibits the NO synthesis by iNOS in a dose-dependent manner. For iNOS, we find a value of $IC_{50} = 12\ \mu M$, even less than for nNOS.[30a]

These results suggest that Fe^{2+}–$(MGD)_2$ complexes should completely inhibit NOS enzymes at Fe^{2+}–$(MGD)_2$ concentrations normally used to trap NO (hundreds of μM). Therefore, adducts should not be formed at all under these conditions. Reality is very different, however. After the addition of 0.6 m*M* Fe^{2+}–$(MGD)_2$ to an active iNOS solution, the formation of EPR detectable adducts was observed easily and unambiguously. In contrast, no adducts are detectable[30] when 1 m*M* of Fe^{3+}–$(MGD)_3$ is added to a solution containing activated nNOS. Clearly, the outcome of a trapping experiment depends crucially on the redox state of the Fe–dithiocarbamates used. By implication, the presence of oxidants such as oxygen is very important.

We propose the following tentative explanation of the results. We expect that the addition of 0.6 m*M* Fe^{2+}–$(MGD)_2$ consumes a large fraction of the oxygen from the solution. We propose that the oxygen concentration is reduced to a level that is high enough to sustain NO synthesis, but too low to initiate the oxidation of enzymatic reductase domain by Fe^{2+}–$(MGD)_2$ complexes. The enzyme

[30] H. Yoneyema, H. Kosaka, T. Ohnishi, T. Kawazoe, K. Mizoguchi, and Y. Ichikawa, *Eur. J. Biochem.* **266,** 771 (1999).

[30a] A. Huisman, I. Vos, E. E. van Faassen, J. A. Joles, H. J. Grone, P. Martasek, A. J. Zonneveld, A. F. Vanin, and T. J. Rabelink, *FASEB J.* **16,** 1135 (2002).

remains functional, and MNIC adducts start to accumulate. If, however, 1 mM of Fe^{3+}–$(MGD)_3$ is added to the nNOS preparation, the oxygen levels remain high, thereby allowing oxidation of the enzymatic reductase domain. Inhibition of NO synthesis is the result, and MNIC adducts are not formed. We stress that this is a tentative explanation and needs to be tested in actual experiments. Clearly, it needs to be verified whether the observed inhibition of NOS is dependent on the oxygen concentration. We also stress that the inhibitory effect of the water-soluble Fe^{2+}–$(MGD)_2$ on various isoforms of NOS does not imply a similar inhibitory effect by lipophylic complexes such as Fe^{2+}–$(DETC)_2$. For example, in cells and tissues, Fe^{2+}–$(MGD)_3$ and Fe^{2+}–$(DETC)_3$ localize in aqueous and membrane compartments, respectively. Therefore, their influence on NOS function can be completely different. To our knowledge, no interference with NOS function has been reported for Fe^{2+}–$(DETC)_2$.[31]

Redox Activity of Nitrosyl–Iron–Dithiocarbamates (MNIC)

Reactions with Superoxide and Peroxynitrite

The previous section dealt with redox chemistry of Fe–dithiocarbamate traps for NO. This section considers the redox chemistry of the paramagnetic MNIC adducts, which are produced in the capture of NO by the Fe–dithiocarbamate trap. Until now, it was usually taken for granted that MNIC–dithiocarbamate complexes were stable, thereby allowing unhindered accumulation in cultured cells and tissues. However, isolated reports going back 30 years suggested that MNIC–dithiocarbamates may be involved in redox chemistry.[5,32–34] It was demonstrated[5] that halogen or nitrogen dioxide molecules can be included in paramagnetic MNIC–dithiocarbamate complexes. The inclusion was in the *cis* position to the NO ligand and resulted in the transformation of MNIC into EPR silent complexes. This transformation could be reversed by strong reductants. More recently, similar data were obtained *in vivo* on injection of paramagnetic MNIC–MGD into LPS-treated mice.[33]

It has also become clear that paramagnetic MNIC–dithiocarbamates can react easily with superoxide[32,34] and peroxynitrite.[34] The reaction between superoxide and MNIC–MGD has been found[34] to be very fast with a reaction rate $k = 3 \times 10^7\ M^{-1}\ s^{-1}$ (37°, pH 7.4). The product of this reaction is an EPR silent nitroso

[31] A. Mülsch, A. Vanin, P. Mordvintcev, S. Hauschildt, and R. Busse, *Biochem. J.* **288,** 597 (1992).

[32] S. Pou, P. Tsai, S. Porasuphatana, H. J. Halpern, G. V. R. Chandramouli, E. D. Barth, and G. M. Rosen, *Biochim. Biophys. Acta* **1427,** 216 (1999).

[33] V. D. Mikoyan, L. N. Kubrina, V. A. Serezhenkov, R. A. Stukan, and A. F. Vanin, *Biochim. Biophys. Acta* **1336,** 225 (1997).

[34] A. F. Vanin, A. Huisman, E. S. G. Stroes, F. C. deRuijter-Heijstek, T. J. Rabelink, and E. E. van Faassen, *Free Radic. Biol. Med.* **30,** 813 (2001).

complex. The reaction constant exceeds that of ascorbate and other superoxide scavengers, such as MnTBAP. It shows that NO retains significant reactivity with superoxide even when ligated to the iron complex. This value should be compared to the rate constant of ca. 7×10^9 M^{-1} s^{-1}, which has been reported[35] for the reaction between free NO molecules and superoxide anions in solution. We note that the latter is more than two orders of magnitude larger than that of the reaction between superoxide and MNIC–MGD.

These findings have strong implications for the interpretation of data from NO trapping experiments. Let us consider effects of superoxide on the trapping yields first. We note that Fe–dithiocarbamate traps are normally used in the millimolar range. Therefore, the endogenous NO in cells and tissues will be trapped as MNIC adducts rather than react with superoxide anions, which are present in living systems at a concentration of not more than a few micromolars. Therefore, the presence of physiological superoxide levels will not affect the yields of the trapping reaction. Let us now consider the implications for the adducts formed in the trapping reaction. Here the situation is very different. These adducts are formed in much lower concentrations, typically as tens of micromolars. As such, the adduct levels will be sensitive to attack by endogenous superoxide, which transforms paramagnetic adducts to the EPR-silent state. A similar loss of MNIC is also observed[34] after the addition of peroxynitrite to a solution of MNIC adducts.

This transformation to EPR silent species can be partially reversed by the addition of a reducing agent such as ascorbate.[34] Nearly complete recovery is obtained by adding the ascorbate 1 min after exposing MNIC–MGD adducts to superoxide. Only partial recovery is achieved when ascorbate is added at later times. The same behavior is also found for the EPR silent species formed when peroxynitrite is added to a solution of MNIC–MGD. To explain these observations, it was proposed[34] that superoxide reacts with the iron-liganded NO to iron-liganded peroxynitrite. The resulting diamagnetic MGD–Fe–peroxynitrite complex is EPR silent. The very same complex can also be formed when peroxynitrite replaces the NO ligand in the MNIC–MGD complex. The proposed mechanism is plausible, as the formation of similar complexes is reported for the reaction of oxygen with nitrosylated heme groups in hemoglobin(myoglobin) or NO with oxygenated hemoglobin.[36,37] Peroxynitrite ligands can bind to the iron in more stable *cis* or less stable *trans* forms, depending on the iron atom binding to nitrogen or oxygen atoms, respectively. Subsequent isomerization of peroxynitrite to nitrate results in decomposition of the complex. An analogous situation seems plausible when peroxynitrite binds to the iron in Fe–dithiocarbamate. Evidently, in Fe–dithiocarbamate complexes the nitrogen atom of NO remains directed toward

[35] M. S. Wolin, *Arterioscler. Thromb. Vasc. Biol.* **20,** 1430 (2000).

[36] S. Herold, *FEBS Letts.* **443,** 81 (1999).

[37] S. Herold, M. Exner, and T. Nauser, *Biochemistry* **40,** 3385 (2001).

the iron, even after the incorporation of the superoxide anion. This complex has the peroxynitrite in the more stable *cis* form and survives for a few minutes, just like the peroxynitrite–hemoglobin complexes with the *cis* form of peroxynitrite. The addition of ascorbate within this time range allows recovery of initial paramagnetic MNIC–dithiocarbamate complexes. On time scales longer than a few minutes, conversion to the unstable *trans* form seems to occur. The latter is continually decomposing by release of nitrate from the complex. Therefore, the complexes with *trans*-peroxynitrite cannot be restored to the paramagnetic state by reductants.

MNIC–Dithiocarbamates as Superoxide Scavengers: The ABC Method

Evidently, the level of MNIC complexes in living systems is ultimately determined by two antagonistic processes: By generation of the MNIC–dithiocarbamates through one of the pathways discussed previously and the rate of its transformation into an EPR silent form. High yields of MNIC adducts require strong generation of MNIC and little decay through redox pathways as described earlier. In living tissues, the dominant loss channels are provided by endogenous superoxide and peroxynitrite. Therefore, yields in living tissues can be expected to be enhanced by the addition of superoxide scavengers. SOD is a very efficient scavenger of superoxide, but is ineffective when added exogenously, as this enzyme will not reach the interior of the cells. Several cell-permeating SOD mimics are known, but these suffer from much lower efficiencies for the removal of superoxide. However, the previous section showed that MNIC–MGD itself is a very potent scavenger of superoxide and can remove peroxynitrite as well. Exploiting these reactions to our advantage, we developed the so-called ABC method to detect and quantify NO formation in tissues.[34] In this method, a known quantity of paramagnetic MNIC–dithiocarbamate complexes is added exogenously to a cell culture or animal tissue. It acts as a very efficient scavenger of superoxide anions and peroxynitrite. Additionally, Fe and dithiocarbamate are also added for trapping endogenous NO. This experiment is repeated in the presence of an NOS inhibitor. The MNIC–dithiocarbamate EPR signals measured in these two experiments are as follows.

$$\begin{aligned} \text{First experiment} &= \mathrm{A} + \mathrm{B} - \mathrm{C} \qquad \text{(with active NOS)} \\ \text{Second experiment} &= \mathrm{A} - \mathrm{C} \qquad \text{(with inhibited NOS)} \end{aligned}$$

where (A) is the signal from exogenous MNIC–dithiocarbamate complexes, (B) is the signal due to endogenous NO production, and the quantity (C) accounts for the MNIC–dithiocarbamate complexes transformed to the EPR silent state by superoxide and/or peroxynitrite.

In an actual experiment, the quantity (A) is known, and the unknown quantities (B) and (C) are to be determined. Of course, (A) should be chosen larger than (C), meaning that the quantity of exogenous traps should be large enough to reduce superoxide and peroxynitrite to insignificant levels. However, (A) should not be chosen too high, as the unknown quantity (B) must be determined from subtracting two intensities. The best value for (A) should be determined empirically by trying a few concentrations in the range of 1–50 μM. The difference (B) should be independent of the choice for (A).

In this way, the subtraction of two EPR intensities allows an estimation of true endogenous production of MNIC–dithiocarbamate complexes (B). Some limitations and assumptions of this ABC method are discussed elsewhere.[34] It is important to note that the endogenous NO is trapped efficiently under conditions of low superoxide and peroxynitrite, as these have been scavenged by the exogenous MNIC complexes. Therefore, endogenously formed complexes will survive and remain EPR detectable. It implies that endogenous NO production may now be detected with higher sensitivity than achieved in a conventional NO trapping experiment. As shown in the final section, practical application of the ABC method to viable cultured endothelial cells has demonstrated an increase in sensitivity by a factor of five.

Release of Iron and Complexation of Cu^{2+} by Dithiocarbamates

In the ideal spin trapping experiment, neither the spin trap nor the adduct interferes in any way with the system under investigation. This idealistic requirement is not satisfied at all by Fe–dithiocarbamates. As discussed earlier, MNIC adducts act as potent scavengers of superoxide. Moreover, the hydrophobic Fe–DETC traps have an additional complication in that they cannot be added to biological tissue. Instead, the iron salt and the DETC chelator have to be added separately, with formation of the Fe–DETC trap actually taking place inside the cells or tissue. Administration of iron as well as DETC can introduce artifacts, which need to be considered carefully.

An extensive literature exists on the physiological pathologies associated with the presence of free iron (for a general review cf. Ref. 38). The majority of publications concentrate on the production of radical oxygen species via redox chemistry involving free iron. Particular attention is given to the very reactive hydroxyl radical, which is produced from hydrogen peroxide via the iron-mediated Fenton reaction. The basic rule is simple: Free iron is dangerous and potentially initiates the artificial generation of oxygen radicals. Therefore, free iron concentrations should be kept as low as possible when Fe–dithiocarbamate trapping is used. In actual experiments, the dithiocarbamate chelator is usually added in excess of iron. This guarantees that hardly any free iron will be released in this way. Experiments[32]

[38] B. Halliwell and J. M. C. Gutteridge, "Free Radicals in Biology and Medicine," 2nd Ed. Oxford Univ. Press, Oxford, 1999.

confirm that dithiocarbamates are among the most efficient chelators in suppressing hydroxyl radical production in buffers contaminated with iron salts.

Let us now consider artifacts resulting from the administration of DETC. This powerful metal chelator is not selective for iron and will form complexes with many divalent and trivalent metal ions. Its chelating power is such that not only free metal ions are liganded, but certain metal ions may be extracted from weak bindings as well. An important example is given by inhibition of the Cu,ZnSOD by free DETC. The binding of copper to DETC is so favorable that the Cu^{2+} is extracted from the SOD enzyme, leaving a disabled nonfunctional ruin.[39] The Cu^{2+}–$(DETC)_2$ complex is paramagnetic ($S = 1/2$) and may be detected by EPR.[40] In practical NO trapping, DETC is usually administered in excess of iron. This has the advantage that most iron will be chelated rapidly without a massive increase of free iron in the system. After formation of the Fe–DETC traps, the remaining DETC will chelate Cu^{2+} and disable a considerable quantity of SOD. The formation of Cu^{2+}–$(DETC)_2$ complexes is observed easily as an obfuscating background signal in NO trapping experiments. In many cases, this copper background is so strong that it becomes very difficult or impossible to detect the MNIC–DETC spectrum. In more favorable situations, the MNIC contribution to the EPR spectrum remains identifiable and may be quantified after the (known) Cu^{2+}–$(DETC)_2$ spectrum has been subtracted.

Colloidal Fe–DETC

To overcome this problem, it has been proposed[41] that to avoid the supple-tion of free DETC, dissolved Fe^{2+} and DETC should be mixed in stoichiometric quantities to combine into Fe^{2+}–$(DETC)_2$. These nonsoluble complexes start precipitating slowly, but for several minutes the microcrystals remain sufficiently small and suspended in the solution. This solution can be handled easily with a syringe. Anaerobic conditions are carefully imposed to keep the complexes in a divalent state. This unstable solution is sometimes referred to as colloidal Fe^{2+}–$(DETC)_2$. It is subsequently applied to the tissue under investigation, and formation of paramagnetic MNIC–DETC will commence. Because the solution contains no free DETC capable of chelating copper, SOD is left intact and no disturbing background from Cu–DETC will appear. This stoichiometric method can be applied to situations in which NO levels in solution are quantified. An important example is detection of NO near the endothelial lining of vascular tissue. This refined NO trapping technique has allowed a more sensitive quantification of basal as well as

[39] D. Cocco, L. Calabrese, A. Rigo, E. Agrese, and G. Rotilio, *J. Biol. Chem.* **256,** 8983 (1981).

[40] Y. Suzuki, S. Fujii, T. Tominaga, T. Yoshimoto, T. Yoshimura, and H. Kamada, *Biochim. Biophys. Acta* **1335,** 242 (1997).

[41] A. L. Kleschyov, H. Mollnau, M. Oelze, T. Meinzert, Y. Huang, D. G. Harrison, and T. Munzel, *Biochem. Biophys. Res. Commun.* **275,** 672 (2000).

stimulated NO levels near vascular tissue. For tissues or cellular cultures, the method is problematic, as the colloidal Fe^{2+}–$(DETC)_2$ complexes are unable to cross cell membranes.

Applications

Evaluation of NO Produced by iNOS

Our first illustration concerns the quantification of NO produced by inducible nitric oxide synthase (iNOS) enzymes in solution. It is instructive to consider this problem, as iNOS is known[42] to produce substantial quantities of superoxide as well as NO. The balance between production of NO and superoxide is strongly affected by the cofactor tetrahydrobiopterin (BH_4). The superoxide may interfere with the NO trapping via two reactions. First, it will scavenge a fraction of superoxide in the direct reaction to peroxynitrite. As discussed earlier, peroxynitrite may also bind to the Fe–dithiocarbamate trap and participate in the redox chemistry of the adducts. Second, the superoxide may react with NO adducts already formed. We have applied the ABC method to quantify total NO production, as well as the fraction of NO adducts, which is transformed into the EPR silent state. The latter fraction would remain unobserved in a conventional trapping experiment.

Experiments are performed at 37° with purified recombinant wild-type BH_4-free iNOS produced as described previously.[43] To the buffer solution [50 mM Tris–HCI (pH 7.4)] is added 10 μg/ml calmodulin, 5 mM $CaCl_2$, 2 mM L-arginine, 0.75 μM iNOS, and 1.5 μM BH_4. A physiological peptide environment is simulated by adding bovine serum albumin (BSA) to a final concentration of 25 μg/ml. After a time interval of 8 min to let BH_4 bind to the iNOS, the NO trap Fe^{2+}–$(MGD)_3$ (0.6 mM final) and NADPH (2.5 mM final) were added. After 30 min of trapping at 37°, the experiment is terminated by snap freezing the sample in liquid nitrogen. These concentrations mimic a conventional NO trapping experiment.

In the framework of the ABC method, this experiment is repeated in the presence of 27 μM exogenous MNIC–MGD. The third experiment is repeated in the presence of 27 μM exogenous MNIC–MGD and 3 mM N^{ω}-nitro-L-arginine (NNLA). NNLA is a selective inhibitor of NOS. In ABC nomenclature, the second experiment provides $A - C$, whereas the third experiment provides $A + B - C$. The amount (B) of endogenously formed MNIC–MGD is subsequently estimated by the subtraction of EPR intensities. Using this subtraction method, MNIC–MGD formation (B) is shown to be threefold higher than the adduct concentration achieved in the conventional experiment.[30a] It shows that

[42] Y. Xia, L. J. Roman, B. S. S. Masters, and J. L. Zweier, *J. Biol. Chem.* **273,** 22635 (1998).

[43] P. Martasek, R. T. Miller, Q. Liu, L. J. Roman, J. C. Salerno, C. T. Migita, C. S. Raman, S. S. Gross, M. Ikeda-Saito, and B. S. S. Masters, *J. Biol. Chem.* **273,** 34799 (1998).

a conventional trapping experiment will fail to detect the majority of endogenous NO as most NO finds itself bound to EPR silent complexes.

Quantification of NO Production in Cultured Endothelial Cells

A higher degree of complexity exists in the biochemistry and metabolism of viable cultured cells. We therefore carried out[34] NO trapping experiments on cultured endothelial BEND3 cells. For these experiments, the hydrophobic trap Fe^{2+}–$(DETC)_2$ is used. "Conventional" NO trapping is performed by adding 2.5 m*M* DETC, followed by ferrous sulfate at a 10 μM final concentration. Cellular NO production is stimulated by incubation with 5 μM calcium ionophore A23187 (CaI) for 15 min. In the framework of the ABC method, exogenous MNIC–DETC is administered to the cell culture. As this complex is hydrophobic, it cannot be directly supplied like the MNIC–MGD of the previous example. Rather, it has to be produced *in situ* by the following procedure. As a first step, the dimeric form of dinitrosyl iron complex (DNIC) with *N*-acetylcysteine is added to a final concentration of [Fe] = 10 μM. After 15 min, the addition of 2.5 m*M* DETC initiates the formation of approximately 10 μM exogenous MNIC–DETC. This exogenous MNIC–DETC plays the role of (A). After another minute, additional free iron is added in the form of 10 μM $FeSO_4$. At this point, formation of the traps commences. After an additional 9 min, cellular NO production is stimulated with 5 μM CaI, and formation of the NO adduct commences. After another 15 min, cell suspensions are snap frozen and stored until EPR assay. The same protocol is followed in a parallel experiment in which the cells have been preincubated with NNLA (1 m*M* for 15 min). Quantification of MNIC–DETC adducts with EPR is done at 77K. Subtracting the EPR intensities from the last two experiments, we find that (B) = 2.5 nmol is synthesized enzymatically by 5×10^6 cells during the 15 min of stimulation. In comparison, the "conventional" method detects 0.5 nmol in these cell suspensions. Just as before, we conclude that the "conventional" trapping method leaves the majority of NO in EPR silent complexes. Only by avoiding the loss of paramagnetic NO adducts can the full extent of NO production be assessed. At the same time, the experimental sensitivity for NO detection has been increased by a factor of five.

A final remark concerns the suppletion of exogenous MNIC–DETC, i.e., the quantity (A) in ABC language. It is produced *in situ* from dimeric DNIC precursors. This procedure does not allow precise control over the amount of exogenous MNIC–DETC set free in the system. In our case, we know that roughly 10 μM of MNIC–DETC is achieved, but precise values cannot be given for (A). However, we do know that this uncertainty does not affect the outcome of our subtraction. After all, the ABC method delivers a value for (B) that does not depend on the precise value of the quantity (A).

NO Evaluation in Pregnant Mice by ABC Method

Experiments are performed on pregnant mongrel mice (third trimester) with acute renal failure (ARF). This pathological state is induced by intravenous injection of X-ray contrast substance (76% solution of verographyn). The aim is to assess the enhanced production of NO under conditions of ARF. Four groups of animals are compared. The first group of animals receives 10 m*M*/kg DETC together with glutathione (30 μM/kg). The second group undergoes an additional pretreatment with 1 m*M*/kg NNLA prior to DETC. The third group receives the dimeric form of DNIC (30 μM/kg). After 30 min, 10 m*M*/kg DETC is administered, thereby initiating the formation of monomeric MNIC–DETC. The fourth group of animals is treated with identical procedures, except for a 20-min pretreatment with NNLA (1 m*M*/kg) prior to the DNIC dose. All groups are terminated 30 min after the administration of DETC. After another 30 min, the animals are terminated by decapitation under ethaminal narcosis. After termination, the liver, kidneys, and placenta are extracted, snap frozen, and kept until the EPR assay. The quantities of MNIC–DETC adducts are determined using EPR at 77K in tissues from the liver, kidney, or placenta. Given these treatments, the first group represents a conventional NO trapping experiment. The third and fourth groups are needed for the ABC method. The results are present in Table I (A. F. Vanin *et al.*, in preparation).

From the ABC method, we find the following quantities of endogenous MNIC–DETC adducts (B): 120 n*M*/kg for liver, 160 n*M*/kg for kidney, and 130 n*M*/kg for placenta. We find that a conventional NO assay (first column) gives good agreement for the liver, but seriously underestimates NO production in kidney and placenta tissues. Obviously, in kidney and placenta the levels of superoxide are sufficient to transform substantial quantities of paramagnetic MNIC–DETC adducts to the EPR silent state. The situation is better in liver tissue where MNIC–DETC adducts are seen to survive in the paramagnetic state. We feel that this reflects the potent antioxidant capacity of liver tissue.

TABLE I

QUANTITIES OF MNIC–DETC ADDUCTS (n*M*/kg) IN TISSUES OF FOUR GROUPS OF PREGNANT MICE WITH ACUTE RENAL FAILURE ($n = 5$)[a]

Tissue	DETC	NNLA + DETC	DNIC + DETC	NNLA + DNIC + DETC
Liver	120 ± 40	50 ± 10	300 ± 60	180 ± 10
Kidney	30 ± 5	10 ± 5	200 ± 50	40 ± 5
Placenta	0	0	190 ± 50	60 ± 10

[a] Within the terminology of the ABC method, the third column to A + B − C and the fourth column corresponds to A − C.

Final Comments and Prospects

This article discussed the pitfalls that may compromise the use of Fe–dithiocarbamate complexes for NO trapping in living systems. We stress that researchers using this method should remain aware of the complex redox chemistry involving the trap as well as the final NO adduct in order to avoid misinterpretation of experimental data. The awareness is particularly needed when NO production is assessed quantitatively. It is our hope to have convinced readers that redox chemistry of the traps or the NO adducts does not make NO trapping an unreliable method. As always, good tools should be handled in the proper way with the appropriate expertise. The reaction pathways discussed here are indeed the basis for further development of the trapping method. MNIC adducts show a striking capacity to scavenge superoxide and can be used profitably as potent antioxidants. This property is exploited in the ABC method in which NO trapping is achieved under conditions of low levels of superoxide and peroxynitrite. We have given examples in which the method is applied to isolated NOS enzyme, cell cultures, and living tissues. The ABC method improves the sensitivity of NO trapping with Fe–dithiocarbamates significantly and gives information on the antioxidant status of the system under investigation.

Acknowledgment

AFV gratefully acknowledges support by the Russian Foundation of Basic Research (Grant 98-04-48455) and from the Netherlands Organization for Scientific Research NWO (Grant NB90-171).

[4] Advanced Spin Trapping of Vascular Nitric Oxide Using Colloid Iron Diethyldithiocarbamate

By ANDREI L. KLESCHYOV and THOMAS MÜNZEL

Introduction

The detection of nitric oxide (NO) in isolated blood vessels is a challenging methodological problem. One of the most reliable approaches for the measurement of biological NO has proved to be electron paramagnetic resonance (EPR) spin trapping.[1–3] This approach implies that the otherwise reactive and versatile

[1] Y. Henry, M. Lepoivre, J. C. Drapier, C. Ducrocq, J. L. Boucher, and A. Guissani, *FASEB J.* **7,** 1124 (1993).

[2] B. Kalyanaraman, *Methods Enzymol.* **268,** 168 (1996).

[3] A. F. Vanin, *Methods Enzymol.* **301,** 269 (1999).

0076-6879/02 $35.00

molecule NO can be caged by a specific trap to form a stable NO adduct measurable by EPR method. From basic chemistry it is known that NO in solution reacts readily with iron (Fe) ions to form a nitrosyl–iron complex. The rate of this reaction, as well as the structure of the complex and its reactivity, is critically dependent on the nature of the ligands, which coordinate Fe, as well as on the Fe redox state and the solvent.[4] It was empirically found many years ago that complexes of Fe^{II} with diethyldithiocarbamate (DETC) react avidly with NO in organic solvent to form the mononitrosyl–iron complexes[5] with a characteristic EPR spectrum.[6] The idea of using lipophilic Fe/DETC to trap NO in biological systems arose in 1984 while studying the metabolism of nitrovasodilators in *in vivo* experiments.[7] After discovery of the L-arginine/NO pathway, the Fe/DETC method received several modifications.[8–10] However, this approach still relies mainly on the chelation of intracellular iron and is not sensitive enough. Moreover it requires toxic concentrations of DETC. There is a common prejudice that the major limitation of the otherwise excellent NO trap, Fe/DETC, is its highly lipophilic character, which makes it difficult to use in physiological experiments.[2,3,8,9] As a result, several alternative water-soluble Fe/DETC analogs, such as Fe/*N*-methyl-*D*-glucamine dithiocarbamate (MGD), have been suggested.[11–13] These NO traps have several advantages but are not free of drawbacks (see later). This article points out that with proper usage, the lipophilicity of the Fe/DETC complex is turned from a limitation to an advantage.

Conventional Fe-Based NO Spin Trapping

Hemoglobin as an NO Spin Trap

Hemoglobin (Hb) reacts rapidly with NO ($k = 2 \times 10^7 M^{-1} s^{-1}$) and forms a relatively stable paramagnetic NO–Hb complex. EPR detection of NO–Hb has

[4] F. T. Bonner and G. Stedman, *in* "Methods in Nitric Oxide Research" (M. Feelisch and J. S. Stamler, eds.), p. 3. Wiley, Chichester, UK, 1996.

[5] H. Reihlen and A. Friedolsheim, *Liebigs Ann. Chem.* **457,** 71 (1927).

[6] J. Gibson, *Nature* **196,** 64 (1962).

[7] A. F. Vanin, P. I. Mordvintcev, and A. L. Kleschyov, *Stud. Biophys.* **102,** 135 (1984).

[8] P. Mordvintcev, A. Mülsch, R. Busse, and A. Vanin, *Anal. Biochem.* **199,** 142 (1991).

[9] A. Mülsch, P. Mordvintcev, and A. Vanin, *Neuroprotocols: A Companion to Methods in Neurosciences* **1,** 165 (1992).

[10] A. F. Vanin and A. L. Kleschyov, *in* "Nitric Oxide in Allograft Rejection and Antitumor Defense" (S. J. Lukiewicz and J. L. Zweier, eds.), p. 49. Kluwer Academic, Norwell, MA, 1998.

[11] A. Komarov, D. Mattson, M. M. Jones, P. K. Singh, and C. S. Lai, *Biochem. Biophys. Res. Commun.* **195,** 1191 (1993).

[12] Y. Kotake, *Methods Enzymol.* **268,** 222 (1996).

[13] T. Yoshimura, H. Yokoyama, S. Fujii, Takayama, K. Oikawa, and H. Kamada, *Nature Biotechnol.* **14,** 992 (1996).

been widely used as unequivocal evidence of NO formation in different experimental settings.[14,15] It is important to note that the reaction of NO with deoxy-Fe^{II}Hb (but not with oxy-Fe^{II}Hb) leads to the formation of the EPR-detectable NO–Hb complex. As in most biological experiments, it is impossible to control oxy-Fe^{II}Hb formation; the stoichiometrical recovery of NO using this method is problematical.[15] Moreover, the EPR signal of NO–Hb is rather broad and the detection limit usually does not exceed 1 μM.[14]

Hydrophobic Fe^{II}–Dithiocarbamates

A representative of this group, the Fe/DETC complex, is widely used in pathophysiological experiments to detect NO both *in vivo* and *in vitro*. However, because of the traditional concern about the low water solubility of Fe/DETC, the animals or isolated tissues are exposed to very high concentrations of DETC (500 mg/kg *in vivo* and 2.5–5 m*M in vitro*). This very high concentration of DETC is necessary to maximally chelate the intracellular Fe, forming the actual trap, $Fe^{II}(DETC)_2$ within the tissues. However, DETC invariably chelates the intracellular copper and probably other metals from the active sites of the enzymes. One of the consequences of such side effects of high DETC concentration is inhibition of Cu,Zn superoxide dismutase (SOD),[15a] the enzyme controlling the superoxide level and consequently the NO level. Apart from the interference with cellular biochemistry, the formation of Cu–DETC is unfortunate because this complex is paramagnetic and at low temperature exhibits the EPR signal, which partially overlaps with the $Fe^{II}(DETC)_2$ signal.[3] One more limitation is that according to several protocols, for "full" recovery of NO–$Fe(DETC)_2$ it is necessary to treat the sample with a strong reducing agent such as sodium dithionite.[8,9,16] Obviously, such treatment can interfere considerably with the natural NO turnover and may even lead to an overestimation of NO formation due to the potential reduction of higher nitrogen oxides to NO.

Hydrophilic Fe^{II}–Dithiocarbamate

Several water-soluble Fe^{II}–dithiocarbamates, such as Fe^{II}–MGD or Fe^{II}-*N*-dithiocarboxysarcosine (Fe^{II}-DTCS), have been used for biological NO spin trapping.[11–13,17] These complexes are not permeable via biological membranes and trap NO in the extracellular space, which is very useful in some experimental

[14] Y. A. Henry, A. Guissani, and B. Ducastel, "Nitric Oxide Research from Chemistry to Biology: EPR Spectroscopy of Nitrosylated Compounds." Landes Bioscience, Austin, TX, 1997.

[15] A. V. Kozlov, A. Bini, A. Iannone, I. Zini, and A. Tomasi, *Methods Enzymol.* **268,** 229 (1996).

[15a] A. L. Kleschyov, H. Mollnau, M. Oelze, T. Meinertz, Y. Huang, D. G. Harrison, and T. Münzel, *Biochem. Biophys. Res. Commun.* **275,** 672 (2000).

[16] K. Tsuchiya, M. Takasugi, K. Minakuchi, and K. Fukuzawa, *Free Radic. Biol. Med.* **21,** 733 (1996).

[17] J. L. Zweier, P. Wang, and P. Kuppusamy, *J. Biol. Chem.* **270,** 304 (1995).

settings. However, because of rapid oxidation,[3,18,19] these complexes are employed at high concentrations (1–2 mM). One of the negative consequences of such high Fe^{II} concentrations can be the decrease of tissue oxygen tension and the generation of superoxide.[18,20] Another problem is that due to instability of both the trap and the corresponding NO adduct, NO formation can be followed for a few minutes only.[12] Again, the quantitative recovery of NO often requires the addition of reducing agents.[21] Finally, it has been noted that the Fe^{II}–MGD complex interacts with inorganic nitrite to produce NO–Fe^{II}–MGD.[20] The Fe^{II}–DTCS complex has been described to inhibit the isolated neuronal NO synthase.[22]

In summary, it is impossible to use the aforementioned NO spin trapping techniques for reliable measurements of NO production in isolated blood vessels, which especially holds true for nonstimulated conditions.

Prerequisites for Successful NO Spin Trapping

The theory[23] predicts that the efficiency of biological spin trapping is determined by (A) the rate constant of the reaction between the trapping agent and a free radical; (B) the concentration of the spin trapping agent; (C) the partition of trapping agent to the same intracellular loci where a free radical is formed or concentrated; (D) the stability of the trapping agent in the intracellular environment; and (E) the stability of the corresponding free radical adduct. Ideally, the trapping agent should minimally modify the biological properties of the tissue and not interfere with free radical generating/degradation processes. When choosing the trapping agent, it is important to remember that the sensitivity of detecting a paramagnetic center is reversibly proportional to the width of its EPR signal.

The free radical NO is well recognized as an intercellular messenger, which can diffuse within tissue for several cell diameters.[24] Nevertheless, because NO is nine times more soluble in a nonpolar organic solvent than in water,[25] it is believed

[18] A. F. Vanin, X. Liu, A. Samouilov, R. A. Stukan, and J. L. Zweier, *Biochim. Biophys. Acta* **1474,** 365 (2000).

[19] A. F. Vanin, A. Huisman, E. S. G. Stroes, F. C. de Ruijter-Heijstek, T. J. Rabelink, and E. E. van Faasen, *Free Radic. Biol. Med.* **30,** 813 (2001).

[20] K. Tsuchiya, M. Yoshizumi, H. Houchi, and R. P. Mason, *J. Biol. Chem.* **275,** 1551 (2000).

[21] T. Münzel, H. Li, H. Mollnau, U. Hink, E. Matheis, M. Hartmann, M. Oelze, M. Skatchkov, A. Warnholtz, L. Duncker, T. Meinertz, and U. Forstermann, *Circ. Res.* **86,** E7 (2000).

[22] H. Yoneyama, H. Kosaka, T. Ohnishi, T. Kawazoe, K. Mizoguchi, and Y. Ichikawa, *Eur. J. Biochem.* **266,** 771 (1999).

[23] G. J. Janzen, M. S. West, Y. Kotake, and C. M. DuBose, *J. Biochem. Biophys. Methods* **32,** 183 (1996).

[24] J. R. Lancaster, Jr., *Proc. Natl. Acad. Sci. U.S.A.* **91,** 8137 (1994).

[25] A. W. Shaw and A. J. Vosper, *J. Chem. Soc. Faraday Trans.* **8,** 1239 (1977).

that NO is concentrated in the hydrophobic interior of lipid membranes.[26,27] Under these circumstances, it is logical to suggest that a hydrophobic trap will react much more efficiently with NO than a hydrophilic one. This speculation should especially hold true in the case of NO produced by endothelial NO synthase, as it is known to be associated with the plasma membrane.

As a trap for biological NO, the $Fe^{II}(DETC)_2$ complex apparently meets most of the aforementioned criteria. It is a highly hydrophobic compound (octanol/water distribution is more than 100). While the rate constant of the reaction between NO and $Fe^{II}(DETC)_2$ in a membrane-like hydrophobic solvent has not yet been determined, it is likely to be in the same range as that calculated for the water-soluble analog, Fe^{II}–proline–dithiocarbamate ($k = 1.1 \times 10^8\ M^{-1}s^{-1}$).[28] Moreover, the reducing intracellular environment is apparently "safe" both for the trap, $Fe^{II}(DETC)_2$, and for the NO adduct, NO–$Fe^{II}(DETC)_2$. This is in contrast to the classical spin traps such as nitronyl nitroxides, for which the endogenous reductants pose a major problem for its biological applications.[2] It is important to note that the NO–$Fe^{II}(DETC)_2$ complex in the organic polar solvent dimethyl sulfoxide remains stable in air for several hours even without the addition of reducing agents.[3] Finally, the EPR signal of NO–$Fe^{II}(DETC)_2$ is specific and relatively narrow (SHF splitting of 1.26 mT). Taking these considerations into account, we hypothesized that the colloidal microparticles formed after mixing aqueous solutions of Fe^{2+} and DETC (molar ration 1 : 2) might be a good tool for the delivery of $Fe(DETC)_2$ into NO producing cells due to their hydrophobic forces.[15a] This methodology can be considered as an advanced NO spin trapping protocol, which has used previously.[9,10,29,30] We simply suggest not being afraid of the $Fe(DETC)_2$ insolubility and increasing Fe^{II} and decreasing DETC concentrations.

Experimental Procedures

Vascular Preparations

Spin trapping of vascular NO using colloidal $Fe^{II}(DETC)_2$ has been validated on several types of blood vessels obtained from different species, including rabbits (aorta, carotic artery, renal artery, vena cava), rats (aorta, vena cava), mice (aorta),

[26] R. J. Singh, N. Hogg, H. S. Mchaourab, and B. Kalyanaraman, *Biochim. Biophys. Acta* **1201,** 437 (1994).

[27] X. Liu, M. J. S. Miller, M. S. Joshi, D. D. Thomas, and J. R. Lancaster, Jr., *Proc. Natl. Acad. Sci. U.S.A.* **95,** 2175 (1998).

[28] S. V. Paschenko, V. V. Khramtsov, M. P. Skatchkov, V. F. Plyusnin, and E. Bassenge, *Biochem. Biophys. Res. Commun.* **225,** 577 (1996).

[29] V. Martin, A. L. Kleschyov, J. P. Klein, and A. Beretz, *Infect. Immun.* **65,** 2074 (1997).

[30] A. L. Kleschyov, B. Muller, T. Keravis, M. E. Stoeckel, and J. C. Stoclet, *Am. J. Physiol.* **279,** H2743 (2000).

and humans (mammarial artery, radial artery, saphenous vein). Additionally, successful experiments have been performed with cultured human umbilical endothelial cells. There are significant differences in NO production among different species and between native and cultured endothelial cells.

Before NO spin trapping experiments, blood vessels must be cleaned carefully but thoroughly of adhering fat and cut into rings or strips. We usually use the ice-cold Krebs solution while cleaning and storing vessels. Usually, a good signal can be obtained from the vessels with an intimal surface of 50–100 mm^2 (about 1–2 $\times 10^5$ of endothelial cells). In contrast, several million endothelial cells are usually necessary to detect NO in cell culture.

Preparation of Colloidal $Fe(DETC)_2$

The concentration of the "stock" $Fe^{II}(DETC)_2$ colloid solution can vary between 0.1 and 0.5 m*M* depending on the experiment. To prepare a 0.4-m*M* solution, Na-DETC (3.6 mg) and $FeSO_4 \cdot 7H_2O$ (2.25 mg) are dissolved separately under argon flow in 2 volumes (10 ml) of filtered and deoxygenated Krebs–HEPES solution. Alternatively, to prevent the possible formation of insoluble complex formation between Fe^{II} and HEPES, $FeSO_4 \cdot 7H_2O$ is dissolved in pure water and DETC in a two times concentrated Krebs solution. Immediately after dissolving $FeSO_4$ (takes about 30 sec), these parent solutions are mixed and aspirated (without air bubbles) into an Eppendorf Combitip. The formed 0.4 m*M* $Fe(DETC)_2$ colloid solution has yellow-brownish color and is slightly opalescent in light. For best results, it is highly recommended to use the $Fe(DETC)_2$ colloid solution immediately after preparation.

Incubation and EPR Spectroscopy

Vascular rings or strips are placed in 24-well clusters filled with ice cold Krebs–HEPES solution. After the specific additions (acethylcholine, A23187, L-NAME, etc.), the samples are treated with colloidal $Fe(DETC)_2$ (final concentration 50–250 μM) and incubated at 37° for 15–60 min. While the efficiency of NO trapping does not change significantly within this concentration range, in general, the higher the incubation volume used in the experiment, the lower the colloidal $Fe(DETC)_2$ concentration should be. The time course of the EPR signal reveals that the accumulation of vascular NO by the trap occurs at a constant rate for 1 hr.[15a] In practice, however, the usual time of incubation is 30 min. Due to its high lipophilicity, the formed NO–$Fe(DETC)_2$ complex is localized exclusively in the vascular tissue and never in the medium. After the incubation period, the vascular strip can either be transferred directly into the EPR tissue flat cell or sampled in a frozen state for EPR measurements at 77K.

EPR spectra can be recorded on any conventional EPR spectrometer (x band). For routine vascular NO spin trapping studies, we recommend a tabletop spectrometer from Magnettech (Germany), which is sufficiently sensitive and easy to operate. For measurements at 77K, the instrument settings are as follows: 10 mW of microwave power, 1 mT of amplitude modulation, 100 kHz of modulation frequency, 60 sec of sweep time, and 10 scans. For quantification of the NO–Fe$(DETC)_2$ formed in vascular tissue, a freshly prepared solution containing a known amount of NO–Fe$(MGD)_2$ can be used. With the EPR spectrometer Magnettech, the detection limit of the NO–Fe$(DETC)_2$ in the sample ($\leq$12 mm length) has been found to be 5 pmol. The absolute amount of NO produced in vascular tissue and trapped in the form of NO–Fe$(DETC)_2$ can be normalized per $cm^2 \times$ min.

NO Spin Trapping Using Colloid Fe$(DETC)_2$

General Discussion

Nonstimulated blood vessels incubated with colloidal Fe$(DETC)_2$ exhibited the triplet EPR signal characteristic of NO–Fe$(DETC)_2$ ($g = 2.035$; $A_N = 1.26$ mT), reflecting NO production (Fig. 1). Quantification reveals that this "basal" NO production usually occurs with a rate of 2–5 pmol/$cm^2 \times$ min. NO–Fe$(DETC)_2$ is found to be associated exclusively with blood vessels, suggesting the "one-way delivery" of the trap into the cellular hydrophobic compartments. Moreover, the NO–Fe$(DETC)_2$ complex does not stick to the vascular surface but is distributed equally within all three tunica. Almost a linear increase of the EPR signal is observed during 1 hr of incubation,[15a] which suggests a high level of availability of Fe$(DETC)_2$ and stability of NO–Fe$(DETC)_2$ within blood vessels as well as the lack of an inhibitory effect on the NO synthase.

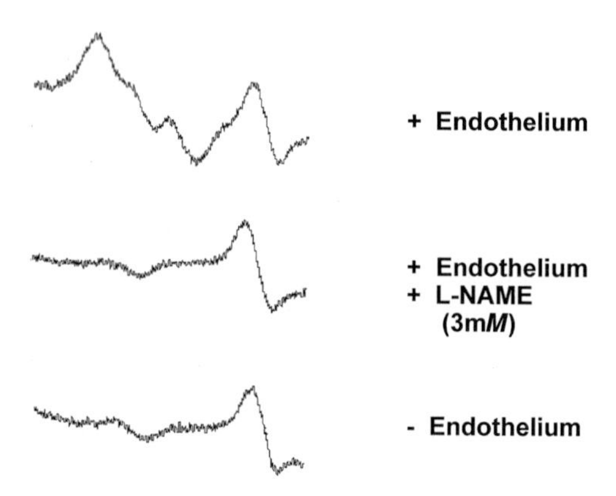

FIG. 1. EPR spectra of NO–$Fe^{II}(DETC)_2$ complexes formed in rabbit aortic strips after a 30-min incubation with 200 μM colloidal $Fe^{II}(DETC)_2$. Removal of endothelium or pretreatment with 3 mM L-NAME prevented formation of the complex.

In colloidal $Fe(DETC)_2$, all molecules of DETC are already occupied by Fe^{2+}, thereby allowing the high efficiency of NO spin trapping in vascular tissue. Another positive point is that because of the absence of free DETC, the chelation of essential intracellular metals such as Cu, Fe, Zn, and Mo is minimized. Indeed, when using colloidal $Fe(DETC)_2$, formation of the notorious Cu–DETC signal[7–10] is negligible. In addition, no inhibition of vascular Cu,ZnSOD was observed with this protocol.[15a] Nevertheless, due to the ligand-exchange reactions, a small Cu–DETC signal is always present. As expected, the stimulation of vascular tissue with calcium ionophore, A231187, or acethylcholine sharply increases the NO–$Fe(DETC)_2$ signal, while removal of endothelium or inhibition of NO synthase abolishes the signal. Interestingly, the more NO–$Fe(DETC)_2$ formed in tissue, the less Cu–DETC signal is observed. These observation may indicate that the $Fe^{+}NO^{+}$ group competes more efficiently with Cu^{2+} for DETC than Fe^{2+} itself.

What Can We Actually Detect?

As was first suggested by Vanin,[3] Fe–dithiocarbamate complexes may react not only with free NO, but also with *S*-nitrosothiols, producing the same NO–$Fe(DETC)_2$ complex. Indeed, when *S*-nitrosoglutathione was added to blood plasma, the treatment of plasma with colloidal $Fe(DETC)_2$ revealed rapid (within 1 min) and quantitative formation of the EPR-detectable NO–$Fe(DETC)_2$ complex. Interestingly, in these experiments colloidal $Fe(DETC)_2$ was found to be a much more efficient "catalyst" of *S*-nitrosoglutathione decomposition than the water-soluble Fe–MGD complex (unpublished data). Thus, using our method, it is impossible to discriminate between the production of NO and the production of low molecular mass *S*-nitrosothiols as both products will react with $Fe(DETC)_2$ and accumulate in the form of NO–$Fe(DETC)_2$. However, because the reaction of $Fe(DETC)_2$ with *S*-nitrosothiols is very fast and does not require long incubation, this approach may be used in the determination of already preformed *S*-nitrosothiols.

Colloidal $Fe(DETC)_2$ can also react with nitrosyl nonheme iron, again giving rise to the same NO–$Fe(DETC)_2$ complex. In biological experiments, this reaction can be carried out with the sodium nitroprusside and dinitrosyl iron complexes.[10,31] However, in this case, one has the possibility of discriminating between NO and Fe–NO species by using, in parallel experiments, either colloidal $Fe(DETC)_2$ or DETC alone.[32]

In contrast to Fe–MGD,[20] no apparent reaction of inorganic nitrite (up to 100 μM) with colloidal $Fe(DETC)_2$ was observed (Fig. 2). This distinction may be attributed to the hydrophobic character of the colloidal microparticle interior, which may prevent the interaction of $Fe(DETC)_2$ with negatively charged nitrite.

[31] A. L. Kleschyov, P. I. Mordvintcev, and A. F. Vanin, *Stud. Biophys.* **105,** 93 (1985).

[32] A. L. Kleschyov, B. Muller, J. C. Stoclet, and T. Münzel, *Circulation.* **104,** 1458 (2001). [Abstract]

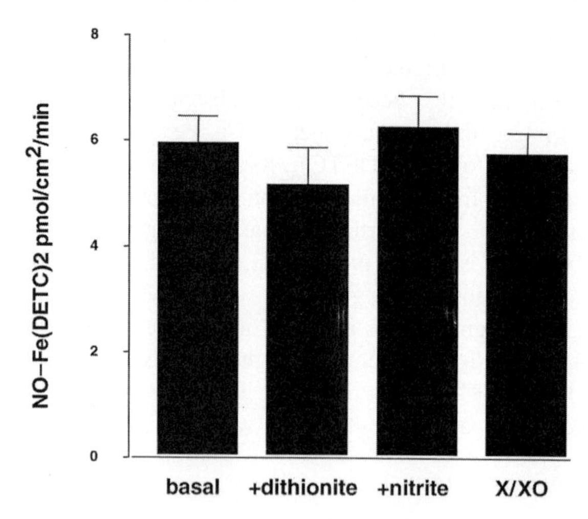

FIG. 2. Effect of treatment with sodium dithionite (1 mg/ml) on the recovery of NO–$Fe^{II}(DETC)_2$ in aorta and of addition of sodium nitrite (100 μM) or xanthine (0.3 mM)/xanthine oxidase (3.2 mU/ml) to medium on the formation of NO–$Fe^{II}(DETC)_2$ in rabbit aorta incubated with colloidal $Fe^{II}(DETC)_2$.

Redox Reactions

In general, $Fe^{II}(MGD)_2$ is sensitive to oxidation by molecular oxygen forming $Fe^{III}(MGD)_3$.[18] It seems that this also holds true for colloidal $Fe^{II}(DETC)_2$, which forms black aggregates during storage exposed to air. However, such an oxidation does not play a significant role in our spin trapping protocol, as $Fe^{III}(DETC)_3$ is also hydrophobic and tends to partition inside cells where it is reduced to $Fe^{II}(DETC)_2$. This conclusion comes from experiments showing that in control vessels, colloidal $Fe^{II}(DETC)_2$ and colloidal $Fe^{III}(DETC)_3$ have the same NO spin trapping capabilities. However, in the later case, a more intensive Cu–DETC signal is observed, which is apparently due to the liberation of one DETC molecule after complex reduction (unpublished data). Additionally, supplementation of the incubation medium with ascorbate (0.5 mM) to prevent the oxidation of $Fe^{II}(DETC)_2$ does not improve vascular NO spin trapping, but rather decreases the EPR signal.

An important issue is whether reactive oxygen species such as superoxide and peroxynitrite interfere with our NO spin trapping assay. It has been shown that in a model system, peroxynitrite reacts rapidly with dinitrosyl iron complexes, converting them to an EPR silent form.[33] Similar data[19] have been obtained for superoxide and NO–Fe^{II}–MGD ($k = 3 \times 10^7\ M^{-1}\text{sec}^{-1}$). In both cases, the addition of sodium dithionite substantially restores the EPR signal. Thus, it might be

[33] I. I. Lobysheva, V. A. Serezhenkov, and A. F. Vanin, *Biochemistry* (*Mosc.*) **64,** 153 (1999).

possible that reactive oxygen species, which are generated in vascular tissue, oxidize NO–$Fe^{II}(DETC)_2$ and cause an underestimation of NO production. However, when blood vessels were thawed after recording EPR spectra, treated with sodium dithionite, and recorded again, no increase of the signal was observed (Fig. 2). These data imply that endogenous oxidants in control vessels are either not accessible to NO–$Fe^{II}(DETC)_2$ or the rate of their generation is not sufficiently high to oxidize NO–$Fe^{II}(DETC)_2$. Moreover, when the superoxide generating system (xanthine/xanthine oxidase) was present in the medium during the incubation of vessels with colloidal $Fe^{II}(DETC)_2$, the efficiency of NO spin trapping was not decreased,[15a] suggesting that even elevated levels of extracellular superoxide cannot significantly compete with the trap for NO and/or do not reach NO–$Fe^{II}(DETC)_2$ (Fig. 2). However, we speculate that the NO–$Fe^{II}(DETC)_2$ signal should be sensitive to superoxide generated by NO synthase itself, as it occurs in the case of "uncoupling" of the NO synthase. Intriguingly, in our study on the model of angiotensin II-induced hypertension in rats, we detected only half of the "normal" vascular NO level, despite the sharp upregulation of endothelial NO synthase.[34]

Concluding Remarks

The colloidal form of $Fe^{II}(DETC)_2$ is an efficient tool for EPR detection of small amounts of NO/S-nitrosothiols in isolated vascular tissue. This method may be of importance for the direct estimation of NO levels in vascular diseases associated with endothelial dysfunction, such as atherosclerosis, hypertension, and diabetes, as well as in searching for new ways of pharmacological intervention. The potential application of colloidal $Fe^{II}(DETC)_2$ for NO spin trapping in other tissues and other experimental settings deserves further study.

Acknowledgments

This work was supported by the Deutsche Forschungsgemeinschaft Mu 1079/3-1, 4-1. We thank Dr. M. Wendt for useful comments.

[34] H. Mollnau, M. Wendt, K. Szöcs, B. Lassègue, E. Schulz, M. Oelze, H. Li, M. Bodenschatz, M. August, A. L. Kleschyov, N. Tsilimingas, U. Walter, U. Förstermann, T. Meinertz, K. Griendling, and T. Münzel, *Circ. Res.* **90,** e58 (2002).

[5] Simultaneous Detection of pO_2 and NO by Electron Paramagnetic Resonance

By PHILIP E. JAMES and HAROLD M. SWARTZ

Introduction

Oxygen is an essential substrate for mitochondrial respiration and many essential synthetic and degradative reactions and it is an essential component in oxidative damage.[1–6] Its supply to tissue and cells is therefore a critical parameter governing normal homeostasis. It is hypothesized that oxygen sensing and control of vessel tone are closely linked and controlled. Nitric oxide (NO) is an important regulator of vascular tone, functioning as a potent vasodilator when produced by endothelial cells lining the blood vessels.[7–9] Nitric oxide synthases (NOS) release NO from the endothelial cells in response to a variety of stimuli, e.g., agonists (such as bradykinin or acetyl choline), shear stress, and flow.[10,11] Oxygen is a critical cofactor for NO biosynthesis and is required at concentrations above 9 μM for normal NO production.[12,13]

Pathological conditions, including heart failure, vascular disease, diabetes, and stroke, all have been linked to a defective oxygen supply or vascular dysfunction as a result of decreased NO production or availability.[14–16] Elevated NO concentrations can also be detrimental. Enzymes of the mitochondrial electron transport

[1] J. L. Zweier, S. Thompson-Gorman, and P. Kuppusamy, *J. Bioenerg. Biomembr.* **23,** 855 (1991).

[2] D. C. H. McBrien and T. F. Slater (eds.), "Free Radicals, Lipid Peroxidation, and Cancer." Academic Press, New York, 1982.

[3] P. J. Simpson and L. R. Lucchesi, *J. Lab. Clin. Med.* **110,** 13 (1987).

[4] J. E. Biaglow, M. E. Varnes, B. Jacobson, and C. J. Koch, *Adv. Exp. Med. Biol.* **159,** 347 (1983).

[5] H. M. Swartz, *Int. Rev. Cytol.* **35,** 321 (1973).

[6] J. E. Johnson, *in* "Biology of Aging" (J. E. Johnson, R. Walford, D. Harman, and J. Miquel, eds.), p. 1. Liss, New York, 1986.

[7] R. M. J. Palmer, A. G. Ferrige, and S. Moncada, *Nature* **327,** 524 (1987).

[8] C. F. Nathan and D. J. Stuehr, *J. Natl. Cancer Inst.* **82,** 726 (1990).

[9] C. Nathan and Q.-W. Xie, *Cell* **78,** 915 (1994).

[10] J. S. Beckman, "Nitric Oxide: Principles and Actions." Academic Press, San Diego, 1996.

[11] R. Joannides, W. E. Haefeli, L. Libden, V. Richard, E. H. Bakkali, C. Thuillez, and T. F. Luscher, *Circulation* **91,** 1314 (1995).

[12] A. Rengasamy and R. A. Johns, *J. Pharmacol. Exp. Ther.* **276,** 30 (1996).

[13] A. R. Whorton, D. B. Simmonds, and C. A. Piantadosi, *Am. J. Physiol.* **272,** L1161 (1997).

[14] C. D. A. Stehower, J. Lambert, A. J. M. Donker, and V. W. M. van Hinsbergh, *Cardiovasc. Res.* **34,** 55 (1997).

[15] M. W. Ramsey, J. Goodfellow, C. J. H. Jones, L. A. Luddington, M. J. Lewis, and A. H. Henderson, *Circulation* **92,** 3212 (1995).

[16] S. H. Kubo, T. S. Rector, A. J. Bank, R. E. Williams, and S. M. Heifetz, *Circulation* **84,** 1589 (1991).

0076-6879/02 $35.00

chain are inhibited directly by NO or some of its metabolites (such as peroxynitrite, ONOO).[10] Hypotension and tissue hypoperfusion associated with a septic shock episode are the direct results of NO overproduction by the inducible form of NOS in response to bacterial products and inflammatory cytokines.[17,18]

Techniques that can monitor either oxygen or NO independently have already provided novel insights into some disease mechanisms. However, it is often desirable (and important) to measure both NO and oxygen and to do this directly from the same site (in cells or in tissue). Dual electrodes have been developed and utilized simultaneously to monitor NO and pO_2 (partial pressure of oxygen), but these systems consume NO (or oxygen) and are extremely brittle and therefore costly. They also cannot be used readily to follow changes over time in the same animal.

For a number of years we have developed and characterized techniques using electron paramagnetic resonance spectroscopy (EPR, or equivalently electron spin resonance, ESR) to measure oxygen.[19–23] More recently, we have extended these studies to combine measurements of oxygen with simultaneous spin trapping and detection of nitric oxide.[18]

Data can be obtained from different types of systems using different frequencies of EPR spectroscopy. For studies in cell suspensions or *ex vivo* samples (from blood or urine), where the sample size is limited to a few millimeters, 9.5 GHz ("X band") is generally used because of the higher sensitivity of higher frequencies. When making measurements in larger samples and *in vivo,* we use lower frequency (1.1-GHz "L band"), which has lower sensitivity but can penetrate more deeply (e.g., throughout the body of a mouse). Other groups are also developing EPR spectrometers operating at even lower frequencies that can make measurements at a depth of several centimeters, but because the drop in frequency results in a consequent further decrease in sensitivity, the application of these frequencies to the measurement of nitric oxide *in vivo* has been limited so far.

Simultaneous measurement of pO_2 (or $[O_2]$) and NO requires that the EPR signals from each can be resolved, and several techniques have been developed that can do this. These are relatively noninvasive and can provide site-specific measurements *in vitro* or from tissue *in vivo*. This article provides an overview of

[17] M. A. Titherage, *Biochim. Biophys. Acta* **1411,** 437 (1999).

[18] P. E. James, M. Miyake, and H. M. Swartz, *Nitric Oxide* **3,** 292 (1999).

[19] P. E. James, K. J. Liu, and H. M. Swartz, *in* "Oxygen Transport to Tissue XX" (A. Hudetz and D. F. Bruley, eds.), p. 181. Plenum, New York, 1997.

[20] P. E. James, O. Y. Grinberg, F. Goda, T. Panz, J. A. O'Hara, and H. M. Swartz, *Magn. Res. Med.* **38,** 48 (1997).

[21] J. F. Glockner and H. M. Swartz, *in* "Oxygen Transport to Tissue XIV" (W. Erdmann and D. F. Bruley, eds.), p. 221. Plenum, New York, 1992.

[22] K. J. Lui, P. Gast, M. Moussavi, S. W. Norby, N. Vahidi, T. Walczak, M. Wu, and H. M. Swartz, *Proc. Natl. Acad. Sci. U.S.A.* **90,** 5438 (1993).

[23] F. Goda, K. J. Liu, T. Walczak, J. A. O'Hara, J. Jiang, and H. M. Swartz, *Magn. Res. Med.* **33,** 237 (1995).

these techniques, illustrating examples of their use and potential advantages and limitations in their application to biological systems.

Measurements of Oxygen

Principles of EPR Oximetry

There are a number of useful methods for measuring oxygen concentration ($[O_2]$) *in vivo,* but all of these techniques also have some significant limitations,[24,25] especially in regard to the ability to make repeated measurements, in the degree of invasiveness required for the methods, or in sensitivity/accuracy.

There has been a very significant amount of progress in EPR oximetry based on developments in several different laboratories.[26,27] This has resulted in the availability of instrumentation and paramagnetic materials capable of measuring pO_2 or $[O_2]$ in tissues with an accuracy and sensitivity comparable or greater than that available by any other method. EPR specifically responds only to molecules with unpaired electrons, including free radicals, free electrons in some types of matrices, and some valence states of metal ions. The phenomenon observed in EPR is the transition (resonance absorption of energy) between the two energy states that can occur in an unpaired electron system in a magnetic field. The magnetic field separates the energy states associated with the two possible spin states of an unpaired electron (spin 1/2 system). The presence of other unpaired electron species can affect the EPR spectrum by introducing a fluctuating magnetic field. Of particular importance to EPR oximetry is the fact that the ground state of molecular oxygen has two unpaired electrons. The two unpaired electrons in oxygen will interact with other unpaired electron species, and the extent of this interaction will be a function of the amount of oxygen that is present. While this effect occurs with all paramagnetic materials, it is much larger in some materials, and these have been selected for use for "EPR oximetry." The EPR signal arising from these materials can be a sensitive reporter of the pO_2 or $[O_2]$ at that location. Typically, the spectral line width (the peak-to-peak splitting along the magnetic field axis) is measured and converted to pO_2 or $[O_2]$ using an appropriate calibration curve.

The choice of material depends on the system being investigated and the parameter to be measured. Both soluble and particulate oxygen-sensitive materials have been used. Each type of preparation has some advantages and potential limitations. Both types share the virtue of relying on a physical interaction with oxygen that does not affect the local concentration of oxygen. This differs from techniques

[24] J. M. Vanderkooi, M. Erecinska, and I. A. Silver, *Am. J. Physiol.* **260,** C1131 (1991).

[25] J. A. Raleigh, M. W. Dewhirst, and D. E. Thrall, *Semin. Radiat. Oncol.* **6,** 37 (1996).

[26] H. M. Swartz and R. B. Clarkson, *Phys. Med. Biol* **43,** 1957 (1998).

[27] H. M. Swartz, *in* "Biological Magnetic Resonance" (L. J. Berliner, ed.), Vol. 20. Plenum, New York, 2002.

such as polarography in which the oxygen is consumed in the process used to measure its concentration.

Soluble Materials. A variety of oxygen-sensitive paramagnetic molecules [especially nitroxide and trityl (triphenyl derivatives) radicals] are available that vary in structure and chemical properties (such as size and hydrophobicity). Most nitroxides exhibit a three-line hyperfine line splitting characteristic of the N^{14} in the N-O group and a line width that shows some dependence on $[O_2]$ (typically around 100 mGauss over the range of 0–20% O_2). Trityls have narrower lines in the absence of oxygen and also broaden by about 100 mGauss over the range of 0–20% O_2. The soluble materials report on $[O_2]$ because their line widths reflect collisions with dissolved oxygen, and the rate of collisions is dependent on the solubility and diffusion in the medium. Nitroxides for oximetry can be synthesized using N^{15} and deuterium substitution, resulting in a narrow intrinsic line width and fewer lines in the EPR spectrum. For example, tempone (4-oxo-2,2,6,6-tetramethylpiperidine-1-oxyl) exhibits a three-line spectrum and a line width; this line width is about 1550 mG in nitrogen to and around 1650 mG in air, whereas per-deuterated ^{15}N tempone (N^{15} PDT) has two spectral lines and a line width of about 160 mG in nitrogen and 260 mG in air. The ability to choose either a two- or a three-line material is particularly important, as it allows the operator to use two probes simultaneously whose EPR lines do not overlap in the spectrum and can be monitored independently and simultaneously.

Particulates. Certain particulate materials have unpaired electrons (especially carbonaceous systems such as coals and chars, and lithium phthalocyanine crystals) and often exhibit a single EPR line. They respond to pO_2 rather than to $[O_2]$. These paramagnetic materials are very inert in biological systems as assayed in both cell cultures and *in vivo*.[19,20,28] Useful features of these materials include noninvasiveness (most approaches do require an initial placement of the paramagnetic materials into the tissues, but after that the measurements are made noninvasively from the surface), and the measurements can be made as frequently as desired over a period of a year or more. The pO_2 measured is the average pO_2 in the tissues, which are in immediate equilibrium with the surface of the paramagnetic particle(s). The characteristics of EPR oximetry, which is based on the use of particulate oxygen-sensitive materials, include the following.

1. Accuracy and sensitivity: ability to make the measurements with an uncertainty of less than 1 mm Hg (pO_2) or 2 μM (oxygen concentration) and to detect values down to 0.5 mm Hg/1 μM

[28] M. Miyake, J. A. O'Hara, P. E. James, T. Panz, and H. M. Swartz, "Proceedings of ESR (EPR) Imaging and *in Vivo* ESR Spectroscopy" (H. Kamada and H. Ohya, eds.), p. 252. Yamagata, Japan, 1998.

2. Localization: ability to make the measurements from a defined volume with a diameter at least as small as 0.5 mm and from within tissue
3. Repeatability: ability to make the measurements as frequently and as long as required for the study—for some purposes, this may mean months or years
4. Rapidity: ability to make the measurements within the time periods that are appropriate for the biological phenomena that are being investigated—in some situations, this may be as short as a few seconds
5. Noninvasiveness: ability to make repeated measurements without compromising the integrity of the biological system and/or the biological process being investigated
6. Very low toxicity: these materials are inert in tissues and remain intact and therefore do not enter cells unless very small particles are used

Measurement of Nitric Oxide

An accurate technique for the direct detection of NO is critical for a complete understanding of the physiological and pathophysiological processes in which NO is implicated. This especially needs to be measured where it has been produced and where it plays an important role. Measurement of NO concentrations, until recently, has been limited to *ex vivo* determinations of complexes in blood or other body fluids,[29,30] with the exception of an electrochemical catheter method that has been used to detect NO released into superficial veins.[31] While these techniques may provide important and complimentary data, they have some potential disadvantages. For example, measurement of nitrite and nitrate (NOx, believed to be the major metabolites of NO in the circulation) by the Griess method is subject to false positives and also cannot distinguish between the various sources of these species (e.g., nitrite derived from peroxynitrite). It also only provides limited time resolution because the half-life of some metabolites is long and there is accumulation in the system being studied. Measurement of NO directly via an electrode certainly has advantages, such as real-time measurement and exquisite sensitivity ($>0.2\ \mu M$), and has been used with good effect *in vitro*. Their use *in vivo* is limited primarily by the fact it is invasive, electrodes are extremely brittle (and consequently expensive), and the baseline is extremely temperature sensitive.

EPR techniques for the measurement of nitric oxide depend on trapping NO to form a more stable NO adduct that has readily measured EPR spectra. The

[29] L. C. Green, D. A. Wagner, J. Glogowski, P. L. Skipper, J. S. Wishnok, and S. R. Tannenbaum, *Anal. Biochem.* **126,** 131 (1982).

[30] P. H. P. Groeneveld, K. M. C. Kwappenberg, J. A. M. Langermans, P. H. Nibbering, and L. Curtis, *Cytokine* **9,** 138 (1997).

[31] P. Vallance, S. Patton, K. Bhagat, R. MacAllister, M. Radomski, S. Moncada, and T. Malinski, *Lancet* **346,** 153 (1995).

principal approach has been to use metal complexes that trap NO and form an EPR detectable product. It may also be feasible to use organic molecules as spin traps for NO. Typically, the amount of adduct present is proportional to the intensity of the signal or, more accurately, the area under peaks of that species in the EPR spectrum.

Metal Chelate Complexes

Diethyldithiocarbamate (DETC) or methyl-D-glucamine dithiocarbamate (MGD) have both been utilized. The technique depends on formation of a trapping complex between the chelate and the ferrous iron [e.g., Fe–$(DETC)_2$] that then forms a complex with NO. The NO–Fe–$(DETC)_2$ formed is stable and exhibits a three-line EPR signal typical of a hyperfine splitting from nitrogen. The precise spectral characteristics vary according to which species is used. *In vivo* detection of NO generation in the liver was first detected by Quaresima *et al.*[32] injecting DETC. The EPR signal generated from the resulting paramagnetic complex, NO-Fe-$(DETC)_2$, was monitored. The signal-to-noise ratio obtained was around 3 : 1 and was very close to the limits of sensitivity of the EPR spectrometer (i.e., 40 μM).[33] Our group and others have since developed several systems utilizing this technique.[18–20,34] Together with developments in EPR instrumentation, the improved sensitivity of L-band EPR spectrometry (linked to a loop gap resonator) has resulted in the detection of improved signals of trapped nitric oxide (by a factor of 10), enabling measurements of tissue NO to be made in other tissues and profiles of tissue NO levels with time. One interesting difference between DETC and MGD is the solubility of the resulting complex with iron: DETC is hydrophobic and consequently enters into cells and tissues readily, whereas MGD is more hydrophilic and therefore tends to remain in the intravascular compartment.[34]

Hemoglobin

Interactions of NO with hemoglobin are among the most extensively studied. Hb meets the criteria for an effective spin trap for NO: it is present in high concentration *in vivo,* displays high affinity for NO, and the HbNO produced following interaction with deoxyHb has a characteristic EPR spectrum.[35] This is particularly attractive for the detection of NO *in vivo* because of the high concentration of Hb. Because oxidation of HbNO may occur in oxygen-rich environments, immediate

[32] V. Quaresima, H. Takehara, K. Tsushima, M. Ferrari, and H. Utsumi, *Biochem. Biophys. Res. Commun.* **221,** 729 (1996).

[33] H. Utsumi, S. Masuda, E. Muto, and A. Hamada, *in* "Oxidative Damage and Repairs" (K. J. A. Davies, ed.), p. 165. Pergamon, New York, 1991.

[34] H. Fujii, J. Koscielniak, and L. J. Berliner, *Magn. Res. Med.* **38,** 565 (1997).

[35] H. Kosaka and T. Shiga, *in* "Methods in Nitric Oxide Research" (M. Freelisch and J. S. Stamler, eds.), p. 373. Wiley, New York, 1996.

freezing of red blood cell samples in liquid nitrogen is desirable. Other small molecular ligands (CO and oxygen) exhibit lesser affinity than NO and form EPR silent complexes.

Nitroxides as Spin Traps for NO

EPR-detectable nitronyl nitroxides have been used to trap NO *in vitro* in solution.[36] They can be used in cell systems with relative ease because they are well tolerated at relatively high concentrations. Carboxy-PTIO (carboxy-2-phenyl-4,4,5,5-tetramethylimidazoline-1-oxyl 3-oxide) exhibits an EPR spectrum with five-line hyperfine splitting. Its NO adduct is characterized by a shift to a nine-line species (however, this is observed as seven lines in the EPR spectrum because of the overlap of some lines). Interestingly, both the spin trap (i.e., the nitroxide) and the spin adduct can be monitored independently under certain circumstances. We have utilized this technique in several studies to detect NO production by cells.[37] As with most nitroxides, they are subject to reduction and cellular "metabolism" with consequent loss of the EPR signal as the nitroxide group is reduced to its corresponding hydroxylamine.[38] This instability limits their use *in vivo,* although direct detection in the blood circulation by EPR spin trapping of NO was demonstrated using nitronyl nitroxides in the tails of septic mice.[39,40]

Simultaneous Measurement of NO and O_2

The technique relies on the fact that the EPR signal(s) arising from the paramagnetic species used for the detection of NO and the oxygen reporting species do not unacceptably overlap in the resulting EPR spectrum. There are several possible combinations from the approaches outlined earlier. For example, particulate materials can generally be used in conjunction with nitroxides or chelates, as the g value of the former (the position in the magnetic field where the center of the EPR signal falls) lies between the mid- and upper-field hyperfine splitting of the latter (Fig. 1). Because both materials are relatively inert, this technique can be applied to a broad range of studies involving NO and oxygen (see later).

We have also undertaken a series of studies that relied on the fact that most nitroxides are oxygen sensitive (as described earlier). It occurred to us that a spin trap suitable for the detection of NO might also show line width dependence on $[O_2]$.

[36] J. Joseph, B. Kalyanaraman, and J. S. Hyde, *Biochem. Biophys. Res. Commun.* **192,** 926 (1993).

[37] A. Barchowsky, L. R. Klei, E. J. Dudek, F. A. Gesek, H. M. Swartz, and P. E. James, *Free Radic. Biol. Med.* **27,** 1405 (1999).

[38] N. Kocherginsky and H. M. Swartz, "Nitroxide Spin Labels: Reactions in Biology and Chemistry," CRC Press, Boca Raton, FL, 1995.

[39] C.-S. Lai and A. M. Komarov, *FEBS Lett.* **345,** 120 (1994).

[40] A. M. Komarov and C.-S. Lai, *Biochem. Biophys. Acta* **1272,** 29 (1995).

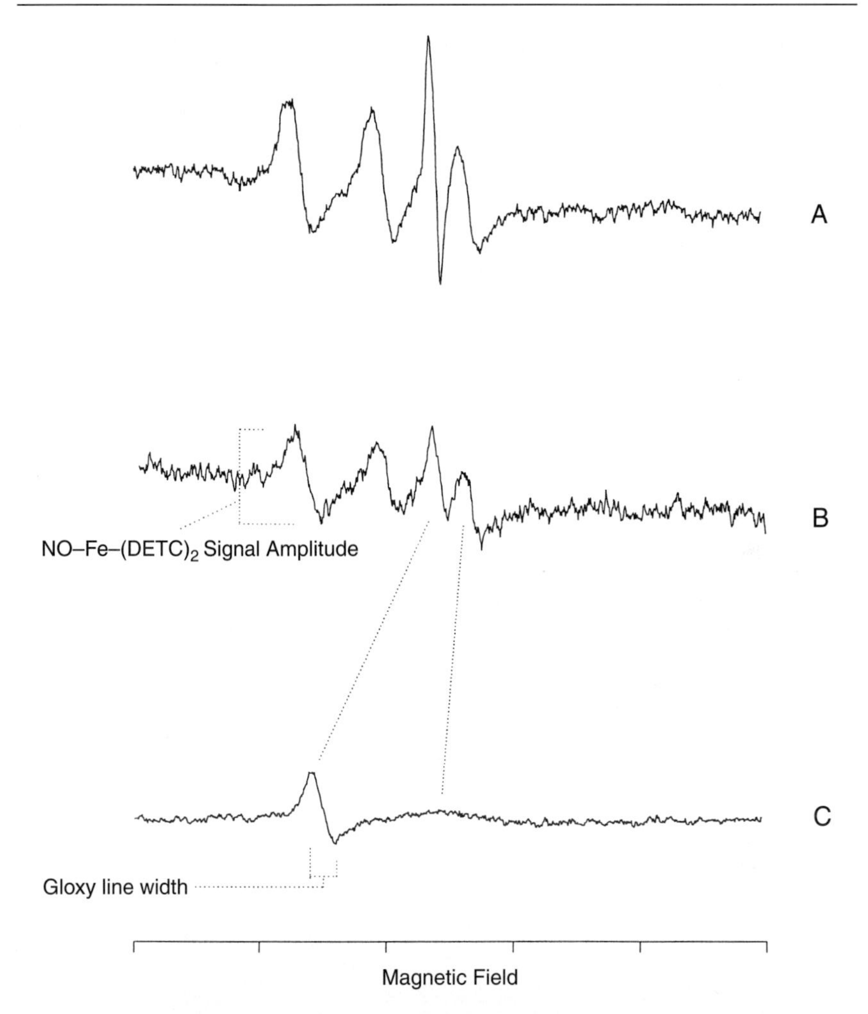

FIG. 1. (A) A typical spectrum observed on addition of an aliquot of a saturated NO solution to saline containing DETC, iron sulfate, sodium citrate, and particles of gloxy (pH 7.4, 37°) *in vitro*. This EPR spectrum is composed of two species: three lines with characteristics typical of NO–Fe–$(DETC)_2$ ($g_{iso} = 2.04$ gauss; $a_N = 13.04$ gauss) and a single line ($g = 2.0028$) arising from gloxy. (B) A spectrum recorded *in vivo* from the liver region of a mouse injected with sodium nitroprusside (SNP). Gloxy had been implanted 7 days earlier, and DETC, iron sulfate, and sodium citrate had also been injected 30 min prior to SNP injection. (C) A typical spectrum recorded from liver [as in (B)] but with spectrometer conditions optimized for recording of the EPR signal from gloxy (reporting on tissue pO_2). Signals from gloxy and NO–Fe–$(DETC)_2$ are clearly distinguishable. The peak-to-peak line width was measured from the gloxy signal (reflecting pO_2) and the signal amplitude of the NO–Fe–$(DETC)_2$ signal. The scale has been normalized to that in spectrum A. The magnetic field sweep was 100 gauss for scans A and B, and 30 gauss for scan C. From P. E. James, M. Miyake, and H. M. Swartz, *Nitric Oxide* **3,** 292 (1999) with permission.

cPTIO is introduced into saline, and the sample is drawn into gas-permeable Teflon tubing and placed into an X-band spectrometer. The [O_2] in the gas perfusing the sample is varied by mixing nitrogen and air and is measured in-line using an oxygen electrode (WPI) immediately prior to the cavity. Spectra are recorded following equilibration at each [O_2]. Typical spectrometer settings are as follows: incident power, 5 mW; scan width, 2 Gauss; and scan time, 60 sec. Modulation amplitude is kept to less than one-third of the spectral line width to avoid artifactual line broadening and overestimation of the [O_2]. A small scan width is chosen, which centers on the low-field hyperfine peak in the spectrum. The spectral line width is measured using a simulation package (EWVoigt software) and is plotted against [O_2] in the in-flowing gas to provide a calibration curve (Fig. 2a). These results demonstrate that cPTIO can also be utilized to measure [O_2] in solution.

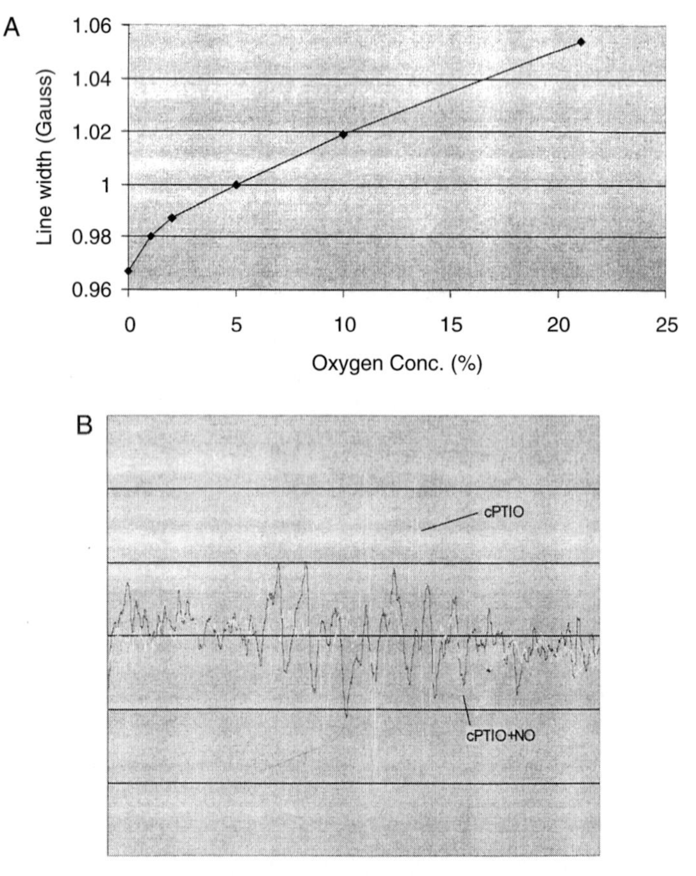

FIG. 2. (a) Spectral line width of cPTIO (low field peak) as oxygen concentration was varied. (b) Typical spectrum of the cPTIO-NO adduct. A burst of NO was produced by the addition of NOC 9 to the system at pH 7.4, and the resulting spectrum was recorded after 5 min.

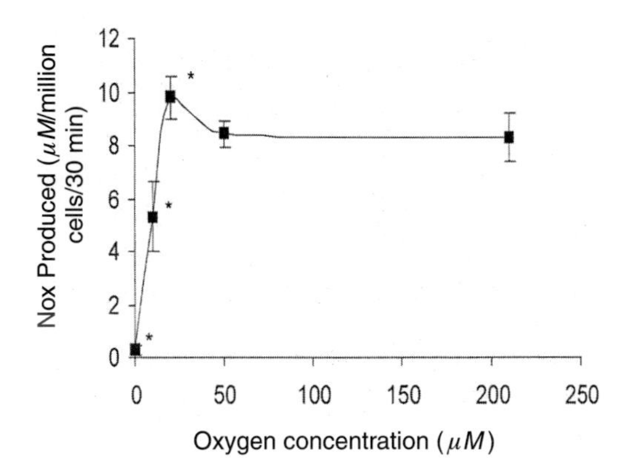

FIG. 3. NOx production by brain endothelial cells exposed to a range of oxygen concentrations in ambient gas. Asterisk denotes where the mean was significantly different from that seen with 210 μM O_2. From Ref. 41.

In Vitro Examples

In Solution. NO and $[O_2]$ can be conveniently measured simultaneously in the same sample using cPTIO (10 μM final concentration). For example, a burst of NO is produced by spiking NOC-9 (Alexis) into saline (pH 7.2); the NOC-9 is kept on ice at pH > 9 until use. EPR spectra can be recorded continuously, and NO production can be monitored as the appearance of the NO adduct (seven hyperfine lines) and simultaneously, $[O_2]$ can be measured conveniently from the spectral line width (Fig. 2b). In practice, however, it is usually more accurate to alternate recording of cPTIO line width (using appropriate modulation amplitude) with optimal recording of the amplitude of the signal of the NO adduct.

In Cells. Production of NO by brain endothelial cells (BEC) is oxygen dependent. This was established by incubating BEC with cPTIO in medium at 37° within the EPR spectrometer cavity. Samples are perfused with gas containing a range of $[O_2]$. NO production and $[O_2]$ are monitored as outlined earlier. Parallel cell samples are coincubated with L-NMMA (to inhibit NOS) or with SOD (dismutation of superoxide and prevention of peroxynitrite formation). These measurements are confirmed in parallel experiments in which total NOx (nitrite and nitrate-stable metabolites of NO) are measured in the culture medium of HUVEC incubated at varied $[O_2]$. Interestingly, these studies yield new and important findings regarding NO availability. Whereas NO production is optimal at $[O_2]$ ranging from 210 to 20 μM, below 20 μM (and >5 μM) NO production appears to increase (Fig. 3).[41] Importantly, similar increases in NO occur following incubation with

[41] P. E. James *et al., in* "Oxygen Transport to Tissue XIX" (D. Wilson, ed.), Plenum, New York, 2002.

SOD. NO production by NOS is unaffected by $[O_2]$ other than below the K_m for oxygen (9 μM),[13] whereas superoxide production (from NADPH-oxidases) typically starts to decrease at higher levels ($K_m = 30–40\ \mu M$, depending on activation state). Taken together, reduced superoxide production results in increased free NO at $[O_2]$ between 5 and 20 μM (despite no change in NOS activity). These findings have important implications for *in vivo* control mechanisms, as increases in available NO may occur *in vivo* as a result of decreasing oxygen, at least within a certain $[O_2]$ range. As expected, NO production declines with decreasing $[O_2]$ below 5 μM $[O_2]$.

In Vivo Examples

In studies where repeated pO_2 measurements are required over longer investigation periods, the oxygen-sensitive material is typically implanted directly into the tissue of interest several days prior to the study. Particles (around 200 μm) are implanted via small gauge needles at the desired tissue depth. The EPR signal arising from the material is then detected noninvasively from the surface using low-frequency EPR spectroscopy. Alternatively, soluble oxygen-sensitive materials (such as trityl) may also be injected directly into tissue or, perhaps more typically, injected iv. In this case, we have found $>80\%$ remains in the bloodstream and therefore reports on blood $[O_2]$. However, probe distribution and pharmacokinetics *in vivo* must be taken into account. When using a NO-detecting metal chelate (DETC or MGD), the chelate is injected ip, and a simultaneous injection of ferrous sulfate and sodium citrate is given sc to facilitate formation of the Fe–$(DETC)_2$ complex. The EPR signal from the pO_2-reporting material does not overlap with that of the NO adduct, permitting simultaneous measurements.

Cerebral pO_2 and NO. Intravenous injection of a NO donor (e.g., sodium nitroprusside) to mice results in vasodilatation and increased cerebral pO_2. *In vivo* EPR is used to monitor pO_2 (from an oxygen-sensitive coal, gloxy, implanted into the brain), and NO is measured simultaneously using the formation of NO–Fe–$(DETC)_2$. Cerebral pO_2 increased with increased levels of NO in the brain (Fig. 4). Similar experiments were undertaken using trityl (administered iv via the lateral tail vein) and DETC in mice. In this case, the trityl reports almost exclusively on pO_2 from the blood, which remained virtually unchanged throughout administration of SNP as described earlier. These results demonstrate the advantage of making site-specific pO_2 measurements simultaneously with NO detection.

Hepatic pO_2 and NO. Overproduction of NO has long been implicated in the pathogenesis and hemodynamic defects associated with septic shock. However, whether NO is harmful in certain locations and beneficial in others remains unclear. Furthermore, questions remain over defects in tissue oxygenation and whether this is the result of an inadequate blood supply (and hence oxygen) or impaired cellular oxygen utilization. We have utilized EPR oximetry to obtain selective measurements of pO_2 from liver sinusoids and from liver parenchyma

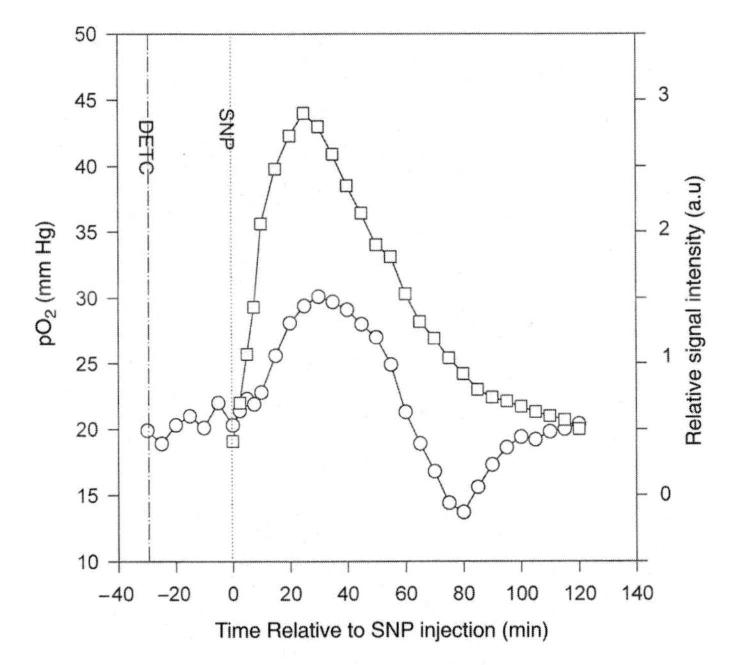

FIG. 4. A graph showing typical pO_2 values and levels of NO from the brain measured prior to and following injection of sodium nitroprusside (SNP), an NO donor, into a mouse.

in vivo. Gloxy was injected either (i) iv as a slurry of fine microparticles that are trapped and reside associated with Kupffer cells lining the sinusoids or (ii) directly as large particles (200 μm) placed into the liver to measure average lobule pO_2. DETC was used in these studies simultaneously to monitor liver NO. Lipopolysaccharide (LPS) was injected ip to induce a septic episode and typical hypotension, as demonstrated previously.[18] We found that the oxygen supply to sinusoids was reduced by 75% in mice treated with LPS. In addition, oxygen utilization across the liver lobule also decreased from 40 to 10 mm Hg. These defects coincide with elevated NO–Fe–$(DETC)_2$ measured directly from the liver lobule (Table I).

Hemoglobin Saturation and HbNO in Vivo. Dependence of the three-line hyperfine structure in Hb_4NO on oxygen saturation was demonstrated by Kosaka and Shiga.[35] A three-line hyperfine structure is associated with α-NO hemes in the low oxygen affinity tense state in which the α-NO is a pentacoordinate. Hbα-NO in the high oxygen affinity-relaxed state is a hexacoordinate and shows little or no hyperfine structure. Rats treated ip with LPS for 6 hr to induce iNOS produce large quantities of NO. Arterial blood is drawn and incubated under 95%N_2/5% CO_2. Samples are taken successively until the deoxygenation of Hb is complete. Spectrophotometric determination of Hb O_2 saturation and HbMet is carried out. At baseline, arterial blood typically shows an EPR spectrum of the hexacoordinate

TABLE I
SUMMARIZING pO_2 DATA OBTAINED FROM LIPOPOLYSACCHARIDE (LPS) OR SALINE-TREATED MICE[a]

	Control pO_2 (mm Hg)	LPS pO_2 (mm Hg)	Decrease (%)
Slurry (measurement of pO_2 at sinusoids)	44.39 ± 5.13	11.22 ± 2.48	74.72
Particles (measurement of pO_2 across the liver)	4.56 ± 1.28	1.16 ± 0.42	74.56
pO_2 difference (slurry pO_2- particles pO_2)	39.83	10.05	74.8

[a] *In vivo* measurements of pO_2 were obtained 6 hr after treatment using L-band EPR spectroscopy. Each value represents the mean pO_2 value $\pm$ SD ($n = 6$ mice) sensed by gloxy (and oxygen-sensitive paramagnetic material) either implanted as particles directly into the liver (reporting on average lobule pO_2) or injected iv as a slurry of fine gloxy particles (mean size = 1.1 μm) that are taken up by Kupffer cells lining the sinusoids.

α-NO heme(s) with only a modest hyperfine structure. As deoxygenation proceeds, a three-line hyperfine signal intensifies, i.e., the intensity of the three-line hyperfine structure of HbNO correlates inversely with the oxygen saturation of Hb (once normalized with respect to total HbNO concentration). In our laboratory, we have also exposed human blood (arterial and separately venous) to a range of NO concentrations (spiked as a NO-saturated solution into the blood sample). Hb O_2 saturation, metHb, and Hb are assessed on a clinical blood gas analyzer. HbNO is assessed by cryogenic EPR and also using chemical cleavage of the NO in acetic acid with subsequent detection of the released NO by an NO electrode (Harvard AMI-700).[42,43] We found that for a given dose of NO, Hb saturation with oxygen in the blood sample could be predicted from the intensity of the three-line hyperfine HbNO. In a carefully controlled system in which total α-HbNO is measured as a fraction of total HbNO, both HbO_2 saturation and NO levels may be predicted (Fig. 5).

General Considerations

There is evidence that certain oxygen-sensitive paramagnetic materials (e.g., fusinite coal) broaden in direct response to NO, which is itself paramagnetic and has

[42] M. T. Gladwin, F. P. Ognibene, L. K. Pannell, J. S. Nichols, M. E. Pease-Fye, J. H. Shelhamer, and A. N. Schechter, *Proc. Natl. Acad. Sci. U.S.A.* **97,** 9943 (2000).
[43] M. T. Gladwin, J. H. Shelhamer, A. N. Schechter, M. E. Pease-Fye, M. A. Waclawiw, J. A. Panza, F. P. Ognibene, and R. O. Cannon III, *Proc. Natl. Acad. Sci. U.S.A.* **97,** 11482 (2000).

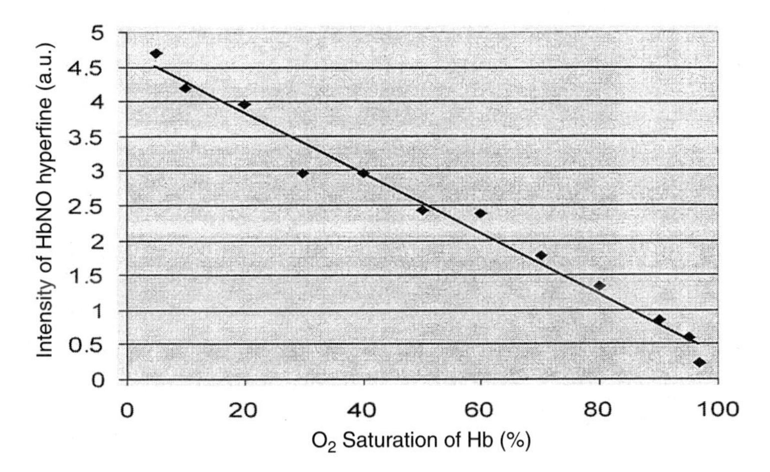

FIG. 5. Total α-HbNO measured as a function of Hb O_2 saturation.

many physical and chemical characteristics in common with oxygen. It is essential, therefore, that adequate controls are performed. The extent of broadening may be of the same order as that expected from oxygen. It must be noted, however, that in biological systems, NO is unlikely to be present at concentrations $>10\ \mu M$. Furthermore, the line width of other substances, such as gloxy, does not respond to [NO]. We have found that exposing some materials to extreme [NO] (e.g., a 10-m*M* saturated solution) in the absence of oxygen may transiently affect their responsiveness to oxygen, which is reversed on reintroducing oxygen into the system.

It is important to note that although cPTIO shows line width responsiveness to [O_2], the line width of cPTIO is the same as that of the NO adduct at a given concentration, providing strong evidence that NO itself does not affect spectral line width.

It is also recognized that limitations of the NO trapping technique include (i) the accumulation of NO adduct is the parameter being measured (rather than rate of NO production per se) and (ii) precise quantitative measurement of NO concentration *in vivo* is not possible (unless the tissue is also removed and X-band EPR is carried out at frozen temperatures to obtain EPR signal/tissue weight). In practice, one observes a progressive increase in the EPR signal followed by a period where the signal is stable. This point reflects a state of equilibrium being established between NO production, NO adduct formation, and NO adduct destruction, and can be considered a parameter that is proportional to the rate of NO formation (assuming the spin trap is in excess).

EPR spectroscopy can detect and quantify NO production only incompletely because some NO will inevitably react with ubiquitous metals (present in hemoglobin and other thiol-rich proteins) or superoxide prior to complexing with the chelating agent. Also, although it is widely believed that only NO can react with MGD and DETC, reports suggest that several NO metabolites (such as nitroxyl anion) can react with MGD.[44–46]

Conclusions

With proper precautions in the use of the technique, combined with the selection of appropriate methods, EPR spectroscopy provides a very effective method for the simultaneous measurement of oxygen and nitric oxide in functioning biological systems, both *in vitro* and *in vivo*. This approach is especially valuable when repeated measurements are needed.

Acknowledgement

The authors acknowledge the Biomedical Technology Research Center at the EPR Center for Viable Systems at Dartmouth Medical School, supported by the National Center for Research Resources, NIH Grant P41 RR11602, Hanover, New Hampshire, and NIH Grant P01 GM51630. The Department of Cardiology, W.H.R.I., is supported by the British Heart Foundation.

[44] K. Tsuchiya, M. Yoshizumi, H. Houchi, and R. Mason, *J. Biol. Chem.* **275,** 1551 (2000).
[45] Y. Xia, A. J. Cardounel, A. F. Vanin, and J. L. Zweier, *Free Radic. Biol. Med.* **29,** 793 (2000).
[46] A. M. Komarov, D. A. Wink, M. Feelish, and H. H. Schmidt, *Free Radic. Biol. Med.* **28,** 739 (2000).

[6] Electron Paramagnetic Resonance Studies of Nitric Oxide in Living Mice

By ANDREI M. KOMAROV

Introduction

Electron paramagnetic resonance (EPR) observation in small living animals is feasible only for stable paramagnetic products, and the problem of free radical stabilization does not have a generic solution. Sulfur ligands are known to stabilize nitrosyliron complexes.[1,2] Vanin and co-workers[3] introduced diethyldithiocarbamate (DETC) as a precursor forming the lipophilic DETC–Fe–nitric oxide (NO) complex in animal tissues. However, DETC forms insoluble complexes with iron

[1] J. F. Gibson, *Nature (Lond.)* **196,** 64 (1962).
[2] C. C. McDonald, W. D. Phillips, and H. F. Mower, *J. Am. Chem. Soc.* **87,** 3319 (1965).
[3] L. N. Kubrina, W. S. Caldwell, P. I. Mordvintcev, I. V. Malenkova, and A. F. Vanin, *Biochim. Biophys. Acta* **1099,** 233 (1992).

0076-6879/02 $35.00

and therefore requires a separate iron injection to produce *in vivo* EPR-visible nitrosyliron complex. This problem has been solved using a water-soluble complex of *N*-methyl-D-glucamine dithiocarbamate (MGD) and iron (MGD–Fe) for EPR detection of endogenous nitric oxide in the circulating tail blood of live conscious mice.[4–6] Dithiocarbamate NO traps yield an intense three-line EPR signal at ambient temperatures, thus allowing the study of *in situ* and real-time nitric oxide generation and spatial distribution of NO in small living animals and isolated organs.[7–11] This article describes a method for tracing NO metabolism and distribution in endotoxin-treated mice using *in vivo* low-frequency EPR spectroscopy in combination with an extracellular MGD–Fe nitric oxide trapping complex.

Properties of Dithiocarbamate–Fe Complex Relevant to *in Vivo* NO Detection

1. Bis(dithiocarbamato)nitrosyliron(II) complexes[12] are tetrasulfur complexes of iron(II) coordinated to a nitrogen atom of nitrogen monoxide. This square pyramidal complex contains a single unpaired electron and a low-spin iron ($S = 1/2$) in the formal oxidation state Fe(I), d^7.[13] The MGD–Fe–NO complex displays EPR spectrum consisting of three lines ($g = 2.04$ and $a_N = 12.5$ G) at room temperature due to the interaction of an unpaired electron with the ^{14}N nucleus (nuclear spin = 1).[4] The MGD–Fe–^{15}NO complex yields a two-line EPR pattern ($g = 2.04$ and $a_N = 17.6$ G) characteristic for the ^{15}N isotope (nuclear spin = 1/2).[14]

2. NO and the water-soluble dithiocarbamate–Fe complex react with a rate constant[15–17] from 10^6 to 10^8 M^{-1} s^{-1} (the rate constant of NO scavenging by

[4] A. Komarov, D. Mattson, M. M. Jones, P. K. Singh, and C.-S. Lai, *Biochem. Biophys. Res. Commun.* **195,** 1191 (1993).

[5] C.-S. Lai and A. M. Komarov, *FEBS Lett.* **345,** 120 (1994).

[6] C.-S. Lai and A. M. Komarov, *in* "Bioradicals Detected by ESR Spectroscopy" (H. Ohya-Nishiguchi and L. Packer, eds.), p. 163. Birkhauser Verlag, Basel, 1995.

[7] V. Quaresima, H. Takehara, K. Tsushima, M. Ferrari, and H. Utsumi, *Biochem. Biophys. Res. Commun.* **221,** 729 (1996).

[8] H. Fujii, J. Koscielniak, and L. J. Berliner, *Magn. Reson. Med.* **38,** 565 (1997).

[9] T. Yohimura, H. Yokoyama, S. Fujii, F. Takayama, K. Oikawa, and H. Kamada, *Nature Biotechnol.* **14,** 992 (1996).

[10] A. M. Komarov, *Cell. Mol. Biol.* **46,** 1329 (2000).

[11] P. Kuppusamy, P. Wang, A. Samoilov, and J. L. Zweier, *Magn. Reson. Med.* **36,** 212 (1996).

[12] G. A. Brewer, R. J. Butcher, B. Letafat, and E. Sinn, *Inorg. Chem.* **22,** 371 (1983).

[13] R. J. Butcher and E. Sinn, *Inorg. Chem.* **19,** 3622 (1980).

[14] Y. Kotake, T. Tanigawa, M. Tanigawa, and I. Ueno, *Free Radic. Res.* **23,** 287 (1995).

[15] S. Pou, P. Tsai, S. Porasuphatana, H. Halpern, G. V. R. Chandramouli, E. D. Barth, and G. M. Rosen, *Biochim. Biophys. Acta* **1427,** 216 (1999).

[16] S. V. Paschenko, V. V. Khramtsov, M. P. Skatchkov, V. F. Plysnin, and E. Bassenge, *Biochem. Biophys. Res. Commun.* **225,** 577 (1996).

[17] V. Misik and P. Riesz, *J. Phys. Chem.* **100,** 17986 (1996).

oxyhemoglobin is $3 \times 10^7\ M^{-1}\ s^{-1}$, and rapid reactions of NO with other radical species are in the range of 10^9–$10^{10}\ M^{-1}s^{-1}$).[18] *In vivo* NO trapping efficiency of the MGD–Fe complex is 50% or higher based on the attenuation of the plasma nitrate/nitrite level following MGD–Fe complex injection in endotoxin-treated mice.[19] The partition coefficients of MGD–Fe and MGD–Fe–NO complexes in octanol/water mixtures are 0.01 and 0.001, respectively.[11] Therefore, based on their physical properties, these complexes should be confined to intravascular and extracellular aqueous compartments. Note that extracellular MGD–Fe and intracellular DETC–Fe complexes bind NO equally in isolated ischemic myocardium when they are applied at the same dose.[20] Such a comparison *in vivo* is complicated by their potentially different clearance rate and distribution.

3. MGD cannot use tissue iron and should be supplemented with exogenous iron.[21] DETC chelates intracellular-free iron in most tissues to yield lipophilic DETC–Fe "traps" However, in brain tissue,[22,23] isolated myocardium,[20] or for *in vivo* EPR studies,[7] DETC requires a separate subcutaneous injection of the Fe–citrate complex (135 μmol/kg of $FeSO_4$ plus 0.64 mmol/kg of sodium citrate)[24] to produce the EPR-visible DETC–Fe–NO complex. Note that iron supplementation enhances the *in vivo* NO trapping efficiency of dithiocarbamate, thus increasing the EPR signal of NO complexes, but it does not change nitric oxide production in tissue when it is given after iNOS activation [i.e., 6 hr after lipopolysaccharide (LPS)].[21]

Experimental Procedures

Materials

MGD and N^G-monomethyl-L-arginine (NMMA) can be obtained from Calbiochem (San Diego, CA). The powder form of MGD should be sealed under nitrogen gas and stored desiccated at 4°.[25] $^{15}N_2$-guanidino-L-arginine (^{15}N-arginine) is purchased from Cambridge Isotope Laboratories (Woburn, MA). Sources of other chemicals are specified in the procedure.

[18] D. A. Wink, M. B. Grisham, J. B. Mitchell, and P. C. Ford, *Methods Enzymol.* **268,** 12 (1996).

[19] A. M. Komarov and C.-S. Lai, *Biochim. Biophys. Acta* **1272,** 29 (1995).

[20] A. M. Komarov, J. H. Kramer, I. T. Mak, and W. B. Weglicki, *Mol. Cell. Biochem.* **175,** 91 (1997).

[21] A. M. Komarov, I. T. Mak, and W. B. Weglicki, *Biochim. Biophys. Acta* **1361,** 229 (1997).

[22] V. D. Mikoyan, N. V. Voevodskaya, L. N. Kubrina, I. V. Malenkova, and A. F. Vanin, *Biochim. Biophys. Acta* **1269,** 19 (1995).

[23] V. D. Mikoyan, L. N. Kubrina, and A. F. Vanin, *Biofizika* **39,** 915 (1994).

[24] L. N. Kubrina, V. D. Mikoyan, P. I. Mordvintcev, and A. F. Vanin, *Biochim. Biophys. Acta* **1176,** 240 (1993).

[25] L. A. Shinobu, S. C. Jones, and M. M. Jones, *Acta Pharmacol. Toxicol.* **54,** 189 (1984).

Animal Procedures

The L-band (1.1 GHz) EPR experiment is performed with 15- to 20-g female ICR mice (Harlan Sprague-Dawley, Indianapolis, IN). The size of animals is restricted by the requirements of L-band EPR measurement (lowering the microwave frequency will allow the use of animals up to 30 g). Mice weighing 25–30 g are suitable for the S-band (3.5 GHz) EPR experiment and are used in our measurements taken from the murine tail. Each mouse is given 4 mg of LPS (*Escherichia coli* 026:B6 from Sigma, St. Louis, MO) via the lateral tail vein (note that none of the LPS-treated mice survive over 24 hr). At 6 hr, VEA mice are anesthetized with methoxyflurane (Pittman-Moore, Mundelein, IL) and then injected subcutaneously with 0.4 ml of the MGD–Fe in water (326 mg/kg MGD and 34 mg/kg of $FeSO_4$).[5] Over time, MGD, like other dithiocarbamate derivatives, decomposes in aqueous solution, yielding toxic carbon disulfide.[26] In addition, the iron incorporated in the MGD–Fe complex can oxidize and precipitate in the air-saturated solution. Thus, it is important to prepare fresh MGD–Fe complex in deoxygenated water (yellow color of solution) maintaining a 5 : 1 MGD-to-iron ratio and aerate it just before injection (brown-colored solution). Measuring absorption at 340, 385, and 520 nm (extinction coefficients 20,000, 15,000, and 3000 M^{-1}, respectively) could discover the presence of Fe(III) in the MGD–Fe complex. This could also be done by EPR of frozen aqueous solutions (77K) at $g = 4.3$ (characteristic EPR signal of ferric iron.)[27] Note that the administration of anaerobic MGD–Fe solution to animals[28] is less desirable, as the exposure to oxygen (and generation of reactive species) will occur *in vivo*. Using mixtures of MGD–Fe and ascorbate to keep the iron in a reduced state is also prone to artifacts, as ascorbate reacts with nitrite to release NO. Note that EPR-visible MGD–Fe–NO complexes can be prepared using Fe(III) salts as an iron source.[29] This may be due to reduction of the EPR-silent Fe(III)–NO complex to a paramagnetic Fe(II)–No derivative as is the case with other nitrosyliron complexes.[27] Therefore, *in vivo* application of Fe(III) NO traps is possible, but it will lead to the loss of reductants such as ascorbate and thiols. Normal mice will tolerate four separate aliquots (0.4 ml each) of MGD–Fe solution prepared as described earlier and injected subcutaneously over 6 hr.[6] For measurement of ^{15}NO production, animals receive a subcutaneous injection of ^{15}N-arginine (10 mg per mouse in saline) and the MGD–Fe injection. For inhibition experiments, at 6 hr after LPS treatment, mice are injected intraperitoneally with an aliquot of NMMA (50 mg/kg in saline).

[26] T. Martens, D. Langevin-Bermond, and M. B. Fleury, *J. Pharmaceut. Sci.* **82,** 379 (1993).

[27] A. F. Vanin, X. Liu, A. Samouilov, R. A. Stukan, and J. L. Zweier, *Biochim. Biophys. Acta* **1474,** 365 (2000).

[28] T. Miyajima and Y. Kotake, *Biochem. Biophys. Res. Commun.* **215,** 114 (1995).

[29] S. Fujii, T. Yoshimura, and H. Kamada, *Chem. Lett.* **9,** 785 (1996).

Measurement of NO in Circulating Blood

Immediately after MGD–Fe complex injection, the mouse still resting in the plexiglass-restraining tube (24-mm diameter, with ventilation holes in the animal chamber) is transferred to a specially built platform on the S-band EPR spectrometer. No anesthetic agent is used and the animal remains conscious without apparent discomfort throughout the entire *in vivo* S-band EPR measurement. The tail of the animal is immobilized by taping it to a thin plexiglass rod and is then placed inside a horizontally oriented loop-gap resonator. A 4-mm loop loop-gap resonator operating at 3.5 GHz[30] is recommended for this *in vivo* experiment because the tip of the mouse tail (2–3 mm) fits well into the diameter of the resonator. The measured unloaded Q of the empty resonator is 3000 and the loaded Q is 400 (with the presence of the mouse tail).[19] EPR spectra are recorded at room temperature using an EPR spectrometer equipped with an S-band microwave bridge. The field modulation amplitude is calibrated using Fremy salt as a standard. Instrument settings for this experiment are as follows: 100 G field scan, 30 sec acquisition time, 0.1 sec time constant, 2.5 G modulation amplitude, 100 KHz modulation frequency, and 25 mW microwave power (each of the spectra recorded is an average of nine individual 30-sec scans).

Measurement of NO in the Murine Body

Immediately after MGD–Fe complex injection, mice are put into a plexiglass-restraining tube for 1 hr, anesthetized with an intraperitoneal injection of Inactin, and placed in a disposable plastic holder inside the loop-gap L-band (1.1 GHz) resonator with a diameter of 25 mm (Fig. 1).[31] Note that anesthetized endotoxin-treated mice are more likely to die during EPR experiments than conscious animals. EPR measurements are carried out at room temperature using an L-band 1- to 2-GHz microwave bridge interfaced to an EPR spectrometer. Instrument settings for this experiment include 84 G field scan, 120 sec acquisition time, 1 sec time constant, 8 G modulation amplitude, 100 KHz modulation frequency, and incident power up to 500 mW (each spectrum is obtained from an average of four individual 120-sec scans). The modulation amplitude suggested for this experiment (8 G) is close to optimal and allows *in vivo* EPR signal measurement from head to tail. It should be noted that the peak-to-peak line width of the EPR signal for MGD–Fe–NO is close to 4.0 G in solution,[14] but in tissue it can increase 50% or more.[20] We recommend reducing modulation amplitude to 4.0 G to observe a clear separation of hyperfine components for ^{14}N and ^{15}N isotopes *in vivo*.

[30] W. Froncisz and J. S. Hyde, *J. Magn. Reson.* **47,** 515 (1982).

[31] W. Froncisz, T. Oles, and J. S. Hyde, *J. Magn. Reson.* **82,** 109 (1989).

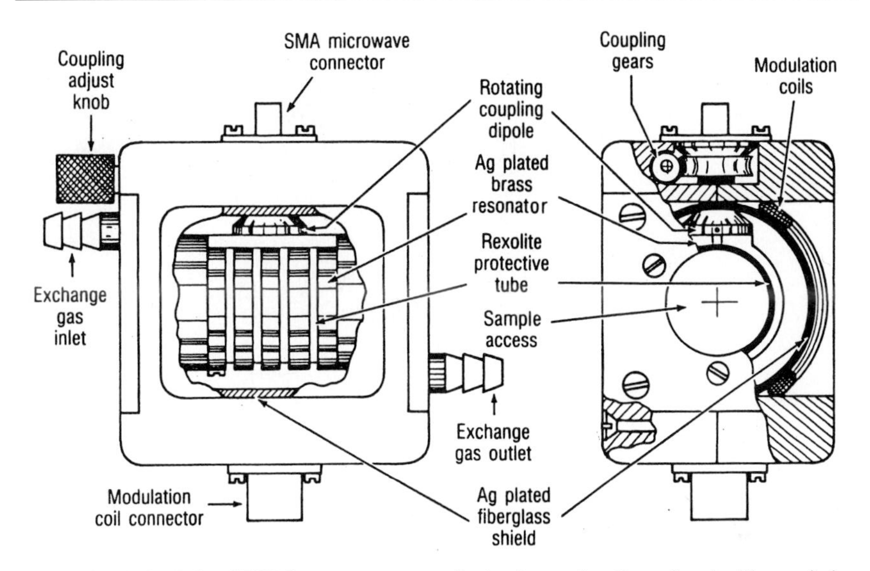

FIG. 1. Murine L-band EPR loop-gap resonator for *in vivo* studies. Reproduced with permission from W. Froncisz, T. Oles, and J. S. Hyde, *J. Magn. Reson.* **82,** 109 (1982).

Ex Vivo Measurement of Nitric Oxide Levels

Some animals are sacrificed 2 hr after MGD–Fe complex injection. Wet tissue samples, including whole blood, liver, and kidney tissue, are transferred into a quartz tube (i.d., 2 mm) for EPR measurement at room temperature. Urine samples are collected from the urinary bladder and are transferred immediately to a quartz flat cell for X-band EPR measurement. The urine sample is dark brown in color, characteristic of the presence of the MGD–Fe complex.[19] The MGD–Fe–NO complex in the urine is found to be stable at 4° for several hours. Reducing agents such as dithionite and ascorbate have been used *ex vivo* to convert diamagnetic EPR-silent derivatives of the dithiocarbamate–nitrosyliron complex into paramagnetic NO complexes.[32] Note, however, that the same procedure may generate NO from nitrite present in the sample.[33] Spectra are recorded at 22° with an X-band EPR spectrometer operating at 9.5 GHz. Instrument settings for this experiment include 100 G field scan, 4 min acquisition time, 0.5 sec time constant, 2.5 G modulation amplitude, 100 KHz modulation frequency, and 100 mW microwave power. We recommend reducing the microwave power to 30 mW to observe clear separation

[32] V. D. Mikoyan, L. N. Kubrina, V. A. Serezhenkov, R. A. Stukan, and A. F. Vanin, *Biochim. Biophys. Acta* **1336,** 225 (1997).

[33] K. Tsuchiya, M. Takasugi, K. Minakuchi, and K. Fukuzawa, *Free Radic. Biol. Med.* **21,** 733 (1996).

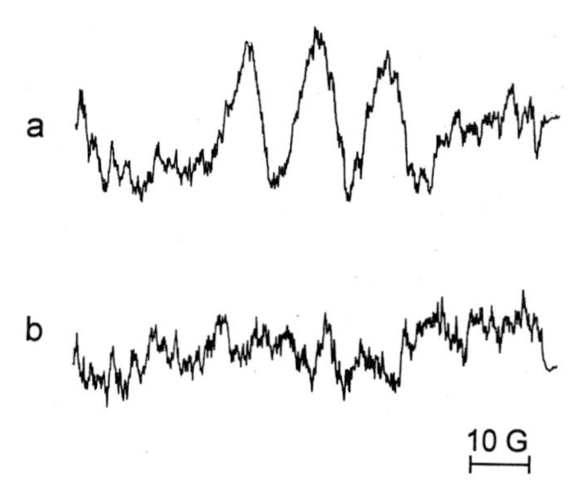

FIG. 2. *In vivo* 1.1-GHz EPR spectra of the MGD–Fe–NO complex in LPS-treated mice. Mice are injected with 0.4 ml of the MGD–Fe complex (a) or the MGD–Fe complex plus NMMA (b) 6 hr after LPS administration. The instrumental settings are as described in the text. Each spectrum presented here is an average of four 120-sec scans. The MGD–Fe–NO complex is formed in the liver by NO produced from endogenous L-arginine. Reproduced with permission from A. M. Komarov, *Cell. Mol. Biol.* **46,** 1329 (2000).

of hyperfine components for ^{14}N and ^{15}N isotopes in the urine sample.[19] Note that the "background signal" of dithiocarbamate–Cu(II) chelates may complicate X-band EPR spectra of NO complexes in tissues and liquid samples.[19,34] MGD forms water-soluble complexes with Cu(II),[19] but it does not form complexes with tissue Cu(II), unlike DETC.[20,21]

Examples of NO Detection in Living Mice

In Vivo EPR Detection of NO Generated from Endogenous L-Arginine

An intensive three-line EPR signal from the MGD–Fe–NO complex in the upper abdomen (liver region) of endotoxin-treated mice is detected by *in vivo* L-band EPR spectroscopy (Fig. 2a). The EPR spectral parameters of the signal are identical to those of the MGD–Fe–NO complex in an aqueous solution of NO ($g = 2.04$ and $a_N = 12.5$ G).[4,10] Note that the *in vivo* EPR signal in the upper abdomen of endotoxin-treated mice is mostly due to the MGD–Fe–NO complex formed *in situ* in the liver tissue of live animals (except the contribution from the blood signal ~7%). It is likely that the MGD–Fe–NO complex formed in the murine liver is excreted and concentrated in the bile, as is the case in rat.[35]

[34] Y. Suzuki, S. Fujii, T. Tominaga, T. Yoshimoto, T. Yoshimura, and H. Kamada, *Biochim. Biophys. Acta* **1335,** 242 (1997).

[35] L. A. Reinke, D. R. Moore, and Y. Kotake, *Anal. Biochem.* **243,** 8 (1996).

Administration of NMMA (50 mg/kg) in septic shock mice decreases the level of the MGD–Fe–NO complex in the upper abdomen below the L-band EPR detectability limit (Fig. 2b). *Ex vivo* examination of isolated tissues and urine shows, however, that NO complex formation is not completely abolished by NMMA administration.[19] This may point to the possibility of the secondary ("nonenzymatic") NO formation from nitrite.

In Vivo EPR Detection of ^{15}NO Generated from ^{15}N-Arginine

In endotoxin-treated mice, ^{15}N-labeled L-arginine competes with the natural analog, producing a mixture of nitric oxide species (^{15}NO and ^{14}NO). Figure 3a exhibits an *in vivo* EPR signal (Fig. 3a) obtained from the upper abdomen of the septic shock mouse after administrating ^{15}N-arginine. The EPR doublet (characteristic of ^{15}N) of the MGD–Fe–^{15}NO complex indicates the origin of nitric oxide from one of the guanidino nitrogens of ^{15}N-labeled L-arginine. The EPR triplet of

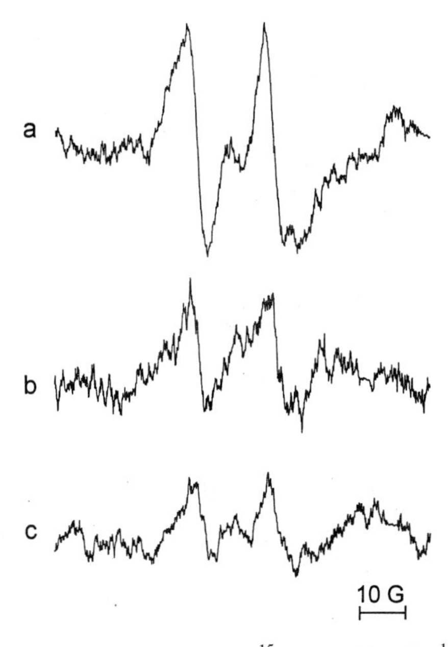

FIG. 3. *In vivo* 1.1-GHz EPR spectra of MGD–Fe–^{15}NO and MGD–Fe–^{14}NO complexes detected in various regions of the body in LPS-treated mice. Mice are injected with 0.4 ml of the MGD–Fe complex and 10 mg of ^{15}N-arginine, and EPR spectra are recorded from the (a) liver region, (b) head region, and (c) lower abdomen region of the body. The instrumental settings are as described in the text. Each spectrum presented is an average of four 120-sec scans, and the receiver gain is the same as for Fig. 2. Note that the doublet EPR pattern is arising from the MGD–Fe–^{15}NO complex, whereas the MGD–Fe–^{14}NO complex is contributing to the signal external "shoulders" and small center field peak. Reproduced with permission from A. M. Komarov, *Cell. Mol. Biol.* **46,** 1329 (2000).

the MGD–Fe–^{14}NO complex (i.e., the complex formed by nitric oxide originated from endogenous L-arginine) contributes external "shoulders" and a small center-field peak in the EPR spectrum (Fig. 3a). *In vivo,* the most intense EPR signal is found in the upper abdomen, which is consistent with the strongest signal having been seen in liver *ex vivo*. Figures 3b and 3c show EPR signals from the head and lower abdomen regions of live septic shock mice that received ^{15}N-arginine; the two-line pattern of the MGD–Fe–^{15}NO complex is evident. Note that there is a pronounced *in vivo* EPR signal in the head region (~60% of the signal intensity in the upper abdomen area). Because the water-soluble MGD–Fe complex is not likely to cross the blood–brain barrier,[36] the circulating MGD–Fe–NO complex may be contributing to the signal found in the head region (Fig. 3b). Nonetheless, the source of NO in the circulating MGD–Fe–NO complexes may be brain tissue where nitric oxide is generated locally by iNOS.[37]

EPR signals from the kidney and urinary bladder containing urine are the most likely contributors to the *in vivo* EPR spectrum in the lower abdomen area (Fig. 3c).

Conclusion

In vivo EPR spectroscopy and the isotopic tracing experiment with ^{15}N-arginine show widespread activation of NO generation in septic shock mice. It is important to emphasize that the NO level detected in the mouse body by the EPR technique may not represent the true "free" NO level existing at any given moment, but rather that NO is trapped and stabilized by the NO-trapping reagent in mice.[19] Note that all redox forms of NO (NO^+, NO, NO^-) *in vivo* may contribute to MGD–Fe–NO complex, as in aerobic conditions the trapping agent is not selective for redox species of NO.[38,39] Note also that the direct reaction of the MGD–Fe trap with nitrite[40] is not likely the source of the MGD–Fe–NO complex *in vivo,* in view of the fact that nitrite levels in tissues and body fluids are low. Complex redox chemistry of dithiocarbamate–iron NO traps is well compensated by their unique and important *in vivo* applications. In brief, recently developed dithiocarbamate–iron nitric oxide-complexing agents allow real-time detection of nitric oxide generation and distribution in small living animals by low-frequency EPR spectroscopy. This new technique significantly extends the capabilities of the traditional spin trapping approach.

[36] T. Yoshimura, S. Fujii, H. Yokoyama, and H. Kamada, *Chem. Lett.* **4,** 309 (1995).

[37] M.-L. Wong, V. Rettori, A. Al-Shekhlee, P. B. Bongiorno, G. Canteros, S. M. McCann, P. W. Gold, and J. Licinio, *Nature Med.* **2,** 581 (1996).

[38] A. M. Komarov, D. A. Wink, M. Feelisch, and H. H. H. W. Schmidt, *Free Radic. Biol. Med.* **28,** 739 (2000).

[39] A. M. Komarov, A. Reif, and H. H. H. W. Schmidt, *Methods Enzymol.* **359,** [2], 2002 (this volume).

[40] K. Tsuchiya and R. P. Mason, *J. Biol. Chem.* **275,** 1551 (2000).

[7] Measuring End Products of Nitric Oxide *in Vivo*

By ANDREAS K. NUSSLER, UWE B. BRUCKNER, JOSEF VOGT, and PETER RADERMACHER

Introduction

The measurement of nitric oxide (NO) in biological fluids has attained increasing importance in order to understand and to develop new therapeutic strategies in (patho)physiological events. In particular, the amount of NO production during inflammatory processes, such as sepsis, has been clearly linked to the severity of diseases. NO, an unstable molecule, is synthesized from L-arginine by three NO synthase (NOS) isoforms that react with oxygen species or other biological molecules to form various end products of the NO pathway, including nitrite (NO_2^-) and nitrate (NO_3^-), as well as various *S*-nitrosothiols (RSNO).[1,2]

Although there are several methods describing NO measurements in biological fluids (for review see Ref. 3), the suitability of these methods to determine NO reaction products in biological fluids has proved to be limited for various reasons. The most frequently used method to measure the stable end products of the NO pathway is with a purple azo dye first described by Griess more than a 130 years ago.[4] This analytical assay can easily be applied in laboratories without a large and expensive experimental setup. However, this method has its limitations with regard to sensitivity and its inability to detect other important physiological products of the NO pathways, such as RSNO, NO by itself, or NO_3^-.

The conventional Griess reaction has been modified frequently; however, to measure nitrite in biological fluids, the detection is limited to 1.0 to 5.0 μM.[5–7] This detection limit has hampered attempts to study precise metabolic pathways of NO in relation to other biological molecules or even in the comparison of NO production among various species. In rodents the measurement of plasma NO_2^-/NO_3^- concentrations certainly is a standard method to assess NO production, but this approach is by far less reliable in higher species such as primates or swine[8] due to the substantially lower endogenous NO formation rate. In fact, other

[1] A. K. Nussler and T. R. Billiar, *J. Leukoc. Biol.* **54,** 171 (1993).
[2] K.-D. Kroncke, K. Fehsel, and V. Kolb-Bachofen, *Clin. Exp. Immunol.* **113,** 147 (1998).
[3] J. P. Griess, *Phil. Trans. R. Soc.* (*Lond.*) **154,** 667 (1864).
[4] M. Feelisch and J. S. Stamler, "Methods in Nitric Oxide Research," p. 1. Wiley, New York, 1996.
[5] L. C. Green, D. A. Wagner, J. Glogowski, P. L. Skipper, J. S. Wishnok, and S. R. Tannenbaum, *Anal. Biochem.* **126,** 131 (1982).
[6] D. L. Granger, R. R. Taintor, K. S. Boockvar, and J. B. Hibbs, Jr., *Methods* **7,** 78 (1995).
[7] M. B. Gilliam, M. P. Sherman, J. M. Griscavage, and L. J. Ignarro, *Anal. Biochem.* **212,** 359 (1993).
[8] L. Traber, *Crit. Care Med.* **27,** 255 (1999).

0076-6879/02 $35.00

authors,[9] as well as our own group,[10,11] reported unchanged plasma NO_2^-/NO_3^- concentrations during porcine endotoxemia, even when the duration of the experiment had been sufficiently long, i.e., at least 5 to 6 hr,[12] to allow NOS-2 activation.[10,13,14] This important species-specific phenomenon is also documented by the extremely differing amount of endogenous NO formation in the upper airways of domestic and zoo animals, which may vary by up to two orders of magnitude.[15] During pathologic conditions, the scarcity[9] and, hence, minor importance of NOS-2 expression/production[10] probably further contributes to these findings.

Therefore, the aim of this article is to summarize our own experiences with NO measurements in biological fluids and to analyze various products of the NO pathway. In addition, we present two different approaches for the estimation of NO formation in higher species: the direct measurement of exhaled NO and the use of stable, nonradioactive ^{15}N-labeled isotopes.

Materials and Methods

Sample Preparation

Venous blood should be drawn in sterile containers from individuals. Red blood cells are removed after clot formation by centrifugation at 2500*g* for 15 min. The serum or plasma is collected, aliquoted, and stored at −70° until further use. Urine samples should also be collected in sterile plastic containers, centrifuged immediately to obtain a cell-free solution, and stored at −70°. For nitrite and nitrate, as well as for the determination of L-citrulline, samples are ultrafiltered (2000*g*/20°/30 min) using a Centrisart-5000 ultrafilter (Satorius AG, Göttingen, Germany) to eliminate high molecular weight particles.[6,16] For RSNO measurement, samples (serum is better than plasma) are not ultrafiltered, but centrifuged at 1000*g* for 4 min at 10° in order to obtain cell-free material.

[9] D. Javeshgani and S. Magder, *Shock* **16,** 320 (2001).

[10] C. M. Pastor, A. Hadengue, and A. Nussler, *Am. J. Physiol. Gastrointest. Liver Physiol.* **278,** G416 (2000).

[11] B. Šantak, P. Radermacher, T. Iber, J. Adler, U. Wachter, D. Vassilev, M. Georgieff, and J. Vogt, *Br. J. Pharmacol.* **122,** 1605 (1997).

[12] M. Villamor, F. Pérez-Vizcaíno, T. Ruiz, J. C. Leza, M. Moro, and J. Tamargo, *Br. J. Pharmacol.* **115,** 261 (1995).

[13] M. Matějovič, P. Radermacher, I. Tugtekin, A. Stehr, M. Theisen, J. Vogt, U. Wachter, F. Ploner, M. Georgieff, and K. Träger, *Shock* **16,** 203 (2001).

[14] F. Ploner, P. Radermacher, M. Theisen, I. Tugtekin, M. Matějovič, A. Stehr, C. Szabó, G. J. Southan, M. Georgieff, U. B. Brücker, and K. Träger, *Shock* **16,** 130 (2001).

[15] U. Schedin, B. O. Röken, G. Nyman, C. Frostell, and G. Gustafsson, *Acta Anaestehsiol. Scand.* **41,** 1133 (1997).

[16] M. Marzinzig, A. K. Nussler, J. Stadler, E. Marzinzig, W. Bartlen, S. M. Morris, Jr., N. C. Nussler, H. G. Beger, and U. B. Brückner, *Nitric Oxide Biol. Chem.* **1,** 177 (1997).

Measurement of Nitrite Based on the Griess Reaction

In the conventional Griess assay, sulfanilamide, H_3PO_4, and *N*-(1-naphthyl)-ethylendiamine (NED) are premixed and then incubated with a study sample in a ratio of 1 : 1 to form a purple azo dye, and the amount of nitrite is determined at a wavelength of 546 nm. Nitrite levels are then calculated from a nitrite standard curve.[4] We use a slightly modified Griess assay in which the reagents are not premixed. Deproteinized samples (150 μl) are mixed with 75 μl ice cold dapsone [4,4′-diaminodiphensylsulfone (Merck, Darmstadt, Germany) 14 m*M* in 2 *N* HCl] and then mixed with 75 μl NED (Merck; 4 m*M* in H_2O). This mixture is incubated at room temperature for 5 min, and light absorption is measured at 550 nm using a microplate reader (EAR 300, SLT, Crailsheim, Germany). Nitrite levels are calculated from a nitrite standard [sodium nitrite (Sigma, Deisenhofen, Germany) in H_2O] curve (0.2 to 100.0 μM) generated in serum, aqueous solutions, or culture medium. Using this assay, a linear range between 0.2 and 80 μM for nitrite and nitrate is achieved, whereas a commercially available kit (Cayman Chemical, Ann Arbor, MI) reaches only a linear range of between 2.0 and 60 μM nitrite.[16]

Measurement of Nitrite Based on Fluorochrome 2,3-Diaminonaphthalene (DAN)

Diluted deproteinized samples (250 μl, diluted 1 : 1 with H_2O) are incubated with 25 μl DAN [(Merck) 633 μM in 0.67 *N* HCl] and incubated at room temperature in the dark for 10 min. Then the mixture is adjusted to pH 11.5–12.0 with 1 *N* NaOH, and fluorescence is measured using a microtiter fluorescence photometer (Fluostar, BMG, Offenburg, Germany) with an excitation of 365 nm and an emission of 405 nm. Nitrite concentrations in samples are calculated from a standard curve using sodium nitrite standards (Sigma) at a linear range of 0.02–3.2 μM diluted in water or serum.[16,17]

Measurement of Nitrite plus Nitrate

In biological systems, NO is oxidized mainly to NO_3^- by the presence of oxyhemoglobin.[18] In order to measure the complete conversion of NO to NO_2^-/NO_3^- in biological fluids, nitrate must first be converted to nitrite in order to use the Griess reaction or the DAN method. With the intention of converting nitrate into nitrite, samples and standards (200 μl) are passed over a copper-coated cadmium column (10×100 mm, Waters, Milford, MA) and connected to a high-performance liquid chromatography (HPLC) system.[5] After samples pass over the cadmium column,

[17] T. P. Misko, R. J. Schilling, D. Salvemini, W. M. Moore, and M. G. Currie, *Anal. Biochem.* **214,** 11 (1993).

[18] P. M. Rhodes, A. M. Leone, P. L. Francis, A. D. Struthers, and S. Moncada, *Biochem. Biophys. Res. Commun.* **209,** 590 (1995).

they are mixed with the modified Griess reagents (dapsone instead of sulfanilamide) in a three-way Tee mixing chamber. Standard curves are generated using defined nitrite or nitrate samples (2.0–100.0 μM) diluted in water or serum. It is noteworthy that samples must be deproteinized before use, as the cadmium column clots very fast. Absorption of standards and study samples is measured using a 485 Tunable UV/visible absorbency detector (Waters) at 543 nm. Using this method, results are expressed in micromoles per liter and represent the total amount of NO_2^- plus NO_3^- in biological fluids. We found a detection limit for nitrite of approximately 3 μM.[16] Disadvantages of this method are the sensitivity, the sample amount needed, and the biohazardous waste of cadmium. More suitable in most laboratories is the use of bacterial nitrate reductase.[6] Deproteinized samples (200 μl) are incubated with 10 μl nitrate reductase (5 U/ml, Boehringer Mannheim, Germany) in the presence of 10 μl FAD [(Merck) 200 μM in H_2O] and 10 μM NADPH [(Merck) 6 mM in H_2O, always prepare fresh and keep at 4° in the dark until use]. This reaction mixture is incubated at 37° for 60 min and is protected from light. Then, excess NADPH is oxidized with 2 U/ml lactic dehydrogenase (Merck) and 5 μmol/ml sodium pyruvate (Merck). Total nitrite levels (representing nitrite plus nitrate) are determined as described earlier using the Griess or DAN method. Using this method, the nitrate concentration is calculated by subtracting from total nitrite the amount of nitrite measured before reducing nitrate to nitrite.

Simultaneous Measurement of Nitrite and Nitrate by Anion-Exchange HPLC

We have shown that this method can be applied easily in a laboratory equipped with a simple HPLC system. Deproteinized samples (10 to 25 μl) are passed over a Polyspher IC AN-1 column (Merck) connected to a HPLC system consisting of a Gynkotek M 480 pump (Gynkotek, Germering, Germany), a Gynkotek-UV detector (210-nm wavelength), and a 3392 integrator (Hewlett-Packard, Waldbronn, Germany). Anions are eluted isocratically by a mobile phase consisting of 2 mM NaCl, 2 mM Na_2SO_4, and 1.5% methanol at pH 3.0 at a flow rate of 2.0 ml/min. The retention times for nitrite and nitrate are about 5 and 10 min, respectively. NO_2^- and NO_3^- concentrations in study samples are correlated to nitrite and nitrate standards (linear range is between 0.5 and 80.0 μM for both ions).[16]

Measurement of S-Nitrosothiols (RSNO)

The determination of RSNO levels in biological fluids is based on a method described elsewhere.[19] Samples (50 μl) are diluted with water at a ratio of 1 : 1 and are mixed with 50 μl ammonium sulfamate [(Merck) 100 μM in H_2O] in order to trap nitrite present. After an incubation period of 10 min at room temperature,

[19] B. Saville, *Analyst* **83,** 670 (1958).

50 μl of the DAN reaction mixture (1 part 1.11 mM $HgCl_2$ + 4 parts 158 μM DAN solved in 0.62 N HCl) is added and incubated at room temperature in the dark for 10 min. Then the mixture is adjusted to pH 11.5–12.0 using 1 N NaOH, and fluorescence is read using a microtiter fluorescence photometer. Preparations of RSNO stocks for generating standard curves are used at a range of 0.15 to 10 μM (glutathione dissolved in 1 N HCl) and mixed with 10 mM sodium nitrite at a ratio 1 : 1 (each 50 μl) at room temperature for 15 min. Then the reaction mixture is incubated with 50 ml amonium sulfamate (100 μM in H_2O), followed by the DAN reaction and the determination of the fluorescence as described earlier.[16]

Measurement of L-*Citrulline*

We measured citrulline, a coproduct of the NO pathway, by modifying the method of Senshu and co-workers.[20] Briefly, samples (100 μl) are incubated with urease [(Merck) 165 U/ml in H_2O] at 37° for 30 min to hydrolyze urea present. Then samples are deproteinized with 50 μl TCA (2.45 M), centrifuged for 5 min at 15,000g, and the collected supernatant (125 μl) is mixed with 375 μl of the reaction mixture: 40% (v/v) diacetylmonoxime (79 mM in 83 mM acetic acid) + 18% (v/v) antipyrine E (47.9 mM in H_2O) + 42% (v/v) 15 N H_2SO_4 and incubated at 96° for 25 min. Once the reaction mixture is cooled to room temperature, the absorption is read at 450 nm using a microplate reader as described. L-Citrulline concentrations in the samples are calculated from standard curves using the range of 1.0 to 500.0 μM generated under identical conditions.[16]

Exhaled NO

As an alternative to the determination of stable end products of NO, the assessment of exhaled NO excretion may serve as a surrogate for the measurement of whole body NO formation. Several techniques have been used for measuring exhaled NO, such as mass spectrometry,[21] gas chromatography–mass spectrometry,[22] or electrochemical sensors,[23] but the most widely used method used to assay NO in exhaled gas is chemiluminescence. This technique is based on the reaction between ozone and nitric oxide producing NO_2 with an electron in an excited state.[24]

$$NO + O_3 \rightarrow NO_2 + O_2 \rightarrow NO_2 + h \cdot \nu$$

[20] T. Senshu, T. Sato, T. Inoue, K. Akiyama, and H. Asaga, *Anal. Biochem.* **203,** 94 (1992).

[21] L. E. Gustafsson, A. M. Leone, M. G. Persson, N. P. Wiklund, and S. Moncada, *Biochem. Biophys. Res. Commun.* **181,** 852 (1991).

[22] M. Leone, L. E. Gustafsson, P. L. Francis, M. G. Persson, N. P. Wiklund, and S. Moncada, *Biochem. Biophys. Res. Commun.* **201,** 883 (1994).

[23] E. P. Purtz, D. Hess, and R. M. Kacmarek, *J. Clin. Monit.* **13,** 25 (1997).

[24] S. L. Archer, *FASEB J.* **7,** 349 (1993).

Emission from the excited NO_2 (600–3000 nm) when it reverts to the ground state is detected by a red-sensitive photomultiplier. In different models of mechanically ventilated pigs[9,13,14,25] and dogs,[26] the amount of exhaled NO has been measured using the following approach: the expired port of the ventilator is connected to a mixing chamber and the mean NO concentration is either measured in repeated aliquots taken in gas-tight syringes or with the gas being aspirated continuously directly from this chamber. This article refers to the NOA 270B or the NOA 280 chemiluminescence analyzer (Sievers Medical Instruments, Boulder, CO), the sensitivity of which is 0.5–1.0 ppb. The NO analyzer is calibrated before each data collection in two steps using both a gas containing <1 ppb ("zero air") and a reference gas of known NO concentration. Meticulous attention has to be paid to avoid any exogenous contamination of the inspired gas by NO from the room air. Therefore, the inspired gas is delivered through a filter consisting of charcoal and potassium permanganate inserted into the respiratory port of the ventilator. NO excretion is then calculated as the product of expired NO concentration and the minute volume, but it should be taken into consideration that changes in the minute volume per se may influence NO excretion.[25] This technique has the advantage of avoiding erroneous measurements due to breath-to-breath variations, but NO breakdown and reaction during the time between sampling and analysis may compromise the accuracy of the measurement.[27] Moreover, "scavenging" of NO due to the presence of oxygen radicals or when bleeding is present in the lung or the airways may lead to underestimation of the NO excretion.[27] Finally, potential inaccuracy is due to the sensitivity of the NO analyser (see earlier discussion) when the measured expired NO levels are low.[25] Nevertheless, measuring NO excretion allowed monitoring the effects of manipulating the endogenous NO formation.[13,14,28]

Stable Isotope Approaches

The use of primed continuous infusions of stable, nonradioactive labeled isotopes for the quantification of whole body and/or organ metabolic pathways is based on the theoretical assumption,[29] that metabolic processes, as well as distribution into the different compartments of the labeled and unlabeled molecule, are not discriminated. Using L-[guanidino-$^{15}N_2$]-arginine, the NO production rate can be determined from the conversion of this tracer to L-[ureido-^{15}N]-citrulline, the

[25] S. Mehta, S. Magder, and R. D. Levy, *Chest* **111,** 1045 (1997).

[26] S. N. A. Hussain, M. N. Abdul-Hussain, and Q. El-Dwairi, *J. Crit. Care* **11,** 167 (1996).

[27] M. Gerlach and H. Gerlach, *Respir. Care* **44,** 349 (1999).

[28] N. Marczin, B. Riedel, D. Royston, and M. Yacoub, *Lancet* **349,** 1742 (1997).

[29] R. R. Wolfe, "Radioactive and Stable Isotopes in Biomedicine: Principle and Practice of Kinetic Analysis." Wiley-Liss, New York, 1992.

coproduct of this reaction being ^{15}NO[30,31] according to the equation:

$$COOH\text{—}HCNH_2\text{—}(CH_2)_3\text{—}NH\text{—}C(\boxed{^{15}N}H_2)_2 \xrightarrow{O_2 + NADPH\ NADP^+} COOH\text{—}HCNH_2\text{—}(CH_2)_3\text{—}NH\text{—}CO\boxed{^{15}N}H_2 + \boxed{^{15}N}O^{\bullet}$$

For this purpose, after a prime of 0.1 μM/kg, the tracer is infused at a rate of 0.1 μM/kg/hr. Steady-state concentrations are achieved within 1 hr. In human 24-hr studies, a bolus of 5 μM/kg followed by a constant intravenous infusion at a rate of 5 μM/kg/hr was used.[30,31] At least four blood samples are taken every 20 min beginning 2 hr after the bolus to ensure state plasma isotope enrichments. Plasma arginine and citrulline concentrations are assessed using standard HPLC procedures. Methylester trifluoroacetyl derivatives of the amino acids are formed first by esterification with acetyl chloride and methanol and subsequent acylation with trifluoroacetic anhydride and dichloroethane.[31] The plasma isotope enrichments $^{15}N_2$-arginine/$^{14}N_2$-arginine and ^{15}N-citrulline/^{14}N-citrulline, defined as enrichments A^{15N} and C^{15N}, are then measured by combined liquid[32] or gas chromatography–mass spectrometry[31] under negative, methane-based chemical ionization. Selective ion monitoring provides an abundance of unlabeled and labeled arginine (Mass 456 vs mass 458), as well as citrulline (Mass 361 vs 362). The citrulline tracer, liberated by NO production, is diluted by citrulline production in the frame of extrahepatic arginine synthesis. Hence, Castillo and co-workers[31] estimated whole body citrulline production by infusing a second citrulline tracer with a labeling pattern that can be separated from ^{15}N-citrulline labeling. This correction is not required if the arginine-to-citrulline conversion model is applied to the organ-specific NO production rate (Q_{NO}). It can be derived from the general equation:

$$Q_{NO} \cdot (\text{arginine precursor enrichment}) = Q_{plasma} \cdot (C^{15N}{}_A \cdot [CIT]_A \text{-} C^{15N}{}_A \cdot [CIT]_V)$$

The right side of the equation reflects the ^{15}N-citrulline organ balance and uses the plasma flow (Q_{plasma}), the arterial and venous plasma citrulline concentrations ($[CIT]_A$ and $[CIT]_V$, respectively), and the isotope enrichments. Because this balance does not correct for ^{15}N-citrulline disposal within the organ, the calculated Q_{NO} underestimates the true NO production rate. In contrast, arterial arginine

[30] L. Castillo, T. C. DeRojas-Walker, T. E. Chapman, J. Vogt, S. R. Tannenbaum, and V. R. Young, *Proc. Natl. Acad. Sci. U.S.A.* **90,** 193 (1993).

[31] L. Castillo, M. Sánchez, J. Vogt, T. E. Chapman, T. C. DeRojas-Walker, S. R. Tannenbaum, A. M. Ajami, and V. R. Young, *Am. J. Physiol. Endocrinol. Metab.* **268,** E360 (1995).

[32] M. H. Van Eijk, D. R. Rooyakkers, and N. E. P. Deutz, *J. Chromatogr.* **620,** 143 (1993).

enrichment is taken as the precursor for the conversion, and any dilution of this precursor enrichment by unlabeled arginine, released within the organ, causes an overestimation of the conversion rate. Probably due to a balance of errors, this approach allowed estimation of the LPS-induced changes of the liver and the portal venous drained viscera in awake pigs during 24 hr of continuous intravenous endotoxemia,[33] as well as the effects of iNOS inhibition with aminoethyl isothiourea.[34]

Based on the conjecture that (1) NO is the main source of NO_3^- in the plasma[18] and (2) that NO is converted rapidly to NO_2^-,[18] which is then completely oxidized to NO_3^-, ^{15}NO will appear in the plasma (and subsequently be excreted into the urine) as $^{15}NO_3^-$:

$$4\,\boxed{^{15}N}O^{\bullet} + O_2 + 2\,H_2O \rightarrow 4\,\boxed{^{15}N}O_2^- + 4\,H^+ \quad \text{and}$$

$$2\,\boxed{^{15}N}O_2^- + O_2 + H_2O \rightarrow 2\,\boxed{^{15}N}O_3^- + H_2O$$

Given the potential underestimation of NO production due to intraorgan citrulline disposal and the conversion of L-[ureido-^{15}N]-citrulline to L-[guanidino-$^{15}N_2$]-arginine, a metabolic pathway that takes place in the kidney,[31] whole body NO formation can also be derived using a primed, continuous infusion of $^{15}NO_3^-$. For this purpose, $Na^{15}NO_3$ is infused continuously (0.5 mg/kg/hr) after a bolus injection (0.75 mg/kg), and arterial blood is sampled before, as well as 5, 10, 15, 30, 60, and 90 min after the bolus and subsequently every 90 min. After centrifugation (3000 U/min, 10 min at 4°), 250 μl of plasma is deproteinized with 500 μl acetonitril, and plasma NO_3^- is converted to 1-nitro-2,4,6-trimethoxybenzene with 1,3,5-trimethoxybenzene. After gas chromatography separation (HP-Ultra capillary column, 12 m, 0.2 mm; temperature program: 80° for 1.5 min, 25°/min up to 150°), the isotope ratio $[^{15}NO_3^-]/[^{14}NO_3^-]$ in plasma is measured in the chemical ionization, selective ion-monitoring mode using helium as a carrier gas comparing the abundance of unlabeled (mass 214) and labeled (mass 215) NO_3^-. A two-compartment model is assumed for the NO_3^- kinetics: one being an active pool in which newly generated NO_3^- appears and from which it is eliminated and the other being an inactive volume of distribution in which only passive exchange takes place. In addition, it is assumed that for any given time point the NO_3^- production rate equals the elimination rate or, in other words, the NO_3^- pool size of the active compartment is constant. Time-dependent changes in the isotope enrichments

[33] M. J. Bruins, P. B. Soeters, and N. E. P. Deutz, *Intensive Care Med.* **25**(suppl. 1), S173 (1999). [Abstract]

[34] M. Poeze, M. J. Bruins, I. J. Vriens, G. Ramsey, and N. E. P. Deutz, *Intensive Care Med.* **27**(suppl. 2), S243 (2001). [Abstract]

$N^{15}{}_{I}$ and $N^{15}{}_{II}$ of the two compartments I and II can then be estimated according to the following equations:

$$\frac{dN^{15}{}_{I}}{dt} = [(N^{15}{}_{II} - N^{15}{}_{I}) * T - N^{15}{}_{I} * P] / Q_{I}; \quad \frac{dN^{15}{}_{II}}{dt} = (N^{15}{}_{I} - N^{15}{}_{II}) * T / Q_{II}$$

where T is the NO_3^- transfer rate between compartment I and II, P is the endogenous NO_3^- production rate, and Q_I and Q_{II} are the NO_3^- pools of the respective compartment. A plasma isotope decay curve is generated by numerical integration and is fitted to the measured plasma $^{15}NO_3^-$ enrichment values based on a nonlinear least-squares regression, which enables calculation of the total NO_3^- pool, as well as the NO_3^- production rate. An increase in NO formation will shift the decay curve of the plasma $^{15}NO_3^-$ enrichment downward due to "dilution" of the continuously infused $^{15}NO_3^-$ by the enhanced formation of $^{14}NO_3^-$ resulting from NO oxidation. In fact, this approach allowed quantification of the endogenous NO_3^- production rate over 9 hr of porcine endotoxemia, and the unchanged decay curve during infusion of the L-arginine analog L-NMMA as an NOS inhibitor confirmed the usefulness of this approach.[11]

Summary

The methods we have presented are suitable for measuring end products of the NO pathway in biological fluids. In particular, the first part of this article demonstrated how standard methods could be improved to increase the sensitivity of NO measurements in biological methods. In addition, we showed two possibilities for the estimation of NO formation in higher species, as NO standard measurements in these species are often not reliable to detect changes of the NO pathway.

[8] Guide for the Use of Nitric Oxide (NO) Donors as Probes of the Chemistry of NO and Related Redox Species in Biological Systems

By DOUGLAS D. THOMAS, KATRINA M. MIRANDA, MICHAEL GRAHAM ESPEY, DEBORAH CITRIN, DAVID JOURD'HEUIL, NAZARENO PAOLOCCI, SANDRA J. HEWETT, CAROL A. COLTON, MATTHEW B. GRISHAM, MARTIN FEELISCH, and DAVID A. WINK

Introduction

Numerous tools are used to decipher the roles nitric oxide (NO) plays in physiological and pathophysiological processes.[1] These approaches include product analysis, inhibition of biosynthetic pathways [e.g., pharmacological and genomic ablation of nitric oxide synthase (NOS) isoforms], and simulation of NO production through application of synthetic donors. Each method provides important information to aid in the understanding of the involvement of NO in cell signaling or cytotoxicity.

Various analytical techniques can be employed to measure the metabolites of NO biosynthesis. These data provide information on the redox environment present during NOS activity. One of the most commonly used indicators for the involvement of NO in a biological process is the presence of the stable end products nitrite (NO_2^-) and nitrate (NO_3^-).[2] The observation that serum nitrite/nitrate levels were increased in rats and in volunteers suffering from infection was a crucial preliminary indication that NO was a participant in the immune response.[3,4] Additionally, the detection of common modified biomolecules, such as *S*-nitrosothiols and 3-nitrotyrosine, is often utilized to delineate the redox environment. Use of competitive inhibitor substrates of NOS is an effective method for probing the involvement of NO in a variety of cellular and physiological processes. The efficacy of methylarginine analogs, initially described by Hibbs and co-workers,[5] resulted in an explosion of biomedical research, which demonstrated that NO was not limited to host defense mechanisms but also impacted the cardiovascular and nervous systems.[6]

[1] M. Feelisch and J. Stamler (eds.), "Methods in Nitric Oxide Research," pp. 303–488. Wiley, New York, 1996.

[2] M. B. Grisham, G. G. Johnson, and J. R. Lancaster, Jr., *Methods Enzymol.* **268,** 237 (1996).

[3] L. C. Green, S. R. Tannenbaum, and P. Goldman, *Science* **212,** 56 (1981).

[4] L. C. Green, K. Ruiz de Luzuriaga, D. A. Wagner, W. Rand, N. Istfan, V. R. Young, and S. R. Tannenbaum, *Proc. Natl. Acad. Sci. U.S.A.* **78,** 7764 (1981).

[5] J. B. Hibbs, Jr., R. R. Taintor, and Z. Vavrin, *Science* **235,** 473 (1987).

[6] S. Moncada, R. M. J. Palmer, and E. A. Higgs, *Pharmacol. Rev.* **43,** 109 (1991).

0076-6879/02 $35.00

One of the seminal discoveries in the field was identification of NO as the endothelium-derived relaxing factor (EDRF). Prior to the 1980s, the mechanism of action of the clinical nitrovasodilators sodium nitroprusside (SNP) and glyceryl trinitrate (nitroglycerin, GTN) was not evident. Metabolism of these compounds was found to be thiol dependent and to result in nitrosylation of the heme in soluble guanylyl cyclase (sGC).[7,8] The determination that free NO activates sGC,[9] in combination with Furchgott's observations on the biochemical properties of EDRF,[10] resulted in the discovery that NO, as the EDRF, mediates vasodilation through sGC activation.

Identification of NO as an important component in a wide variety of biological functions has led to a demand for compounds that consistently deliver defined fluxes and types of nitrogen oxides under specific conditions. This article discusses the pros and cons of the three most commonly used classes of NO donors and their usefulness in determining the outcome of NO biosynthesis in specific cellular microenvironments.

Types of NO Donors

Synthetic NO donors differ remarkably in their ability to release NO or other nitrogen oxides, their cofactor requirements for tissue/cell metabolism, and their decomposition by-products. An in-depth discussion of the biochemical characteristics and pharmacological properties of all available NO donors is beyond the scope of this article and has been reviewed previously.[11,12] Choosing an appropriate NO donor for a particular investigation is not a trivial issue, and it is important to realize that full chemical characterization is not available for all donor classes. Unfortunately, an incomplete understanding of nitrogen oxide release from certain commonly used agents has lead to misconceptions concerning the biological role of NO. Further, interpretation and comparison of experimental results are complicated by differential rates of release and amounts of free NO produced, as well as alterations in redox chemistry with varied donor concentrations. Seemingly contradictory results, particularly with the same donor, can be reconciled with an understanding of the environment generated by application of the donor.

[7] L. J. Ignarro, J. C. Edwards, D. Y. Gruetter, B. K. Barry, and C. A. Gruetter, *FEBS Lett.* **110,** 275 (1980).

[8] L. J. Ignarro, H. Lippton, J. C. Edwards, W. H. Baricos, A. L. Hyman, P. J. Kadowitz, and C. A. Gruetter, *J. Pharm. Exp. Ther.* **218,** 739 (1981).

[9] F. Murad, C. K. Mittal, W. P. Arnold, S. Katsuki, and H. Kimura, *Adv. Cyclic Nucleotide Res.* **9,** 145 (1978).

[10] R. F. Furchgott and J. V. Zawadzki, *Nature* **288,** 373 (1980).

[11] M. Feelisch and J. S. Stamler (eds.), "Methods in Nitric Oxide Research," pp. 71–115. Wiley, New York, 1996.

[12] M. Feelisch, *Naunyn Schmiedebergs Arch. Pharmacol.* **358,** 111 (1998).

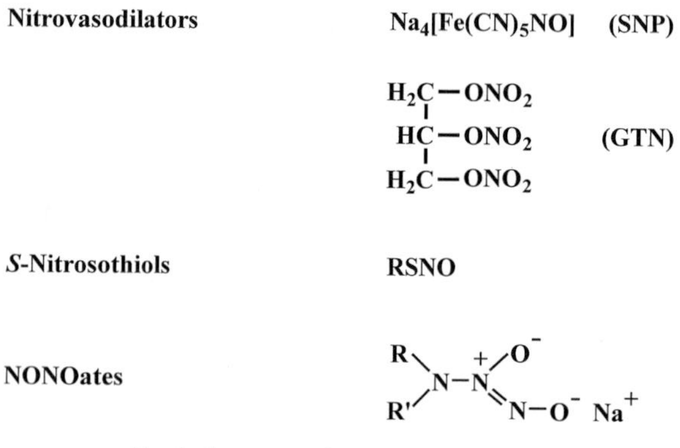

Nitrovasodilators	$Na_4[Fe(CN)_5NO]$	**(SNP)**
	H_2C-ONO_2 $HC-ONO_2$ H_2C-ONO_2	**(GTN)**
***S*-Nitrosothiols**	**RSNO**	
NONOates	$RR'N-N^{+}(O^{-})=N-O^{-}\ Na^{+}$	

FIG. 1. Structures of common NO donors.

The most commonly used NO donors fall into three broad categories: clinical nitrovasodilators, *S*-nitrosothiols, and diazeniumdiolates (NONOates) (Fig. 1). In addition, compounds are available that generate NO redox chemistry (i.e., other related nitrogen oxides) and thus provide important insights into NO biology. These include synthetic peroxynitrite ($ONOO^-$), SIN-1 (3-morpholinosydnonimine), which cogenerates NO and superoxide (O_2^-), and Angeli's salt ($Na_2N_2O_3$), which produces HNO.[11] The kinetics of product release from a given compound also has an important impact on the experimental outcome, as it determines, together with other environmental factors, the extent of biomolecule modification. The conditions required for nitrogen oxide release are also donor dependent. For example, some compounds decompose spontaneously under physiological condition, whereas others require oxidative or reductive metabolism. Finally, the effects of end products and environmental alterations, such as consumption of O_2, must be considered.

Clinical Nitrovasodilators

Initially, SNP and GTN were the agents commonly used to investigate physiological processes mediated by NO. Initial reports suggested that the administration of SNP was representative of NO biosynthesis; however, SNP was later shown to function through a complex set of actions with reductive metabolism required for the generation of NO in cells, tissue, or media.[13] The extent of SNP reduction and NO release depends on a number of factors, including reductant bioavailability. Additionally, the NO^+ moiety of SNP can bind directly to protein amine and thiol residues,[14] producing compounds that decompose slowly. Further complicating

[13] L. J. Ignarro, B. K. Barry, D. Y. Gruetter, J. C. Edwards, E. H. Ohlstein, C. A. Gruetter, and W. H. Baricos, *Biochem. Biophys. Res. Commun.* **94,** 93 (1980).

[14] D. L. H. Williams, *in* "Nitrosation," pp. 201–206. Cambridge Univ. Press, Cambridge, 1988.

clear interpretation of results is the potential for release of five equivalents of cyanide (CN^-) upon reduction, formation of toxic iron complexes in the presence of hydrogen peroxide,[15,16] and photochemical degradation.[17]

Reductive metabolism of organic nitrates such as GTN appears to be similarly complex and likely involves both nonenzymatic and enzymatic pathways.[18] Desensitization to nitrovasodilators can develop, which is a phenomenon known as nitrate tolerance. Studies utilizing SNP and GTN therefore do not definitively indicate the involvement of free NO. Nevertheless, positive results with these compounds may indicate a role for sGC in the process under examination. Verification of sGC participation can be achieved through use of non-NO donor activators such as YC-1 [3-(5′-hydroxymethyl-2′-furyl)-1-benzylindazole],[19] which activates guanylyl cyclase, or 8-bromo cGMP, which stimulates cGMP-dependent protein kinases.[20] Additionally, sGC inhibitors, such as 1H-[1,2,4]oxadiazolo[4,3-α]quinoxalin-1-one (ODQ),[21] are available.

S-Nitrosothiols

The discovery that the active intermediate of nitrovasodilator metabolism may often be *S*-nitrosothiols[7,13] has led to frequent use of these compounds as NO donors. Although isolation in the crystalline form is somewhat difficult, *S*-nitrosothiols are relatively straightforward to generate *in situ*. As an example, *S*-nitrosothiol adducts of cysteine, *N*-acetylcysteine, or glutathione (GSH) are prepared in water by exposure of the parent thiol to equimolar sodium nitrite. The resulting acidic solution converts nitrite to the nitrosating species N_2O_3, and the deep red product is formed rapidly.[22] The yield can generally be enhanced by the addition of HCl. The concentrations of stock solutions are determined easily by measuring the absorbance at 334 nm ($\varepsilon = 908\ M^{-1}\ cm^{-1}$).[22,23] Dilution in 100 m*M* phosphate buffer rather than phosphate-buffered saline (PBS) is recommended to provide adequate buffer capacity. Isolation of these compounds requires several recrystallization steps with organic solvent and often results in preparations with

[15] M. Yamada, K. Momose, E. Richelson, and M. Yamada, *J. Pharmacol. Toxicol. Methods* **35,** 11 (1996).

[16] D. A. Wink, J. A. Cook, R. Pacelli, W. DeGraff, J. Gamson, J. Liebmann, M. C. Krishna, and J. B. Mitchell, *Arch. Biochem. Biophys.* **331,** 241 (1996).

[17] S. Kudo, J. L. Bourassa, S. E. Bogg, Y. Sato, and P. C. Ford, *Anal. Biochem.* **247,** 193 (1997).

[18] E. Noack and M. Feelisch, *Basic Res. Cardiol.* **86,** 37 (1991).

[19] E. Martin, Y. C. Lee, and F. Murad, *Proc. Natl. Acad. Sci. U.S.A.* **98,** 12938 (2001).

[20] R. M. Rapoport, M. B. Draznin, and F. Murad, *Proc. Natl. Acad. Sci. U.S.A.* **79,** 6470 (1982).

[21] F. Brunner, H. Stessel, and W. R. Kukovetz, *FEBS Lett* **376,** 262 (1995).

[22] T. W. Hart, *Tetrahedron Lett.* **26,** 2013 (1957).

[23] J. A. Cook, S. Y. Kim, D. Teague, M. C. Krishna, R. Pacelli, J. B. Mitchell, Y. Vodovotz, R. W. Nims, D. Christodoulou, A. M. Miles, M. B. Grisham, and D. A. Wink, *Anal. Biochem.* **238,** 150 (1996).

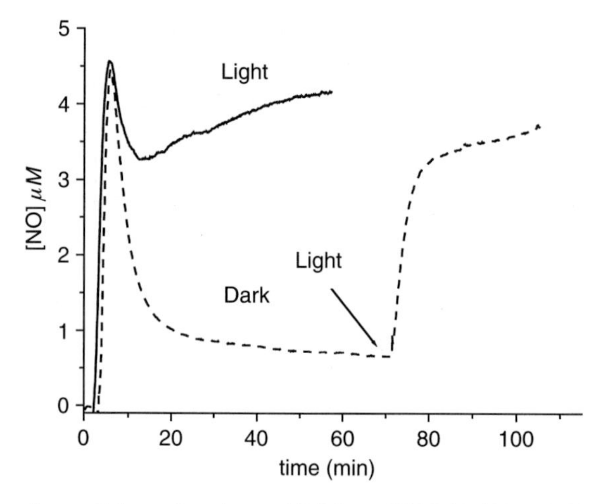

FIG. 2. Effect of room light on the release of NO from GSNO (1 mM) in PBS + 50 μM DTPA at 37° as measured with a NO-sensitive electrode. Initial peaks are a function of equilibrium of the stock solution with the buffer. In the dark, a steady state of NO is achieved after autoxidation of the initial NO peak (dashed line). In the presence of room light, significant enhancement of NO release is apparent (solid line).

varied thiol disulfide content. Therefore, fresh *in situ* preparations are more convenient for donor use. However, as these solutions can be cytotoxic, isolated solid may be more appropriate for *in vitro* use. Limited storage time for solid *S*-nitrosothiols, in the dark at $<0°$, is recommended.

The rate of NO release from *S*-nitrosothiols is dependent on the type of thiol and the buffer composition, in particular the concentration of contaminating transition metals such as Cu^{2+}.[24] Addition of the metal chelator diethylenetriaminepentaacetic acid (DTPA) will reduce the effect of metals greatly, but EDTA is generally not recommended as chelated metals remain redox active. The complex relationship between *S*-nitrosothiols and metals makes it difficult to extrapolate experimental results to *in vivo* conditions. Another often overlooked factor to consider when working with *S*-nitrosothiols is the effect of light. As seen in Fig. 2, room light has a dramatic effect on the rate of NO release.

Transnitrosation, or transfer of the NO^+ group, to another thiol, amine, alcohol, or metal can result in the formation of new nitrosated species, which will exhibit different properties and half-lives. For example, exposure of excess GSH to the *S*-nitrosothiol *S*-nitrosyl-*N*-acetyl-DL-penicillamine (SNAP) produces *S*-nitrosoglutathione (GSNO). Under certain conditions, transnitrosation may be

[24] D. L. Williams, *Methods Enzymol.* **268,** 299 (1996).

more efficient than release of free NO from *S*-nitrosothiols.[25] Transnitrosation from thiols is generally accelerated in the presence of metals via an exchange reaction involving a metal–thiol complex. It has been proposed that a reversible equilibrium may exist between nitrosated compounds and these metal complexes in tissues.[26] These alternate reactions, as well as metal status and light exposure, must be considered when interpreting experimental results obtained with *S*-nitrosothiols, and electrochemical measurement of NO release becomes essential for the comparison of results under varied conditions. These donors may be ideal for mimicking a biological response that involves a nitrosation reaction, but less beneficial if direct NO signaling predominates.

NONOates

NO biosynthesis can be simulated with diazeniumdiolates, which are commonly known as NONOates. Drago[27] first showed in the early 1960s that exposure of amines to several atmospheres of NO produces zwitterionic salts that contain two NO groups (Fig. 1). Keefer and co-workers[28] expanded this work into development of a series of NO donors:

$$\mathrm{RR'NN(O)NO} + \mathrm{H}^+ \rightarrow \mathrm{RR'NH}^+ + 2\mathrm{NO}$$

The kinetics of NO release from these compounds has been characterized under a variety of biological conditions.[16,28] The decomposition rate is dependent on the amine component, but, in contrast to *S*-nitrosothiols, is a function of pH and temperature with little influence from thiols, metals, light, or other media constituents.

The half-life for NO release is NONOate specific, ranging from several seconds to days at physiological pH and temperature. The most commonly used NONOates are diethylamine (DEA/NO), *N*-propyl-1,3-propanediamine (PAPA/NO), spermine (SPER/NO), and diethylenetriamine (DETA/NO) in which the first part of the name refers to the amine component. The amine and half-lives are given in Table I.[29,30] The pH and temperature dependence on decomposition rate allow for convenient storage of both solids and stock solutions in 10 m*M* NaOH at -20°. Concentration can be determined by measuring the absorbance at 250 nm ($\varepsilon = 8000\ M^{-1}\ \mathrm{cm}^{-1}$ for all NONOates)[29] shortly before use. Generally, long-term storage of stock solutions under 10 m*M* is not recommended.

[25] D. R. Arnelle and J. S. Stamler, *Arch. Biochem. Biophys.* **318,** 279 (1995).

[26] A. F. Vanin, *Biochemistry (Moscow)* **60,** 308 (1995).

[27] R. S. Drago, *Adv. Chem. Ser.* **36,** 143 (1962).

[28] L. K. Keefer, R. W. Nims, K. M. Davies, and D. A. Wink, *Methods Enzymol.* **268,** 281 (1996).

[29] C. M. Maragos, D. Morley, D. A. Wink, T. M. Dunams, J. E. Saavedra, A. Hoffman, A. A. Bove, L. Isaac, J. A. Hrabie, and L. K. Keefer, *J. Med. Chem.* **34,** 3242 (1991).

[30] J. A. Hrabie, J. R. Klose, D. A. Wink, and L. K. Keefer, *J. Org. Chem.* **58,** 1472 (1993).

TABLE I
NONOate Amine Components and their Half-Lives

Abbreviation	Amine	Half-life (37°, pH 7.4)
DEA/NO	Diethylamine	2–4 min
PAPA/NO	*N*-Propyl-1,3-propanediamine	15 min
SPER/NO	Spermine	39 min
DETA/NO	Diethylenetriamine	20 hr

Although insensitive to most buffer constituents and contaminants, NONOate decomposition is extremely sensitive to small changes in pH.[31] These compounds decompose in a pH-dependent manner, which often has a linear relationship to proton concentration. As a result, a change in pH between 7.2 and 7.6 can alter the rate of NO release by two- to threefold. To appropriately buffer the 10 m*M* NaOH solvent, 100 m*M* phosphate buffer should be used for diluting $<100\times$ stock solutions, otherwise PBS is sufficient. As a general rule, the NONOate should be added to the assay solution or media as the initiating reactant.

NONOates decompose to release NO and the parent amine. Autoxidation of NO will produce nitrite and potentially nitrosamines, particularly at high concentrations (millimolar) of NONOate. To control for potential confounding effects (e.g., amine backbone, nitrite, NaOH vehicle), a decomposed NONOate solution should be included as a control. Overnight decomposition at 37° in assay buffer or media is sufficient for all the NONOates in Table I except DETA/NO, which has a longer decay time.

Each NONOate produces a different flux of NO over varying time periods, thereby providing a convenient means to mimic the low, short bursts of NO synthesized by eNOS and nNOS or the higher, sustained concentrations produced by iNOS. Figure 3 shows the NO profiles that can be achieved with various NONOates. Significantly different concentrations are required to obtain comparable maximal production. Additionally, a steady state can be achieved with the longer lived NONOates.

Because NO is a gas, loss of NO from solution through volatilization must be considered. The amount of NO that escapes from solution is dependent on several factors, including NO concentration, temperature, vessel size and shape, surface area, degree of stirring, and type and concentration of scavengers. Figure 4 illustrates the effect of four different vessels of varying size and shape on NO detection following 1 hr decomposition of 100 μM PAPA/NO at 37°. An experiment

[31] K. M. Davies, D. A. Wink, J. E. Saavedra, and L. K. Keefer, *J. Am. Chem. Soc.* **123,** 5473 (2001).

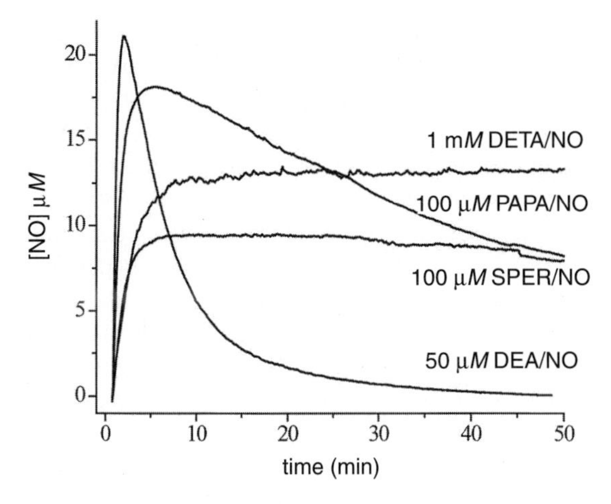

FIG. 3. NO electrode measurements demonstrating NO profiles for various NONOates in PBS + 50 μM DTPA at 37°. NO donor concentrations were chosen to give approximately equivalent maxima.

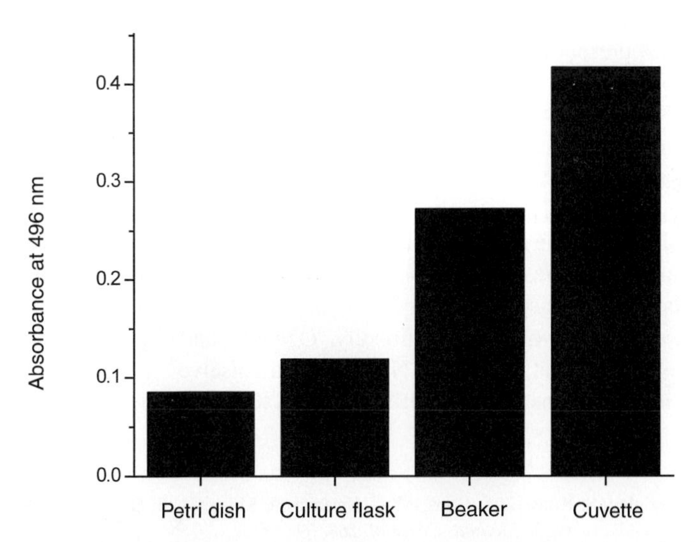

FIG. 4. NO concentration as a function of vessel size and shape. Each vessel contained 100 μM PAPA/NO in PBS + 50 μM DTPA with neutral Griess reagents. Solutions were incubated for 1 hr at 37°, after which 1 ml was extracted and the resultant absorbance was read at 496 nm. Fluid volumes: 100 × 20-mm petri dish, 10 ml; 250-ml culture flask, 20 ml; 20-ml beaker, 10 ml (stirred); and 4-ml cuvette, 2 ml (stirred).

requiring a defined concentration of NO calls for careful consideration of the type of vessel used. Alternately, a sealed vessel without headspace will ensure that NO will remain in solution. However, under these conditions, the solution can become anaerobic rapidly, particularly with high concentrations of NONOate.

NONOate decay can be monitored by measuring the rate of decrease in absorbance of the parent compound ($\lambda_{max} = 250$ nm, $\varepsilon = 8000\ M^{-1}\ cm^{-1}$).[29] However, this may not be possible in biological media, which often contain species that absorb strongly in this region. This coupled with vessel-dependent volatilization, pH sensitivity, and the fact that NONOates can release less than the theoretical amount of two NO per NONOate dictates the need for alternate means of quantification. Direct measurement of NO can be accomplished by electrochemical detection (Fig. 3).[32] Numerous conventional methods are also available,[1,33] such as gas-phase chemiluminescence or the oxymyoglobin assay (Fig. 5A). Alternatively, nitrosation of reporter substances in aerobic solution via NO autoxidation can be used as an indirect index of NO production. For instance, nirosation of sulfanilamide followed by reaction with *N*-(1-naphthyl)ethylenediamine leads to diazonium production in the classic Griess reaction. At neutral pH, this reaction is specific for NO (orange product with $\lambda_{max} = 496$ nm, $\varepsilon = 26{,}500\ M^{-1}\ cm^{-1}$,[34] and was the data acquisition method for Fig. 5C. Nitrosation of 2,3-diaminonaphthalene (DAN) results in formation of a fluorescent naphthaltriazole ($\lambda_{ex} = 375$, $\lambda_{em} = 450$ nm;[34] Fig. 5B). Strong fluorescence from endogenous flavins precludes monitoring of DAN nitrosation in the presence of cells. Diaminofluorescein (DAF) can be nitrosated to form a fluorescent triazole with spectral qualities suitable for intracellular imaging ($\lambda_{ex} = 488$ nm, $\lambda_{em} = 512$ nm).[35]

Techniques that use oxidation reactions for NO detection from NONOates can be problematic. Although oxidation of 2-2′-azinobis(3-ethylbenzthiazoline-6-sulfonic acid) (ABTS) is an excellent technique to determine NO concentrations in solutions,[34] the oxidized ABTS radical can react directly with NONOates through an as yet unresolved mechanism. Interestingly, there is no appreciable reaction between ferricyanide and NONOates, indicating that a redox reaction is not necessarily involved. Additionally, $ONOO^-$ and O_2^-[36] do not oxidize NONOates. In general, however, NONOates themselves are not reactive, and observed modifications are a result strictly of NO release. In fact, oxidation of

[32] D. Christodoulou, S. Kudo, J. A. Cook, M. C. Krishna, A. Miles, M. B. Grisham, M. Murugesan, P. C. Ford, and D. A. Wink, *Methods Enzymol.* **268,** 69 (1996).

[33] L. Packer (ed.), *Methods Enzymol.* **268,** 58–259 (1996).

[34] R. W. Nims, J. C. Cook, M. C. Krishna, D. Christodoulou, C. M. B. Poore, A. M. Miles, M. B. Grisham, and D. A. Winks, *Methods Enzymol.* **268,** 93 (1996).

[35] M. G. Espey, K. M. Miranda, D. D. Thomas, and D. A. Wink, *J. Biol. Chem.* **276,** 30085 (2001).

[36] A. M. Miles, D. S. Bohle, P. A. Glassbrenner, B. Hansert, D. A. Wink, and M. B. Grisham, *J. Biol. Chem.* **271,** 40 (1996).

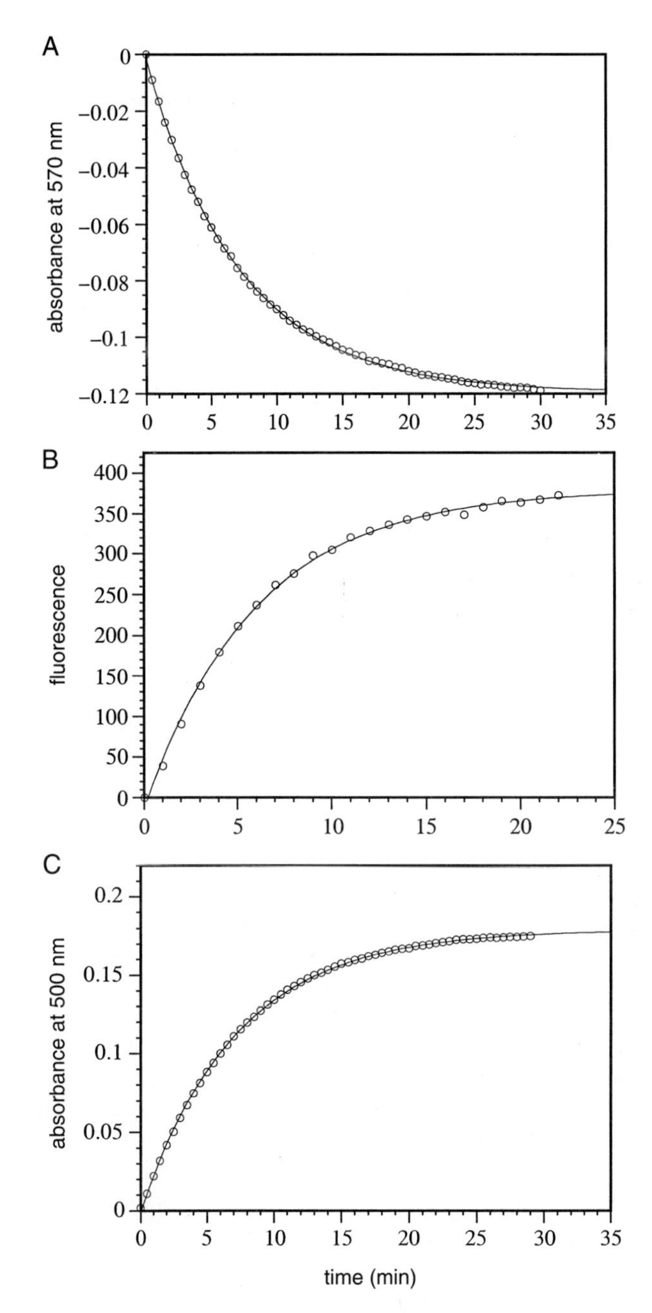

FIG. 5. Profiles of NO fluxes using the oxymyoglobin, DAN, and neutral Griess assays. Each reaction was monitored for 30 min at room temperature in PBS: (A) 10 μM DEA/NO + 20 mg/ml $Mb(O_2)$ at 570 nm; (B) 10 μM DEA/NO + 200 μM DAN with excitation at 375 nm and emission at 450 nm (slit widths 2.5 mm); and (C) 100 μM DEA/NO + Griess at 500 nm.

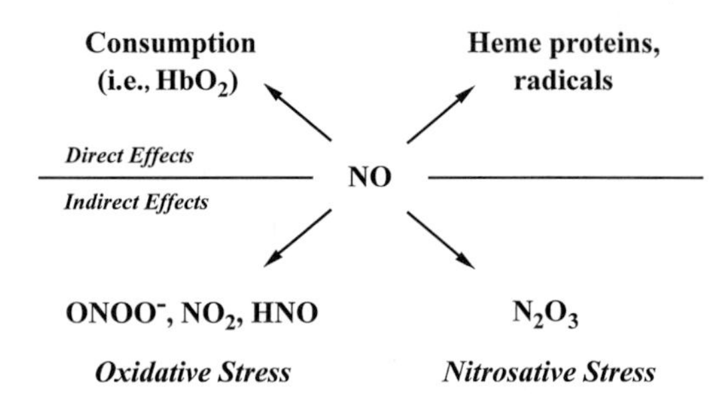

FIG. 6. The chemical biology of NO.

dihydrorhodamine (DHR) to the fluorescent dye rhodamine is a viable technique for NO measurement from NONOates. In summary, NONOates currently provide the most reliable source of exogenous NO and are ideal for most *in vitro* and *in vivo* experiments.

Producing NO Redox Chemistry *in Situ*

The redox chemistry of NO is the single most important factor to consider in mechanistic interpretations of the biological role of NO. Because NO can undergo a variety of chemical reactions, it is important to evaluate these reactions in different biological systems.[37] In an attempt to simplify the complexity of NO chemistry in biology, we have developed the concept of the *chemical biology of NO*[37,38] in which kinetically viable reactions are placed into the two classifications of direct and indirect effects (Fig. 6).

Direct chemical reactions are defined as those in which NO itself interacts with biological targets, most commonly heme proteins. Such reactions are generally rapid and likely account for the majority of the physiological effects of NO in biological systems as well as control of NO concentration. Conversely, indirect effects involve reactive nitrogen oxide species (RNOS) formed from the reaction between NO and either O_2 or O_2^-. Unlike NO, these reactive species can interact with a variety of cellular constituents, potentially leading to post-translational modification of critical moieties in macromolecules. Direct effects predominate when NO is produced at low concentrations for short periods of time, whereas prolonged exposure to higher local NO concentrations can initiate indirect effects.

[37] D. A. Wink and J. B. Mitchell, *Free Radic. Biol. Med.* **25,** 434 (1998).

[38] K. M. Miranda, M. G. Espey, D. Jourd'heuil, M. B. Grisham, J. Fukuto, M. Feelisch, and D. A. Wink, *in* "Nitric Oxide: Biology and Pathobiology" (L. Ignarro, ed.), pp. 41–55. Academic Press, San Diego, 2000.

The reactive intermediates responsible for indirect effects mediate either nitrosative or oxidative stress depending on the species involved.[37,38] Nitrosation occurs primarily through N_2O_3, whereas oxidation can be affected by HNO, $ONOO^-$, and NO_2. Nitrosative stress is orthogonal to oxidative stress with respect to RNOS chemistry.[37–39] On the one hand, nitrosative stress can be abated by scavenging of NO with oxidants such as O_2^- or HNO. On the other hand, NO can intercept reactive oxygen species (ROS) such as O_2^- and iron–oxo complexes. In addition, NO can convert oxidants such as $ONOO^-$ into nitrosating agents, nicely demonstrating the balance that exists between nitrosative and oxidative stress and the dependence on reactant concentration and thus cellular environment. Because biological outcome varies under each stress type, a shift in this balance may determine the eventual responses, which range from signal transduction to cell death.

Oxidative Stress: The NO/O_2^- Reaction

One of the first clues as to the chemical identity of EDRF was that its activity is attenuated by O_2^-, suggesting a radical species,[10] and the physiological activity of NO can be counteracted by reaction with O_2^-.[40] The reaction of NO with O_2^- and the subsequent chemistry of $ONOO^-$ is now one of the most prolific and controversial topics in NO research. $ONOO^-$ is a potent oxidant, which has led to speculation that it is one of the primary mediators of oxidative stress *in vivo*.[41] Synthetic $ONOO^-$ is capable of oxidizing or nitrating numerous biological components,[42] which led to the conclusion that the simultaneous production of NO and O_2^- was toxic and mediated tissue injury.[43] However, many of the reported biochemical and biological experiments utilized high concentrations of $ONOO^-$ and may not have significant physiological bearing.

NO is a powerful antioxidant that protects cells from H_2O_2/O_2^- cytotoxicity[44] and also downregulates the oxidant-mediated adhesion of leukocytes.[45] The question thus arises as to the factors that determine whether NO biosynthesis results in cytotoxicity, antioxidant and thus protective effects, or simply acts as a signaling agent. Closer examination of the chemistry of NO under different conditions provides insights into these very different effects. Miles *et al.*[36] first showed that the rates of NO and O_2^- production determine the degree of redox chemistry produced

[39] M. B. Grisham, D. Jourd'heuil, and D. A. Wink, *Am. J. Physiol.* **276,** G315 (1999).

[40] R. F. Furchgott and P. M. Vanhoutte, *FASEB J.* **3,** 2007 (1989).

[41] J. S. Beckman, T. W. Beckman, J. Chen, P. A. Marshall, and B. A. Freeman, *Proc. Natl. Acad. Sci. U.S.A.* **87,** 1620 (1990).

[42] W. A. Pryor and G. L. Squadrito, *Am. J. Phys.* **268,** L699 (1995).

[43] J. S. Beckman, *Nature* **345,** 27 (1990).

[44] D. A. Wink, I. Hanbauer, M. C. Krishna, W. DeGraff, J. Gamson, and J. B. Mitchell, *Proc. Natl. Acad. Sci. U.S.A.* **90,** 9813 (1993).

[45] P. Kubes, M. Suzuki, and D. N. Granger, *Proc. Natl. Acad. Sci. U.S.A.* **88,** 4651 (1991).

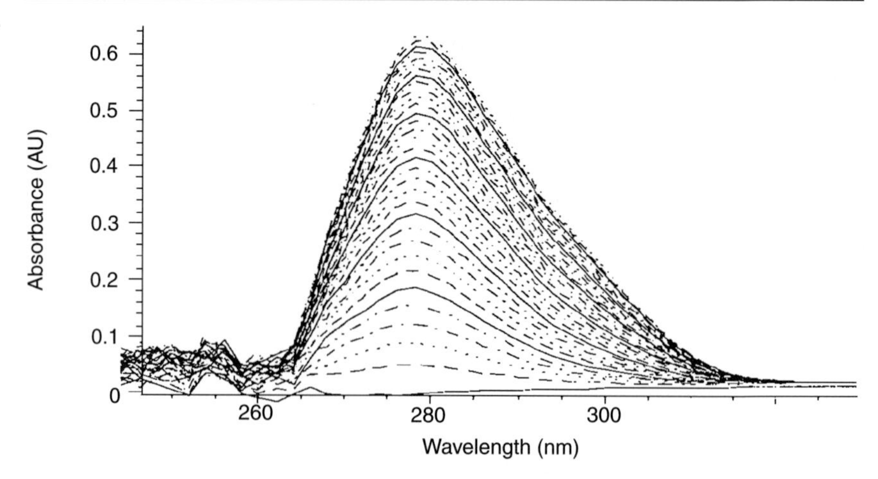

FIG. 7. Formation of xanthine (λ_{max} 272 nm) and urate (λ_{max} 292 nm, ε 1.1×10^4 M^{-1} cm^{-1}) from hypoxanthine (500 μM) by xanthine oxidase (1.5 μM O_2^-/min). The rate of urate formation under these conditions is 15 μM/hr.

from $ONOO^-$. If either radical is produced in excess, the respective oxidation or nitration reaction is quenched.[36,46] The secondary reactions of $ONOO^-$ with NO and O_2^- dramatically limit direct participation of $ONOO^-$ in cells and tissue. Furthermore, an imbalance in radical stoichiometry can lead to formation of a nitrosating rather than oxidizing species.[46,47] Therefore, in order to critically evaluate the biological relevance of the NO/O_2^- interaction, bolus additions of synthetic $ONOO^-$ should be avoided; rather, systems utilizing defined production rates of both radicals under controlled conditions are desirable.

Generating and Analyzing the Redox Chemistry of the NO/O_2^- Reaction

Separate sources of NO and O_2^- can be used to simulate the reactions relevant to *in vivo* conditions *in vitro*. The most commonly used paradigm is a combination of NONOate with the O_2^- generating system, hypoxanthine/xanthine oxidase (HX/XO). There has been some debate about the usefulness of this reaction, in particular the possible accumulation of urate, which is a potent scavenger of $ONOO^-$.[48] However, at <1 mM HX, which can be added directly to the assay buffer, the formation of urate is minimal, as XO preferentially utilizes HX rather than xanthine as the substrate (Fig. 7). Another potential pitfall of XO use is oxygen

[46] S. Pfeiffer and B. J. Mayer, *Biol. Chem.* **273,** 27280 (1998).

[47] A. M. Miles, M. Gibson, M. Krishna, J. C. Cook, R. Pacelli, D. A. Wink, and M. B. Grisham, *Free Radic. Res.* **23,** 379 (1995).

[48] C. D. Reiter, R. J. Teng, and J. S. Beckman, *J. Biol. Chem.* **275,** 32460 (2000).

consumption, particularly at high concentrations (>10 mU) or in low surface area vessels, such as test tubes. In general, the O_2 concentration should be measured regularly where possible, as even the relatively slow NO autoxidation can lead to considerable O_2 consumption during long incubations. As trace metals and H_2O_2 can elicit redox chemistry under these conditions, a metal chelator such as DTPA should be used to prevent Fenton-type reactions.

In addition, catalase can be added to scavenge the unavoidable production of H_2O_2 to ensure examination of the effect of NO/O_2^- chemistry alone. Catalase should be used with caution, however, as it can react with NO, resulting in scavenging of NO and reduction of enzyme activity. An effect attributed to H_2O_2 production may therefore actually be a result of reduced NO. Generally, only small amounts of catalase are necessary to remove H_2O_2 due to the fast reaction rate. Because the reaction of NO with O_2^- is several orders of magnitude faster than with catalase, H_2O_2 can be removed effectively with the careful use of catalase under most experimental conditions utilizing the NO/O_2^- generating system.

If the presence of urate, hypoxanthine, or xanthine presents a problem, then an alternative substrate can be used such as lumazine.[49] However, because the rate of O_2^- production from lumazine is approximately seven-fold lower compared to HX, the fluxes of NO or XO must be adjusted accordingly.

The rate of NO release from the NONOate can be measured as described earlier, whereas the flux of O_2^- can be determined by the ferricytochrome *c* assay. Reduction of ferricytochrome *c* (15 mg/ml) by O_2^- is zero order in the presence of XO and can be followed at 550 nm ($\varepsilon = 21{,}000\ M^{-1}\ \text{cm}^{-1}$).[50] A unique method developed by Kelm *et al.*[51] is also available to measure both of these radicals simultaneously. It should be pointed out again that small changes in buffer composition, such as addition or elimination of CO_2, may result in minor pH changes, which will alter the decay of the NONOate. Therefore, it is mandatory to measure each of these parameters for every buffer preparation.

It should also be noted that NO is metabolized by cells in an O_2-dependent manner.[52] Therefore, when using NONOates in the presence of cells, especially in suspension, the steady-state flux of NO may be much lower than that observed in solution alone. Although differences may not be significant when working with monolayer cell cultures, cellular suspensions can have substantial effects. In general, the steady-state NO concentration will be a function of cell density and O_2 concentration.

[49] D. Jourd'heuil, F. L. Jourd'heuil, P. S. Kutchukian, R. A. Musah, D. A. Wink, and M. B. Grisham, *J. Biol. Chem.* **276,** 28799 (2001).

[50] I. Fridovich, "Handbook of Methods for Oxygen Radical Research," p. 121. CRC Press, Boca Raton, FL, 1986.

[51] M. Kelm, R. Dahmann, D. A. Wink, and M. Feelisch, *J. Biol. Chem.* **272,** 9922 (1997).

[52] D. D. Thomas, X. Liu, S. P. Kantrow, and J. R. Lancaster, Jr., *Proc. Natl. Acad. Sci. U.S.A.* **98,** 355 (2001).

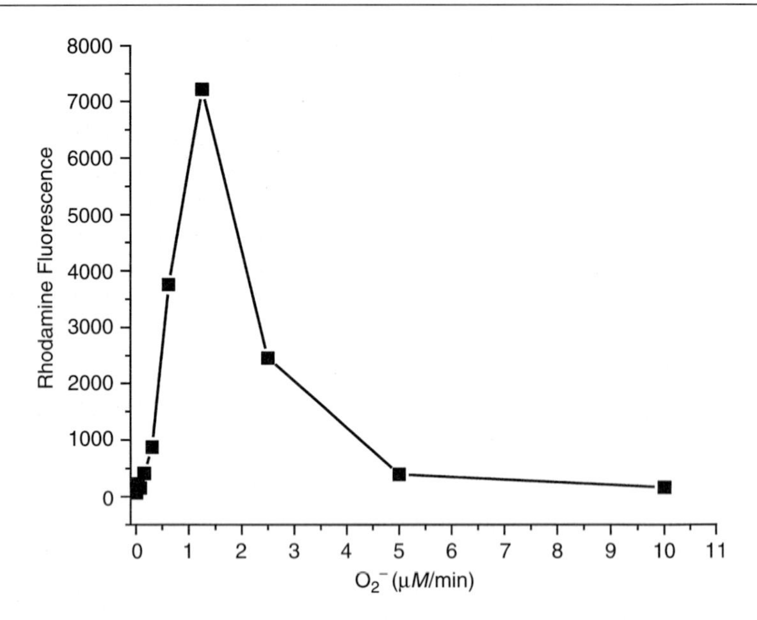

FIG. 8. Rhodamine production from oxidation of DHR under varied fluxes of O_2^- with constant NO. A PBS buffer containing 500 μM HX, 50 μM DTPA, and 100 μM SPER/NO was exposed to increasing concentrations of XO (0–10 μM O_2^-/min) for 1 hr at 37°. Samples in a 96-well plate were read at $\lambda_{ex/em}$ 500/570 nm.

The extent of nitrosation and oxidation under specific reaction conditions can be quantified through the modification of DAN or DHR, respectively, to fluorescent products.[36,47] Generally, stock solutions are made in dimethyl formamide (DMF) or dimethy sulfoxide (DMSO) at 100× to minimize organic solvent effects. Figure 8 demonstrates the effects on DHR oxidation as a result of modifying the redox environment with varied O_2^- flux (varied XO) at constant NO (100 μM SPER/NO). Figure 8 also illustrates the dependence of $ONOO^-$-mediated chemistry on an optimal and narrow range of reactant ratios.

The sydnonimine SIN-1 is also used commonly to examine the effects of the NO/O_2^- reaction.[53] This compound hydrolyzes to the open ring nitrosamine SIN-1A, which is oxidized by molecular oxygen, producing O_2^- and an unstable radical intermediate. A stable product is then formed by the elimination of NO, thereby providing a single donor of both NO and O_2^- (Scheme 1). Many studies have reported similar results for SIN-1 and synthetic $ONOO^-$. However, there are discrepancies, such as the observed potent vasodilatory activity of SIN-1. In general, oxidation of SIN-1 results in NO production; however, its oxidation can

[53] M. Feelisch, J. Ostrowski, and E. Noack, *J. Cardiovasc. Pharmacol.* **14,** S13 (1989).

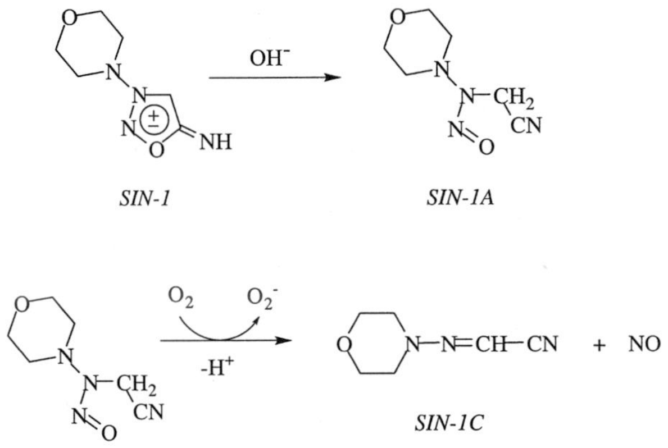

OH^-

SIN-1 *SIN-1A*

O_2 O_2^- $-H^+$

O N—N=CH—CH + NO

SIN-1C

SCHEME 1

occur by species other than O_2, including $ONOO^-$ itself. SIN-1 may therefore act primarily as an NO donor under certain conditions, particularly *in vivo*[54] or with high initial donor concentrations.

Nitrosative Stress

Nitrosation is defined as electrophilic addition of the nitrosonium ion (NO^+), generally donated by N_2O_3, to a nucleophile such as thiols, amines, or hydroxyl groups. Nitrosation should not be confused with nitrosylation, which is the addition of NO without any formal change in the redox state. However, nitrosylation often occurs through a multistep mechanism, forming the same product as nitrosation. For example, DAF can be converted to the fluorescent trizole by nitrosation via NO auoxidation or by oxidative nitrosylation, which involves substrate oxidation, such as by an intermediate of the NO/O_2^- reaction, prior to the addition of NO.[55] Obviously, the redox conditions for product formations by these pathways are quite different. Reductive nitrosylation can also occur with NO itself acting as the reducing agent in the presence of a nucleophile[56]:

$$Fe^{III}(\text{porphyrin})Cl + NO \rightarrow Fe^{II}(\text{porphyrin})(NO^+)Cl$$
$$Fe^{II}(\text{porphyrin})(NO^+)Cl + ROH \rightarrow Fe^{II}(\text{porphyrin}) + RONO + H^+ + Cl^-$$
$$Fe^{II}(\text{porphyrin}) + NO \rightarrow Fe^{II}(\text{porphyrin})NO$$

[54] R. J. Singh, N. Hogg, J. Joseph, E. Konorev, and B. Kalyanaraman, *Arch. Biochem. Biophys.* **361,** 331 (1999).

[55] H. Kojima, N. Nakatsubo, K. Kikuchi, S. Kawahara, Y. Kirino, H. Nagoshi, Y. Hirata, and T. Nagano, *Anal. Chem.* **70,** 2446 (1998).

[56] B. B. Wayland and L. W. Olson, *J. Am. Chem. Soc.* **96,** 6037 (1974).

Further, HNO, the one electron reduction product of NO, can mediate reductive nitrosylation, particularly of oxidized heme proteins:

$$Fe^{III}(heme) + HNO \rightarrow Fe^{II}(heme)(NO)$$

Nitrosation and oxidative nitrosylation can be distinguished through specific scavenging of N_2O_3 by azide (N_3^-).[57] DAF, DAN, and thiols can be modified by either process, and azide again can be used as a convenient method to quantify the respective pathway contributions.

As discussed previously, nitrosation can result from N_2O_3 production or by exchange of a NO^+ group from, e.g., a *S*-nitrosothiol to another nitrosatable nucleophile. N_2O_3 is an intermediate of NO autoxidation.[58] This reaction is second order in NO, and in aqueous solution is thus relatively slow compared to other consumptive pathways. However, because both NO and O_2 are approximately 10-fold more soluble in hydrophobic solvents, these reactants will tend to accumulate and concentrate in hydrophobic regions.[59] As a result, in a system containing both aqueous and hydrophobic phases, e.g., lipid membranes, the autoxidation reaction will be approximately 300 times faster in the hydrophobic phase than in the corresponding aqueous solution.[60] This suggests that NO autoxidation under biological conditions most likely occurs within lipids and other hydrophobic compartments of tissue.

The reaction medium also alters the mechanism of NO autoxidation. In the gas phase and in hydrophobic solvents, NO autoxidation produces N_2O_3 through the intermediacy of NO_2[61]:

$$2NO + O_2 \rightarrow 2NO_2$$
$$NO + NO_2 \rightleftharpoons N_2O_3$$

Despite the similarity in rate constants, the mechanism of NO autoxidation in aqueous solution is altered. Rather than production of free NO_2, N_2O_4 appears to be the intermediate that leads to N_2O_3,[35,58] which is further hydrolyzed to nitrite:

$$2NO + O_2 \rightarrow N_2O_4$$
$$NO + NO_2 \rightleftharpoons N_2O_3$$
$$N_2O_3 + H_2O \rightarrow 2H^+ + 2NO_{2-}$$

[57] M. G. Espey, D. D. Thomas, K. M. Miranda, and D. A. Wink, *Proc. Natl. Acad. Sci. U.S.A.,* in press.

[58] D. A. Wink, J. F. Darbyshire, R. W. Nims, J. E. Saveedra, and P. C. Ford, *Chem. Res. Toxicol.* **6,** 23 (1993).

[59] A. W. Shaw and A. J. Vosper, *J. Chem. Soc. Faraday Trans.* **8,** 1238 (1977).

[60] X. Liu, M. J. S. Miller, M. S. Joshi, D. D. Thomas, and J. R. Lancaster, Jr., *Proc. Natl. Acad. Sci. U.S.A.* **95,** 2175 (1998).

[61] S. E. Schwartz and W. H. White, *in* "Trace Atmospheric Constituents: Properties, Transformation and Fates" (S. E. Schwartz, ed.), pp. 1–117. Wiley, New York, 1983.

Additionally, the two autoxidation mechanisms appear to produce different isomers of N_2O_3. In aqueous solution the nitrosating species is significantly less reactive than that produced from the reaction of NO with NO_2.[35] A comparison of these two reactants can be made in aqueous solution by the conversion of NO to NO_2 via oxygen atom transfer from the nitroxide 2-phenyl-4,4,5,5-tetramethylimidazoline-1-oxyl 3-oxide (PTIO).[62,63]

Determining NO Concentration in the Microenvironment of NOS Containing Cells and Tissues

The concentration of NO that is physiologically relevant has sparked much debate. The critical NO concentration will be dependent on the physiological process and the type of cell or tissue. For example, modulation of enzyme activity by iron–nitrosyl complex formation will generally require a significantly lower concentration of NO than nitrosation of a critical thiol by NO autoxidation. However, the relative distance of the target from the NO source and the concentration of scavengers can complicate even this seemingly straightforward example. Speculated values for physiological NO concentrations range from as low as 10 n*M* to as high as 5 μM, and each may be an accurate representation of the concentration required to elicit the biological effect in question. The concentrations of the major NO decomposition products, nitrite and nitrate, in drained venous effluent of the perfused organ or in serum and plasma is reported to be a reasonable approximation of NO levels *in vivo*.[64] Similarly, evaluation of the concentration of a wide varity of NO donors required to elicit a biological effect *in vitro* has been used as quantification of *in vivo* levels. Whether exposure of an exogenous NO donor, dispersed homogeneously in solution, corresponds to the flux of NO produced endogenously by NOS in the heterogeneous environment of a cell or tissue is questionable. However, extrapolation of *in vitro* data to an *in vivo* situation can provide at least a preliminary indication of the required NO concentration.

The major criterion for this type of experiment is use of a NO donor that releases NO chemically, rather than through a metabolic pathway. The two types of NO donor used most commonly for this purpose are *S*-nitrosothiols and NONOates; however, NONOates are recommended due to their stability to redox-active metals. It is important to recall that the concentration of donor does not translate linearly to the concentration of free NO, which should be measured directly in the cell culture medium with a NO-specific electrode (Fig. 3).

[62] T. Akaike and H. Maeda, *Methods Enzymol.* **268,** 211 (1996).

[63] T. Akaike, M. Yoshida, Y. Miyamoto, K. Sato, M. Kohno, K. Sasamoto, K. Miyazaki, S. Ueda, and H. Maeda, *Biochemistry* **32,** 827 (1993).

[64] D. L. Granger, R. R. Taintor, K. S. Boockvar, and J. B. Hibbs, Jr., *Methods Enzymol.* **268,** 142 (1996).

Maragos *et al.*[29] were the first to estimate the amount of NO required to stimulate sGC in vascular smooth muscle cells of isolated aortic rings. Complete relaxation of a norepinepherine-constricted aortic ring required 1 μM DEA/NO (EC_{50} of 0.15 μM). Under similar conditions,[65] this concentration of DEA/NO produces 50–100 nM steady-state NO, suggesting synthesis of submicromolar NO concentrations by eNOS in the vascular smooth muscle, assuming the absence of major consumption processes in the intracellular microenvironment.

Production of NO from nNOS was examined by a similar method, and the EC_{50} for dopamine release by DEA/NO was determined to be 50 μM.[66,67] Electrochemical measurements in later investigations suggest that the level of NO required to stimulate dopamine release in the synaptic region is in the micromolar range.[68] These values are considerably higher than those obtained for sGC activation in vascular smooth muscle cells, suggesting that the concentration of NO synthesized by eNOS would not result in significant dopamine release. NO produced by nNOS in the striatum will diffuse 10–100 μm depending on the cell type and O_2 concentration.[52,69] The concentration of NO therefore is likely to fall below the threshold for sGC activation after it diffuses a short distance from its source of production, effectively confining the effect of nNOS to the synaptic region. This example illustrates nicely the importance of both NO concentration and tissue architecture on physiological regulation. Additionally, the differential concentrations estimated by the *in vitro* models exemplify the difficulties in establishing physiologically relevant concentrations of NO. Indeed, the EC_{50} values only represent the concentration required for the specific measured response and may be significantly below the amount actually synthesized by the enzyme.

Considerably more nitrite/nitrate is produced by activated leukocytes, such as macrophages, and hepatocytes. Activation of these cells results in the expression of iNOS, which can generate NO over longer periods of time than either eNOS or nNOS. However, it may not be assumed that the NO concentration produced by iNOS is substantially higher than that generated by either of the constitutive isoenzymes. The specific activities of all three NOS isoforms are quite similar.[70] In addition to the amount of enzyme present, each cell type may differ in the posttranslational regulation of NOS.

[65] D. Morley, C. M. Maragos, X. Y. Zhang, M. Boignon, D. A. Wink, and L. K. Keefer, *J. Cardiovasc. Pharm.* **21,** 670 (1993).

[66] I. Hanbauer, D. Wink, Y. Osawa, G. M. Edelman, and J. A. Gally, *Neuroreport* **3,** 409 (1992).

[67] I. Hanbauer, G. W. Cox, and D. A. Wink, *Ann. N.Y. Acad. Sci.* **738,** 173 (1994).

[68] A. M. Lin, L. S. Kao, and C. Y. Chai, *J. Neurochem.* **65,** 2043 (1995).

[69] J. R. Lancaster, Jr., *Proc. Natl. Acad. Sci. U.S.A.* **91,** 8137 (1994).

[70] D. J. Stuehr, *Annu. Rev. Pharmacol. Toxicol.* **37,** 339 (1997).

The mode by which iNOS expression is stimulated can determine the ultimate redox chemistry of NO.[71] Sequential cytokine activation of murine macrophages with interferon (IFN)-γ followed by either tumor necrosis factor (TNF)-α or interleukin (IL)-1β resulted in induction of iNOS and production of nitrite (20 n*M*/min), but failed to elicit nitrosation of extracellular DAN. An approximately two-fold higher level of iNOS and nitrite production was observed with secondary stimulation by bacterial lipopolysaccharide (LPS), rather than TNF-α or IL-1β. The concentration of NO released into the culture medium and nitrosation of DAN were increased substantially in the LPS case ($\geq$30-fold) relative to the TNF-α or IL-1β cytokine treatment.[71] A similar activation pattern was observed in macrophages exposed to cytokines and other pathogen products, such as gram-positive antigen or virus. Although rat hepatocytes showed a similar pattern, the highest nitrosative capacity was observed in rat vascular smooth muscle cells induced by IL-1β alone relative to any combination of IFN-γ, TNF-α, and LPS (unpublished observations). Thus, the NO redox profiles derived from iNOS can be quite differential and will depend on the cell type and the signaling cascades involved in iNOS induction.

An important issue is the mechanism by which NO formed by an iNOS expressing cell translates into functional changes. Lewis and colleagues[72] presented a model for the diffusion of NO from LPS-activated RAW cell macrophages *in vitro*. Several comparisons have been performed on the responses elicited by coculture with activated iNOS-containing cells and by exposure to NO donors. Human lung adenocarcinoma A-549 cells coincubated with LPS-stimulated ANA-1 macrophages resulted in the release of latent TGF-β comparable to exposure to millimolar DEA/NO.[73] Similar results for the inhibition of dopamine β-hydroxylase activity were obtained in human neuroblastoma cells.[74] Cytostasis of intracellular *Listeria monocytogenes* also required millimolar concentrations of NONOates to achieve an effect analogous to activated macrophages.[75] Activation of human astrocytic NOS results in an inhibition of HIV-1 replication similar to 1 m*M* SNAP.[76] Under *in vitro* conditions, the functional changes elicited by the generation of NO from iNOS can be comparable to what is observed with millimolar concentrations of NO donors.

[71] M. G. Espey, K. M. Miranda, R. M. Pluta, and D. A. Wink, *J. Biol. Chem.* **275,** 11341 (2000).

[72] R. S. Lewis, S. Tamir, S. R. Tannenbaum, and W. H. Deen, *J. Biol. Chem.* **270,** 29350 (1995).

[73] Y. Vodovotz, L. Chesler, H. Chong, S. J. Kim, J. T. Simpson, W. DeGraff, G. W. Cox, A. B. Roberts, D. A. Wink, and M. H. Barcellos-Hoff, *Cancer Res.* **59,** 2142 (1999).

[74] X. Zhou, M. G. Espey, J. X. Chen, L. J. Hofseth, K. M. Miranda, S. P. Hussain, D. A. Wink, and C. C. Harris, *J. Biol. Chem.* **275,** 21241 (2000).

[75] R. Ogawa, R. Pacelli, M. G. Espey, K. M. Miranda, N. Friedman, S. M. Kim, G. Cox, J. B. Mitchell, D. A. Wink, and A. Russo, *Free Radic. Biol. Med.* **30,** 268 (2001).

[76] K. Hori, P. R. Burd, K. Furuke, J. Kutza, K. A. Weih, and K. A. Clouse, *Blood* **93,** 1843 (1999).

Dependence on the cellular conditions, such as rate and duration of formation, scavengers, and diffusion distance, may mask the actual concentration produced by each NOS isoenzyme to a significant extent. Electrochemical measurements can provide information on the concentration of NO required to mimic a NOS-dependent response. Selection of the appropriate NO donor is critical. NONOates provide the greatest flexibility for the delivery of various fluxes of NO (Table I, Fig. 3). One can infer that low doses of NO *at the reaction site* will initiate direct reactions, such as the activation of sGC, whereas prolonged *exposure* to NO will result in indirect effects, such as oxidation and nitrosation.

NADPH oxidase activation in endothelial cells by processes such as shear stress can result in expression of a number of factors, such as ICAM-1 and MCP-1.[77] These proteins are produced downstream of ras and MAPK activation, which can be blocked by the administration of SNAP (100 μM). Under these conditions, NO scavenges reactive oxygen species formed by NADPH oxidase, thereby abating ras activation. Cells transfected with eNOS also exhibited similar inhibition of MAPK-related pathway activation. Thus, NO abatement of NADPH oxidase activity in endothelial cells can be mimicked by low micromolar SNAP.

NO donors have been used to decipher the redox chemistry of the glutamate subtype NMDA receptor. Colton *et al.*[78] were the first to show that exposure to H_2O_2 and other reactive oxygen species results in closure of the NMDA channel. Six years later, Lipton *et al.*[79,80] reported that various reactive nitrogen oxide species also affected this channel. These studies were performed using SIN-1, SNP, and several *S*-nitrosothiols prior to the elucidation of NO donor nuances. It was later learned that the combination of SIN-1 and HEPES, which is a normal component of the buffers used in patch clamp and other cell culture experiments, can produce significant H_2O_2.[81] Therefore, it is difficult to interpret the relative contribution of oxidation and/or nitrosation reactions from these data. In addition, SNP has the potential to bind to thiols and thus might serve to block the channel itself. An examination of NMDA-induced neurotoxicity using NONOates found significant protection with 0.3–1 mM exposures.[82] These effects were independent of the presence of ascorbate, suggesting that the effect was not due to receptor nitrosation by N_2O_3. The effect of NO on the conductance properties of the NMDA

[77] B. S. Wung, J. J. Cheng, Y. J. Chao, H. J. Hsieh, and D. L. Wang, *Circ. Res.* **84,** 804 (1999).

[78] C. A. Colton, J. S. Colton, and D. L. Gilbert, *Free Radic. Biol. Med.* **2,** 141 (1986).

[79] S. Z. Lei, Z. H. Pan, S. K. Aggarwal, H. S. Chen, J. Hartman, N. J. Sucher, and S. A. Lipton, *Neuron* **8,** 1087 (1992).

[80] S. A. Lipton, Y. B. Choi, Z. H. Pan, S. Z. Lel, H. S. Chen, N. J. Sucker, J. Loscalzo, D. J. Singel, and J. S. Stamler, *Nature* **364,** 626 (1993).

[81] M. Kirsch, E. E. Lomonosova, H. G. Korth, R. Sustmann, and H. de Groot, *J. Biol. Chem.* **273,** 12716 (1998).

[82] A. S. Vidwans, S. Kim, D. O. Coffin, D. A. Wink, and S. J. Hewett, *J. Neurochem.* **72,** 1843 (1999).

receptor is highly dependent on oxygen tension.[83] These examples demonstrate the importance of NO donor selection and the necessity of complete understanding of the potential chemistry.

Conclusion

Nitric oxide donors can provide important and useful tools for deciphering complex biological mechanisms. Using the appropriate controls, a variety of different redox conditions can be created to aid in identifying the chemistry responsible for specific responses. These methods will provide a framework for future investigation of such mechanisms and offer insight into the possible NO redox environment seen in biologic settings.

[83] M. Gbadegesin, S. Vicini, S. J. Hewett, D. A. Wink, M. Espey, R. M. Pluta, and C. A. Colton, *Am. J. Physiol.* **277,** C673 (1999).

[9] Amperometric Measurement of Nitric Oxide Using a Polydemethylsiloxane-Coated Electrode

By FUMIO MIZUTANI

Introduction

The aim of this article is to describe a polymer-coated electrode that can be prepared easily and applied to the measurement of nitric oxide (NO) in physiological concentrations. The concentration of NO released from endothelial cells is, for example, initially of submicromolar level and is diminished within several minutes due to the high reactivity of the analyte with various molecules.[1,2] Hence a sensitive and rapid measuring method is required for *in situ* NO monitoring. Amperometric determinations can accomplish this function; NO is oxidized on a metal or catalyst-attached electrode around 0.9 V vs Ag/AgCl. However, such anodic detection systems often suffer from electrochemical interference by oxidizable species (e.g., L-ascorbic acid, uric acid, acetaminophen, catechol amines, active oxygens, and nitrite) in a complex matrix. The electrode oxidizes these

[1] R. F. Furgott and J. V. Zawadski, *Nature* **299,** 373 (1980).

[2] T. Malinski, Z. Taha, G. Grunfeld, S. Patton, M. Kaputurczak, and P. Tomboulian, *Biochem. Biophys. Res. Commun.* **193,** 1076 (1993).

0076-6879/02 $35.00

interferants as well as NO, which brings about a current response with positive error. Therefore the electrode system for measuring NO usually uses a coating layer that allows the passage of NO but prevents interferants from reaching the electrode surface. Nafion has been widely used for the suppression of interference response.[3–10] However, the anionic polymer is ineffective in preventing neutral (e.g., acetaminophen) and cationic (e.g., dopamine) interferants from reaching the electrode surface. Although Friedman *et al.*[11] have employed an electrochemically prepared poly(phenylenediamine) layer, it is well established that the polymer does not form an effective barrier to the interference from hydrogen peroxide, a kind of active oxygen.

For the discrimination between interferants and NO, the use of a hydrophobic coating layer would be the suitable approach: a gaseous compound such as NO would easily pass through the hydrophobic layer to reach the electrode surface,[12] whereas the transport of hydrophilic interferant molecules is strongly restricted. Commercially available NO sensors[13] use a hydrophobic polymer layer, although the detail of the polymer is not known. We have found that a polymer-coated electrode prepared by dip coating from an emulsion of polydimethylsiloxane (PDMS) is useful for the rapid, sensitive, and selective determination of NO.[14,15]

Experimental

Materials

An emulsion of PDMS is obtained from Toray Dow Corning Silicone, Tokyo [type, BY22-826; 45% (w/v)]. A standard solution of NO is prepared by bubbling it (99.7%, Sumitomo Seika Chemicals, Tokyo) into pure water whose temperature is kept at $25.0 \pm 0.2^\circ$. The concentration of the saturated solution is reported to

[3] T. Malinski and Z. Taha, *Nature* **358,** 676 (1992).

[4] F. Pariente, J. L. Alonso, and H. D. Abruna, *J. Electroanal. Chem.* **379,** 191 (1994).

[5] F. Lantione, S. Trevin, F. Bedioui, and J. Devynck, *J. Electroanal. Chem.* **392,** 85 (1995).

[6] F. Bedioui, S. Trevin, J. Devynck, F. Lantione, A. Brunet, and M.-A. Devynck, *Biosens. Bioelectron.* **12,** 205 (1997).

[7] K. M. Mitchell and E. K. Michaelis, *Electroanalysis* **10,** 81 (1998).

[8] T. Haruyama, S. Shiino, Y. Yanagida, E. Kobatake, and M. Aizawa, *Biosens. Bioelectron.* **13,** 763 (1998).

[9] J. Jin, T. Miwa, L. Mao, H. Tu, and L. Jin, *Talanta* **48,** 1005 (1999).

[10] L. Mao, K. Yamamoto, W. Zhou, and L. Jin, *Electroanalysis* **12,** 72 (2001).

[11] M. N. Friedmann, S. W. Robinson, and G. A. Gerhardt, *Anal. Chem.* **68,** 2621 (1996).

[12] K. Shibuki, *Neurosci. Res.* **9,** 69 (1990).

[13] "World Precision Instruments Homepage," www.wpiinc.com

[14] F. Mizutani, Y. Hirata, S. Yabuki, and S. Iijima, *Chem. Lett.,* 802 (2000).

[15] F. Mizutani, S. Yabuki, T. Sawaguchi, Y. Hirata, Y. Sato, and S. Iijima, *Sens. Actuat.* **76,** 489 (2001).

be 1.88 mM.[16] Other reagents are of analytical reagent grade (Sigma, St. Louis, MO). Deionized, doubly distilled water is used throughout.

Polymer-Modified Electrode

A platinum electrode (diameter, 1.6 mm; Bioanalytical Systems, West Lafyette, IN) is polished with 0.05 μm alumina slurry, rinsed with water, sonicated in water for 2 min, and then cleaned by an electrochemical oxidation–reduction treatment [-0.19 to 1.16 V vs Ag/AgCl (saturated with KCl), 0.1 V s^{-1}] in 0.5 M H_2SO_4 for 30 min or until the voltammetric characteristics of clean platinum electrode is obtained.

The platinum electrode thus cleaned is dipped into an emulsion of PDMS, which is diluted to 4% (w/v) with water just before the dipping procedure, and allowed to dry with the surface facing up for 4 hr at room temperature. The electrode is placed under a vacuum (1 Pa) for another hour. The thickness of the PDMS layer is ca. 5 μm.

Measurements

A potentiostat (HA-150, Hokuto Denko, Tokyo) equipped with a function generator (HB-104, Hokuto Denko) is used in a three-electrode configuration for electrochemical measurements. The polymer-coated electrode, an Ag/AgCl electrode (saturated with KCl, Bioanalytical Systems), and a platinum wire serve as working, reference, and auxiliary electrodes, respectively. The test solution usually used is an argon-saturated potassium phosphate buffer (0.1 M, pH 7.0, 20 ml), and its temperature is kept at $25.0 \pm 0.2°$. The solution is stirred with a magnetic bar during the amperometric measurements.

Results and Discussion

Characterization of PDMS as a Sensor Material

Linear sweep voltammograms for bare and PDMS-coated electrodes ($\nu = 10$ mV s^{-1}) show an anodic peak for the oxidation of NO at 0.82 V vs Ag/AgCl for each electrode, although the polymer coating results in a decrease in the peak height. Then the permeabilities of a variety of electroactive species, including NO into the PDMS layer, are evaluated as follows. Current responses on a platinum electrode for a variety of electroactive species are measured at 0.85 V vs Ag/AgCl before and after PDMS coating. The ratio of current response on the PDMS-coated electrode to that on the bare electrode is recorded for each species. The ratios for NO and well-known interferants are given in Table I. A high ratio (0.3) is obtained for NO. In contrast, ratios are very low for L-ascorbic acid,

[16] J. A. Dean (ed.), "Lange's Handbook of Chemistry," 12th Ed., p. 105. McGraw-Hill, New York, 1978.

TABLE I
PDMS-COATED ELECTRODE RESPONSE TO BARE ELECTRODE RESPONSE RATIOS FOR OXIDIZABLE SPECIES[a]

Species used	Ratio
NO	3×10^{-1}
L-Ascorbic acid	$<10^{-4}$
Uric acid	$<10^{-4}$
L-Cysteine	$<10^{-4}$
Acetaminophen	5×10^{-3}
Dopamine	5×10^{-3}
Hydrogen peroxide	$<10^{-4}$
Nitrite	$<10^{-4}$

[a] The anodic current response for PDMS-coated electrodes and bare electrodes was measured after the addition of each species at 0.85 V vs Ag/AgCl, and the ratio of the current for the former electrode to that for the latter was calculated. Each ratio was averaged over three measurements and was reproducible within ±40% in the measurements using three different electrodes. The concentrations of the species were 10 μM for NO and 0.1 mM for the others.

uric acid, L-cysteine, acetaminophen, dopamine, hydrogen peroxide, and nitrite. This means that the PDMS-coated electrode can be used for the anodic detection of NO without serious error from such interferants. As expected, the hydrophilic property of PDMS is suitable for discriminating gaseous compound such as NO from hydrophilic interferants. Further investigations are required to elucidate the detailed mechanism responsible for the permselectivity of the layer.

Electrode Response for NO

Figure 1 shows a current–time curve for the PDMS-coated electrode on the successive addition of NO in 10 μM steps. The steady-state current increase, which was obtained within 3 sec after the addition of NO, was proportional to the analyte concentration up to 50 μM. The inset in Fig. 1 shows a current–time curve for the addition of 20 nM NO, which suggests that such a low concentration of analyte can be measured on the electrode without the use of a Faraday cage (signal-to-noise ratio, >5). The relative standard deviation for 10 successive measurements of 2 μM NO was 1.5%.

Then the current response for NO was measured in the testing buffer solution saturated with air. The addition of NO brought about an increase in the anodic current, but a steady-state current response was not observed; the current reached a peak ca. 3 sec after the addition of the analyte and returned to the baseline within 8 min (Fig. 2). The current decrease to revert the baseline was attributable to the

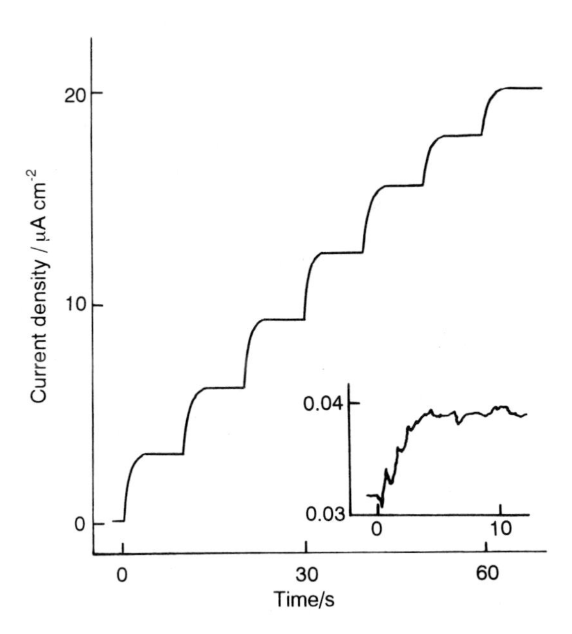

FIG. 1. Current–time curve for the PDMS-coated electrode obtained on increasing NO concentration in 10 μM steps under an argon atmosphere at 0.85 V vs Ag/AgCl. The inset shows current–time curve obtained after the addition of 20 nM NO.

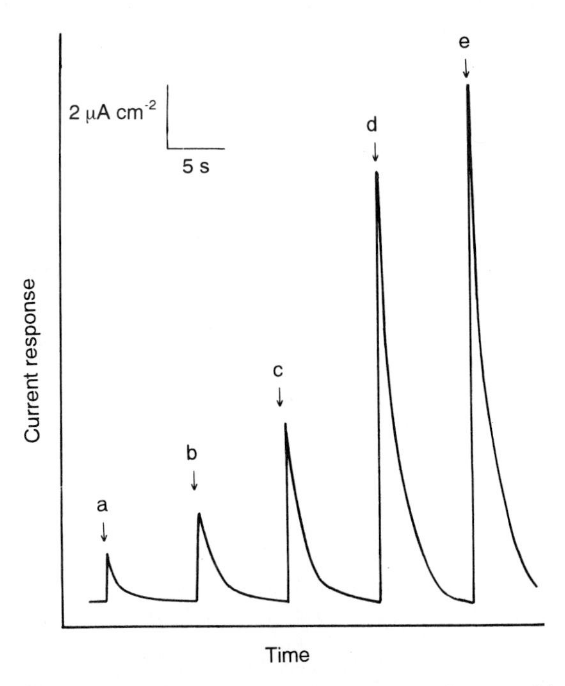

FIG. 2. Current–time curve for the PDMS-coated electrode in an air-saturated buffer solution for 5 μM (a), 10 μM (b), 25 μM (c), 50 μM (d), and 0.1 mM (e) NO.

consumption of NO through the reaction with oxygen. However, the peak height for NO was equal to the magnitude of the steady-state current response obtained in the argon-saturated buffer solution on the addition of the same concentration of analyte (see Figs. 1 and 2); i.e., the concentration of NO (20 nM–50 μM) can also be measured in air-saturated media.

The long-term stability of the PDMS-coated electrode was examined; measurements of the current response for 2 μM NO and 0.1 mM uric acid were, respectively, carried out three times a day, each day, for a month. Average values of the electrode response for both compounds in the respective three measurements, as well as the baseline, did not change for a month. This indicates that the PDMS-coated electrode could be used for the selective determination of NO for long periods of time.

Conclusion

The PDMS-coated electrode has proved to show high-performance characteristics in terms of sensitivity, selectivity, and response time, enough for monitoring NO-releasing processes in biological systems. The dip coating of PDMS is useful as a simple and reproducible procedure for preparing NO-sensing electrodes; the coating procedure is of particular interest because it is highly suitable for microsensors. We have prepared an NO-sensing microelectrode with the use of a platinum microelectrode (diameter, 5 μm) as the base transducer. For the mass production of microplanar sensors, the spin coating of the polymer would be a useful procedure.

The PDMS-coated electrode could be used for the cathodic detection of oxygen, when the electrode potential is at a negative potential, e.g., around −0.3 V vs Ag/AgCl. By combining the oxygen electrode with immobilized enzyme layers, a variety of enzyme sensors with low interference levels have been prepared.[17–21]

[17] F. Mizutani, T. Sawaguchi, S. Yabuki, and S. Iijima, *Electrochemistry* **67,** 1138 (1999).
[18] F. Mizutani, S. Yabuki, and S. Iijima, *Electroanalysis* **13,** 370 (2001).
[19] F. Mizutani, T. Sawaguchi, and S. Iijima, *Chem. Lett.,* 556 (2001).
[20] F. Mizutani, Y. Sato, Y. Hirata, and S. Iijima, *Anal. Chim. Acta* **441,** 175 (2001).
[21] F. Mizutani, T. Sawaguchi, Y. Sato, S. Yabuki, and S. Iijima, *Anal. Chem.* **73,** 5738 (2001).

[10] Nitric Oxide Monitoring in Hippocampal Brain Slices Using Electrochemical Methods

By ANA LEDO, RUI M. BARBOSA, JOÃO FRADE, and JOÃO LARANJINHA

Introduction

Nitric oxide (•NO) is a ubiquitous, gaseous, signaling molecule in the central nervous system involved in an impressive number of diverse functions, ranging from neurotransmission to neurodegeneration.[1,2] Due to its small size, hydrophobic nature, and reactivity, •NO may act within the cell from where it originated, or diffuse to underlying cells, in the opposite direction of normal neurotransmission, thus linking the activities of neurons in a local volume of tissue, regardless of whether the neurons are connected directly by synapses.[3]

It is therefore likely that •NO must convey information by its concentration, which implies that the half-life of •NO is an important determinant for its biological function.[4] However, the mechanisms that control •NO diffusion in the brain are still unclear, and cellular sinks for •NO, independent of its reaction with oxygen, superoxide anion, and hemoglobin, have been proposed.[5] Hence, the evanescent nature of nitric oxide allied with the multifaceted nature of its biological activity makes the evaluation of the dynamics of •NO production and removal/degradation an important approach to gain critical insights of its role in physiological and pathological processes.

Among the methods currently available to measure •NO, electrochemical methods, using modified carbon fiber microelectrodes, offer several advantages over other approaches,[6–16] including, as the most relevant for an *in vivo* situation, high

[1] J. Garthwaite and C. L. Boulton, *Annu. Rev. Physiol.* **57,** 683 (1995).
[2] V. L. Dawson and T. M. Dawson, *Prog. Brain Res.* **118,** 215 (1998).
[3] J. A. Gally, P. R. Montague, G. N. Reeke, Jr., and G. M. Edelman, *Proc. Natl. Acad. Sci. U.S.A.* **87,** 3547 (1990).
[4] J. S. Beckman, *in* "Nitric Oxide: Principles and Actions" (J. Lancaster, ed.), p. 1. Academic Press, San Diego, 1996.
[5] C. Griffiths and J. Garthwaite, *J. Physiol.* **536,** 855 (2001).
[6] K. Shibuki and D. Okada, *Nature* **349,** 326 (1991).
[7] T. Malinski and Z. Taha, *Nature* **358,** 676 (1992).
[8] T. Malinski and L. Czuchajowski, *in* "Methods in Nitric Oxide Research" (M. Feelish and J. Stamler, eds.), p. 19. Wiley, New York, 1996.
[9] T. Malinski, S. Mesaros, and P. Tomboulian, *Methods Enzymol.* **268,** 58 (1996).
[10] D. A. Wink, D. Christodoulou, M. Ho, M. C. Krishna, J. A. Cook, H. Haut, J. K. Randolph, M. Sullivan, G. Coia, R. Murray, and T. Meyer, *Methods: Companion Methods Enzymol.* **7,** 71 (1995).
[11] D. Christodoulou, S. Kudo, J. A. Cook, M. C. Krishna, A. Miles, M. B. Grisham, M. Murugesan, P. C. Ford, and D. A. Wink, *Methods Enzymol.* **268,** 69 (1996).

0076-6879/02 $35.00

sensitivity, real-time response, and high spatial resolution (e.g., measurement at the single cell and single mitochondria[17] levels). One of the main drawbacks of the electrochemical approach relates to selectivity because the surface of the electrode is in contact with the biological medium, subject to interference of a variety of redox active biological compounds. This problem has been partially solved by covering electrode active surfaces with semipermeable membranes and polymers.[7,12,18,19] In this regard, the porphyrinic sensor covered with Nafion, a negative charged film that repels anions introduced by Malinski and Taha[7] in 1992, has been a fundamental advance for the measurement of ·NO in a biological setting.

An additional strategy to improve selectivity is to complement the commonly used amperometric detection (current monitored as a function of time at a constant potential) with differential pulse voltammetry (DPV).[9,10,14] However, despite its high sensitivity, this technique lacks time resolution for the real-time measuring of ·NO, as it may require at least 10 sec for the voltammogram to be run.

Fast cyclic voltammetry (FCV) measures currents while the potential is scanned toward cathodic and anodic directions; i.e., at the end of the oxidation sweep, the potential cycles back in the reduction direction. The reduction and oxidation peak potentials thus obtained provide the electrochemical signature of the redox species. Therefore, FCV is typically applied in biological systems (using carbon fiber microelectrodes with $\approx 8\ \mu m$ diameter) when chemical identification of the substance being detected is required.[20,21] Additionally, by scanning the potential at very high speed (up to 1000 V/sec), one can obtain a signature of the redox species in a few milliseconds,[22] an important feature for the measurement of an evanescent molecule such as ·NO in biological systems. Despite the high scan rate of the applied voltage waveform, FCV does not affect neuronal activity.[23] Therefore, we attempted to apply FCV to monitor levels of ·NO in brain slices.

---

[12] J. K. Park, P. H. Tran, J. K. T. Chao, R. Ghodadra, R. Rangarajan, and N. V. Thakor, *Biosens. Bioelectr.* **13,** 1187 (1998).

[13] N. Villeneuve, F. Bedioui, K. Voituriez, S. Avaro, and J. P. Vilaine, *J. Pharmacol. Toxicol. Methods* **40,** 95 (1998).

[14] F. Bedioui and S. Tévin, *Biosens. Biolelectr.* **13,** 227 (1998).

[15] Y. Xian, W. Zhang, J. Xue, X. Ying, and L. Jin, *Anal. Chim. Acta* **415,** 127 (2000).

[16] N. Diab and W. Schuhmann, *Electrochim. Acta* **47,** 265 (2001).

[17] A. J. Kanai, L. L. Pearce, P. R. Clemens, L. A. Birder, M. M. VanBibber, S. Y. Choi, W. C. de Groat, and J. Peterson, *Proc. Natl. Acad. Sci. U.S.A.* **98,** 14126 (2001).

[18] M. N. Friedemann, S. W. Robinson, and G. A. Gerhardt, *Anal. Chem.* **68,** 2621 (1996).

[19] M. Pontié, C. Gobin, T. Pauporté, F. Bedioui, and J. Devynnck, *Anal. Chim. Acta* **411,** 175 (2000).

[20] J. A. Stamford, F. Crespi, and C. A. Marsden, *in* "Monitoring Neuronal Activity: A Practical Approach" (J. A. Stamford, ed.), p. 113. Oxford Univ. Press, New York, 1992.

[21] P. S. Cahill, Q. D. Walker, J. M. Finnegan, G. E. Michelson, E. R. Travis, and R. M. Wightman, *Anal. Chem.* **68,** 3180 (1996).

[22] J. A. Stamford and J. B. Justice, Jr., *Anal. Chem.* 359A (1996).

[23] J. A. Stanford, *in* "Brain Slices in Basic and Clinical Research" (A. Schurr and B. M. Rigor, eds.), p. 66. CRC Press, Boca Raton, FL, 1995.

In vitro brain slice preparations, widely used in neurosciences, exhibit significant merits.[24] They retain a high degree of structural integrity and functionality of neuronal circuitry and, when perfused, allow control of features of the physiological environment (e.g., composition, redox status, and stimuli concentration) in a manner not possible *in vivo*. Also, insertion of the electrodes in slices is done under visual control, which allows small regions involved in specific functions to be studied (e.g., CA1 versus CA3 regions of the hippocampus). These characteristics are pertinent for measuring ·NO dynamics during functionally induced changes in the tissue slice. This is particularly relevant in glutamate-induced ·NO production through the *N*-methyl-D-aspartate (NMDA) glutamate receptor.[25] The glutamate–NMDA receptor in the hippocampus has been strongly implicated in learning and development processes, and ·NO has been shown to play an essential role in the induction of long-term potentiation in the hippocampus, the most widely studied neuronal equivalent of learning.[26] However, ·NO has also been implicated in the neuronal excitoxicity associated with excessive stimulation of the glutamate NMDA receptor.[27]

This article describes the use of a ·NO microsensor (single 8 μm diameter carbon fiber microelectrode modified with nickel-porphyrin and Nafion) to monitor the production of ·NO by FCV and amperometry in perfused rat hippocampal brain slices stimulated with glutamate and NMDA.

Preparation of Nitric Oxide Microsensors

Electrochemical methods for the measurement of nitric oxide have been reviewed in detail.[7,9,11,28] We have used a homemade single nickel porphyrinic/Nafion carbon fiber electrode based on the one described by Malinski and Taha.[7] Whereas Nafion, a thin anionic film consisting of SO_3^- functional groups that prevent the diffusion of nitrite and other anions to the catalytic surface of the sensor, increases the selectivity of the sensor, the nickel porphyrin polymerized at the carbon surface catalyzes the oxidation of ·NO to NO^+, enhancing the sensitivity of the sensor to ·NO. Of note, the small dimensions of the sensor (diameter 8 μm, tip length $\approx$ 100 μm) are important for placement into the brain slices with minimal damage to the tissue.

The microelectrodes are made essentially as described elsewhere.[29] Briefly, single carbon fibers (Courtaulds, Ltd, UK; a gift from J. Stamford, Royal London

[24] A. Schurr and B. M. Rigor, "Brain Slices in Basic and Clinical Research." CRC Press, Boca Raton, FL, 1995.

[25] J. Garthwaite, G. Garthwaite, R. M. J. Palmer, and S. Moncada, *Eur. J. Pharmacol.* **172,** 413 (1989).

[26] E. M. Schuman, *in* "Nitric Oxide in the Nervous System" (S. Vincent, ed.), p. 125. Academic Press, New York, 1995.

[27] S. A. Lipton, *Cell Death Differ.* **6,** 943 (1999).

[28] J. P. Rivot, A. Sousa, J. Montagne-Clavel, and J. M. Besson, *Brain Res.* **821,** 101 (1999).

[29] R. M. Barbosa, A. M. Silva, A. R. Tome, J. A. Stamford, R. M. Santos, and L. M. Rosario, *J. Physiol.* **510,** 135 (1998).

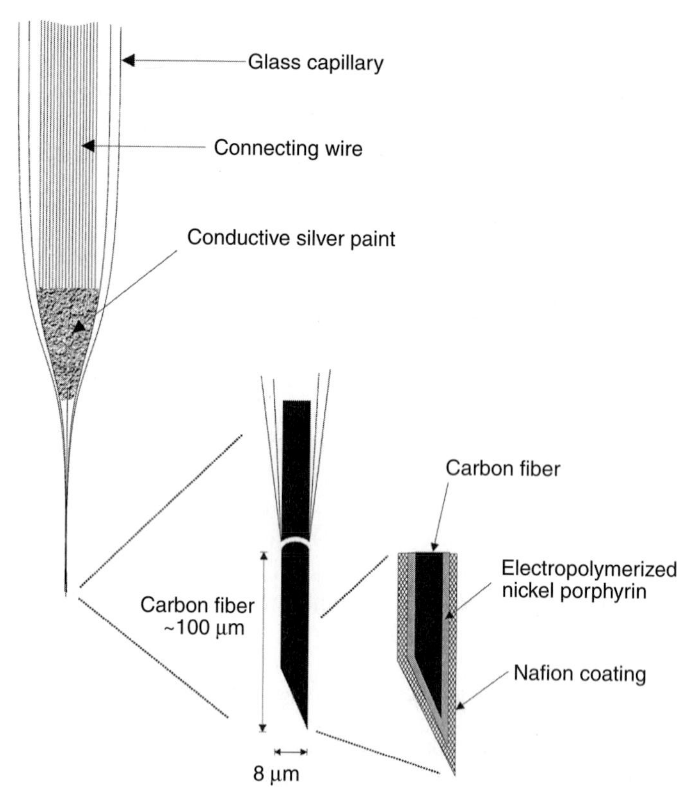

FIG. 1. Schematic representation of the nitric oxide microsensor showing the details of tip geometry and surface modifications.

Hospital) are inserted into borosilicate glass capillaries (1.16 mm i.d. × 2.0 mm o.d.; Clark Electromedical Instruments, UK). The capillaries are then pulled on a horizontal puller (P-77, Sutter Instruments Company, CA), after which the protruding carbon fibers are cut to the required size and beveled at 30° on a grinder (EG-4, Narishige, Japan). Electrical contact between the carbon fiber and the wire is provided by electrically conductive paint (RS, UK) (Fig. 1). Afterward, the carbon tip surface is coated with nickel(II) tetrakis(3-methoxy-4-hydroxyphenyl)porphyrin (Frontier Scientific, Logan, UT) (Ni-TMPP) and Nafion (Aldrich, Milwaukee, WI) according to the procedures described by Malinski and Taha.[7] The polymeric film of Ni-TMPP is deposited on the surface of the carbon fiber electrode by electropolymerization of the porphyrin (500 μM) dissolved in 0.1 M NaOH by means of cyclic voltammetry (scanning potential between −0.2 and 1.0 V vs Ag/AgCl applied at 100 mV/sec). The electrode coverage takes about 10 min and is monitored by following the oxidation/reduction peaks of Ni^{II}/Ni^{III} until a steady intensity is reached.

The porphyrin carbon fiber is then coated with Nafion by dipping the tip in a 5% aliphatic alcohol solution for 10 sec (twice) and is then dried for 10–15 min at 80°.

From our experiments and those described in the literature,[7,8,11] we have established two layers of Nafion as the optimal number to coat the electrode surface because of the following reasons. First, it excludes most of the potentially interfering compounds found in the brain environment at high concentrations, including ascorbate, dihydroxyphenylacetic acid, glutamate, glucose, and amino acids.[8,11] However, electroactive neutral or positively charged compounds, including dopamine, noradrenaline, and serotonin, are potentially interfering compounds in amperometric detections when present in the low micromolar range. However, such interferences may be overcome by fast cyclic voltammetry measurements because oxidation and reduction peaks of ·NO are distinguished readily from those of catecholamines and indols (see discussion later); e.g., Iravani *et al.,*[30] using a bare carbon electrode, reported the simultaneous selective measurement of ·NO and dopamine in rat caudate putamen slices. Second, the two layers of Nafion did not compromise the sensitivity of the microsensor for the expected concentrations of ·NO *in vivo* (see calibration later).

Microsensor Testing Procedures

Each microsensor is calibrated individually with a fresh saturated solution of ·NO (1.7 m*M* at 25°)[31] prepared by bubbling purified ·NO gas (removed from higher nitrogen oxides by two consecutive passages through a 10 *M* deaerated KOH solution) through ultrapure deionized water (Milli-Q system, Millipore Company, Bedford, MA), degassed previously with argon in a stoppered glass tube for 30 min. Aliquots of the saturated solution are transferred to phosphate-buffered saline (PBS; 140 m*M* NaCl, 2.7 m*M* KCl, 8.1 m*M* $Na_2HPO_4 \cdot 12H_2O$, 1.8 m*M* KH_2PO_4, 100 μM diethylenetriaminepentacetic acid), pH 7.4, with gas-tight syringes in order to achieve the required concentration. Evaluation of the microsensors, except for the determination of the oxidation peak potential of ·NO, is assessed by flow injection analysis using a homemade flow cell prior to the experiments. Such dynamic evaluation is important for determining not only the sensitivity, but also how the microsensors reproduce the response to repeated ·NO injections (see later).

Square Wave Voltammetry (SWV) and Differential Pulse Voltammetry (DPV) in Solution

The oxidation potential for detecting ·NO was determined from SWV and DPV assays using an μAutolab type II running with GPES version 4.9 software (Echo

[30] M. M. Iravani, J. Millar, and Z. L. Kruk, *J. Neurochem.* **71,** 1969 (1998).

[31] F. T. Bonner, *Methods Enzymol.* **268,** 50 (1996).

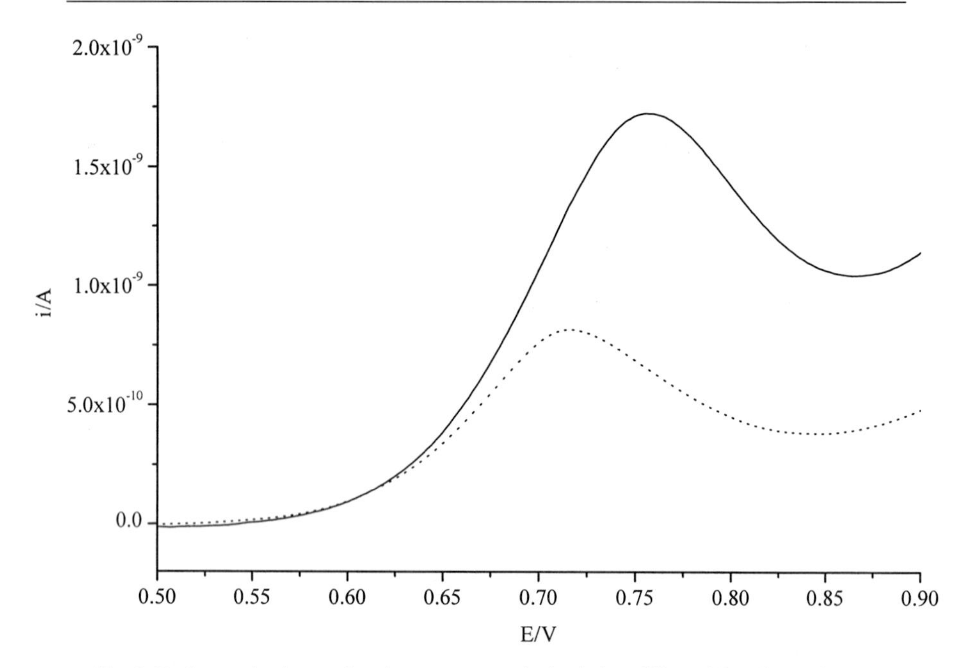

FIG. 2. Background-subtracted voltammograms obtained by differential pulse voltammetry (dashed line) and square wave voltammetry (full line) of 10 μM nitric oxide in a deaerated PBS solution, pH 7.4. Peak potentials are 713 and 749 mV vs Ag/AgCl, respectively. Experimental conditions: DPV, pulse amplitude 25 mV, pulse width 50 ms, scan rate 10 mV/sec, SWV, pulse amplitude 25 mV, frequency 25 Hz, step potential 2 mV, and scan rate 50 mV/sec.

Chimie, NL). SWV is an electrochemical technique that allows a higher scan rate than DPV while retaining good resolution and sensitivity.

Figure 2 illustrates typical voltammograms of 10 μM ·NO in PBS. ·NO detected by SWV exhibits an oxidation potential of 0.75 V versus Ag/AgCl (3 *M*), which is ca. 40 mV higher as compared with that obtained with the DPV approach (likely due to the higher scan rate of SWV, 50 mV/sec relatively to DPV, 10 mV/sec). Under the conditions used, the current obtained with SWV doubles that obtained with DPV for the same concentration of ·NO.

Usually, the ability of the porphyrin sensors to selectively measure ·NO in *in vitro* or *ex vivo* biological preparations requires consideration of nitrite interference. Because the oxidation peak potential of nitrite in aqueous solution (0.7–0.8 V) is only slightly more positive than that of ·NO, several authors[8,32,33]

[32] S. Mesaros, *Methods Enzymol.* **301,** 160 (1999).

[33] F. Crespi, M. Campagnola, A. Neudeck, K. McMillan, Z. Rossetti, A. Pastorino, U. Garbin, A. Fratta-Pasini, A. Reggiani, G. Gaviraghi, and L. Cominacini, *J. Neurosci. Methods* **109,** 59 (2001).

have addressed the risk of confounding the peak of ·NO with that of nitrite. The relevance of this problem to the *in vitro* experiments with brain slices relies on the fact that, during the perfusion of the slices, the medium is saturated continuously with Carbox (5% CO_2, 95% O_2) and, under such high O_2 tension, the reaction of ·NO with O_2 producing NO_2^- [reaction (1)], although slow,[34] may occur to a certain extent.

$$4{\cdot}NO + O_2 + 2H_2O \rightarrow 4NO_2^- + 4H^+ \tag{1}$$

As already pointed out, Malinski and Taha[7] have introduced a nickel porphyrin carbon fiber sensor covered with Nafion and were able to selectively monitor ·NO at 0.64 V (versus saturated calomel reference electrode). The negatively charged Nafion coating is highly permeable to ·NO and prevents nitrite from gaining access to the catalytic porphyrin film at the surface of the sensor. However, one may consider that different coating procedures influence nitrite (and other anions as well) access to the electrode surface and that changes in the chemical modifications of carbon surfaces alter the oxidation potential for ·NO. Also, for identical electrodes, the oxidation potential of ·NO varies with the electrochemical technique used. Figure 2 documents this conclusion as discussed previously. These notions explain, at least in part, that subsequent to Malinski and Taha's pioneer work,[7] different oxidation potentials to measure ·NO, ranging from 0.55 to 0.95 V, have been reported.[17,28,32,35–37] In particular, Crespi *et al.*[33] were able to monitor the oxidation of ·NO *in vitro* and *in vivo* at 0.55 V with electrically pretreated 30-μm bare carbon fiber electrodes, thus avoiding the interference of nitrite above 0.7 V.

Under the conditions we have chosen to coat the sensor with Nafion (see earlier discussion) and at the oxidation peaks reported in Fig. 2, 100 μM of nitrite produces a current that is only 17% of the current intensity obtained with 10 μM of ·NO (not shown). It is therefore reasonable to conclude that currents due to oxidation of ·NO at the surface of the microsensor we have prepared do not suffer a significant contribution from the oxidation of nitrite for identical concentrations of the redox species. Also, as discussed later, nitrite is readily distinguished from ·NO by the analysis of FCV voltammograms.

Finally, one may consider that the difficulty in having healthy brain slices thicker than 400 μm because of oxygen diffusion problems may be taken as evidence for a low physiological concentration of oxygen in the innermost cell layers (2–10 μM); i.e., although the surface of the slices are bathed with artificial cerebrospinal fluid (aCSF) saturated with 95% oxygen, a concentration gradient

[34] P. C. Ford, D. A. Wink, and D. M. Stanbury, *FEBS Lett.* **326,** 1 (1993).
[35] A. Meulemans, *Neurosci. Lett.* **171,** 89 (1994).
[36] F. Lantoine, S. Trevin, F. Bedioui, and J. Devynck, *Electroanal. Chem.* **392,** 85 (1995).
[37] S. Burlet and R. Cespuglio, *Neurosci. Lett.* **226,** 131 (1997).

of O_2 is operative, decreasing from the surface to the deep layers of the slice where the sensor tip is located, thereby making the reaction of ·NO with oxygen to form nitrite nonsignificant. That is, under such conditions, ·NO is likely to decay (be removed) by mechanisms other than the reaction with molecular oxygen.

Amperometry

Amperometric currents originating from the oxidation of ·NO to NO^+ at the porphyrinic sensor interface were measured using an amperometer (AMU 130 Tacussel/Radiometer, France; sensitivity range 0.1 pA–20 nA), with the ·NO microsensor held at a potential of 0.8 V vs Ag/AgCl. The current was low-pass filtered (100 Hz) and recorded using a chart recorder.

Figure 3 depicts a flow calibration of the microsensor recorded for different concentrations of ·NO (0.1 to 1.0 μM) injected repeatedly and in sequence at a flow rate of ca. 2 ml/min (Fig. 3A) and the respective calibration curve (Fig. 3B). The high analytical performances of the microsensor are apparent from Fig. 3, as inferred from linearity ($R = 0.999$), sensitivity (312 pA/μM), and detection limit (20 nM).

Fast Cyclic Voltammetry

As mentioned earlier, FCV is a highly selective technique for monitoring redox species in biological media. In order to evaluate the feasibility of this approach for monitoring ·NO in hippocampal slice preparations, the responsiveness of the microsensor is first assessed in solution by FCV in flux.

FCV measurements are performed using a three-electrode potentiostatic system involving the ·NO sensor as working electrode, a platinum wire as the auxiliary electrode, and an Ag/AgCl reference electrode. An EI-400 potentiostat (Ensman Instruments, Bloomington, IN), provided with an internal triangle generator and a sample and hold amplifier circuits, is employed. The scan rate is 200 V/sec, ranging from −0.6 to 1.4 V and sweeping initially in the anodic direction. Each triangle wave is repeated at 20-ms intervals. The voltammetric signal is filtered at 2 kHz using the internal filter of the potentiostat. The current signals are monitored on a digital storage oscilloscope (Tektronix TDS 220, Portland, OR) and stored on a microcomputer via RS232 using WaveStar software version 2.3.

Figure 4 depicts the fast cyclic voltammogram of 10 μM ·NO in PBS solution consisting of one oxidation peak at about 1.2 V and one reduction peak at about −0.4 V. A flow calibration is run by sampling the current at the oxidation peak of ·NO for concentrations between 0.2 and 1.0 μM injected repeatedly and in sequence. Data are fitted to a linear regression curve showing good linearity ($r = 0.994$).

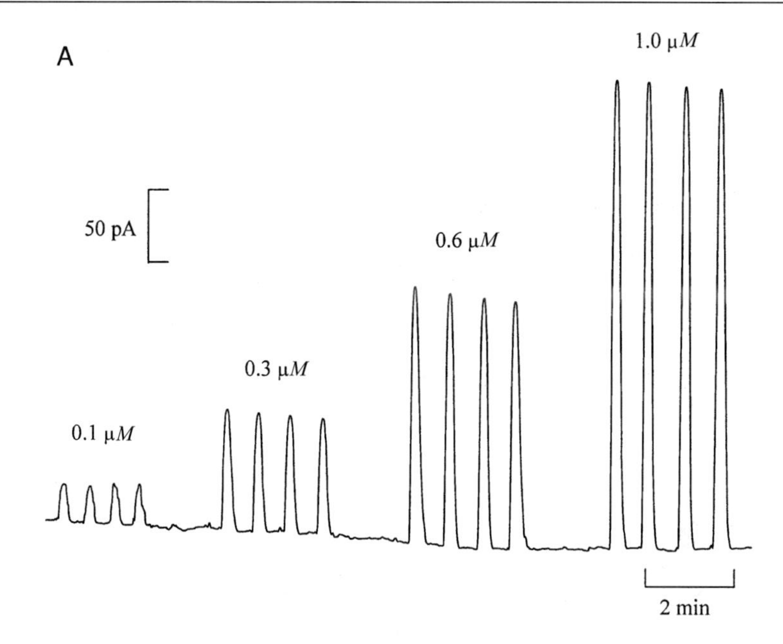

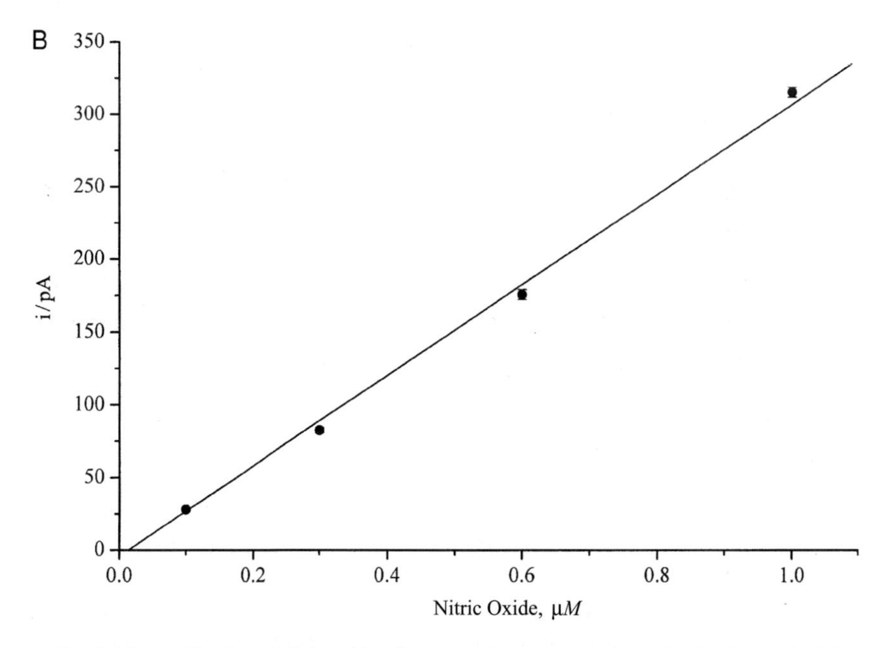

FIG. 3. Flow calibration of nitric oxide microsensor in amperometric mode. (A) Sequential injections of different nitric oxide concentrations, as indicated on top of the respective group of signals. (B) Calibration curve taken from data obtained in A.

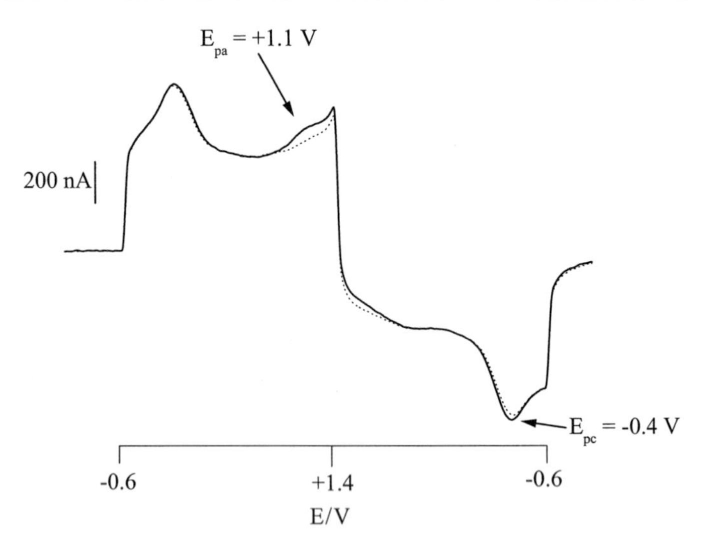

FIG. 4. Fast cyclic voltammogram of 10 μM ·NO in PBS solution, pH 7.4, obtained under flow conditions. Oxidation and reduction peak potentials of nitric oxide (full line) are shown against a background (dotted line).

Expectedly, the detection limit obtained with the FCV technique is at least two- to threefold higher relative to the amperometric detection (Fig. 3). However, it should be emphasized that the ·NO concentrations FCV detects are within the range of the concentrations of ·NO expected to be found *in vivo,* following stimulation of nitric oxide synthasc.[1,2]

Catecholamines (e.g., dopamine) and indoles (e.g., 5-hydroxytryptamine), compounds that are potential interfering species in the amperometric mode because of their partition in the Nafion layer, produced peak oxidation potentials by FCV between 0.5 and 0.6 V, well resolved from the oxidation peak of ·NO (not shown). In agreement with previous reports[30] using bare, not-covered electrodes, high concentrations of nitrite (>200 μM) produced an oxidation current at the same potential of ·NO, but, at variance with ·NO, no reduction peak of nitrite was noticed (not shown).

In summary, by sampling the current at the oxidation peak potential of ·NO, determined from fast cyclic voltammograms with the porphyrin/Nafion microsensor, the major problems associated with its selective detection in brain slices are solved: (1) the interference of nitrite is prevented by the Nafion layer and its presence may be checked by the absence of a reduction peak in the cyclic voltammograms and (2) the interference of Nafion-permeable neutral and cationic species is solved because the redox signatures of these species differ from that of ·NO.

Hippocampal Slices

There are many different protocols for preparing hippocampal slices that, in part, are influenced by the nature of the investigation to be carried out.[38,39] The technical apparatus also varies, including different approaches for cutting the slices and different recording chambers as well. As a general rule, the process must be fast, physical trauma must be avoided, and the whole procedure should be performed in ice cold aCSF continuously oxygenated. A standard aCSF solution gassed with 95% O_2/5% CO_2 contains 2 m*M* KCl, 124 m*M* NaCl, 25 m*M* $NaHCO_3 \cdot 10H_2O$, 1.25 m*M* KH_2PO_4, 2 m*M* $CaCl_2$, 2 m*M* $MgSO_4$, and 10 m*M* D-glucose. In the experiments described in Fig. 6, Mg^{2+} was removed from the medium. We used the following methodology to prepare transverse slices (across the longitudinal axis) of rat brain hippocampus.

One rat (4- to 6-week old male Wistar rats) is decapitated, and the brain is removed rapidly and placed in ice cold aCSF for 2 min. The two hippocampi are dissected out and placed on the stage of the McIlwain chopper (WPI) with the long axis at right angles to the cutting blade. Sections of 400 μm thickness are cut, and the slices, submerged in ice cold aCSF, are separated gently under a dissecting microscope with glass tools made from drawn-out Pasteur pipettes with fire-polished tips. The slices are allowed to recover in oxygenated aCSF at room temperature for 1 hr prior to electrochemical recordings. The slices are then submerged in the recording chamber at 32° (Harvard Medical Systems) and perfused with aCSF flowing at 2 ml/min (temperature controller TC-202A; Harvard Medical Systems).

A typical experiment starts with the insertion of the microsensor (mounted in a micromanipulator) approximately 150–200 μm deep into a selected region of the hippocampus. Then, a stable background signal is recorded and the effects of agonists, added either locally by microinjection or included in the perfusion medium, are studied.

The preparation and use of hippocampal slices to measure glutamate-induced •NO production deserve additional comments: the mechanisms underlying opening and closing of the NMDA receptor channel are extremely complex, and a vast array of factors and conditions modulate its function.[40] From the methodological point of view, during the preparation of the slices, one should consider the influence of pH, the presence of oxidizing compounds, oxygen tension, Ca^{2+} concentration, allosteric effects of glycine, histamine, and polyamines, mechanosensitivity, voltage-dependent blockage by Mg^{2+}, and feedback modulation by nitric oxide.

[38] Z. I. Bashir and M. Vignes, *in* "Neuroscience LabFax" (M. A. Lynch and S. M. O'Mara, eds.), p. 13. Academic Press, London, 1997.

[39] P. Lipton, P. G. Aitken, F. E. Dudek, K. Eskessen, M. T. Espanol, P. A. Ferchmin, J. B. Kelly, N. R. Kreisman, P. W. Landfield, and P. M. Larkman, *J. Neurosci. Methods* **59,** 151 (1995).

[40] C. J. McBain and M. L. Mayer, *Physiol. Rev.* **74,** 723 (1994).

Finally, one should consider that the fast reaction of ·NO with oxyhemoglobin is unlikely to occur, as it is expected that hemoglobin is washed out during the perfusion of the brain slices with aCSF.

Nitric Oxide Production in Response to Glutamate and NMDA in Perfused Hippocampal Slices: Amperometry and Fast Cyclic Voltammetry Monitoring

Glutamate, the major excitatory neurotransmitter in the brain, has been shown to mediate neuronal injury by a mechanism involving a massive calcium influx through a subtype of the glutamate receptor, the NMDA receptor.[41] Current evidence implicates ·NO in the mechanisms of cell injury and death, following overstimulation of NMDA receptors by glutamate.[27] Additionally, the production of free radicals other than ·NO is a natural consequence of the activation of glutamate receptors.[42] For instance, the superoxide radical increases following the exposure of hippocampal slices to NMDA[43] and, whereas the formation of superoxide radical was shown to modulate the induction of LTP in the CA1 region of the hippocampus,[44] its extremely fast reaction with ·NO ($K = 1 \times 10^{10}\ M^{-1}s^{-1}$),[45] producing the strong oxidant peroxynitrite, may have deleterious cellular consequences.

However, the involvement of ·NO in the mechanisms of glutamate-dependent injury may be more subtle because ·NO-related species modulate NMDA receptor function by modifying its redox status.[27] Glutamate transporters (glutamate is not metabolized by extracellular enzymes) also exhibit cysteine residues vulnerable to the action of biological oxidants, resulting in reduced glutamate uptake and contributing to the build-up of neurotoxic extracellular glutamate levels.[46]

The aforementioned observations stress the relevance in measuring ·NO in the hippocampus in relation to glutamate stimulus. The production of ·NO in hippocampal slices in response to nicotine and acetylcholine stimuli has already been documented.[47]

It is accepted that ·NO is produced continuously within the central nervous system *in vivo* and so, considering the high degree of structural integrity and functionality of the neuronal circuitry retained by the hippocampal slices, a basal ·NO level in the low nanomolar range during the perfusion of the hippocampal

[41] J. T. Coyle and P. Puttarcken, *Science* **262,** 689 (1993).

[42] J. A. Dickens, *in* "Mitochondria and Free Radicals in Neurodegenerative Diseases" (M. F. Beal, N. Howel, and I. Bodis-Wollner, eds.), p. 29. Wiley-Liss, New York, 1997.

[43] V. P. Bindokas, J. Jordán, C. C. Lee, and R. J. Miller, *J. Neurosci.* **16,** 1324 (1996).

[44] C. M. Atkin and J. D. Sweatt, *J. Neurosci.* **19,** 7241 (1999).

[45] R. E. Huie and S. Padmaja, *Free Radic. Res. Commun.* **18,** 195 (1993).

[46] D. Trotti, N. C. Danbolt, and A. Volterra, *Trends Pharmacol. Sci.* **19,** 328 (1998).

[47] D. A. Smith, A. F. Hoffman, D. J. David, C. E. Adams, and G. A. Gerhardt, *Neurosci. Lett.* **255,** 127 (1998).

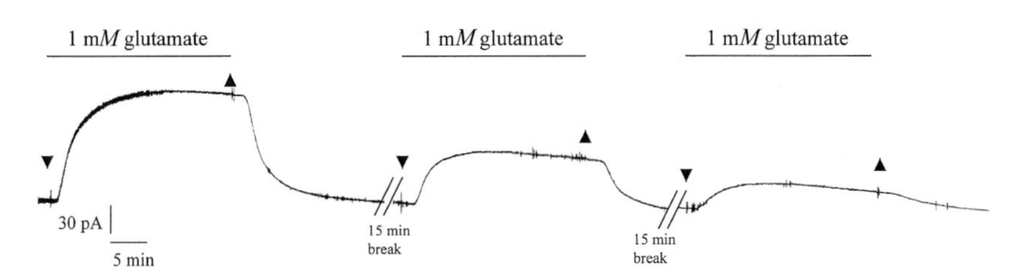

FIG. 5. Glutamate-induced production of nitric oxide in the CA3 region of the hippocampus monitored by amperometry in perfused hippocampal slices. Glutamate was introduced in and removed from the perfusion medium at the time indicated by upward and downward arrows, respectively.

slices is expected. Figure 5 documents the increased production of ·NO in the CA3 region of the hippocampus by exogenous glutamate (1 m*M*) added to the perfusion aCSF medium. Likewise, following supplementation of perfusion aCSF with NMDA (50 μM), a synthetic analog of aspartic acid that selectively activates the glutamate NMDA receptor ion channel, an increased production of ·NO could be observed (Fig. 6).

When comparing glutamate- and NMDA-induced ·NO production, one can find similarities: (a) It is apparent that in both cases a decrease in peak intensity is observed for sequential stimulations. It is accepted that physiologically, at the majority of synapses, the glutamate peaks at $\approx$1 m*M* and decays exponentially with a time constant of $\approx$1 ms.[48] Therefore, the conditions used (perfusion of 1 m*M* of glutamate for 30 min) are likely to mimic excitotoxic conditions to a certain extent. While desensitization of the glutamate receptor cannot be discarded, the decreased signal intensity observed in Figs. 5 and 6 may also be supported by mechanisms related to ·NO itself or with a massive Ca^{2+} entry in the cells, occurring during the sustained stimulation by glutamate or NMDA. (b) The time course of ·NO production and removal/degradation is similar in both cases, an expected observation, as the underlying mechanism for the production of ·NO is similar for glutamate and NMDA: stimulation of nitric oxide synthase by Ca^{2+} influx through the NMDA channel. The increase of ·NO started soon after glutamate or NMDA reached the slices and reached a maximum in about 10 min. So long as glutamate or NMDA is perfused, ·NO is maintained at high but slowly decreasing concentration. Removal of the stimuli results in a fast decay of ·NO concentration to the basal levels.

Figure 7 documents the use of fast cyclic voltammetry to measure glutamate-evoked ·NO production in the CA1 region of the hippocampus in brain slices under perfusion. The current was sampled at the oxidation peak of ·NO (see Fig. 4). It is apparent from Fig. 7 that the time course of ·NO currents in the CA1 region is slower than that obtained in amperometric monitoring in the CA3 region (Fig. 5) on introduction of 1 m*M* glutamate in the perfusion. Also, after ca. 25 min of glutamate

[48] J. D. Clements, R. A. Lester, G. Tong, C. E. Jahr, and G. L. Westbrook, *Science* **258,** 498 (1992).

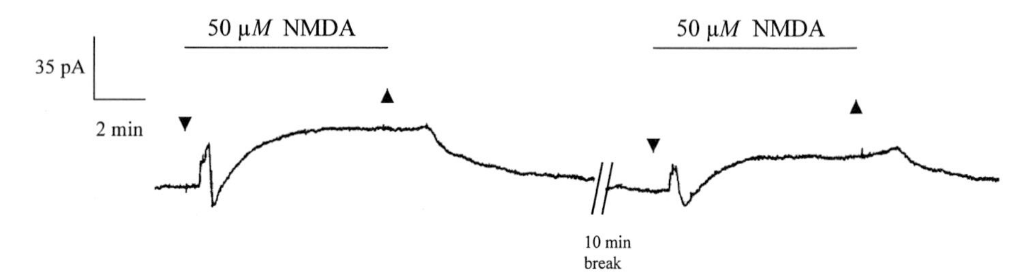

FIG. 6. NMDA-induced production of nitric oxide in the CA3 region of the hippocampus monitored by amperometry in perfused hippocampal slices. NMDA was introduced in and removed from the perfusion medium at the time indicated by upward and downward arrows, respectively. In these experiments, Mg^{2+} was removed from the perfusion medium.

perfusion, the levels of ·NO decay to a value close to the basal signal without removing the stimulus. Several mechanisms may contribute to this phenomenon, including, among others, a ·NO-dependent feedback inhibition of nitric oxide synthase or glutamate NMDA receptor.

Taken together, data shown provide strong support for the electrochemical monitoring of ·NO in hippocampal slices functionally dependent on glutamate stimulus.

Concluding Remarks

The difficulty in measuring ·NO in a real-time and selective way has been a major limitation to the understanding of its role in brain pathophysiology. The use of Nafion-coated porphyrinic microsensors, coupled with the use of fast cyclic voltammetry, affords a high degree of selectivity and spatial resolution for the real-time measurement of ·NO in brain slices. Therefore, this approach may contribute to the elucidation of the mechanisms by which ·NO exerts its biological effects in

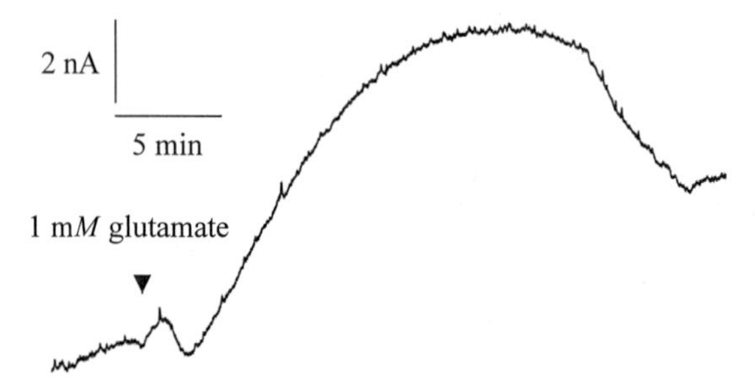

FIG. 7. Glutamate-induced production of nitric oxide in the CA1 region of the hippocampus monitored by fast cyclic voltammetry in perfused hippocampal slices by sampling the current at the oxidation peak potential of nitric oxide.

the brain. In particular, it is well suited to answer questions related to (1) the time course of agonist-induced ·NO production in different brain areas; (2) the type of relationship that may be established between ·NO concentration and cellular effects; (3) how the balance between oxidants and antioxidants (redox state) of cells and tissue modulates the time course of ·NO release and spatial distribution, and (4) the development of therapeutic strategies to modulate the activity of ·NO in the brain.

Acknowledgments

These studies were supported by Fundação Ciência e Tecnologia (Grant PRAXIS/P/BIA/1395/1998), Portugal. Brain slice preparation facilities and expertise provided by Drs. Catarina Oliveira and João Malva are gratefully acknowledged.

[11] Electrochemical Detection of Physiological Nitric Oxide: Materials and Methods

By BARRY W. ALLEN, LOUIS A. COURY, JR., and CLAUDE A. PIANTADOSI

Measuring Nitric Oxide in Living Systems

An eagerly sought but elusive goal of contemporary biology is the ability to study nitric oxide (NO) signaling at physiological levels. Physiological levels are those needed to maintain normal function in an organism, tissue, cell, or organelle free of injury or disease. These levels have been estimated to be in the picomolar range and are likely to be brief releases followed by rapid disappearance due to the presence of biological scavenging molecules. The half-life of nitric oxide in biological systems is variously estimated to be a few seconds to one-tenth of a second.[1,2] Therefore, any method for measuring physiological NO, in addition to exquisite sensitivity, must operate with a short time constant. Although a great deal of progress has been made in the technical aspects of real-time nitric oxide detection, no device presently exists that seamlessly accomplishes these dual objectives.

Of the techniques available for measuring nitric oxide, *ex situ* methods, those requiring that a sample be removed from a biological system, in some cases undergo secondary processing, and finally be transferred to a laboratory analytical device, seem less promising. Sample extraction, transfer, processing, and analytical

[1] D. A. Wink and P. C. Ford, *Methods: Companion Methods Enzymol.* **7,** 14 (1995).

[2] J. R. Lancaster, *in* "Nitric Oxide: Biology and Pathobiology" (L. J. Ignarro, ed.), p. 209. Academic Press, San Diego, 2000.

0076-6879/02 $35.00

signal analysis may take longer than nitric oxide persists. Some methods seek to circumvent this limitation by measuring the more persistent oxidation products of NO, such as nitrite (NO_2^-), nitrate (NO_3^-), or other nitrogen oxides, collectively NO_X. However, total NO_X levels in biological systems are much larger than moment-to-moment NO production and are regulated by many physiological systems independent of NO production. In fact, NO levels and total NO_X levels can, in principle, change in opposing directions at the same time.

However, techniques that permit analytical chemistry to be performed at the tip of a slender probe inserted directly into a site of nitric oxide release seem more promising. To date, electrochemistry (using electrodes) and photochemistry (using optodes) have both been accomplished at the tip of a miniature probe. The very interesting bridging technology of microdialysis puts the sampling process in a probe and couples it to an *ex situ* detector but does not qualify here because more than a few seconds are required for the sample to reach the detector.

An optode, a term coined from "optical" and "electrode," is a device in which the reagents for photochemistry are deposited at the tip of an optical fiber. A number of nitric oxide-sensitive photochemical systems are available, and some may significantly decrease the micromolar limit of detection demonstrated so far.[3] Electrodes, however, can now detect nanomolar nitric oxide during short-term release (tens of seconds) and picomolar *in vitro* during steady-state production by long-acting nitric oxide-releasing compounds. If electrochemical devices can be refined further, it is likely that they will be the first to reliably detect physiological levels in living systems.

Electrodes behind Hydrophobic Membranes

Electroanalytical chemistry is unique in that it simultaneously can detect nitric oxide itself (not nitrite or nitrate), operate in the liquid phase (and thus in biological fluids), and provide real-time data continuously. Amperometry is an electrochemical technique well suited to detecting NO *in vivo,* as it produces a continuous record and is among the most sensitive of electroanalytical methods.[4] In this technique, a constant potential is applied between the working electrode (the electrode at which nitric oxide is detected) and a reference electrode (an electrode providing a stable reference potential). Current is the measured variable reflecting NO activity.

Electroanalytical devices embodied in slender probes suitable for insertion into a site of biological nitric oxide release were initially based on the Clark electrode (1953) for the electrochemical detection of oxygen.[5] Briefly, this electrode electrochemically reduces dioxygen (O_2). A platinum (Pt) working electrode is held at

[3] S. L. R. Barker, R. Kopelman, T. E. Meyer, and M. A. Cusanovich, *Anal. Chem.* **70,** 971 (1998).
[4] A. J. Bard and L. R. Faulkner, "Electrochemical Methods." Wiley, New York, 1980.
[5] L. C. Clark, R. Wolf, D. Granger, and Z. Taylor, *J. Appl. Physiol.* **6,** 189 (1953).

a fixed potential of approximately −700 mV with respect to a silver/silver chloride (Ag/AgCl) reference electrode. The electrodes are immersed together in a strong electrolyte solution. The critical innovation of the Clark electrode is the use of a membrane permeable only to gases that separates the electrolyte surrounding the Pt and the Ag/AgCl electrodes from the analyte fluid. This membrane prevents most interfering substances, which are not gases, from reaching the sensor and allows a strong electrolyte solution to bathe the electrodes (and thus increase their electrochemical efficiency) without altering the composition of the biological fluid being studied. In 1990, Shubuki[6] described a nitric oxide electrode based on Clark's technology. Shibuki's inventive step was to replace the negative (reducing) potential that Clark had applied between his electrodes with a positive potential, 800–900 mV, to oxidize nitric oxide. This change was necessitated by the fact that both gases are reduced at the same potential and therefore cannot be distinguished by electrochemical reduction; however, only NO can be oxidized in aqueous solution. The electrons stripped from nitric oxide are measured as current. Two widely used commercially available nitric oxide sensors employ Clark/Shibuki-type sensors that are sheathed in silicone or fluropolymer gas-permeable membranes that confine a gel electrolyte; a positive potential of approximately 800 mV is applied between working and reference electrodes.[7,8] Both devices use a metallic working electrode, presumably Pt, for their larger probes and a graphite fiber for their smaller probes.

Gas-permeable membranes and enclosed electrolytes such as those used in the Clark, Shubuki, and commercial sensors described earlier are partial barriers to gas diffusion and therefore increase the time required for the analyte gas to reach the electrodes. In the case of O_2, which can be stable indefinitely and has a high concentration in tissues and physiological fluids, this barrier simply introduces a short delay in the recorded data, a temporal offset that does not alter the essential dynamics of the gas. However, because NO has a very short half-life and acts at much lower physiological concentrations, a short delay can diminish its apparent concentration to levels below the limit of detection of an analytical method. At the very least, a significant delay in detection will alter the measured dynamic behavior of the gas. This problem is far less apparent when the much higher pathological levels of nitric oxide are present.

Electrodes behind Hydrophilic Membranes

The first published report of a nitric oxide electrode that did not use a membrane permeable only to gases was that of Malinski and Taha.[9] This electrode was coated

[6] K. Shibuki, *Neurosci. Res.* **9,** 69 (1990).

[7] World Precision Instruments, Inc., Sarasota, FL.

[8] Harvard Apparatus Company, Holliston, MA.

[9] T. Malinski and Z. Taha, *Nature* **358,** 676 (1992).

with Nafion (Dupont, Wilmington, DE), a fluorocarbon polymer comprising a backbone of polytetrafluoroethylene (PTFE, Teflon) with pendant side chains of sulfonic acid. When in thick layers (tens of micrometers) that are only partially hydrated, this membrane is sparingly permeable to gases. However, when applied in thin layers (tens of nanometers) that are fully hydrated, and the sulfonate groups are neutralized with a cation such as Na^+, Nafion admits gases as well as small cations and neutral species in aqueous solution.[10,11] Anions, however, are excluded by the negative charges on the sulfonate anions. When employed as an anion-exclusion coating on electrodes in aqueous media, Nafion is used in thin layers and is fully hydrated.[9,12]

Catalytic Electrodes

Of the electrochemical devices described earlier, all but one use as a working electrode a material that is generally regarded as chemically inert and serves mainly as a conductor of electrons (either a platinum wire or an unmodified graphite fiber). The electrode of Malinski and Taha was the first published attempt to make a catalytic sensor for the oxidation of nitric oxide by using a metalloporphyrin in an effort to promote the oxidation reaction. In fact, catalysis may indeed occur on this electrode, as nitric oxide is oxidized at a lower potential (630 mV)[13] than used on the platinum electrode in Shibuki's sensor (800–900 mV). Malinski and Taha used a working electrode consisting of a flame-sharpened graphite fiber modified with a coating of a nickel-bearing porphyrin.[13,14] Subsequent work has shown that porphyrin coating is not necessary for the ability of the graphite fiber to sensitively detect nitric oxide.[15] However, the performance of this electrode may have been enhanced by the thermal treatment, which is known to increase the surface area of graphite electrodes,[16] and by the cycling between high and low potentials that was used to apply the porphyrin coating and which is also known to increase the surface area of graphite electrodes.[16] Both of these effects would increase the response to NO. In 2000, we described a catalytic electrode using ruthenium, a metal of the platinum group (elements 44–46 and 76–78) that operates at a lower potential and exhibits a higher current density than seven other electrode materials

[10] T. Sakai, H. Takenaka, and E. Torikai, *J. Electrochem. Soci.* 88 (1986).

[11] J. S. Chou and D. R. Paul, *Ind. Eng. Chem. Res.* **27,** 2161 (1988).

[12] B. W. Allen, C. A. Piantadosi, and L. A. Coury, *Nitric Oxide* **4,** 75 (2000).

[13] T. Malinski and L. Czuchajowski, *in* "Methods in Nitric Oxide Research" (M. Feelisch and J. S. Stamler, eds.), p. 19. Wiley, Chichester, 1996.

[14] T. Malinski, E. Ubaszewski, and F. Kiechle, *in* "Nitric Oxide Synthase: Characterization and Functional Analysis" (M. D. Maines, ed.), p. 14. Academic Press, San Diego.

[15] M. M. Irvani, Z. L. Kruk, and J. Millar, *J. Physiol.* **467,** 48P (1993).

[16] K. Kinoshita, "Carbon, Electrochemical and Physiochemical Properties," p. 251. Wiley, New York, 1988.

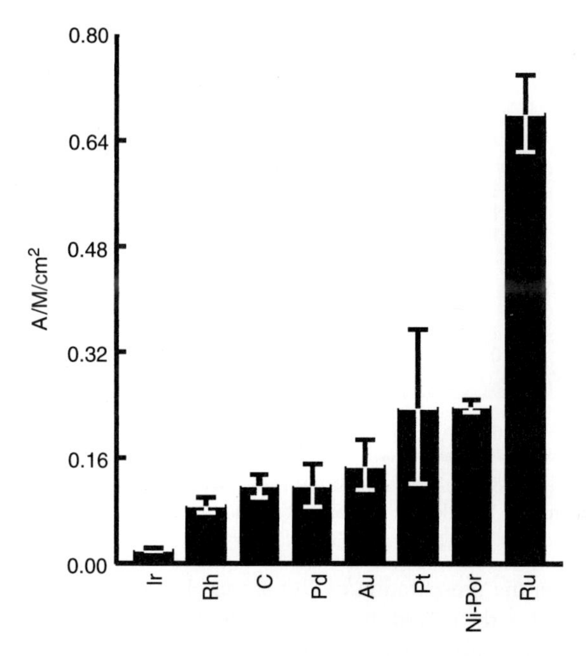

FIG. 1. Electrocatalysis at a ruthenium electrode. The ruthenium NO electrode demonstrates electrocatalysis for NO oxidation and a threefold increase in the NO signal compared to the next best material. Shown here are average current densities, normalized for electrode surface area, observed during NO oxidation for a variety of tested electrode materials: iridium (Ir), rhodium (Rh), carbon (C), palladium (Pd), gold (Au), platinum (Pt), a nickel-porphyrin on carbon (Ni-Por), and ruthenium (Ru). Data from B. W. Allen, C. A. Piantadosi, and L. A. Coury, *Nitric Oxide* **4,** 75 (2000).

tested, including those used in the other nitric oxide electrodes described earlier[12] (see Figs. 1 and 2). Enhancement of NO oxidation on this electrode did not depend on an increased surface area.

Measuring Physiological NO Levels

Optimal electrode performance is key to measuring physiological levels of nitric oxide; however, other technical problems must also be overcome: both electrical and temporal resolution of the other components of the complete detection system must be refined. These, in fact, are related issues, as the conventional method of increasing the limit of detection for an electrochemical sensor is to incorporate a low pass filter that attenuates signals whose frequencies are above a chosen cut-off level, without regard to whether these signals originate from extraneous sources, as "noise," or represent a real signal; i.e., both "noise" and some real data are removed by this strategy. Electronic filtering also introduces a significant

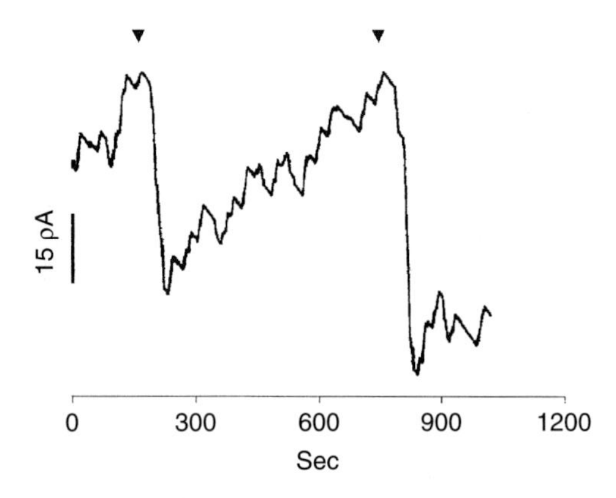

FIG. 2. Limit of detection at ruthenium electrode. The amperometric current at the Ru electrode in response to two successive injections of 3 n*M* NO (injected at arrows) is shown. The signal is significantly above electrical background noise. Data were maximally filtered electronically during the experiment. Data from B. W. Allen, C. A. Piantadosi, and L. A. Coury, *Nitric Oxide* **4,** 75 (2000).

time lag between the presentation of NO at the electrode surface and the registration of a signal at the output device. This blunts the amplitude of an authentic signal while increasing its duration, even for signals below the cut-off frequency. Therefore, the smallest signals that can be distinguished on most electroanalytical instruments actually reflect levels of NO that are higher than they appear, and even smaller signals disappear into the electrochemical baseline. That physiological releases of NO may be very small events means that filtering strategies work against a device intended to record physiological NO release. A useful instrument would be one with minimal electronic filtering and that operates in ordinary laboratory environments, without a Faraday cage, and yet still provides low noise output.

Two further obstacles must be overcome in order to measure physiological NO release: the influences of temperature and hydrodynamics. Changes in both temperature and flow occur *in vivo* and cause changes in the amplitude of an electrochemical signal that is impossible to distinguish from altered NO activity. Again, because physiological NO release is small, these confounding variables can overwhelm the real signal. One commercially available device has temperature compensation[8]; however, because this system uses a temperature sensor that is separate from and larger than its nitric oxide-sensitive electrochemical probe, it serves more as an acknowledgment that temperature compensation is needed than as a practical solution to the problem. No available instrument addresses the issue of sensitivity to fluid flow, although this has been recognized to be a problem.[17]

[17] K. Schmidt and B. Mayer, *in* "Nitric Oxide Protocols" (M. A. Titheradge, ed.), p. 101. Humana Press, Totowa, NJ, 1998.

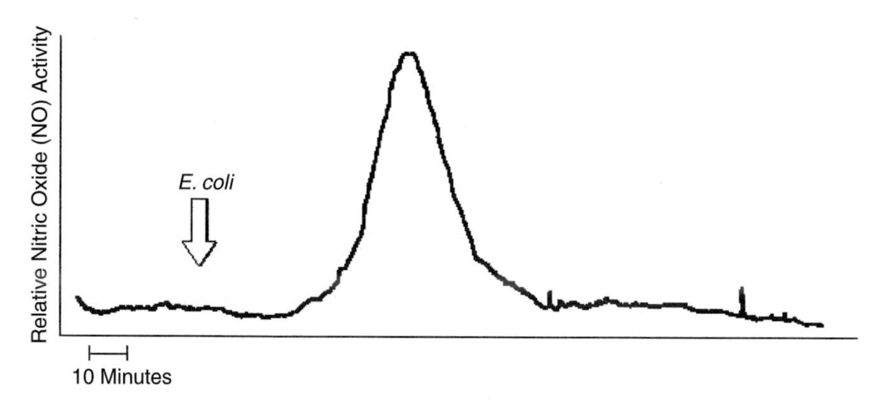

FIG. 3. Detecting pathological levels of NO *in vivo*. An intravascular nitric oxide sensor was inserted in a femoral vein in an adult male baboon (*Papio cynocephlus*) that was anesthetized and ventilated mechanically, according to an approved protocol. Heat-killed bacteria (*Escherichia coli*) were infused into a brachial vein. Approximately 20 min later, the output of a ruthenium electrode demonstrated a strong signal consistent with nitric oxide production by the animal in response to the experimental bacteremia. Other *in vivo* experiments confirmed this as a true nitric oxide response, as it was abolished by an inhibitor of biological nitric oxide production, L-nitroarginine methyl ester. Unpublished data of B. W. Allen and C. A. Piantadosi.

In fact, all readily available publications that report nitric oxide levels *in vivo* do so in pathological states, after pharmacological interventions, or in the special case in which an intense exogenous or supraphysiological stimulus, such as injected or infused bradykinin,[9,18] acetylcholine,[18] lipopolysacharride (Fig. 3), or interferon-γ[19] is used to elicit an NO response. The pathological release of NO can produce levels several orders of magnitude greater than physiological levels. For example, Porterfield *et al.*[19] reported a steady-state nitric oxide flux in a region 1 μm from the surface of a stimulated macrophage to be 1 pmol/cm^2/sec persisting over more than 100 sec. This translates into a local NO concentration of at least 10 μM.

The lack of an analytical tool to measure nitric oxide in the picomolar range and at time constants of less than 1 sec has severely limited the ability of investigators to answer fundamental questions of nitric oxide biology. Such questions include the following: How much NO is free and how much is bound in normal physiological states? What are baseline NO levels in various parts of the circulation and in different tissues? What changes in NO levels may indicate early, asymptomatic disease states? Is the NO release required to maintain normal physiological function continuous or pulsatile? What biochemical variables modulate physiological

[18] Z. G. Liu, X. C. Liu, A. P. Yim, and G. W. He, *Ann. Thorac. Surg.* **71,** 133 (2001).

[19] D. M. Porterfield, J. D. Laskin, S.-K. Jung, R. P. Malchow, B. Billack, P. J. S. Smith, and D. E. Heck, *Am. J. Physiol. Lung Cell Mol. Physiol.* **281,** L904 (2001).

NO release, e.g., O_2, CO, and what other variables govern its disappearance under normal conditions?

Criteria for Physiological Measurement

New analytical approaches are needed to provide sensitive and specific detection of NO at levels now beyond reach. Electrochemistry applied at picomolar resolution is difficult, and variables that are insignificant at nanomolar and micromolar levels can mask or confound the NO signal at picromolar levels. We, therefore, propose the following criteria by which to facilitate and evaluate experiments that attempt to measure physiological levels of nitric oxide.

Control for Interventional Artifacts

Control experiments should show the effect on the electrode system of changed conditions or of the addition of reagents or drugs to elicit the nitric oxide response being studied. Therefore, if acetylcholine, bradykinin, tumor necrosis factor, interferon, and so on are used to stimulate NO release, then the effect of these agents must be tested on the electrochemical detection system in the absence of the biological system being studied and published along with the biological results. For example, Malinski and Taha did this by demonstrating that their electrode did not respond to the bradykinin that they subsequently used to stimulate nitric oxide production in a biological system. Also, putative NO responses should be confirmed with specific inhibitors of the nitric oxide synthases. Likewise, the biological system must be tested with inactive or placebo agents to reveal the effects of the mechanical aspects of the intervention on that system. For example, insertion of a needle into a blood vessel, injection of a bolus of fluid, and rapid removal of the needle can, in and of themselves, cause the release of a nitric oxide by disturbing the vascular endothelium. Extensive surgical preparation is a trauma that can cause a delayed release of nitric oxide, and the introduction of microbes inadvertently resulting from surgery or other invasive manipulations can also produce a significant response after a delay of 20 min or more, particularly in experiments lasting hours.

Control for Changes in the Fluid Path between Electrodes

The pathway between working and reference electrodes has an electrical resistance. If this resistance changes, the signal from the electrode system will change too and can be misinterpreted as a change in nitric oxide concentration. Thus, if a tissue changes in volume during an experiment (and NO itself does modulate vascular and tissue volume), the working electrode may be moved relative to the nitric oxide source, or the distance between working and reference electrodes may be altered. Therefore, all electrodes should be positioned as close to each other as possible and fixed securely together. Mechanical movement itself may produce a change in electrical current (triboelectric current).

Control for Changes in Electrochemical Background Oxidation

This problem becomes extreme when working and reference electrodes are on opposite sides of an excitable membrane. In this case, one may actually measure an action potential, not NO release.

Calibrate in Accord with Experimental Conditions

The electrode system should be calibrated in conditions as similar as possible to the experimental system, in estimated nitric oxide concentration, osmotic strength, temperature, and hydrodynamic flow.

Control for Changes in Temperature

Changes in temperature will change the electrochemical baseline and may be indistinguishable from an authentic signal. *In vivo* experiments should be done with constant temperature thermal pads for animal studies and with close thermostatic control of tissue perfusion or cell culture solutions. All injections, infusions, and other interventions should be at the same temperature as the experimental preparation.

Control for Changes in Hydrodynamics

Because the rate of a chemical reaction changes with the rate at which reactants are supplied to the reaction site, alterations in stirring speed *in vitro,* or in blood flow *in vivo,* will produce spurious signals. For example, significant differences in fluid flows do occur *in vivo,* as blood flow can vary up to 60-fold between rest and exercise.

Select an Appropriate Time Constant

The investigator should be aware of the time constant of the total detection system, and this information should be included along with published data. Electronic filtering as well as excessive or hindered diffusion paths increase the interval between the biological release of NO and the appearance of a signal, and they also flatten the response curve. An instructive experiment is to increase the electronic filtering to the maximum available on the apparatus being used and then to gently disturb the electrode system, say by adding a drop of cold solution to a warm electrochemical cell. At high levels of filtering, the effect of this disturbance will register seconds after the intervention. Under these conditions, one could easily be convinced that the observed signal is real, especially if the added agent is known to stimulate NO production. This problem is avoided by testing for these effects in control experiments. Control data should be shown along with "real" data in published results to avoid the criticism that the purported data are artifacts.

Summary

Advances in the electroanalytical technology of NO detection make it possible to detect the release of robust concentrations of NO from living systems under pathological or pharmacological conditions. However, technical improvements should enable the construction of research instruments with one or two orders of magnitude improvement in both detection limit and temporal resolution. Such instruments would be capable of revealing physiological NO production and could help quantify the correlations between NO levels and health or disease, ultimately leading to important applications in biomedical research and clinical medicine.

Acknowledgment

This work was supported by a grant from the Office of Naval Research.

[12] Detection of Nitric Oxide Production by Fluorescent Indicators

By JEAN-YVES CHATTON and MARIE-CHRISTINE BROILLET

Introduction

The design of detection methods for nitric oxide (NO) is complicated by the unstable nature of the NO molecule, as well as by its low cellular production.[1,2] Several methods for detecting NO have been developed, such as trapping NO with hemoglobin,[3] chemiluminescence assays,[4] electron paramagnetic resonance spectroscopy,[5] or measurements with different electrochemical electrodes.[6] These

[1] K. Shibuki and D. Okada, *Nature* **349,** 326 (1991).

[2] T. Malinski, Z. Taha, S. Grunfeld, S. Patton, M. Kapturczak, and P. Tomboulian, *Biochem. Biophys. Res. Commun.* **193,** 1076 (1993).

[3] S. R. Vincent, J. A. Williams, P. B. Reiner, and A. E.-D. El-Husseini, *Prog. Brain Res.* **118,** 27 (1998).

[4] A. M. Leone, V. W. Furst, N. A. Foxwell, S. Cellek, and S. Moncada, *Biochem. Biophys. Res. Commun.* **221,** 37 (1996).

[5] T. Yoshimura, H. Yokoyama, S. Fujii, F. Takayama, K. Oikawa, and H. Kamada, *Nature Biotechnol.* **14,** 992 (1996).

[6] T. Malinski, S. Mesaros, and P. Tomboulian, *Methods Enzymol.* **268,** 58 (1996).

0076-6879/02 $35.00

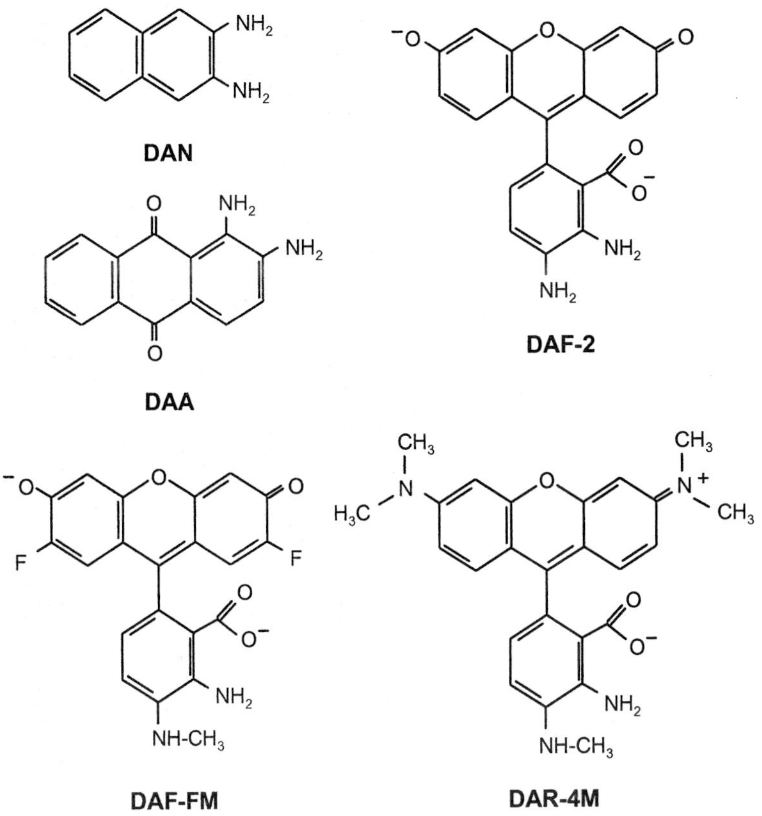

DAN **DAF-2**

DAA

DAF-FM **DAR-4M**

FIG. 1. Chemical structure of diamino aromatic compounds used as fluorescent NO indicators: 2,3-diaminonaphthalene (DAN), 1,2-diaminoanthraquinone (DAA), 4,5-diaminofluorescein (DAF-2), 4-amino-5-methylamino-2′,7′-difluorofluorescein (DAF-FM), and diaminorhodamine-4M (DAR-4M).

detection methods are limited by their relatively low sensitivity and provide little information about the source of NO or the identification of NO-synthesizing cells.

NO fluorescent indicators have been developed[7–10] (Fig. 1), which should allow the detection and real-time biological imaging of NO.

[7] A. M. Miles, D. A. Wink, J. C. Cook, and M. B. Grisham, *Methods Enzymol.* **268,** 105 (1996).

[8] H. Kojima, N. Nakatsubo, K. Kikuchi, S. Kawahara, Y. Kirino, H. Nagoshi, Y. Hirata, and T. Nagano, *Anal. Chem.* **70,** 2446 (1998).

[9] H. Kojima, K. Sakurai, K. Kikuchi, S. Kawahara, Y. Kirino, H. Nagoshi, Y. Hirata, and T. Nagano, *Chem. Pharm. Bull.* **46,** 373 (1998).

[10] H. Kojima, M. Hirotani, N. Nakatsubo, K. Kikuchi, Y. Urano, T. Higuchi, Y. Hirata, and T. Nagano, *Anal. Chem.* **73,** 1967 (2001).

This article outlines the spectrofluorimetric and microscopic methods used in our laboratories to detect NO generated by NO-releasing compounds and by cultured cells with these fluorescent NO indicators. The critical evaluation presented also highlights some peculiar properties of these dyes.

Principle of NO Detection by Diamino Derivatives of Fluorophores

Several fluorimetric methods for detecting NO have been developed based on aromatic diamino derivatives of fluorescent chromophores.[7–10] Successful examples (Fig. 1) are 2,3-diaminonaphthalene[7] (DAN, Sigma, St. Louis, MO), 1,2-diaminoanthraquinone[11] (DAA, Sigma), 4,5-diaminofluorescein[9] (DAF-2, Sigma), 4-amino-5-methylamino-2′,7′-difluorofluorescein (DAF-FM, Molecular Probes, Eugene, OR), and diaminorhodamine-4M[10] (DAR-4M Alexis Biochemicals, Lausen, Switzerland). These poorly fluorescent diamino compounds react rapidly with NO to yield the highly fluorescent benzotriazole products. DAF-FM and DAF-2 are observed using standard fluorescein filter sets ($\lambda_{ex} = 495$ nm, $\lambda_{em} = 515$ nm), whereas DAR-4M can be used adequately with rhodamine filter sets ($\lambda_{ex} = 560$ nm, $\lambda_{em} = 575$ nm).

Among the five dyes, DAN ($\lambda_{ex} = 365$ nm, $\lambda_{em} = 415$ nm) would be suited for NO measurement in extracellular fluids using a spectrofluorimeter, whereas the other indicators are membrane permeant (DAA) or are also available commercially as membrane-permeant ester derivatives (DAF-2 diacetate, DAR-4M acetoxymethyl ester, and DAF-FM diacetate). Stock solutions of the membrane-permeant derivatives should be prepared in anhydrous dimethyl sulfoxide (DMSO) at 5–10 m*M* and kept in small aliquots at −20° in the dark until dissolved immediately before use in physiological medium at a final concentration of 1–20 μM. Cell loading is typically performed at 37° for 30 min; however, depending on the cell type, longer times may be required. The technique for dye loading via membrane-permeable derivatives[12] is used routinely for other compounds, such as Ca^{2+} and pH indicators, and is based on the ability of intracellular esterases to hydrolyze such ester moieties to leave the negatively charged, membrane-impermeable form of the dye trapped in the cell (Fig. 2). On reaction with NO, the fluorescent triazole product accumulates, yielding an increasingly bright fluorescence.

These indicators have different sensitivities that extend to the low nanomolar range. DAR-4M, for example, should allow the detection of 7 n*M* NO[10] and DAF-2 of 4 n*M* NO.[13] These indicators are to be used with two main experimental

[11] X. Chen, C. Sheng, and X. Zheng, *Cell Biol. Int.* **25,** 593 (2001).

[12] R. Y. Tsien, *Nature* **290,** 527 (1981).

[13] H. Kojima, N. Nakatsubo, K. Kikuchi, Y. Urano, T. Higuchi, J. Tanaka, Y. Kudo, and T. Nagano, *NeuroReport* **26,** 3345 (1998).

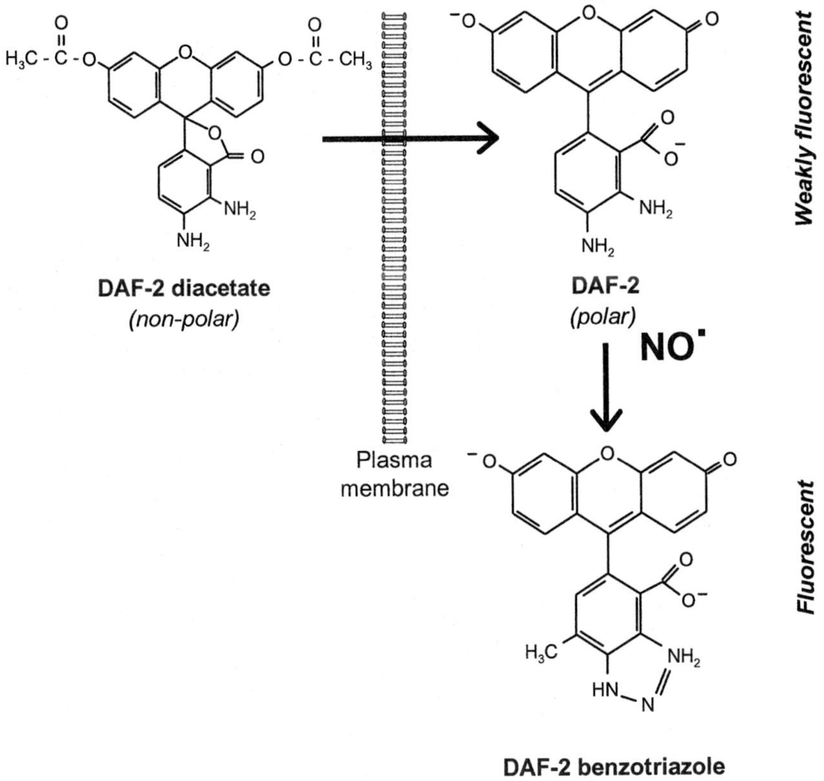

FIG. 2. Principle of intracellular NO detection. In the example shown, the membrane-permeable diacetate form of DAF-2 undergoes passive diffusion across the cell membrane. Once in the cytosol, cellular esterases hydrolyze the ester form to yield the negatively charged and membrane-impermeant form trapped in the cell. In their diamino form, these compounds are only weakly fluorescent. In the presence of NO, they rapidly form a benzotriazole compound that is highly fluorescent. Accumulation of this fluorescent product is indicative of the presence of NO either delivered by means of donors or locally produced by the nitric oxide synthases. The same scheme applies to DAF-2-FM diacetate and DAR-4M acetoxymethyl ester.

techniques: spectrofluorimetery and microscopy. Using spectrofluorimetry, they should have useful applications in the detection of NO in the serum, plasma, urine, and other body fluids,[14,15] as well as in culture medium in order to evaluate, for example, the activities of eNOS and nNOS isoforms during *in vitro* cultures.

[14] T. Yoshioka, N. Iwamoto, F. Tsukahara, K. Irie, I. Urakawa, and T. Muraki, *Br. J. Pharmacol.* **129,** 1530 (2000).

[15] Y. Itoh, F. H. Ma, H. Hoshi, M. Oka, K. Noda, Y. Ukai, H. Kojima, T. Nagano, and N. Toda, *Anal. Biochem.* **287,** 203 (2000).

Indeed, cells expressing these low output enzymes produce submicromolar levels of nitrite and nitrate when cultured *in vitro* and such a method could definitely be useful. With microscopy, these indicators should allow the real-time detection of NO production in single living cells.

NO Detection in Solution: Extracellular Fluids

It is very important to first test the sensitivity of the chosen indicator in the specific experimental conditions. In order to do that, substances releasing NO such as NO donors can be used. Indeed, NO donors or releasing agents are used increasingly in *in vivo* or *in vitro* studies to mimic the effects of NO. A number of them are available commercially (Cayman Chemicals and Alexis Corporation, Lausen, Switzerland). We generally use the *S*-nitrosothiol *S*-nitrosocysteine (SNC, stock solution 100 m*M*), as a NO donor because it is inexpensive, easy to prepare, and reliable. Moreover, it probably represents a common *in vivo* donor of NO, as the most abundant source of free thiol groups in mammalian tissue is cysteine (in either the free or peptide form). SNC has to be prepared on ice from the combination in acid solution of equimolar amounts of L-cysteine hydrochloride and sodium nitrite[16] (Sigma). SNC is formed quantitatively within 1 min. The result of this acid-catalyzed reaction is a red liquid. The stock SNC solution remains effective for approximately 2 hr. NO donors, such as sodium nitroprusside (SNP, stock solution: 20 m*M* in H_2O) (Sigma-Aldrich, Steinheim, Germany)[17] or DETA NONOate (DETA/NO, stock solution: 6 m*M* in the selected Ringer solution) (Alexis Biochemicals, San Diego, CA),[18] can also be used.

These three donors are known to release NO with different time course characteristics.[19,20] Their features should be found again as a control when monitoring NO release with the chosen fluorescent dye (Fig. 3).

The NO produced by these donors should be detected easily by the chosen fluorescent indicator diaminofluorescein or diaminorhodamine. Figure 3A shows how diaminofluorescein (DAF-2; 5 m*M* stock solution) (Calbiochem, Foster City, CA) detects the presence of NO produced by three different NO donors.

The NO donors, SNC (500 μM), SNP (5 m*M*), and DETA/NO (1.5 m*M*), are added with DAF-2 (15 μM) immediately before the beginning of each experiment

[16] S. Z. Lei, Z. H. Pan, S. K. Aggarwal, H. S. V. Chen, J. Hartman, N. J. Sucher, and S. A. Lipton, *Neuron* **8,** 1087 (1992).

[17] B. Brune and E. G. Lapetina, *J. Biol. Chem.* **264,** 8455 (1989).

[18] D. L. Mooradian, T. C. Hutsell, and L. K. Keefer, *J. Cardiovasc. Pharmacol.* **25,** 674 (1995).

[19] L. K. Keefer, R. W. Nims, K. M. Davies, and D. A. Wink, *in* "NONOates" (1-Substituted Diazen-1-ium-1,2-diolates) as Nitric Oxide Donors: Convenient Nitric Oxide Dosage Forms" (L. Packer, ed.), p. 281. Academic Press, San Diego, 1996.

[20] J. A. Bauer, B. P. Booth, and H.-L. Fung, *in* "Nitric Oxide Donors: Biochemical Pharmacology and Therapeutics" (L. Ignarro and F. Murad, eds.), p. 361. Academic Press, San Diego, 1995.

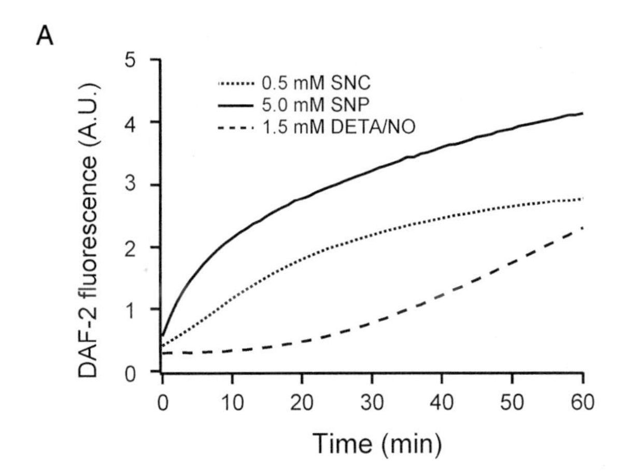

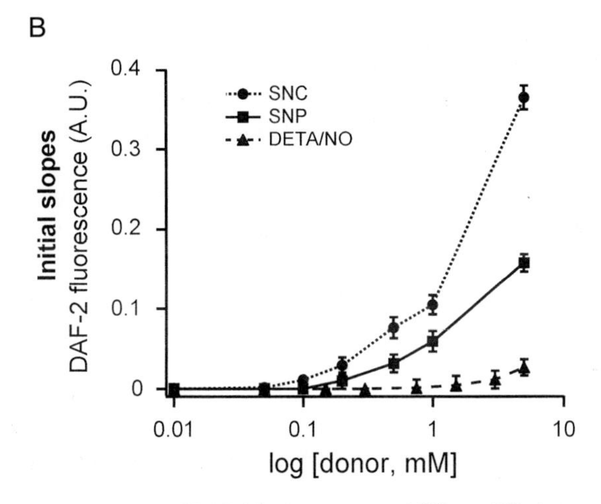

FIG. 3. Fluorescence response of DAF-2 in the presence of different NO donors. (A) Representative experiments showing the time courses of DAF-2 (15 μM) fluorescence intensity depending on NO generated by three different NO donors: *S*-nitrosocysteine (SNC, 0.5 mM), sodium nitroprusside (SNP, 5 mM), and DETA/NO (1.5 mM) during 60 min. (B) Initial slopes of DAF-2 fluorescence calculated on the first 10 min of NO released by SNC, SNP, and DETA/NO. The concentrations of NO donors were increased from 10 μM to 10 mM in a Ca^{2+}/Mg^{2+}-free Ringer solution (averages of five experiments for each NO donor).

to a Ca^{2+}/Mg^{2+}-free Ringer solution containing (in mM) NaCl, 145; EGTA, 0.5; EDTA, 0.5; and HEPES, 20. The pH of the solution is 7.6. The reactions are performed in Falcon Microtest 96-well plates (Becton Dickinson, NJ). The volume of the reaction medium is 200 μl. Measurements are performed at 37° with a fluorimeter (Perceptive Biosystems, Cytofluor multiwell plate reader 4000) with excitation wavelength at 485 nm (20-nm bandwidth) and emission wavelength at 530 nm (25-nm bandwidth). Continuous data acquisition is performed for 60 min on a PC Pentium Pro (333 MHz). Data are presented as mean values ± SEM. Differences between means are analyzed using the Student's *t* test for paired experiments.

Different time courses of DAF-2 fluorescence increase can indeed be recorded (Fig. 3A). After a fast initial increase, a slow down in kinetics is observed for both SNP and SNC (after 14 ± 2 min for 0.5 m*M* SNC and after 11 ± 3 min for 5 m*M* SNP; $n = 5$ for each donor). DETA/NO (1.5 m*M*) increases the fluorescent signal slowly but continuously for a longer period of time without showing any decrease in the rate of fluorescent product formation in this time frame. These data are consistent with previous findings showing that the time course of NO release is strongly dependent on the type of NO donor.[19,20] Indeed, SNC is a fast and efficient short-lived NO-releasing compound, SNP has an intermediate behavior, and DETA/NO releases NO slowly but can be used if a long-lasting presence of NO is necessary. If we consider the initial slopes of the responses calculated in the first 10 min of the reactions, concentration–response curves for the different NO donors can be obtained (Fig. 3B). As shown by Kojima *et al.*[9], the intensity of DAF-2 fluorescence increased according to the half-life and concentration of the added NO donors.

These results show that the use of DAF-2 in a Ca^{2+}/Mg^{2+}-free Ringer solution allows the continuous monitoring of NO released by all three donors and reveals an increase in fluorescence corresponding with the intrinsic characteristics of NO release by each donor. Such measurements/calibration can also be performed with NO gas.[21]

While performing these fluorimetric experiments, a series of factors appeared to interfere with our measurements. These factors and their influence on the detection of NO are listed in the next paragraphs. They should be taken into account if one wants to use the fluorescent indicators for the spectrofluorimetric detection of NO.

Effect of Light on NO Detection by Fluorescent Indicators

An unexpected and major increase in the fluorescence intensity of DAF-2 is observed when the number of readings per minute of the fluorimeter is increased from one read (0.400 ± 0.005) to three reads (0.590 ± 0.009) or five reads (1.220 ± 0.005) ($n = 4$) in the presence of 0.1 m*M* SNC and 15 μM DAF-2 (Fig. 4[22]). Each read has a duration of 0.2 sec. The same effect was observed in the presence of NO released by SNP or by DETA/NO (data not shown). This phenomenon is never observed in the absence of NO donors in the Ringer solution. These measurements show that a NO-dependent enhancement of DAF-2

[21] N. Suzuki, H. Kojima, K. Kikuchi, Y. Hirata, and T. Nagano, *JBC* **227,** 47 (2002).
[22] M.-C. Broillet, O. Randin, and J.-Y. Chatton, *FEBS Lett.* **491,** 227 (2001).

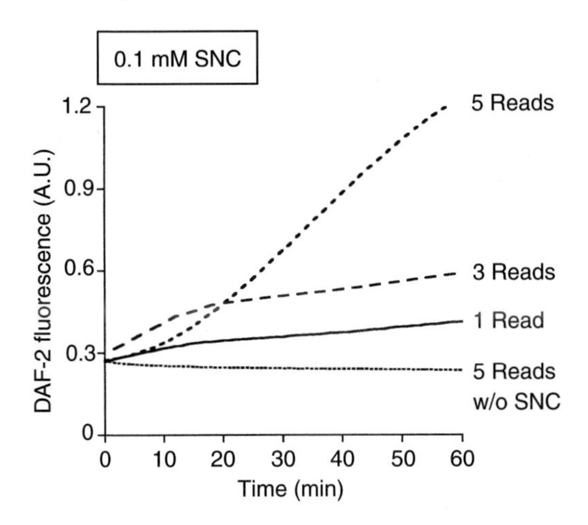

FIG. 4. Photosensitivity of DAF-2 fluorescence response. Representative recordings of DAF-2 fluorescence made with increasing duration of illumination on the spectrofluorimeter (1, 3, and 5 reads/min of 0.2 sec each) as indicated in the graph in the presence of 0.1 m*M* SNC. In comparison, a curve is presented for the longer light exposure time (5 reads, 1 sec) but in the absence of SNC (DAF-2 alone).

fluorescence proportional with the duration of light exposure occurs in a cell-free environment.

As a consequence, precise and controlled light conditions have to be used, and when comparing data, one has to be sure they were acquired under similar conditions.

Effect of pH on NO Detection by Fluorescent Indicators

The diaminofluoresceins and rhodamines have a marked pH sensitivity, which will affect the NO detection level. The applicable range of pH for these fluorescent dyes is above pH 6 for DAF (DAF-2, DAF-FM)[9] and above pH 4 for DAR-4.[10]

Effect of Divalent Cations on NO Detection by Fluorescent Indicators

Extracellular environments, such as interstitial or biological fluids, typically contain millimolar concentrations of Ca^{2+}, whereas physiological saline solutions contain a wide range of Ca^{2+} and/or Mg^{2+} concentrations depending on the experimental conditions.

We therefore examined the effects of Ca^{2+} and Mg^{2+} on DAF-2 fluorescence using Ringer solutions containing different concentrations of these divalent cations. The free Ca^{2+} and Mg^{2+} concentrations can be determined using the software WEBMAXCv2.10.[23] With the help of a fluorimeter, we monitored the increase

[23] D. M. Bers, C. W. Patton, and R. Nuccitelli, *Methods Cell Biol.* **40,** 3 (1994).

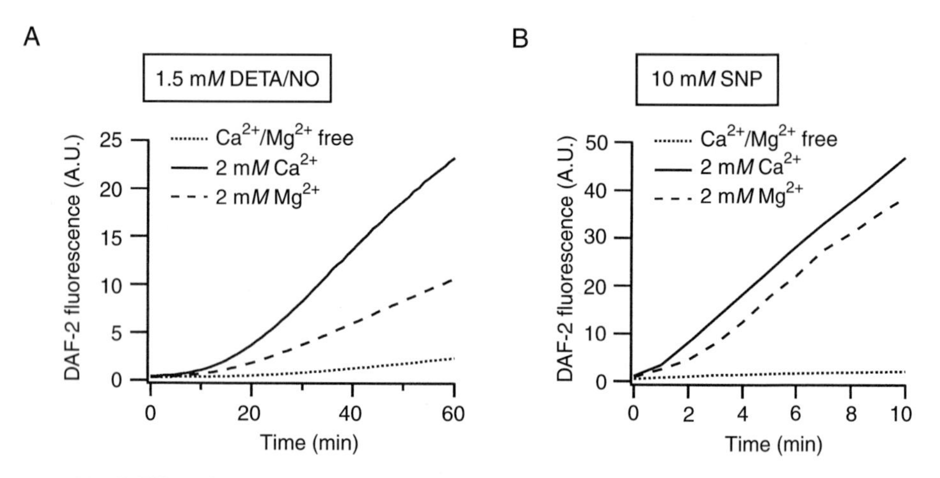

FIG. 5. Effect of divalent cations on the fluorescence response of DAF-2. Representative experiments where DAF-2 (15 μM) fluorescence was monitered for 60 min in the presence of NO generated by (A) DETA/NO (1.5 mM) or (B) SNP (10 mM) in Ca^{2+}/Mg^{2+}-free solutions or in the solutions containing either 2 mM Ca^{2+} or 2 mM Mg^{2+}.

of fluorescence levels over a period of 60 min following the addition of 15 μM DAF-2 to a Ringer solution containing, respectively, 1.5 mM DETA NONOate (DETA/NO) or 10 mM sodium nitroprusside (SNP), two NO donors of different chemical structure and reactivity (Fig. 5[22]).

Different time courses of DAF-2 fluorescence increase were recorded in a Ca^{2+}/Mg^{2+}-free Ringer solution versus solutions containing 2 mM Ca^{2+} or Mg^{2+}. Figure 5 shows that the ionic composition of the Ringer solution strongly influenced the fluorescence level of DAF-2: DAF-2 fluorescence intensity was 10 ± 1-fold increased in the presence of 2 mM Ca^{2+} with NO generated by 1.5 mM DETA/NO and 22 ± 2-fold with 10 mM SNP compared to a divalent cation-free solution ($n = 5$ for each donor and experimental conditions). The increase in DAF-2 fluorescence in the presence of Ca^{2+} is not caused by an increase in NO production by the donors themselves. NO production measured with the Griess reaction measuring the nitrite concentration (NO_2^-) on a spectrophotometer (540 nm)[24] showed that 10 mM SNP produces 17.60 ± 0.51 versus 16.73 ± 1.11 μM nitrites (n = 5, NS) in the presence or absence of 2 mM Ca^{2+}, respectively, and 1.5 mM DETA/NO produces 48.54 ± 3.75 versus 49.98 ± 2.39 μM nitrites ($n = 5$, NS) in the presence or absence of 2 mM Ca^{2+}, respectively.

An increase in DAF-2 fluorescence is also observed in the presence of Mg^{2+} in the Ringer solution (Fig. 5[22]). Indeed, when 2 mM Mg^{2+} is present in the 200-μl

[24] H. H. H. W. Schmidt and M. Kelm, *in* "Determination of Nitrite and Nitrate by Griess Reaction" (M. Feelisch and J. S. Stamler, eds.), p. 491. Wiley, New York, 1996.

reaction Ringer solution, DAF-2 fluorescence intensity is increased by a factor of 5 ± 1- and 19 ± 3-fold in the presence of NO generated, respectively, by 1.5 m*M* DETA/NO or 10 m*M* SNP (Fig. 5[22]). As for Ca^{2+}, Mg^{2+} does not alter the ability of the donors to release NO (data not shown).

The effect of Ca^{2+} and Mg^{2+} does not seem to be on background DAF-2 fluorescence but on its sensitivity to NO, with the presence of NO being required to observe an increase in fluorescence intensity as in the absence of NO donors, Ca^{2+} at concentrations of up to 20 m*M* had no effect on DAF-2 fluorescence intensity [0.215 ± 0.005 versus 0.204 ± 0.008 with 20 m*M* Ca^{2+} ($n = 5$, NS)]. Therefore the presence of Ca^{2+} or Mg^{2+} in the reaction medium appears to enhance the conversion of DAF-2 into its fluorescent product in the presence of NO regardless of the type of NO donor.

The strong effect of Ca^{2+} on DAF-2 fluorescence described earlier for the extracellular environment is also relevant for intracellular NO measurements. We performed fluorimetric measurements in conditions where Mg^{2+} was set at 1 m*M* and Ca^{2+} varied over a range of 50 n*M*–1 m*M*, which apply to the cytosolic, mitochondria, and endoplasmic reticulum environments.[25–27] Under these conditions, Ca^{2+} also enhances the NO-evoked DAF-2 fluorescence response markedly (Fig. 6[22]). For example, the change in Ca^{2+} from 50 n*M* to 1 μM enhances DAF-2 fluorescence by a factor of 4.45-fold in the presence of NO released by 0.5 m*M* SNC.

In summary, these dyes can indeed detect NO with relatively high sensitivity; however, for routine laboratory measurements, these fluorimetric methods can be expensive and time-consuming. For example, quantification of NO concentrations in solutions and determination of the amounts and rates of NO release from various donor agents or from cells with DAF-2 can be cumbersome as the reactions are complicated by multiple interfering agents. The interferences described earlier for DAF-2 (e.g., light and Ca^{2+}) are also found with DAR-4M.[28] Therefore, each time an assay has to be performed, it is necessary to consider the experimental conditions very critically to avoid misinterpretations and to make certain to use the same conditions systematically for NO monitoring.

Intracellular NO Detection

With the ability to incorporate NO-sensitive indicators in cells, researchers envisioned realizing the ultimate assay of imaging local production of NO with cellular or even subcellular resolution. Experiments using cells expressing the inducible NO synthase (iNOS), the Ca^{2+}-activatable neuronal (nNOS), or endothelial-type

[25] R. Rizzuto, P. Bernardi, and T. Pozzan, *J. Physiol.* **529** (Pt. 1), 37 (2000).
[26] F. L. Bygrave and A. Benedetti, *Cell Calcium* **19,** 547 (1996).
[27] J. Y. Chatton, H. Liu, and J. W. Stucki, *FEBS Lett.* **368,** 165 (1995).
[28] J. Y. Chatton, C. Pélofi, O. Randin, and M.-C. Broillet, submitted for publication.

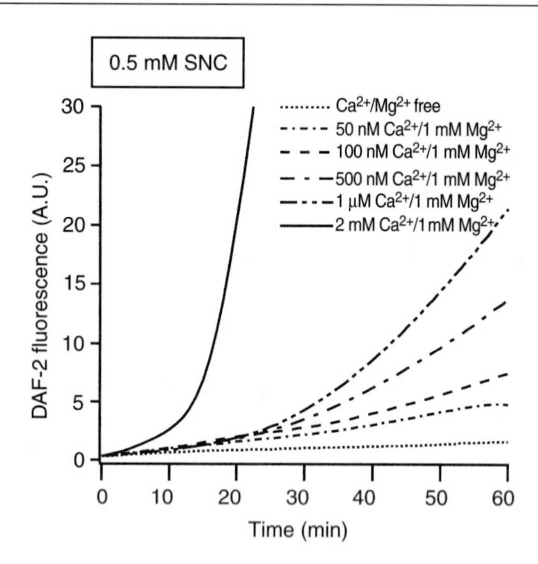

FIG. 6. Enhancement of DAF-2 fluorescence by physiological calcium concentrations. Representative experiment where DAF-2 (15 μM) fluorescence was monitored for 60 min in the presence of NO generated by SNC (0.5 mM) in Ca^{2+}/Mg^{2+}-free solution or in solutions containing different Ca^{2+} concentrations (from 50 nM to 2 mM) in the presence of 1 mM Mg^{2+}.

(eNOS) NO synthases have been attempted by several groups with mixed success (see, for example reference[29]).

Because of the properties of DAF-2[22] and DAR-4M[28] described earlier, it is advisable to first critically consider the imaging setup. The following experiments were carried out with primary mouse cortical astrocytes cultured on glass coverslips[30] loaded for 25 min at 37° with 10 μM DAR-4M/AM in Ringer and imaged on an inverted microscope equipped with a filter wheel for fluorescent excitation (Sutter Instruments, Novato, CA). Fluorescence images were acquired by a Gen III+-intensified CCD camera (VideoScope, Sterling, VA) at a rate of 0.5–0.1 Hz, with intermittent light exposure at 540 nm and an incident light intensity decreased by means of neutral density filters to ~5 μW (low light level mode), as measured at the back focal plane of the 40 × 1.3 NA objective using an optical power meter (LaserCheck, Coherent, Auburn, CA).

In Fig. 7A, an experimental trace is indicated where cultured mouse primary astrocytes loaded using DAR-4M/AM were subjected to a superfusion of sodium nitroprusside (SNP, 10 mM). Figure 7A shows that under low light level conditions, SNP barely affected the baseline DAR-4M fluorescent signal. However,

[29] J. F. Leikert, T. R. Rathel, C. Muller, A. M. Vollmar, and V. M. Dirsch, *FEBS Lett.* **506,** 131 (2001).
[30] J. Y. Chatton, P. Marquet, and P. J. Magistretti, *Eur. J. Neurosci.* **12,** 3843 (2000).

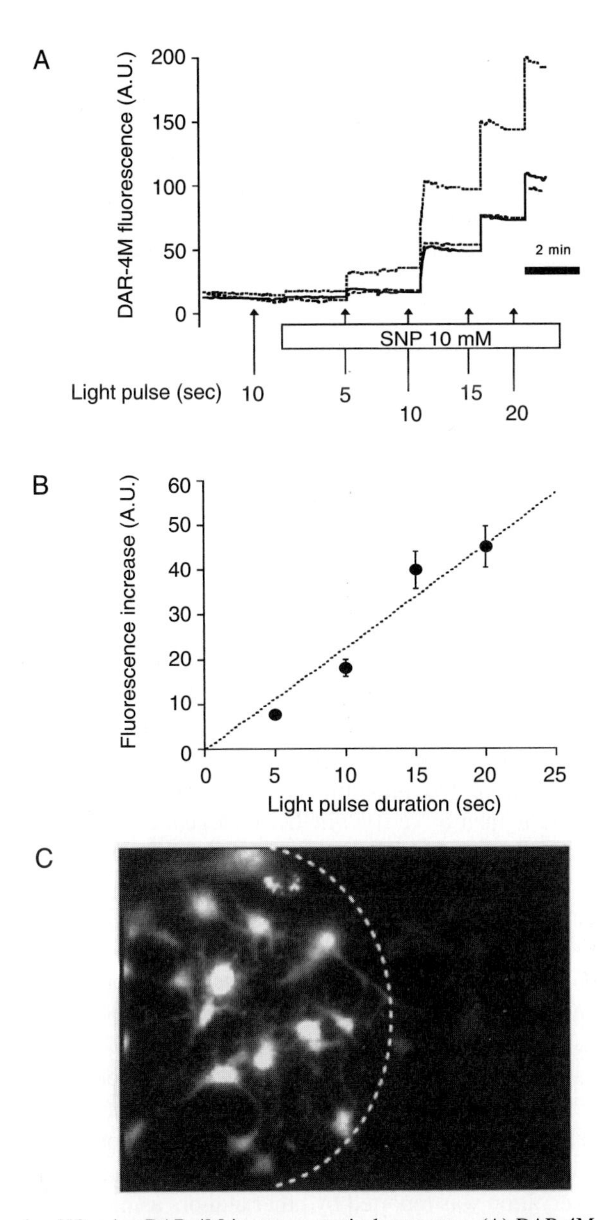

FIG. 7. Imaging NO using DAR-4M in mouse cortical astrocytes. (A) DAR-4M fluorescence in single astrocytes is plotted against time. NO was delivered by means of superfused sodium nitroprusside (SNP, 10 m*M*), and repetitive pulses of higher intensity light (~500 μW) at 540 nm were applied as indicated in the graph. (B) Relationship between light pulse duration (sec) and incremental DAR-4M intensity increase. Data are means ± SEM of 6 cells in the field of view. (C) Image of the astrocytes loaded with DAR-4M at the end of the experiments shown in (A). The field of view was shifted to the left, revealing cells that had not been illuminated during the experiment but had been nevertheless exposed to the same SNP perfusion and were almost devoid of detectable fluorescence.

when brief pulses of higher intensity light were applied, DAR-4M fluorescence increased in a stepwise manner. The fluorescence increase, which did not occur in the absence of the NO donor, was directly proportional to the duration of the light pulse applied (Fig. 7B) and indicated that formation of the fluorescent benzotriazole compound is strongly dependent on incident light. The same photoactivation was observed with the other NO dye, DAF-2, as well as with other NO donors (e.g., *S*-nitrosocysteine)[22] and was found in all cell types tested (i.e., HEK-293 cells, RIN m5f cells, EMT-6 cells, and primary cultured mouse astrocytes), which indicates that the phenomenon is not cell type or NO donor dependent. The effect of light is seen most strikingly in Fig. 7C, where the field of view has been moved sideways at the end of the experiment to reveal the neighboring cells that had not been illuminated during the experiment and remained extremely weakly fluorescent, despite having experienced the same NO application.

This unexpected behavior of DAF-2 and DAR-4M has to be considered carefully when designing experiments in the sense that the very light utilized for image formation catalyzes the formation of the fluorescent product. We observed that in low light level conditions (incident light power of 5–10 μW at the back focal plane of the objective), both DAF-2 (at 490 nm) and DAR-4M (at 540 nm) failed to reveal a significant fluorescence increase in the presence of NO and that light levels of $\sim$0.5–1.5 mW used for the light pulses activated the dye effectively. These relatively high intensities are typically achieved in confocal microscopy[31] and are most probably associated with phototoxic side effects on cells.

The next issue is then to verify whether endogenous NO production can be detected adequately using these dyes. Figure 8 shows an experiment performed on insulinoma RIN 5mf cells, loaded with DAR-4M, that had been stimulated overnight using interleukin 1β (10 U/ml) plus interferon-γ (150 U/ml). This treatment has been shown[32] to maximally activate the iNOS in these cells. Observation of these cells for 10–15 min under low light level conditions did not reveal any fluorescence increase, despite a high rate of NO production (9.21 $\pm$ 0.31 μM nitrites as measured with the Griess reaction[24]).

However, application of light pulses under the same conditions as in Fig. 7 showed an incremental increase in fluorescence that did not occur in nonstimulated cells (data not shown). This observation indicates that the photoactivation of DAR-4M not only occurs with NO donors, but also with NO produced endogenously by iNOS. This observation was reported by other authors using NO-producing COS cells and DAF-2 DA.[33] We observed that the ability of NO-producing cells to increase the DAR-4M signal decreased with repeated high-intensity light pulses. The disappearance of response to light pulses was not due to the depletion of an

[31] R. Nitschke and K. R. Spring, *J. Microsc. Soc. Am.* **1,** 1 (1995).

[32] M. R. Heitmeier, A. L. Scarim, and J. A. Corbett, *J. Biol. Chem.* **272,** 13697 (1997).

[33] M. O. Lopez-Figueroa, C. Caamano, R. Marin, B. Guerra, R. Alonso, M. I. Morano, H. Akil, and S. J. Watson, *Biochim. Biophys. Acta* **1540,** 253 (2001).

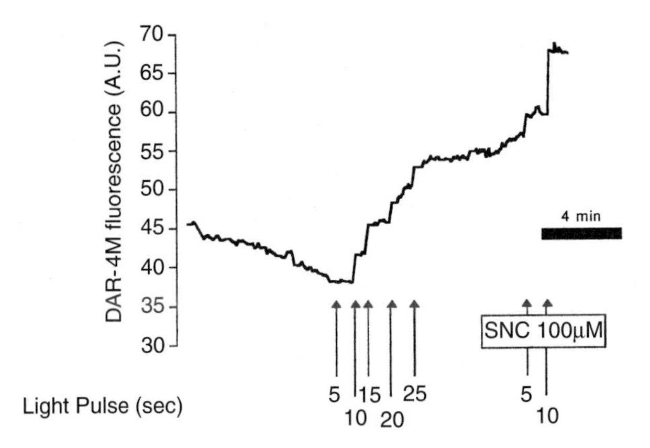

FIG. 8. Imaging endogenously produced NO. Stimulated insulinoma RIN m5f cells were loaded using DAR-4M/AM, and a baseline fluorescent signal was acquired under low-light level imaging conditions in Ca^{2+}-containing Ringer solution complemented with 5 m*M* glucose and 1 m*M* arginine. Instead of increasing, the signal gradually decreased may be because of transporter-mediated extrusion of the indicator. When pulses of higher intensity light (see Fig. 7) were applied, stepwise signal increases were produced, revealing endogenously produced NO. At the end of the experiment, the NO donor SNC (100 μM) was perfused and yielded comparable photoactivation by light pulses.

available fluorescent precursor because NO delivery from SNC at the end of the experiment produced a clear fluorescence response, but probably, rather, revealed the deleterious effect of light on cellular enzymes, a well-known problem with live cell fluorescence imaging.

Cuvette measurements (see earlier discussion) indicated that divalent cations, particularly Ca^{2+}, might significantly enhance the response of both DAF-2 and DAR-4M to NO. Evidence was reported[22] that this was the case at Ca^{2+} concentrations relevant for cytosolic environment and that it also applied to intracellularly loaded DAF-2.

This issue should be tested in experimental protocols in which intracellular Ca^{2+} is increased in a controlled way (e.g., using Ca^{2+} ionophores). However, the difficulty is to sort out an effect of increased intracellular Ca^{2+} on Ca^{2+}-stimulated NOS enzymes from an effect on the indicator itself, which is sometimes not possible. If the Ca^{2+} sensitivity of DAF-2 and DAR-4M is confirmed for endogenously produced NO, it would mean, e.g., that an intercellular Ca^{2+} wave increasing the sensitivity of the intracellular indicator to NO could be interpreted incorrectly as a NO wave.

Conclusions

The development of diamino fluorescent indicators capable of sensing the ambient NO concentration was considered a major advance because it appeared to be the long-awaited method for the detection of NO with cellular resolution. Even though the indicators described in this article effectively detect the presence of NO, they remain difficult experimental tools to use because they are influenced strongly by solution composition and by the very light used to elicit their fluorescence. Moreover, the high incident light flux necessary for dye activation will undoubtedly have side effects on living cells and may compromise the success of experimental protocols.

Thus, the use of these dyes should be evaluated critically and applied in situations in which experimental parameters are under very careful control.

Acknowledgments

We sincerely thank Dr. L. Juillerat for fruitful discussions and for providing us access to the fluorimeter. We gratefully acknowledge the excellent technical assistance of O. Randin and C. Pélofi. This work was supported by Grants 31-51061.97 and 3130-051920.97 (to M.-C. Broillet) and 31-55786.98 (to J.-Y. Chatton) from the Swiss National Foundation for Scientific Research.

[13] Specific Analysis of Nitrate and Nitrite by Gas Chromatography/Mass Spectrometry

By GEORGE A. SMYTHE and GABRIJELA MATANOVIC

Introduction

The fact that nitrogen monoxide (nitric oxide, NO) plays major roles in a wide range of pathological and physiological states has necessitated the development of reliable methods of measurement in order to investigate fully its role and to assess aspects of its production and activity. The measurement of nitric oxide per se is difficult due to its short half-life, low concentration in biological samples, and its reactive nature.[1] As a consequence, indirect methods of assessing nitric oxide production have generally been employed; these usually rely on the measurement of the circulating or excreted levels of its end products, nitrate and nitrite ions (NO_3^- and NO_2^-). Provided dietary and other exogenous sources are taken into account, the measurement of nitrate (or nitrite after reduction of nitrate) has proved a useful index of nitric oxide production in many clinical situations.[2] The high degree of specificity and accuracy endowed by isotope dilution-based gas chromatography/mass spectrometry (GC/MS) enables the employment of definitive

[1] J. R. Lancaster, Jr., *Nitric Oxide* **1,** 18 (1997).

[2] G. Ellis, I. Adatia, M. Yazdanpanah, and S. K. Makela, *Clin. Biochem.* **31,** 195 (1998).

0076-6879/02 $35.00

analyses for both nitrate and nitrite. This article describes the GC/MS analytical methods employed in our laboratory for the assay of nitrate and nitrite in physiological media. We also outline potential sources of analytical interference that can be avoided.

Methods

Stable Isotopes and Chemical Reagents

Potassium [^{15}N]nitrate (98+ %, Cambridge Isotope Laboratories, Inc., Andover, MA); sodium [^{15}N]nitrite (99.4%) purchased from Isotec., Inc. (Miamisburg, OH); trifluoroacetic anhydride (TFAA), and α-bromo-2,3,4,5,6-pentafluorotoluene (pentafluorobenzylbromide) obtained from Sigma-Aldrich; sodium nitrate and sodium nitrite purchased from Fluka, and reagent grade organic solvents and other chemicals used are obtained from commercial suppliers and are used without treatment. Distilled water (MilliQ, 18m $\Omega.cm^{-1}$ at 25°) is used in all work.

General Principles and Procedures

Analysis of Nitrate as Nitrotoluene

The nitration of aromatic compounds by electrophilic substitution is often utilized in analyses of nitrate concentrations in physiological samples by gas chromatographic methods. Problems associated with the use of concentrated sulfuric acid, which is normally used to catalyze this reaction,[3] led us to develop an alternative method using TFAA as catalyst.[4] In this technique, toluene acts as both reaction solvent and electrophile to react with nitrate in the presence of TFAA. In the presence of excess toluene, the nitrate ion is quantitatively converted to nitrotoluene, producing the three nitrotoluene isomers (ratio *o*- : *m*- : *p*- approximately 57 : 3 : 40).[4] Samples are prepared for analysis as follows.

An aliquot of an aqueous solution of potassium [^{15}N]nitrate internal standard (100 μM, 50 μl) is mixed in a screw-top glass tissue culture vial (100 × 13 mm) with plasma, urine, or other physiological sample (generally, 50 μl) or, for the standard curve, calibrated dilutions of sodium [^{14}N]nitrate. The mixture is then dried under a stream of nitrogen [or a Speed-Vac rotary evaporator (Savant Instruments)]. Trifluoroacetic anhydride (200 μl) and toluene (1 ml) are added to the residue, and the tubes are capped and heated at 70° for 60 min. After cooling to room temperature, the toluene solution is washed sequentially with water (1 ml), aqueous sodium bicarbonate (1%, 1 ml), and water (1 ml). The toluene solution is dried

[3] L. C. Green, D. A. Wagner, J. Glogowski, P. L. Skipper, J. S. Wishnok, and S. R. Tannenbaum, *Anal. Biochem.* **126,** 131 (1982).

[4] G. A. Smythe, G. Matanovic, D. Yi, and M. W. Duncan, *Nitric Oxide* **3,** 67 (1999).

over anhydrous sodium sulfate (500 mg) and is transferred to a GC autosampler vial for analysis.

The nitrotoluene isomers, particularly *o*- and *p*-, are readily assayed with the following GC/MS methods utilizing different ionization techniques.

Electron Ionization (EI) GC/MS

In this method, EI GC/MS is performed on a Hewlett-Packard 5890 gas chromatograph interfaced to either a Hewlett-Packard 5989B mass spectrometer or a Hewlett-Packard 5971A MSD. Chromatographic separations are performed in splitless mode using an HP-5MS capillary column (30 m $\times$ 0.25 mm i.d. with a 0.25-μm stationary phase film thickness, Agilent Technologies) with the following temperature program: 70° constant for 2 min and then 20° per minute to 150°. The GC/MS interface heater, the ion source, quadrupole, and injection port temperatures are maintained at 280°, 250°, 100°, and 240°, respectively. In this and all of the GC/MS methods described, the GC inlet port was fitted with a 4-mm deactivated liner packed with 140 mg of 1% SP-2100 on 100/120 mesh Supelcoport (Sigma-Aldrich). This technique has the effect of extending capillary column life and avoiding some nonspecific interferences without significantly affecting chromatographic peaks. These liners are generally changed on a weekly basis.

Nitrotoluene analyses are performed in the selected ion monitoring (SIM) mode and ions are generated by electron ionization. Mass spectra of the three isomers of nitrotoluene differ in one important respect: the *m*- and *p*-isomers yield strong signals (base peaks) for the molecular ion at *m/z* 137; the *o*-isomer does not. This is due to the "ortho effect," whereby the *o*-nitrotoluene favorably dissociates under electron ionization, losing OH to form, as the base peak, the ion *m/z* 120.[5] Figure 1A shows the selected ion chromatograms of the three nitrotoluene isomers in an equimolar mixture, and Fig. 1B shows the selected ion chromatograms obtained from ^{15}N-labeled nitrate used as an internal standard. Analysis of *o*-nitrotoluene is based on the *m/z* 120 ion, whereas the molecular ion (*m/z* 137) is used for *p*-nitrotoluene. The low-yielding *m*-nitrotoluene isomer (see Fig. 1B) is not used for quantification. For analysis, the (*m/z*) peak areas are determined and the area ratios of the ion pairs are calculated. Calibration curves deriving from the *o*- and *p*-isomers are virtually identical.[4]

Electron-capture Negative Ionization (ECNI) GC/MS

ECNI GC/MS is performed on an Agilent Technologies 6890 gas chromatograph interfaced to an Agilent 5973 mass selective detector (Agilent Technologies). Chromatographic separations are performed in splitless mode using HP-5MS capillary columns (30 m $\times$ 0.25 mm i.d. with a 1.0-μm stationary phase film thickness,

[5] D. Tsikas, R. H. Boger, B. S. Bode, F. M. Gutzki, and J. C. Frolich, *J. Chromatogr. B Biomed. Appl.* **661,** 185 (1999).

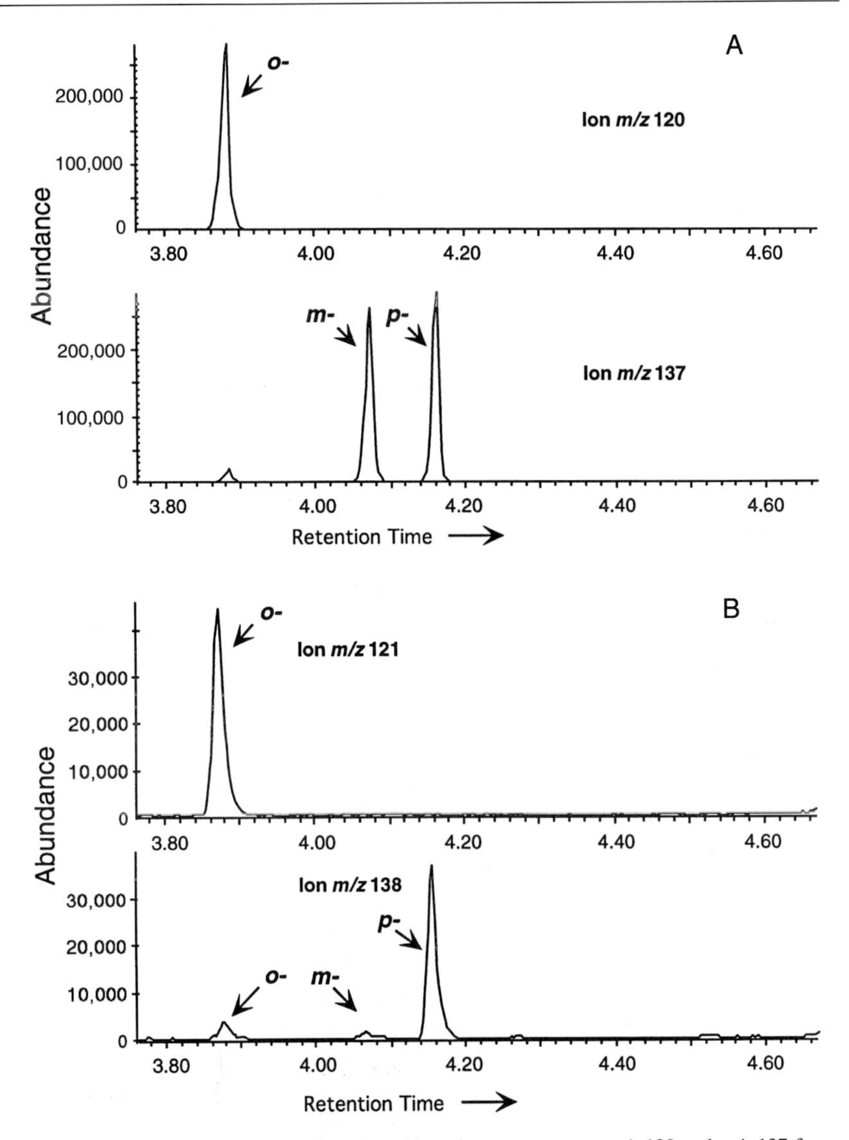

FIG. 1. (A) Electron ionization (EI)-selected ion chromatograms at *m/z* 120 and *m/z* 137 for an equimolar mixture of *ortho-*, *meta-* and *para-* (*o-*, *m-*, and *p-*) isomers of nitrotoluene. (B) EI-selected ion chromatograms at *m/z* 121 and *m/z* 138 for *o-*, *m-*, and *p-*isomeric nitration products derived from ^{15}N-labeled nitrate. Reproduced from G. A. Smythe, G. Matanovic, D. Yi, and M. W. Duncan, *Nitric Oxide* **3,** 67 (1999) with permission.

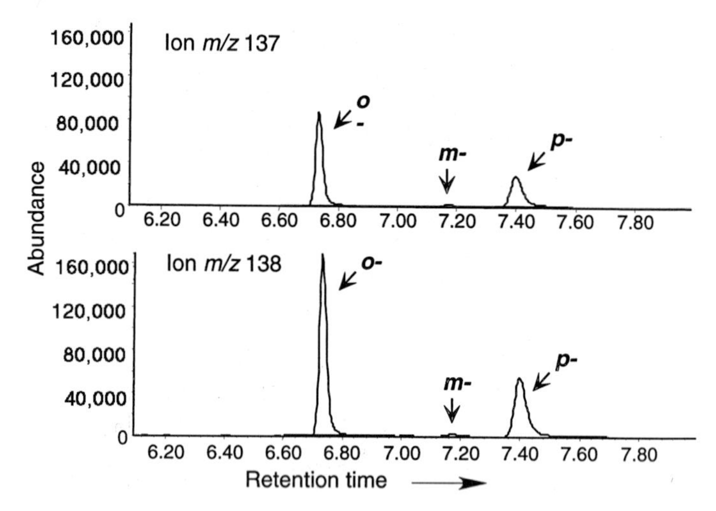

FIG. 2. Electron-capture negative ionization-selected ion chromatograms for *ortho*-, *meta*-, and *para*- (*o*-, *m*-, and *p*-) isomers of [^{14}N]nitrotoluene (m/z 137) and the internal standard stable isotope [^{15}N]nitrotoluene (m/z 138).

Agilent Technologies). GC oven temperature programs are 50° constant for 1 min and then 30° per minute to 180°. The GC/MS interface heater, the ion source, quadrupole, and injection port temperatures are maintained at 280°, 150°, 106°, and 240°, respectively. MS analyses are performed with an ECNI source, using methane as the reagent gas (ultrapure grade, Matheson Gas Products, Inc.). Unlike the situation with EI, the nitrotoluene molecular ion exhibits no significant fragmentation under ECNI conditions and essentially the only ion observed is the molecular anion at m/z 137 (and m/z 138 for the ^{15}N-nitrotoluene internal standard). Typical in chromatograms are shown in Fig. 2.

For both EI and ECNI assays, calibration curves are generated after correction for (a) the contribution to the ^{15}N-nitrotoluene peak areas due to the natural abundance of ^{13}C present in endogenous or unlabeled nitrotoluene and (b) the contribution to the calculated ^{14}N peak areas arising due to the small amount of unlabeled nitrate present (<2%) in the original [^{15}N]nitrate used as internal standard. These corrections are made for each assay by the analysis of separate samples containing unlabeled and ^{15}N-labeled nitrate alone. Regardless of the method of ionization or isomer chosen for analysis, calibration curves for nitrotoluene consistently show an R^2 of 0.996 or better over concentration ranges from 0 to 970 pmol (injected on column). The limits of quantification for nitrate analysis using the aforementioned assays are less than 100 fmol on column (s/n; 40 : 1); note that this level of sensitivity is obtained when injecting only 1 μl of 1000 μl of the total sample volume.

Specific Analysis of Nitrite as α-Nitropentafluorotoluene

The analysis of nitrite by ECNI GC/MS relies on its facile conversion to α-nitropentafluorotoluene by reaction with pentafluorobenzylbromide, and the method used is a modification of that described by Tsikas *et al.* We have found the assay suitable for analysis of sample types including plasma, urine, tissue culture medium, or tissue extracts. To glass tissue culture vials (100 × 13 mm) are added sample (200 μl), an aliquot of sodium [^{15}N]nitrite internal standard (20 μ*M*, 50 μl), aqueous potassium carbonate [10% (w/v), 100 μl], and an aliquot of pentafluorobenzyl bromide solution in acetone [2.5% (w/v), 200 μl]. The vials are vortexed briefly, sealed with Teflon-lined caps, and heated at 50° for 60 min. After cooling to room temperature, water (1 ml) and toluene (1 ml) are added, the samples are vortexed and centrifuged, and the toluene layer is transferred to GC autosampler vials for GC/MS analysis. At the same time, calibration curves are prepared using the same procedure but substituting prepared aliquots of sodium nitrite standard solutions. Under the ECNI conditions utilized here, fragmentation of α-nitropentafluorotoluene molecular anion occurs and the predominant ion produced is the nitrite anion (NO_2^-) at *m/z* 46. Because the NO_2^- ion is the monitored species, no correction for the natural abundance of the ^{13}C isotope in the α-nitropentafluorotoluene molecule is necessary in the analysis of nitrite. Correction still needs to be made for unlabeled nitrite present (<2%) in the original [^{15}N]nitrite used as the internal standard (Fig. 3).

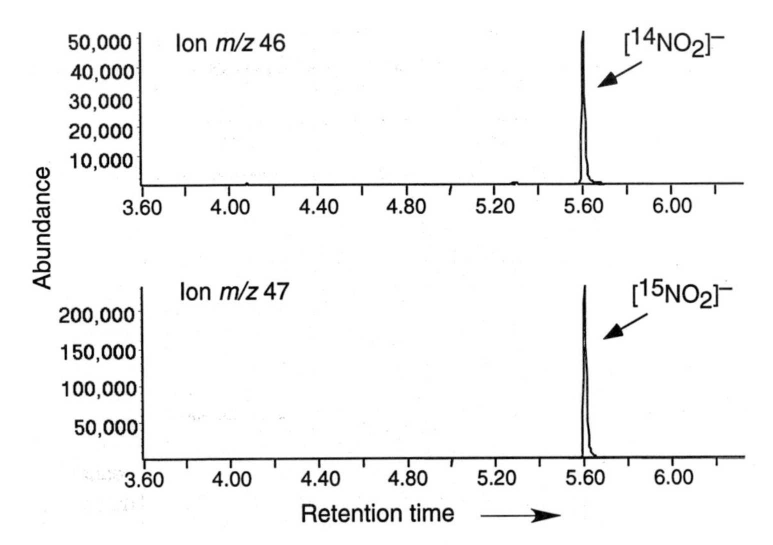

FIG. 3. Selected ion chromatograms for [^{14}N]- and [^{15}N]nitrite anions derived from the electron-capture negative ionization of α-nitropentafluorotoluene. The ion chromatograms at *m/z* 46 and *m/z* 47 correspond to endogenous and stable isotope-labeled nitrite, respectively.

The analysis of α-nitropentafluorotoluene by ECNI GC/MS readily enables quantification of the low levels of nitrite (<10 fmol on column) in biological media. The assay offers a method capable of accurately measuring nitrite in the presence of a high excess of nitrate, and this high specificity has important implications (see later).

Avoiding Artifact and Specificity Problems in the Analysis of Nitrate and Nitrite

Artifactual Nitrate Derived from the Use of Nitric Oxide Synthase (NOS) Inhibitors

We have shown that nitrate can be nearly quantitatively generated from samples containing the frequently used NOS inhibitors N^{ω}-nitro-L-arginine (L-NNA) and N-nitro-L-arginine methyl ester (L-NAME) when sulfuric acid is used to catalyze the nitration reaction.[7] This complication is avoided through the use of TFAA as a catalyst. Table I shows the results of the comparison between sulfuric acid and TFAA-catalyzed nitrate analysis in plasma or urine samples containing NOS inhibitors. These data indicate that nitrate is generated artifactually, and nearly quantitatively, from samples containing L-NNA and L-NAME when sulfuric acid is used to catalyze the reaction. When TFAA was used, there was little or no artifactual generation of nitrate from L-NNA, but a small degree of degradation of L-NAME occurred (less than 5%). In the case of the NOS inhibitor canavanine, which does not contain a nitro group, there was no artifactual generation of nitrate. These *in vitro* data have been confirmed in an *in vivo* study.[4] As shown in Fig. 4, nitrate levels in L-NNA-treated sheep plasma samples analyzed by either TFAA or sulfuric acid-catalyzed methods gave significantly different results. In this example, nitrate levels derived using TFAA catalysis were not significantly different from those of basal samples, but were up to 50 times higher when estimated using sulfuric acid catalysis.

Specific Analysis of Nitrite in the Presence of a Large Excess of Nitrate

The specific assessment of nitrite in the presence of high exogenous concentrations of nitrate can enable the underlying activity and production of nitric oxide to be revealed. An example of this is an experiment in which a mouse macrophage cell line (RAW 264.7) was stimulated by interferon (IFN)-γ, in the presence and absence of the nitric oxide synthase (NOS) inhibitor L-NAME.[6] As shown in Fig. 5, very high, but not significantly different, levels of nitrate were found in each of the experimental groups. Subsequent analysis of the culture medium used showed it to contain millimolar concentrations of nitrate. In contrast, specific analysis of nitrite

[6] G. Matanovic, Ph.D. Thesis, University of New South Wales, UNSW, Sydney, 2000.

TABLE I

COMPARATIVE ANALYSES OF FOUR DIFFERENT SAMPLES (A–D) CONTAINING ADDED N^{ω}-NITRO-L-ARGININE (L-NNA), N-NITRO-L-ARGININE METHYL ESTER (L-NAME), CANAVANINE, OR SODIUM NITRATE USING SULFURIC ACID OR TFAA AS NITRATION CATALYST[a]

Catalyst	Matrix	Compound added	Amount added (μmol/liter)	Nitrate found (μmol/liter) (mean ± SE)	Molar % of added compound detected as nitrate
Sulfuric acid	Plasma (a)	—	—	13.95 ± 0.5	—
	Plasma (a)	L-NNA	259.36	217.0 ± 15.3	78.3
	Plasma (a)	Nitrate	24.47	38.1 ± 0.3	100.0
	Plasma (b)	—	—	8.4 ± 0.4	—
	Plasma (b)	L-NAME	504.8	526 ± 3.9	100
	Plasma (b)	Canavanine	252.4	8.3 ± 0.35	0.0
	Urine (a)	—	—	159.16 ± 1.7	—
	Urine (a)	L-NNA	518.72	601.54 ± 16.9	85.3
	Urine (a)	Nitrate	150.44	318.59 ± 1.2	100.0
	Urine (b)	—	—	163.9 ± 4.1	—
	Urine (b)	L-NAME	1090	1040.9 ± 4.3	80.5
	Urine (b)	Canavanine	1009.5	168.4 ± 7.3	0.0
TFAA	Plasma (c)	—	—	20.04 ± 0.52	—
	Plasma (c)	L-NNA	259.36	22.79 ± 1.5	0.0
	Plasma (c)	Nitrate	24.47	44.39 ± 0.5	100.0
	Plasma (d)	—	—	10.4 ± 0.2	—
	Plasma (d)	L-NAME	504.8	32.1 ± 0.9	4.3
	Plasma (d)	Canavanine	252.4	12.0 ± 0.3	<1
	Urine (c)	—	—	147.22 ± 4.1	—
	Urine (c)	L-NNA	518.7	157.65 ± 3.9	2.0
	Urine (c)	Nitrate	150.44	301.61 ± 2.5	100.0
	Urine (d)	—	—	229.1 ± 6.3	—
	Urine (d)	L-NAME	1090	268.0 ± 8.4	3.6
	Urine (d)	Canavanine	1009.5	239.2 ± 4.1	0.0

[a] Reproduced from G. A. Smythe, G. Matanovic, D. Yi, and M. W. Duncan, *Nitric Oxide* **3,** 67 (1999) with permission.

in the same samples showed IFN-γ stimulation to increase nitrite concentration significantly, a response blocked when NOS was inhibited (Fig. 5).

Laboratory Glassware Contamination

Laboratory glassware represents a serious source of sample contamination with both nitrate and nitrite.[7] We have found that glass tissue culture vials received direct from the supplier can contain up to 6 nmol of nitrate, which can vary considerably

[7] T. Ishibashi, M. Himeno, N. Imaizumi, K. Maejima, S. Nakano, K. Uchida, J. Yoshida, and M. Nishio, *Nitric Oxide* **4,** 516 (2000).

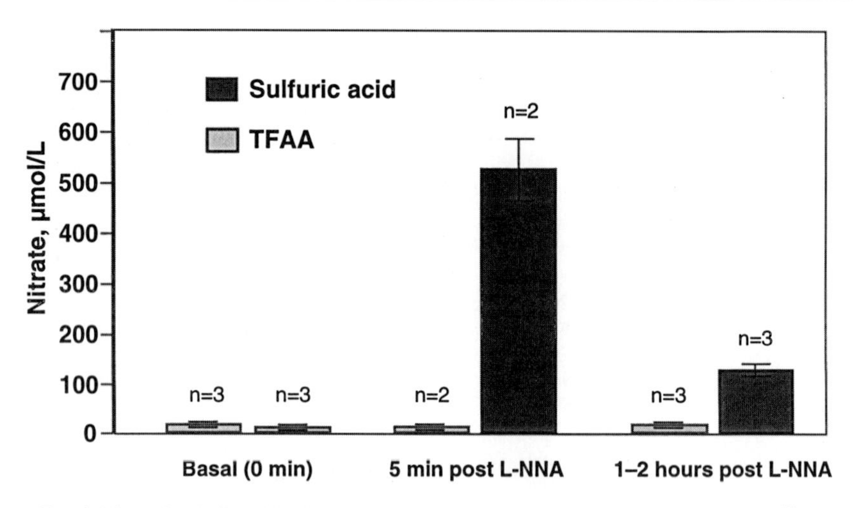

FIG. 4. Nitrate levels found in sheep plasma following treatment with a bolus dose of N^{W}-nitro-L-arginine (L-NNA, 10 mg/kg body weight using either TFAA or sulfuric acid-catalyzed nitration of toluene to assay nitrate. Means ± SE are shown; n = number of sheep in each sample group. Reproduced from G. A. Smythe, G. Matanovic, D. Yi, and M. W. Duncan, *Nitric Oxide* **3,** 67 (1999) with permission.

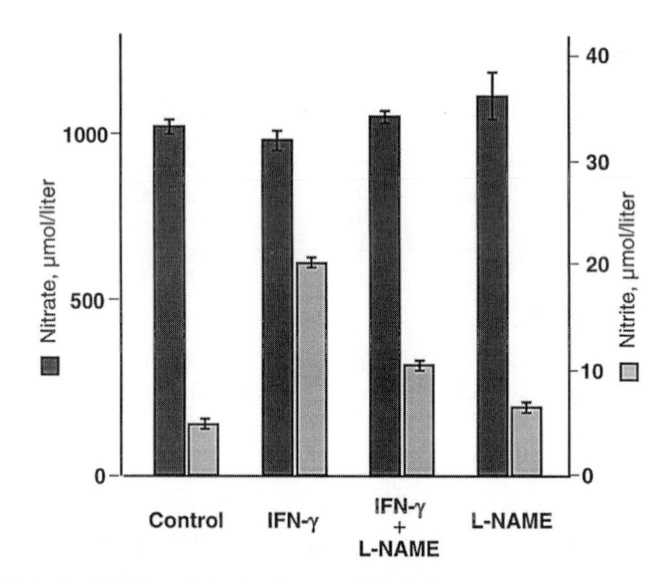

FIG. 5. Selective GC/MS analysis of endogenous nitrite in the presence of excess exogenous nitrate. High levels of native nitrate present in the tissue culture medium (see text) preclude any conclusions about NOS activity in this experiment. The specific analysis of nitrite shows that IFN-γ causes a significant increase in NOS activity partly blocked by the NOS inhibitor L-NAME. Means ± SEM are shown.

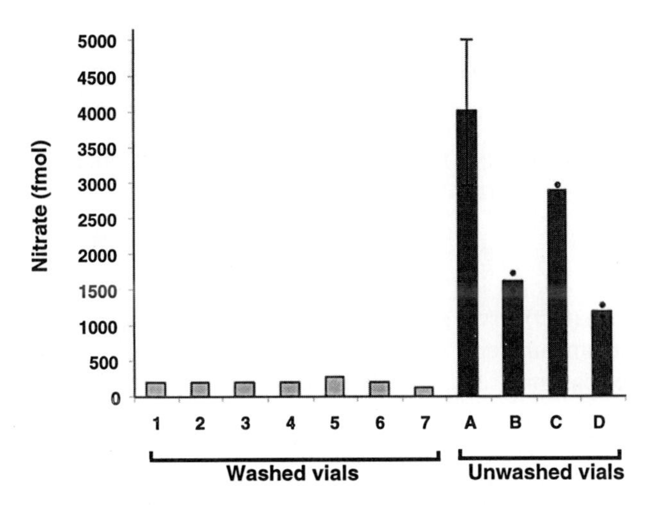

FIG. 6. Contaminant nitrate present in new glass tissue culture vials is largely removed by washing with deionized water. Amounts of nitrate present after washing individual vials (1–7) taken from the same batch of vials (A) that were untreated and for which contaminant nitrate levels were highest (A, mean ± SEM, $n = 10$). Histograms B–D ($n = 2$ per group) show levels of nitrate in different batches of the same vials.

from batch to batch. This contamination, however, can largely be removed by washing the vials three times with cold deionized water (Fig. 6).

Conclusion

The high specificity and accuracy endowed through the use of a stable isotope ^{15}N-labeled internal standard enable GC/MS methods to be considered definitive for the analysis of nitrate or nitrite, and these methods are able to provide an effective estimation of endogenous NO production.

[14] Griess Method for Nitrite Measurement of Aqueous and Protein-Containing Samples

By PETRA KLEINBONGARD, TIENUSH RASSAF, ANDRÉ DEJAM, STEFAN KERBER, and MALTE KELM

Introduction

Nitric oxide (NO) serves a variety of biological functions, such as regulation of vascular tone, immunologic response, coagulation, and neurotransmission.[1] NO is synthesized enzymatically from the amino acid L-arginine by the NO synthase (NOS).[2] NOS is expressed in three isoforms (designated as types I–III of which type III is mainly expressed in endothelial cells).[3] Due to its ultrashort life span, determination of NO itself in biological samples such as plasma appears to be a complicated task. Nitrite and nitrate are the products of oxidative NO chemistry in biological fluids and tissue. The ratio of both N oxides formed depends critically on the surrounding redox conditions. Because of this, the concentration of either nitrite or nitrate or both (the sum of N oxides, NOx) has been used frequently as potential markers of eNOS activity. The high background concentration of nitrate and its relatively long half-life in comparison to nitrite raise the question as to the sensitivity of NOx for measuring eNOS activity in mammalians *in vivo*. It has been demonstrated that the measurement of nitrite reflects acute changes in eNOS activity in human forearm circulation.[4] This stresses the necessity for a reliable method for analyzing nitrite in the physiological concentration range.

Several methods have been developed to fulfill this task, such as the hemoglobin assay,[5] chemiluminescence (CL),[6,7] capillary electrophoresis,[8] and high pressure liquid chromatography (HPLC).[9] These methods do not appear to be applicable

[1] S. Moncada, E. A. Higgs, H. F. Hodson, R. G. Knowles, P. Lopez-Jaramillo, T. McCall, R. M. J. Palmer, M. W. Radomski, D. D. Rees, and R. Schulz, *J. Cardiovasc. Pharmacol.* **17,** 1 (1991).

[2] I. J. Ignarro, *Sci. Am.* 535 (1990).

[3] U. Förstermann, E. I. P. J. S. Closs, M. Nakane, P. Schwarz, I. Gath, and H. Kleiner, *Hypertension.* **23,** 1121 (1994).

[4] T. Lauer, M. Preik, T. Rassaf, B. E. Strauer, A. Deußen, M. Feelisch, and M. Kelm, *Proc. Natl. Acad. Sci. U.S.A.* **98,** 12814 (2001).

[5] M. Feelisch, D. Kubitzek, and J. Werringloer, *in* "Methods in Nitric Oxide Research" (M. Feelisch and J. S. Stamler, eds.), p. 455. Wiley, Chichester, 1996.

[6] V. Hampel, C. L. Walters, and S. L. Archer, *in* "Methods in Nitric Oxide Research" (M. Feelisch and J. S. Stamler, eds.), p. 310. Wiley, Chichester, 1996.

[7] M. T. Gladwin, J. H. Shelhamer, A. N. Schechter, M. E. Pease-Fye, M. A. Waclawiw, J. A. Panza, F. P. Ognibene, and R. O. Cannon III, *Proc. Natl. Acad. Sci. U.S.A.* **97,** 11482 (2000).

[8] A. M. Leone and M. Kelm, *in* "Methods in Nitric Oxide Research" (M. Feelisch and J. S. Stamler, eds.), p. 499. Wiley, Chichester, 1996.

[9] Y. Michigami, Y. Yamamoto, and K. Ueda, *Analyst* **114,** 1201 (1998).

0076-6879/02 $35.00

to routine analysis in clinical settings. The first attempt in this direction was made by combining HPLC with an electrochemical detection device.[10] Another way to measure nitrite is accomplished via a photometric assay comprising the 2,3-diaminonaphthalene DAN assay[11] and the Griess reaction.[12] The former detects a change of color that occurs when NO reacts directly with an indicator compound. The latter is the most commonly used spectrophotometric method for the determination of nitrite based on the Griess reaction. The commercially available colorimetric tests do not possess the sensitivity for a precise measurement in the physiological concentration range.[13] This article describes an analytical method using the technique of flow injection analysis (FIA) in combination with the Griess reagent that has a sensitivity down to nanomolar concentration, with a higher reproducibility and a lower susceptibility to interference of various matrices in comparison to the commercially available colorimetric kits.[14] FIA has been around since the early 1970s and is characterized especially by its simple design and great versatility.[15,16] Initially used only in pharmaceutical applications and food chemistry, this method has become increasingly prevalent in clinical settings.[17–19]

In this article, meticulous effort has been made to characterize potential interferences to proteins and pitfalls due to varying redox conditions and contamination during probe sampling. In order to demonstrate the accuracy and efficacy of our method to sensitively quantify nitrite even in complex protein-containing biological matrices, data on the quantification of nitrite in plasma from various mammalian species relevant to experimental and clinical research (such as in humans, dogs, pigs, rabbits and mice) are presented.

Analytical Setup

A schematic diagram of the FIA setup to determine nitrite via the Griess reaction is shown in Fig. 1. The system consists of a degasser (Sunchrom, Germany), a HPLC pump (Sunchrom), an automatic injector (Triathlon, Netherlands), and an UV/Vis detector (Sykam, Germany). All capillaries are 0.5 mm in diameter and

[10] H. Preik-Steinhoff and M. Kelm, *J. Chromatogr. B* **685,** 348 (1996).

[11] A. M. Miles, D. A. Wink, J. P. Cook, and M. B. Grisham, *Methods Enzymol.* **268,** 105 (1996).

[12] X. Liu, M. J. S. Miller, M. S. Joshi, D. D. Thomas, and J. R. Lancaster, *Proc. Natl. Acad. Sci. U.S.A.* **95,** 2175 (1998).

[13] D. L. Granger, N. M. Anstey, W. C. Miller, and J. B. Weinberg, *Methods Enzymol.* **301,** 49 (1999).

[14] K. Schulz, S. Kerber, and M. Kelm, *Nitric Oxide* **3,** 225 (1999).

[15] J. Ruzicka and E. H. Hansen, *Anal. Chim. Acta* **78,** 145 (1975).

[16] E. H. Hansen and J. Ruzicka, *Anal. Chim. Acta* **148,** 111 (1983).

[17] I. Walcerz, S. Glab, and R. Konicki, *Anal. Chim. Acta* **369,** 129 (1998).

[18] M. Frojanowicz, P. W. Alexander, and D. B. Hibbert, *Anal. Chim. Acta* **366,** 23 (1998).

[19] P. F. Ruhn, J. D. Taylor, and D. S. Hage, *Anal. Chem.* **66,** 4265 (2001).

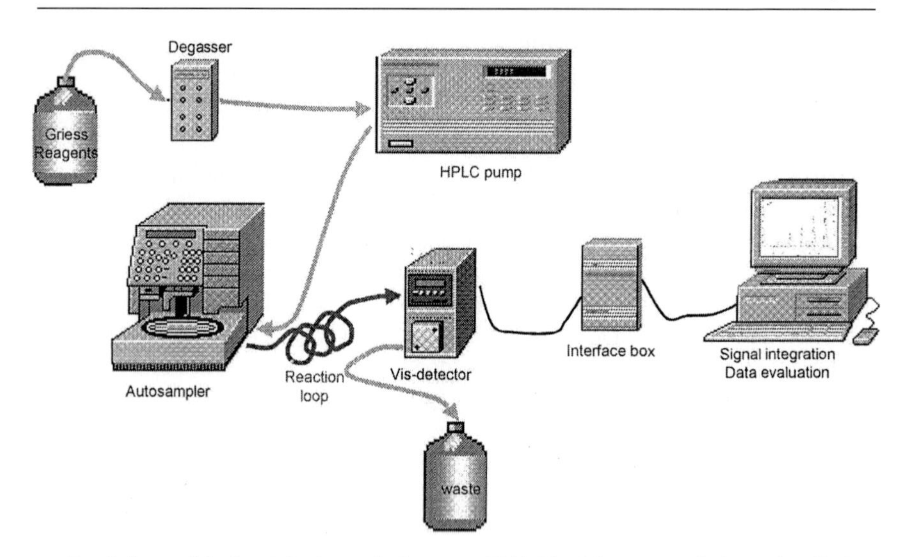

FIG. 1. Setup of the flow injection analysis system (FIA). The Griess reagent is degased and is then pumped continuously through the system via a HPLC pump with a constant rate of 1 ml/min. Twenty microliters of sample volumes is injected into the system through an injection valve by means of an automatic sampler. The reaction between nitrite and the Griess reagent takes place in the loop between the autosampler and the detector with a standardized length and diameter. A visible detector measures the absorption units at 545 nm and sends data via an analog–digital converter to the integrating HPLC software. Peak height serves as the reference for the evaluation of data.

are made of polyether ether ketone (PEEK). The length of capillary tubes is kept as short as possible; only the reaction loop is 90 cm from autosampler to detector. Data are transmitted by an interface box (Knauer, Germany) and are integrated by HPLC evaluation software (ChromGate 2.55, Knauer, Germany). The carrier solution consists of the Griess reagent: 40 g/liter sulfanilamide (Sigma) dissolved in a 1% HCl solution (stable at room temperature in a lightproof container) and 2 g/liter *N*-(1-naphthyl)ethylenediamine (Sigma) dissolved in water (LiChrosolv, Merck, Germany) (stable at 4°). The two components are mixed and used as carrier solution. The solution is degassed and then pumped through the system with a constant flow rate of 1 ml/min. Samples with a volume of 20 μl are injected into the measuring system and are transported to the detector with a stream of dye. A color reaction takes place on the way through the reaction loop. The sample is then analyzed in the ultraviolet detector in a measuring cell (6 mm, 9 μl, Kel-F, Linear-Instruments) at the characteristic absorption spectrum for this diazotization reaction (545 nm). The detected absorption units are transformed by the interface box into processable signals for HPLC software. The evaluation of the peak is done in reference to the peak height (Fig. 2).

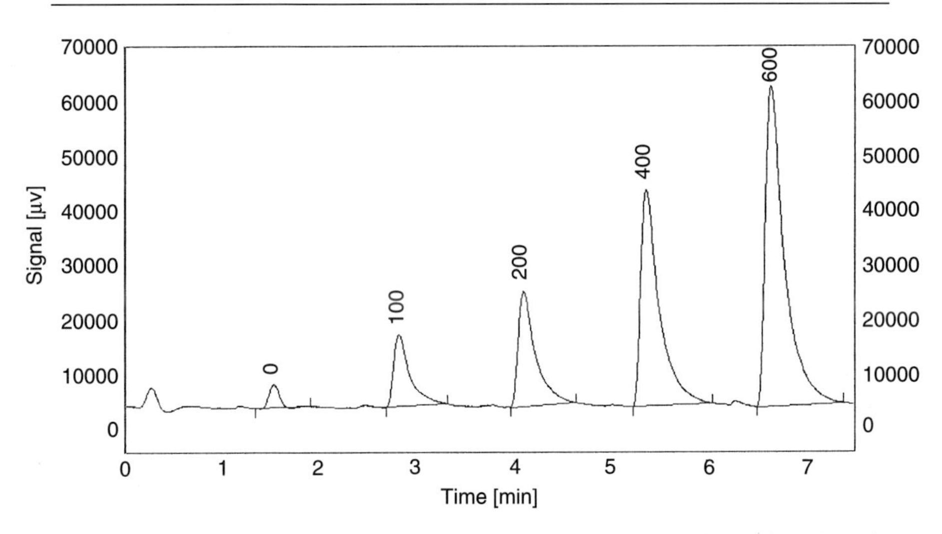

FIG. 2. Original recording of an analytical run of a blood sample. Five individual injections of a sample to which nitrite was added in indicated (over the peaks) concentrations are introduced into the system at intervals of 1.2 min via an automatic injection device. The first peak in this tracing is caused by cleaning the injection system. The intensity of the signals is measured photometrically at 540 nm.

Standard Addition Method

In order to analyze the nitrite concentration in the plasma sample, the standard addition procedure is used. Ten microliters of increasing concentrations of nitrite (0, 2, 4, 8, and 12 μmol/liter) is added to five cups, each containing 190 μl of the sample ultrafiltrate. This results in the addition of 0, 100, 200, 400, and 600 nmol/liter nitrite to the basal nitrite concentration of the sample.

Figure 3 shows the calculation of regression lines. A regression line is drawn through the five data points obtained. The intersection between the straight lines and the X axis at point $y = 0$ represents the nitrite concentration in the plain sample. The existing nitrite in the buffer used for dilution (20–50 nmol/liter on average) is subtracted from the plasma. This corrected concentration has to be multiplied by the dilution factor (5) in order to obtain the final concentration (sample preparation vide infra). All analyses have been performed in triplicate.

Determination of Aqueous Samples

Aqueous samples can be used directly for nitrite determination by the standard addition method. Nevertheless, cautious handling of probes must be exercised, considering the potential for contamination by the surrounding base load N oxide content (see later).

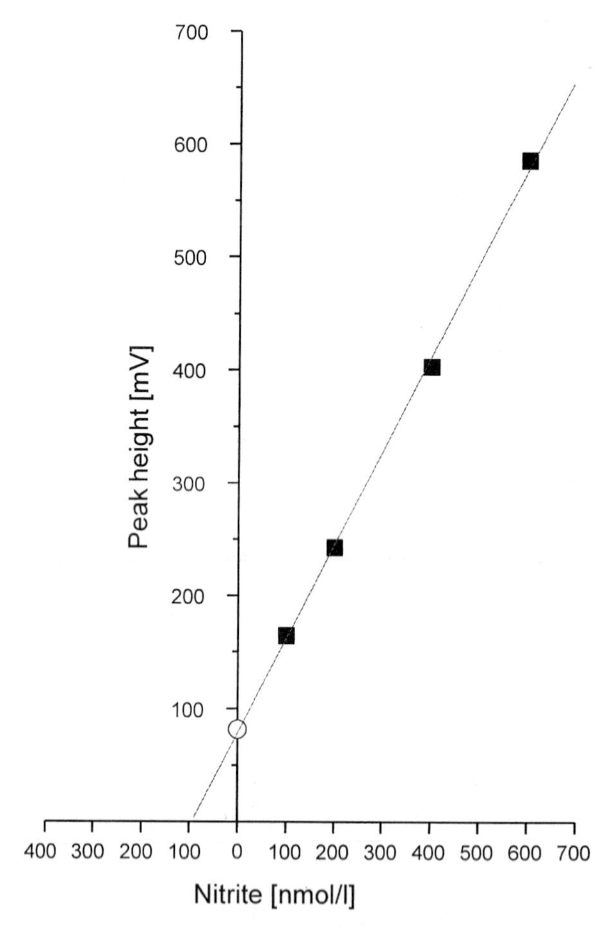

FIG. 3. Principle of the standard addition method for determining nitrite. A biological sample is divided into five aliquots. Increasing concentrations of nitrite are added to samples (■). Intersection of the straight lines and the *X* axis determines the nitrite concentration in the plain sample (○). The nitrite content in the sample is calculated considering the dilution factor.

Processing of Protein-Containing Sample (e.g., Human Plasma)

For the determination of nitrite, venous blood from a fasting patient should be drawn into a 5-ml syringe with a butterfly needle (internal diameter 0.8; 19 gauge). Needle diameter and volume of the syringe are optimized to avoid hemolysis. The first sample (vènous blood) should be rejected to reduce the contamination of subsequent samples.

Drawn blood must be diluted 1 : 5 in 0.9% NaCl (Braun, Germany) with 20 IU/ml heparin to prevent coagulation and to stabilize the nitrite concentration.

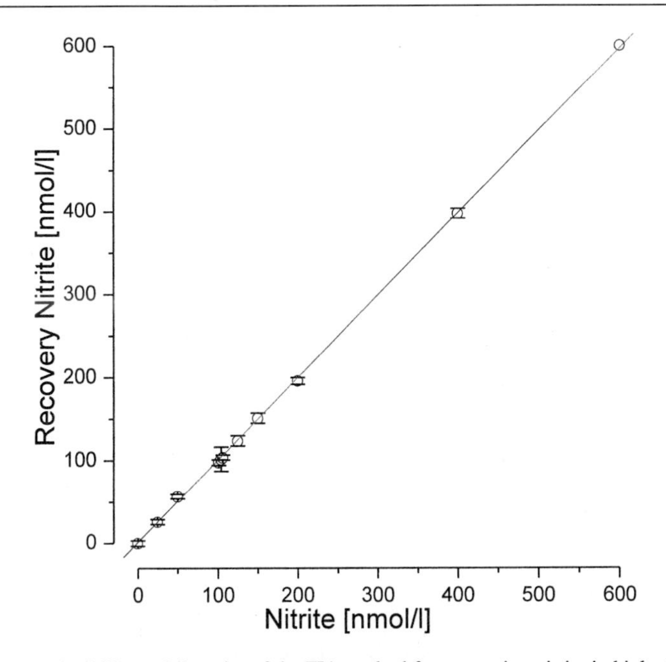

FIG. 4. Reproducibility and linearity of the FIA method for measuring nitrite in biological fluids. Nitrite was added to probes of human plasma, and recovery for added nitrite was determined after correction for baseline levels. Reproducibility is indicated by the low standard error of the mean of three separate measurements at each concentration of nitrite tested.

This step is followed by centrifugation at 800*g* at 4° for 10 to 15 min to separate the cellular compartment of blood. To avoid the mixing of different fractions, the centrifuge should not be stopped abruptly. In order to minimize protein concentration in the plasma sample, the supernatant is transferred to ultrafiltration tubes (Centricon, Millipore) with a pore size of 10 kDa and centrifuged at 4° for 2 hr at 6000*g*. Immediately after ultrafiltration, the nitrite concentration of the plasma sample is measured using the standard addition procedure in triplicate (Fig. 4).

Assay Characteristics

The detection limit for nitrite in aqueous phases is 10 nmol/liter at a 3 : 1 signal-to-noise ratio. There is a linear correlation (without dilution steps) from 10 to 600 nmol/liter ($r = 0.998$). A linear correlation between the nitrite concentration and the signal intensity has been demonstrated for protein-containing matrices over a range of 50 to 3000 nmol/liter. Even in complex matrices, such as processed blood samples (as described), the detection threshold for nitrite is determined to be 50 nmol/liter. Reproducibility, including sample processing and analysis, is found to be 9.8% (CV). During the Griess reaction, only nitrite reacts in a diazotization of sulfanilic acid and coupled with diamine, it builds a measurable metabolite.

Potential Pitfalls during Probe Sampling

Interactions are known to occur in the Griess reaction with hemoglobin and alcohol, leading to false-positive determinations and in a false-negative manner with citrate (unpublished data). These interactions should be considered when choosing the solution for diluting the sample. While using solutions that potentially cause hemolyses, e.g., 0.1 *N* NaOH, interference of free hemoglobin with analysis should be controlled carefully. According to the interaction described earlier, citrate cannot be used for anticoagulation. In addition, EDTA poses significant problems due to its baseline contamination of nitrite. Heparin is one of the only noninteracting anticoagulants that can be applied in the aforementioned concentration range. Many more substances are described that interact with the Griess reaction, e.g., biogenic amines, zinc sulfate, cadmium, manganese, iron, zinc, and urea.[20–22] To exclude these possible interferences of the critical substance, the linearity of the Griess reaction should be tested with increasing concentrations of nitrite in aqueous solutions. This can be tested easily by comparing the slopes of the calibration curves. In order to obtain a high sensitivity baseline, oscillations should be kept to a minimum. Disturbance may be caused by irregular flow of the carrier solution evoked by fluctuations of the pump performance or leakages of capillary connections. The measuring cell can be contaminated by crystalline sedimentation of the Griess reactant. To avoid its sequelae, the system must be flushed regularly with isopropanol (99.7% Merck, Germany) and water (LiChrosolv, Merck, Germany), one after another. Also in this setting the slope of the calibration curve is a useful parameter for detecting gradual contamination of the measuring cell. Moreover, sensitivity is influenced by the nitrite contamination of the Griess reactant by itself, which is caused mainly by water that dissolves the reactants. Lack of reproducibility may also be related to a deficient accuracy of the autoinjector during sample application. To avoid loss of reproducibility due to unpredictable nitrite contamination acquired during sample preparation, rinsing of containers with distilled water and keeping the contact time of sample with air as minimal as possible are necessary.

[20] G. Giovannoni, J. M. Land, G. Keir, E. J. Thompson, and S. J. R. Heales, *Ann. Clin. Biochem.* **34,** 193 (1997).

[21] D. Tsikas, I. Fuchs, F.-M. Gutzki, and J. C. Frölich, *J. Chromatogr. B* **715,** 441 (1998).

[22] Deutsche Einheitsverfahren zur Wasser-Abwasser- und schlammuntersuchung Anionen (GruppeD), D10 DIN 38405

Examples of Analytical Application in Mammalians

The aim of this study was to demonstrate the characteristics of the analyzing method presented when applied to different species. This study showed the ability of FIA to measure nanomolar concentrations of nitrite in plasma samples conveniently and quickly enough so that it can be applied in routine settings. We studied the nitrite concentration of humans, pigs, dogs, rabbits, and mice. Each group consisted of nine subjects. Figure 5 shows the different nitrite concentrations of the corresponding species. The mean $\pm$ SEM nitrite concentration amounts to 225 ± 22, 312 ± 37, 265 ± 39, 926 ± 121, and 466 ± 70 nmol/liter, respectively.

In another experiment, we critically evaluated the diagnostic accuracy of nitrite measurement for the assessment of NOS activity *in vivo*. To prove the hypothesis that nitrite reflects acute changes in eNOS activity accurately, forearm venous

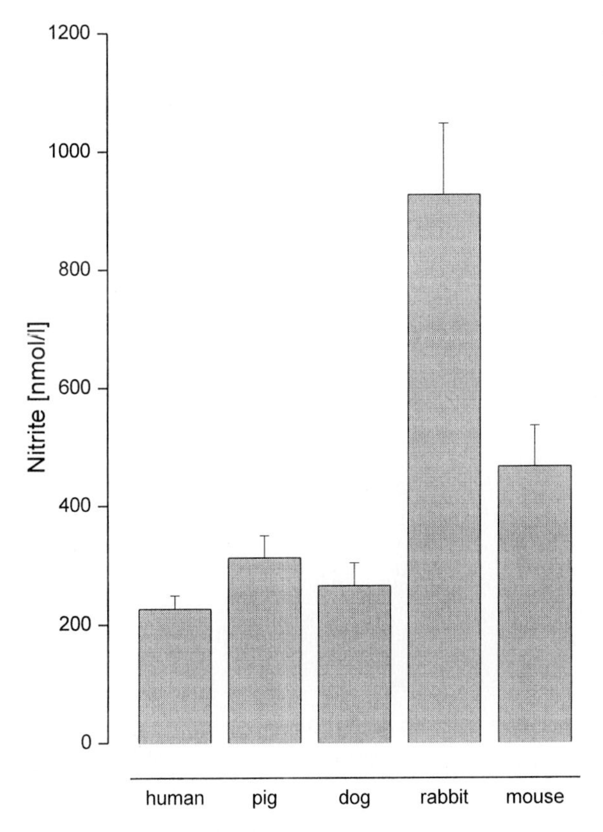

FIG. 5. Comparison of basal levels of nitrite in venous plasma samples drawn from healthy human volunteers, pigs, dogs, rabbits, or mice (for each $n = 9$; means $\pm$ SEM).

plasma levels of nitrite in humans were measured at rest and during inhibition and stimulation of NO synthesis. To verify our results, we compared the subjects to a placebo group that received physiological saline solution instead of the NOS inhibitor L-NMMA in concentrations (12 μmol/liter) or stimulator of NOS acetylcholine (ACh) (10 μg/min). Each group consisted of three subjects. During L-NMMA infusion, venous plasma nitrite concentration decreased from 315 ± 57 nmol/liter to a minimum of 136 ± 21 nmol/liter at 12 μmol/min L-NMMA in a dose-dependent relationship. Intraarterial infusion of ACh increased venous plasma nitrite from 327 ± 39 to 593 ± 112 nmol/liter.

Comparison between nitrite concentrations of treated and nontreated individuals showed statistically significant differences. The specificity of plasma nitrite as an indicator of regional NO formation was shown under basal and stimulated conditions. This demonstrated that plasma nitrite levels change with eNOS stimulation and inhibition. Findings of this study show that plasma concentrations of nitrite sensitively reflect acute changes in regional eNOS activity (Fig. 6).

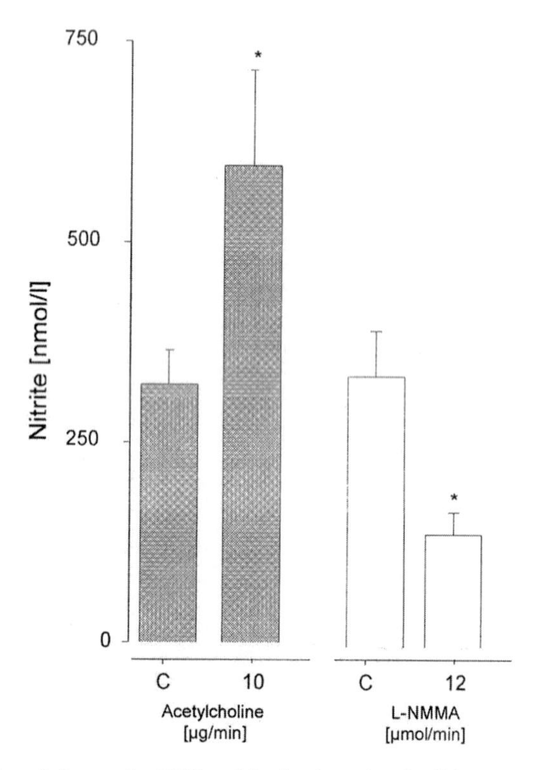

FIG. 6. Detection of changes in eNOS activity *in vivo* using the Griess reagent within the FIA method. Venous plasma nitrite level in human at baseline (C, control) and during intraarterial applicated doses of L-NMMA (12 μmol/min) and acetylcholine (10 μg/min); means $\pm$ SEM; $n = 3$. Asterisk indicates significant difference from control ($p < 0.05$).

TABLE I
COMPARISON OF ANALYTICAL KEY FEATURES OF FLOW INJECTION ANALYSIS AND A COMMERCIALLY AVAILABLE COLORIMETRIC ASSAY USING THE GRIESS REACTION[a]

	Griess reaction within FIA	Griess reaction within colorimetric assay
Linearity	0–10 μmol/liter	0–35 μmol/liter
Recovery	100 ± 10%	100 ± 10%
Reproducibility (CV)		
Nannomolar range	9.8%	Not measurable
Micromolar range	1.9%	5%
Sensitivity	50 nmol/liter	12.5 μmol/liter
Sample volume ($n = 3$)	3000 μl	240 μl
Analysis time; data evaluation ($n = 3$)	1 hr	0.5 hr

[a] Values were obtained in samples of human plasma, as a representative example of protein-containing biological fluids. CV, coefficient of variation. For details, see text.

Comparison of Nitrite Determination within Methods Based on Griess Reaction and Other Analytical Setups

In comparison to nitrite determination with methods based on the Griess reaction, the point-to-point discrimination of the FIA method is 250 times higher than that of the commercially available colorimetric assay (Cayman Chemicals). Recovery (within the characteristic analysis ranges) of the compared methods is 100%. In the micromolar range, reproducibility of the FIA is twofold higher than that of the colorimetric assay (Table I).

The difference in our results on circulating plasma levels (225 nmol/liter) in mammalians compared with those obtained by other techniques (130–3600 nmol/liter) is explicable by emerging confounders and variations due to probe sampling as well as analytical setup. Alternative analytical approaches, which are also valid for nitrite determination, include DAN assay, CLD, HPLC, and gas chromatography-mass spectrometry (GC-MS).[7,11,23–25]

Conclusion

FIA, in combination with the Griess reaction, can be used universally to determine nitrite in aqueous and protein-containing samples. Correct sample processing

[23] A. M. Leone, P. L. Francis, P. Rhodes, and S. Moncada, *Biochem. Biophys. Res. Commun.* **200,** 951 (1994).

[24] M. Gorenflo, C. Zehng, A. Poge, M. Bettendorf, E. Werle, W. Fiehn, and H. E. Ulmer, *Clin. Lab.* **47,** 441 (2001).

[25] D. Tsikas, R. H. Boeger, S. M. Bode-Böger, F.-M. Gutzki, and J. C. Frölich, *J. Chromatogr. B* **661,** 185 (1994).

and the use of a standard addition method in combination with the assay make it possible to use this method to obtain valid nitrite measurements even in complex matrices with nitrite content in the nanomolar range (Fig. 5).

Acknowledgment

A part of this work is supported by the Deutsche Forschungsgemeinschaft (DFG): Ke 405/4-3, Sonderforschungsbereich 1919. TR is a research fellow funded by the DFG (Ra969/1-1).

[15] Gas-Phase Oxidation and Disproportionation of Nitric Oxide

By HIROKAZU TSUKAHARA, TAKANOBU ISHIDA, and MITSUFUMI MAYUMI

Introduction

Nitrogen and oxygen together comprise over 98% of the air we breathe. Nitric oxide (NO) is a simple molecule, consisting of a single nitrogen bonded to one oxygen atom, which makes its chemistry accessible for study in great detail.[1,2] However, it is only recently that mammalian cells were discovered to produce NO as a short-lived intercellular messenger.[3,4] NO participates in blood pressure control, neurotransmission, and inflammation. Moreover, NO is the biologically active species released from a variety of cardiovascular drugs, such as nitroglycerin and isosorbide dinitrate.[5] One important function of NO is in the macrophage-dependent killing of invaders, and possibly cancer cells, indicating the potential of this free radical to mediate cytotoxic and pathological effects.[6]

When inhaled, NO acts as a selective pulmonary vasodilator. There is intense clinical interest in inhalation of low doses of NO (less than 1 to 80 ppm) in the treatment of diseases characterized by pulmonary hypertension and hypoxemia

[1] J. Laane and J. R. Ohlsen, *Prog. Inorg. Chem.* **27,** 465 (1980).

[2] F. T. Bonner, *Methods Enzymol.* **268,** 50 (1996).

[3] C. Nathan, *FASEB J.* **6,** 3051 (1992).

[4] L. J. Ignarro, G. M. Buga, K. S. Wood, R. E. Byrns, and G. Chaudhuri, *Proc. Natl. Acad. Sci. U.S.A.* **84,** 9265 (1987).

[5] D. R. Janero, *Free Radic. Biol. Med.* **28,** 1495 (2000).

[6] B. Gaston, J. M. Drazen, J. Loscalzo, and J. S. Stamler, *Am. J. Respir. Crit. Care Med.* **149,** 538 (1994).

0076-6879/02 $35.00

TABLE I
POTENTIAL CLINICAL USES FOR INHALED NO

Persistent pulmonary hypertension of the newborn
Primary pulmonary hypertension
Acute respiratory distress syndrome
Pulmonary hypertension following cardiac surgery
Cardiac transplantation
Lung transplantation
Congenital heart disease
Congenital diaphragmatic hernia
Chronic obstructive pulmonary disease
Bronchospasm

(Table I).[7,8] The inhaled NO therapy is fairly inexpensive, but it seems that it is not indicated for everybody with regard to the paradigm of its efficiency and potential toxicity. NO reacts with O_2 in the gas phase to form nitrogen dioxide (NO_2)[1,2,9,10]:

$$2NO\ (g) + O_2\ (g) = 2NO_2\ (g) \tag{1}$$

In this therapy, the inhaled NO_2 level must be kept as low as possible because of its toxic pulmonary effects.[7,8]

Commercial NO, which is stored under compression, invariably contains impurities. Formation of the two main contaminants, NO_2 and nitrous oxide (N_2O), is due to a disproportionation reaction, whose rate is immaterially slow at 1 atm and below, but becomes alarmingly high at elevated pressures[2]:

$$3NO\ (g) = N_2O\ (g) + NO_2\ (g) \tag{2}$$

Potential adverse effects of N_2O are also a cause of concern in NO inhalation. N_2O, known as laughing gas, works both as an anesthetic and as an analgesic. Under most conditions, N_2O appears to be inert, but prolonged exposure may produce myelotoxic, neurotoxic, and reproductive adverse effects.[11] Therefore, a proper assessment of the chemical kinetics, i.e., rate law and the relevant rate constants, of reactions (1) and (2) is indeed necessary in the medical community.

Reaction (2) must also be borne in mind when compressed NO in commercial cylinders is employed in high-precision experiments. The presence of NO_2 in the

[7] C. G. Frostell and W. M. Zapol, *in* "Methods in Nitric Oxide Research" (M. Feelisch and J. S. Stamler, eds.), p. 645. Wiley, Chichester, 1996.

[8] E. Troncy, M. Francoeur, and G. Blaise, *Can. J. Anaesth.* **44,** 973 (1997).

[9] D. L. Baulch, D. D. Drysdale, and D. G. Horne, "Evaluated Kinetic Data for High Temperature Reactions," Vol. 2, p. 285. Butterworths, London, 1973.

[10] H. Tsukahara, T. Ishida, and M. Mayumi, *Nitric Oxide Biol. Chem.* **3,** 191 (1999).

[11] R. T. Louis-Ferdinand, *Adverse Drug React. Toxicol. Rev.* **13,** 193 (1994).

compressed NO clearly creates difficulties in experimental studies of NO reactions and can pose an explosion hazard when the gas is brought into contact with hydrocarbons.[2]

Our goal is to provide an introduction to the reaction kinetics of gas-phase oxidation and disproportionation of NO, while emphasizing the prediction of NO_2 formation from reactions (1) and (2).

Gas-Phase Oxidation of NO

Chemical Kinetics of the Reaction of NO with O_2

In oxygenated gaseous environments, the reaction of NO with O_2 (reaction 1) proceeds as a third-order reaction[1,2,9,10]:

$$-d[\mathrm{NO}]/dt = +d[\mathrm{NO_2}]/dt = 2k[\mathrm{NO}]^2[\mathrm{O_2}] \tag{3}$$

Since the first chemical kinetic study of this reaction by Bodenstein and Wackenheim,[12,13] a majority of published experimental results support the differential rate law (3) and the view that reaction (1) is a homogeneous gas-phase reaction and does not require participation of solid surfaces.[9,10] These results include older data obtained with NO pressures ranging from a few torrs to a few hundred torrs and more recent ones involving low NO concentrations down to the ppm level. This section briefly reviews the reaction mechanisms that have been proposed to explain the rate law (3).

Three mechanisms have been proposed.[10]

Termolecular Reaction. According to this view, two molecules of NO and one O_2 molecule collide simultaneously and form a transient complex, which thus takes a single step to form two molecules of NO_2. Although this reaction has long served as a classical example of termolecular reaction in chemistry textbooks, a majority of researchers in recent years prefer the two-step mechanisms, as described later.

Preequilibrium Mechanism with Dimer of NO as an Intermediate.

$$\mathrm{NO} + \mathrm{NO} \rightleftarrows (\mathrm{NO})_2 \quad \text{[fast]} \tag{4a}$$

$$(\mathrm{NO})_2 + \mathrm{O_2} \rightarrow 2\mathrm{NO_2} \quad \text{[slow and rate determining]} \tag{4b}$$

$$-d[\mathrm{NO}]/dt = k_1k_2\,[\mathrm{NO}]^2[\mathrm{O_2}]/(k_{-1} + k_2[\mathrm{O_2}]) \tag{4c}$$

Equation (4c) results from the assumption that the rate constants k_1 and k_{-1} for the forward and reverse reactions of Eq. (4a) are significantly greater than that (k_2) of the second step, Eq. (4b). Because $[O_2]$ is only 0.041 $\mathrm{mol.liter^{-1}}$ even in pure O_2 gas at 1 atm and 298 K, the relation $k_{-1} \gg k_2[O_2]$ is assumed valid, thus making Eq. (4c) take the form of Eq. (3).

[12] M. Bodenstein and L. Wachenheim, *Z. Elektrochem.* **24,** 183 (1918).
[13] M. Bodenstein, *Z. Physik. Chem.* **100,** 68 (1922).

Preequilibrium Mechanism with NO_3 as an Intermediate.

$$NO + O_2 \rightleftarrows NO_3 \quad \text{[fast]} \tag{5a}$$

$$NO_3 + NO \rightarrow 2NO_2 \quad \text{[slow and rate determining]} \tag{5b}$$

$$-d[\mathrm{NO}]/dt = k_3 k_4 [\mathrm{NO}]^2[\mathrm{O_2}]/(k_{-3} + k_4[\mathrm{NO}]) \tag{5c}$$

In Eq. (5c), k_3 and k_{-3} are, respectively, the rate constants of the forward and reverse directions of Eq. (5a) and k_4 is the constant for Eq. (5b). The molar concentration of NO is low even with the NO pressures at a few hundred torrs, and expression (5c) converges to the form of Eq. (3). NO_3 is a well-known intermediate in the decomposition of dinitrogen pentoxide to NO_2 and O_2.[14] Two structures for NO_3 are in the literature: (a) nitrogen trioxide, a nitrate-like planar trigonal structure, in which the N atom is surrounded by three equivalent O atoms[1,15] and (b) peroxynitrite radical, a bent-chain structure with an atomic sequence of O-O-N-O.[1,16] Although peroxynitrite radical is more persuasive than nitrogen trioxide as an intermediate of the third mechanism, there is insufficient evidence to definitely favor one over the other. The formal oxidation state of the N atom in NO_3 is +6. At 298K the standard reduction potential E° for $ONOO^{\bullet} + e^- \rightarrow ONOO^-$ is 0.4 V[17]; i.e., peroxynitrite radical is more oxidative than peroxynitrite anion ($ONOO^-$), which itself is a strong oxidant: E° for $ONOO^- + 2H^+ + e^- \rightarrow NO_2 + H_2O$ is 1.4 V at pH 7.[17] For nitrogen trioxide, various estimates for E° of $NO_3 + e^- \rightarrow NO_3^-$ fall within a limit of 2.5 ± 0.2 V,[18] again indicating a highly oxidative nature of the intermediate of the third mechanism. Thus, the intermediate of the third pathway could be biologically at least as damaging as $ONOO^-$ and more damaging than NO_2. However, reactivity and toxicity studies of NO_3 have largely escaped biologists' attention.

Bodenstein and Wachenheim[12] and Bodenstein[13] considered the second and third mechanisms as possible explanations of their experimental results.[12,13] More recent investigations compared the two mechanisms and concluded that both have equal merits.[9,19,20]

Rate Constant for Oxidation of NO to NO_2

Equation (3) defines the rate constant *k* for reaction (1), which is used in this section. Experimental results from 22 investigations on the rate constant *k* for the oxidation of NO have been summarized.[10] In these investigations, the concentration of

[14] R. A. Ogg, Jr., *J. Chem. Phys.* **15,** 337 (1947).
[15] I. C. Hisatsune, *J. Phys. Chem.* **65,** 2249 (1961).
[16] W. A. Guillory and H. S. Johnston, *J. Am. Chem. Soc.* **85,** 1695 (1963).
[17] W. H. Koppenol, *Methods Enzymol.* **268,** 7 (1996).
[18] D. M. Stanbury, *Adv. Inorg. Chem.* **33,** 69 (1989).
[19] J. Heicklen and N. Cohen, *Adv. Photochem.* **5,** 157 (1968).
[20] J. Olbregts, *Int. J. Chem. Kinet.* **17,** 835 (1985).

NO ranged over five orders of magnitude from 8.2×10^{-8} to 1.8×10^{-2} mol·liter^{-1} and the temperature from 225 to 843K. In many studies, N_2 gas was used as a diluent. Most of the kinetic studies agree that the rate constant k is independent of variations in the total pressure (from less than 20 torr to atmospheric pressure) and the absolute NO concentration or the NO:O_2 ratio (up to a thousandfold variation) and that the logarithm of the rate constant is inversely proportional to the absolute temperature. The addition of N_2, CO_2, methane, and olefins or moisture (up to 90% relative humidity) has no apparent effect on the order and rate constant for the reaction of NO and O_2. Because water reacts with NO_2 but not with NO in the absence of dissolved O_2, it is expected that the presence of water in the NO/O_2 reaction systems will shift the final equilibrium toward the products, including oxyacids of nitrogen and their anions, but it will not change the rate with which NO is oxidized.

The best accepted temperature dependence of k in T = 273 to 600K is represented by an Arrhenius-type equation[9,10]:

$$k(\text{liter}^2\ \text{mol}^{-2}\ \text{s}^{-1}) = 1.2 \times 10^3\ e^{530/T} = 1.2 \times 10^3 \times 10^{230/T} \quad (6)$$

where T is absolute temperature. It is noted that the rate constant of reaction (1) has a negative temperature coefficient. In summary, $k = 7.0 \times 10^3$ liter^2 mol^{-2} s^{-1} near the ambient temperature (~298K) and decreases by only 0.04×10^3 liter^2 mol^{-2} s^{-1} per degree of rise in the temperature. Thus, the rate constant is rather insensitive to changes of temperature over a wide range.

Prediction of NO_2 Formation Using Rate Constant k

In the setting of NO inhalation environment, as well as of ambient atmospheres, NO is present in much lower concentrations than O_2, and it can be assumed that $[O_2]$ remains unchanged. Therefore, a pseudo-second-order kinetics can be applied to reaction (1). Integration of Eq. (3) yields

$$1/[NO]_t - 1/[NO]_0 = 2k\,[O_2]\,t \quad (7a)$$

where $[NO]_t$ represents the molar concentration of NO after a residence time t and $[NO]_0$ is the initial NO concentration. Because the difference between $[NO]_0$ and $[NO]_t$ is $[NO_2]$,

$$t = \{1/(2k\,[O_2]\,[NO]_0)\}\,[NO_2]/([NO]_0 - [NO_2]) \quad (7b)$$

$$t = \{1/(2k\,[O_2]\,[NO]_0)\}\,f/(1 - f) \quad (7c)$$

where f is the fraction of NO that has become NO_2 in time t (i.e., $f = [NO_2]/[NO]_0$) and $k = 7.0 \times 10^3$ liter^2 mol^{-2} s^{-1} near the ambient temperature. It is evident from Eq. (7c) that the time required for the conversion of a certain fraction of NO to NO_2 is inversely proportional to the initial NO concentration and the rate constant k. Equation (7c) can be used to calculate the time to reach a certain concentration of NO_2 for any concentrations of initial NO in O_2 (Table II).

TABLE II
TIME (min) TO YIELD 5 ppm NO_2 FROM REACTION $2NO + O_2 = 2NO_2$ AT $T = 298K$ UNDER A CONSTANT TOTAL VOLUME[a]

	NO (ppm)						
O_2 (%)	10	20	40	80	200	400	800
20	355.83	59.31	12.71	2.97	0.456	0.113	0.028
30	236.64	39.44	8.45	1.97	0.303	0.075	0.019
40	177.48	29.58	6.34	1.48	0.228	0.056	0.014
50	141.99	23.66	5.07	1.18	0.182	0.045	0.011
60	118.80	19.80	4.24	0.99	0.152	0.038	0.009
70	101.77	16.96	3.64	0.85	0.130	0.032	0.008
80	89.01	14.84	3.18	0.74	0.114	0.028	0.007
90	79.10	13.18	2.83	0.66	0.101	0.025	0.006
100	71.17	11.86	2.54	0.59	0.091	0.023	0.006

[a] Threshold values of NO_2 have been set at 5 ppm or lower for a permissible exposure limit for workers.[7,8] Therefore, we have calculated the time to reach 5 ppm NO_2 from NO concentrations of 10 to 800 ppm in O_2 concentrations of 20 to 100%.

The initial rate of production of NO_2 is proportional to the square of the initial concentration of NO. Thus, gas-phase oxidation of NO is slow at very low concentrations of NO in ambient atmospheres. For example, about 6 hr is required for 50% oxidation of 10 ppm NO. Conversely, with high-inspired O_2, a small error in the initial timing of NO inhalation could lead to a larger error in the exposure to the undesirable NO_2. Under a condition of inspired O_2 fraction = 1.0, the concentration of NO_2 reaches 52, 8.8, 2.3, 0.57, and 0.023 ppm in the first 10 sec of inhalation of 500, 200, 100, 50, and 10 ppm NO, respectively. At high NO concentrations, even a 1-sec error in the timing could cause serious consequences.

Gas-Phase Disproportionation of NO

Chemical Kinetics and Rate Constant of the Disproportionation of NO

NO is thermodynamically unstable, as indicated by its large positive Gibbs energy of formation ($\Delta_f G^o_{298} = 86.32$ kJ mol^{-1}).[1,2] Despite the thermodynamic instability of NO, its decomposition and disproportionation are hindered kinetically near the ambient temperature. Reaction (2) does not occur to any appreciable extent near the ambient temperature and near 1 atm pressure.[2,21] However, the rate of this reaction can be increased by the use of catalysts[22] and elevated pressures.[23]

[21] H. Tsukahara, T. Ishida, Y. Todoroki, M. Hiraoka, and M. Mayumi, submitted for publication.
[22] W. E. Addison and R. M. Barrer, *J. Chem. Soc.* 757 (1955).
[23] T. P. Melia, *J. Inorg. Nucl. Chem.* **27,** 95 (1965).

Melia[23] reported that after having been stored in a cylinder under pressures between 50 and 100 atm and near the ambient temperature, freshly purified NO showed an increase of NO_2 and N_2O contents at a rate of 2–3% per month. Although the aforementioned observations are the only quantitative experimental data available until now and the experiments were performed at temperatures above the ambient and under relatively high pressures, data of Melia[23] have been generally referred to in the contemporary review papers on the chemistry of NO gas.[2,10,21]

It has been proposed that under elevated pressures, NO undergoes disproportionation proceeded by the following preequilibrium mechanism[2,21,23]:

$$2NO \rightleftarrows (NO)_2 \quad \text{[fast]} \tag{8a}$$

$$NO + (NO)_2 \rightarrow NO_2 + N_2O \quad \text{[slow and rate determining]} \tag{8b}$$

and the overall rate law:

$$-d[NO]/dt = k' \, [NO][(NO)_2] = Kk' \, [NO]^3 \tag{8c}$$

where K is the equilibrium constant for reaction (8a) and k' is the rate constant of reaction (8b). Melia[23] concluded that in the temperature range, 303 to 323K, and at pressures of up to 400 atm, the overall reaction for the disproportionation of NO obeys third-order kinetics. Melia gave the overall rate constant $Kk' = 2.6 \times 10^{-5}$ liter^2 mol^{-2} hr^{-1}at T = 303K and $Kk' = 2.7 \times 10^{-5}$ liter^2 mol^{-2} hr^{-1} at T = 323K, with an initial pressure $P_0 = 200$ atm. According to Melia, the insensitivity of Kk' to changes of temperature in this range appears to result from the increase in the rate constant k' being offset by the decrease in concentration of $(NO)_2$ brought about by its thermal decomposition.

Buildup of NO_2 and N_2O in Pressurized NO Gas

At very low pressures, the behavior of a real gas approaches the ideal. At higher pressures, however, the physical properties of a real gas may deviate substantially from the ideal. In Melia's equations (8a–8c), [NO] is the molar concentration of NO as a real gas (NO_{real}) and not an ideal gas (NO_{ideal}). Keeping in mind that the nonideality of NO plays a central role in the evaluation of quantitative implication of Melia's experimental results on reaction (2), we calculated the compressibility factor, defined as $z = [NO_{ideal}]/[NO_{real}]$, of NO at 298K and at 12 values of pressure (from 1 to 200 atm) using the ideal gas law and van der Waals equation, and then least squares fitted these z to a cubic function of pressure:

$$z(P) = 0.998 - 0.001438P - 0.000008488P^2 + 0.00000003735P^3$$

where P is the pressure (in atm).

Because $[NO] = (1/RT)(P/z)$ in which R is the gas constant (0.08206 liter atm mol^{-1} K^{-1}), T is the absolute temperature, and (P/z) is a function of P, Eq. (8c) becomes

$$-\{1/(P/z)^3\}d(P/z) = \{Kk'/(RT)^2\}\,dt$$

Integration of this equation between $t = 0$ and $t = $ t yields

$$(1/2)\{(z_f/P_f)^2 - (z_0/P_0)^2\} = \{Kk'/(RT)^2\}\,t \quad (9)$$

where the subscripts "0" and "f" refer to the initial and final states, respectively. For a given P_0, T, and t, Eq. (9) can be solved by finding the values of P_f (and the corresponding z_f) that satisfy Eq. (9). Between $t = 0$ and $t = $ t, the number of moles of NO decreases from $n_0 = \{V/(RT)\} \cdot (P_0/z_0)$ to $n_f = \{V/(RT)\} \cdot (P_f/z_f)$, where V is the internal volume of a container that has been assumed unchanged. Then, from the stoichiometry of reaction (2), the number of moles of NO_2 and N_2O at residence time t is $1/3\,\{V/(RT)\}\,(P_0/z_0 - P_f/z_f)$ and the total number of moles of NO, NO_2, and N_2O is $1/3\,\{V/(RT)\}\,\{2(P_0/z_0) + (P_f/z_f)\}$. Consequently, the mole fractions, x, of the three gases at time t are

$$x_{NO} = 3(P_f/z_f)/\{2(P_0/z_0) + (P_f/z_f)\} \quad (10a)$$

$$x_{NO_2} = x_{N_2O} = \{P_0/z_0 - P_f/z_f\}/\{2(P_0/z_0) + (P_f/z_f)\} \quad (10b)$$

These computations have been carried out at $T = 298K$ for eight initial NO pressures (from 1 to 200 atm) and for eight reaction periods (from 1 to 8760 hr). Results on the final mole fractions of NO_2 and N_2O based on Melia's data, $Kk' = 2.6 \times 10^{-5}$ liter^2 mol^{-2} hr^{-1}, are summarized in Table III.[21] This value of Kk' is for $T = 303K$, but Kk' is very insensitive to the changes in temperature. A similar calculation done at $T = 303K$ has yielded practically the same result.[21]

TABLE III

MOLE FRACTIONS OF NO_2 AND N_2O FORMATION FROM REACTION $3NO = N_2O + NO_2$ AT $T = 298K$ UNDER A CONSTANT TOTAL VOLUME

	Initial pressure (atm)							
Time (hr)	200	100	50	20	10	5	2	1
1	0.0008	0.0002	0	0	0	0	0	0
3	0.0024	0.0005	0.0001	0	0	0	0	0
6	0.0048	0.0011	0.0002	0	0	0	0	0
24	0.0181	0.0042	0.0010	0.0002	0	0	0	0
72	0.0484	0.0123	0.0029	0.0004	0.0001	0	0	0
240	0.1192	0.0373	0.0094	0.0014	0.0004	0.0001	0	0
720 (30 days)	0.2133	0.0903	0.0264	0.0043	0.0011	0.0003	0	0
8760 (1 year)	0.3941	0.2984	0.1689	0.0449	0.0124	0.0032	0.0001	0.0001

As seen from Table III, the formation of NO_2 and N_2O due to the disproportionation of NO is generally negligible at low initial pressures, but becomes higher at elevated pressures. After 10 days of storage, the mole fractions of these contaminants are 0.01% or lower under an initial pressure of 5 atm or below, while they can become as high as 12% in the same period when $P_0 = 200$ atm. Even when pure NO is stored under an initial pressure of 50 atm, the mole fractions of NO_2 and N_2O can reach as high as 17% after 1 year.

Threshold for the Exposure to NO_2

The United States Occupational Safety and Health Administration has set the permissible limits for NO to not exceed 25 ppm for an 8-hr time-weighted average period and NO_2 to not exceed 5 ppm during any part of the working day.[7,8] However, the United States National Institute for Occupational Safety and Health has set the recommended exposure limit for NO_2 to not exceed 1 ppm for a 15-min exposure and a maximum inhaled NO_2 level of 5 ppm. Public safety agencies from other countries have adopted similar or lower levels. However, these recommendations are intended for healthy workers and not for patients (especially neonates) with serious diseases, although some guidelines regarding the use of NO in patients have been based on these recommendations.[24,25] In addition, inhaled NO therapy must frequently be continued for prolonged periods ranging from several hours to several weeks.

The national ambient air quality standard for the annual NO_2 level in the United States is 0.05 ppm.[26] A "safe" level of NO_2 is difficult to determine. NO_2 levels of less than 0.5 ppm may enhance human airway hyperreactivity. Toxic pulmonary effects of breathing NO_2 at or below 5 ppm have been reported.[7,26] These include altered surfactant chemistry and metabolism, epithelial hyperplasia of terminal bronchioles, and increased cellularity of alveoli in rats, as well as diffuse inflammation and hyperreactivity. At higher inhaled doses, pulmonary edema and death have been reported.

Preparation, Storage, and Application of Ultrahigh Purity NO

Disproportionation of NO under pressure must be kept in mind when compressed NO in commercial cylinders is employed, as described earlier. For this reason, purification of commercial NO, most particularly for the removal of NO_2, is essential to its use in either research or clinically.[2]

[24] L. Foubert, B. Fleming, R. Latimer, M. Jonas, A. Oduro, C. Borland, and T. Higenbottam, *Lancet* **339,** 1615 (1992).

[25] M. Bouchet, M. H. Renaudin, C. Raveau, J. C. Mercier, M. Dehan, and V. Zupan, *Lancet* **341,** 968 (1993).

[26] H. Gong, Jr., *Clin. Chest Med.* **13,** 201 (1992).

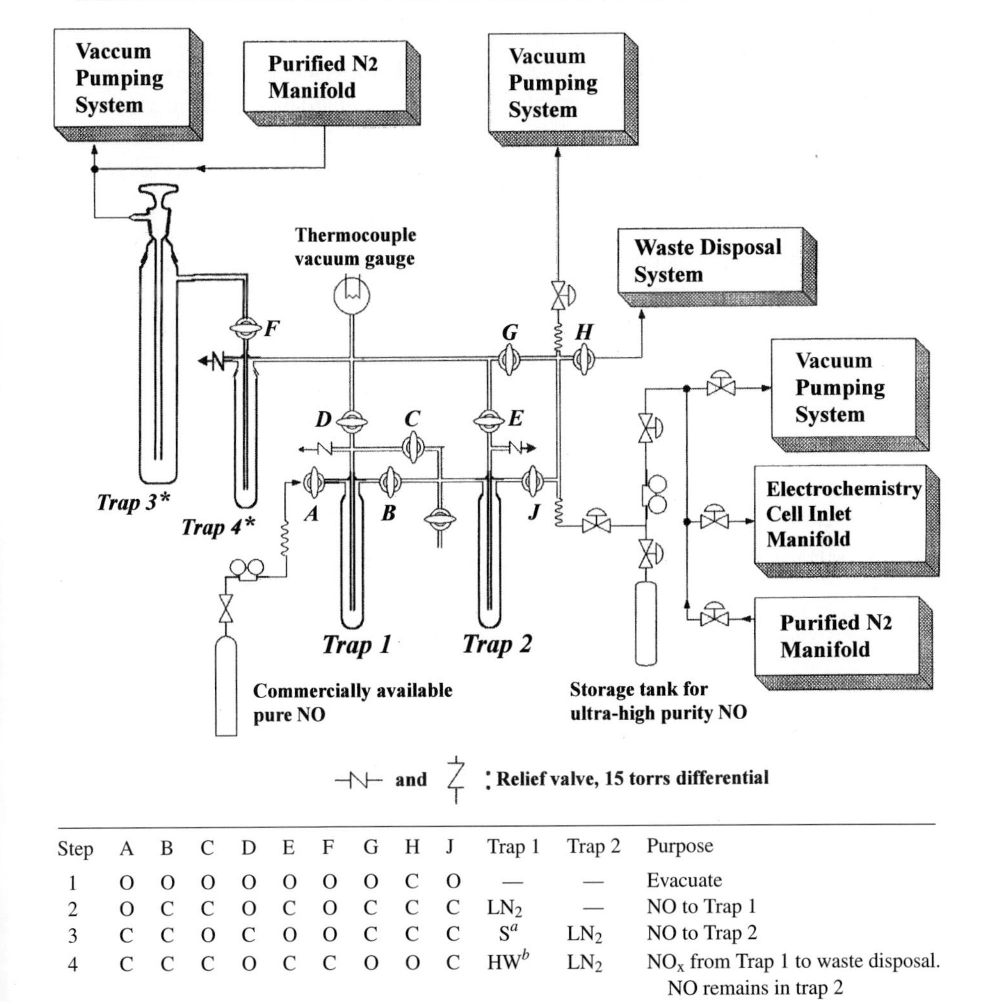

Step	A	B	C	D	E	F	G	H	J	Trap 1	Trap 2	Purpose
1	O	O	O	O	O	O	O	C	O	—	—	Evacuate
2	O	C	C	O	C	O	C	C	C	LN_2	—	NO to Trap 1
3	C	C	O	C	O	O	C	C	C	S[a]	LN_2	NO to Trap 2
4	C	C	C	O	C	C	O	O	C	HW[b]	LN_2	NO_x from Trap 1 to waste disposal. NO remains in trap 2
5	C	C	C	O	C	O	O	C	C	—	LN_2	Degas Trap 1
6	C	O	C	O	C	O	C	C	C	LN_2	S	NO to Trap 1
7	C	C	C	C	O	C	O	O	C	LN_2	HW	NO_x from Trap 2 to waste disposal. NO remains in Trap 1
8	C	C	C	C	O	O	O	C	C	LN_2	—	Degas Trap 2
9	C	C	O	C	O	O	C	C	C	S	LN_2	NO to Trap 2
10	(Go back to Step 4, if necessary)											

Abbreviations: O, open; C, closed; LN_2, liquid nitrogen. [a]Slurry of organic solid in its own liquid. Choose a preferably nonflammable compound whose melting temperature is above the boiling temperature of NO but below the melting temperatures of other NO_x. [b]Hot water ($\sim 50°$), which will evaporate all NO_x. Traps 3* and 4* are always kept in LN_2 during the procedure.

FIG. 1. Procedure for bulb-to-bulb distillation of pure NO.

The normal (1 atm) melting and boiling temperatures are 109K and 121K for NO, 262K and 294K for NO_2, and 182K and 185K for N_2O, respectively.[1] In our laboratory, NO is purified by a repeated bulb-to-bulb distillation at the cryogenic temperatures as follows. The all-borosilicate glass purification system has been designed to provide the bulb-to-bulb distillation in a most effective manner (Fig. 1). When the gas mixture is evaporated from trap 1 to trap 2, NO_x impurities, which are all higher boiling than NO, tend to condense on the walls of trap 2 near the entrance to the trap, whereas NO tends to condense on the walls deeper down the trap. When the condensates are evaporated from trap 2 back to (degassed) trap 1, the pathway between the two traps is built in such a way that NO leaves trap 2 while the majority of NO_x impurities remain trapped in trap 2. If the impurities ever leave trap 2 for trap 1, they must be the last ones to leave trap 2. This procedure can be repeated to remove completely the impurities and to obtain ultrahigh purity NO. The final ultrahigh purity NO can be transported through the high vacuum

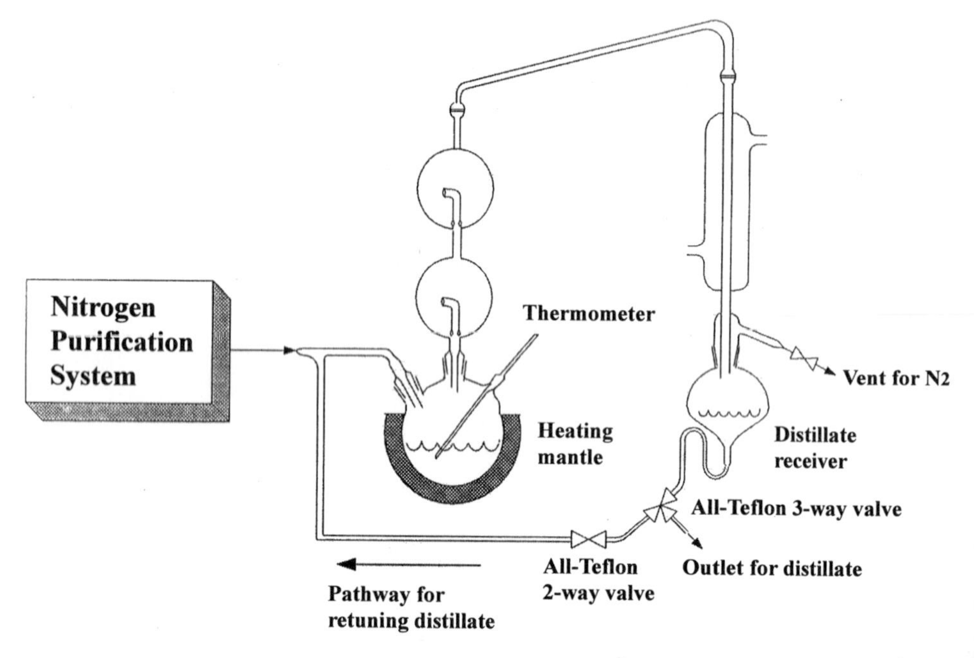

FIG. 2. Water distillation apparatus. Water is distilled from 10^{-3} M permanganate solution in 10^{-2} M H_2SO_4 at a rate of 150 ml/hr for at least 24 hr, during which the distillate is continuously fed back to the still by gravity. This is then followed by a period of withdrawal of product water at a rate of 150 ml/hr. The still is a 5-liter round bottom flask with three tapered ground points. Water vapor from the still passes through two liquid traps, which catch spatters of the permanganate solution. The system is purged continuously by a stream of specially purified N_2 (see text) to carry gaseous impurities out of the apparatus through a "vent," a poly(perfluoroethylene) check valve.

system from trap 2 to the storage tank (Type-316 stainless steel), which is put in a liquid N_2 trap during the transfer. Once purified, NO can be stored safely in the storage tank at low pressures (e.g., 5 atm or below).

NO gas is colorless, and the solid and liquid forms exhibit a very faint blue, which can be taken as being colorless. Because the condensates of dinitrogen trioxide (N_2O_3) are bright blue, some regard the faint blue color of condensed NO as evidence of NO_2, as NO and NO_2 react readily to form N_2O_3 at cryogenic temperatures.[1] The melting and boiling temperatures of N_2O_3 are 162K and 275K, respectively. When more NO_2 is present, the condensates color becomes greenish.

In the study of aqueous NO systems, it is important that the solvent be thoroughly deoxygenated. NO reacts readily with O_2 to form NO_2 and, further, N_2O_3 and dinitrogen tetroxide (N_2O_4). NO_2, N_2O_3, and N_2O_4 will form the nitroxy anions on interaction with water, causing side reactions, some of them catalytic. Inert gas bubble purging is one of the most effective deoxygenation methods, so long as the gas is sufficiently pure. In our laboratory, N_2 is purified by passage through activated charcoal and silica gel and then through copper metal turnings heated red hot in a silica tube furnace to remove NO_x (e.g., $2NO + 2Cu \rightarrow N_2 + 2CuO$) and O_2 (i.e., $O_2 + 2Cu \rightarrow 2CuO$). Preparation of ultrapure water involves a two-step purification process. The first step yields the water, which we may call acid-permanganate distillation water (Fig. 2).[27] The product is free of all impurities, with the exception of O_2. The second step is the removal of dissolved O_2 by purging with the ultrapure N_2 in a specially built all-borosilicate glass apparatus. This water is used throughout our investigation in aqueous solutions.

[27] L. C. Krebs and T. Ishida, "Characterization of Electrochemically Modified Polycrystalline Platinum Surfaces," p. 21. U.S. Department of Energy Report No. DOE/ER/13855-6, 1991.

[16] Detection and Quantification of Nitric Oxide (NO) Synthase-Independent Generation of NO

By DIANA A. LEPORE

Introduction

Nitric oxide (NO) is considered to play a key role in "ischemia–reperfusion injury," a major limiting factor in the survival of tissues involved in trauma, transfers, or infarcts.[1,2] The development of methods for the direct detection and quantification of NO is important to fully characterize the role of NO in ischemia–reperfusion injury. Classically, the generation of NO has been described as deriving from the enzymatic pathway in which L-arginine is converted to L-citrulline via nitric oxide synthase (NOS).[3] The generation of NOS-dependent NO during ischemia–reperfusion injury has been reported in numerous studies.[4–19] However, in more recent years, significant generation of NO from a NOS-independent

[1] P. Grace, *Br. J. Surg.* **81,** 637 (1994).

[2] A. G. Stewart, J. E. Barker, and M. J. Hickey, *in* "Ischaemia–Reperfusion Injury" (P. A. M. Grace and R. T. Mathie, eds.), p. 180. Blackwell Science, Oxford, 1998.

[3] S. Moncada and A. Higgs, *N. Eng. J. Med.* **329,** 2002 (1993).

[4] A. Seekamp, M. S. Mulligan, G. O. Till, and P. A. Ward, *Am. J. Pathol.* **142,** 1217 (1993).

[5] B. Zhang, K. R. Knight, B. Dowsing, E. Guida, L. H. Phan, M. J. Hickey, W. A. Morrison, and A. G. Stewart, *Clin. Sci.* **93,** 167 (1997).

[6] J. E. Barker, K. R. Knight, R. Romeo, J. V. Hurley, W. A. Morrison, and A. G. Stewart, *J. Pathol.* **194,** 105 (2001).

[7] L. H. Phan, M. J. Hickey, Z. B. M. Niazi, and A. G. Stewart, *Microsurgery* **15,** 703 (1994).

[8] J. L. Zweier, P. Wang, and P. Kuppusamy, *J. Biol. Chem.* **270,** 304 (1995).

[9] V. C. Patel, D. M. Yellon, K. J. Singh, G. H. Neild, and R. G. Woolfson, *Biochem. Biophys. Res. Commun.* **194,** 234 (1993).

[10] G. Matheis, M. P. Sherman, G. D. Buckberg, D. M. Haybron, H. H. Young, and L. J. Ignarro, *Am. J. Physiol.* **262,** H616 (1992).

[11] M. W. Williams, C. S. Taft, S. Ramnauth, Z. Q. Zhao, and J. Vinten-Johansen, *Cardiovasc. Res.* **30,** 79 (1995).

[12] S. Hoshida, N. Yamashita, J. Igarashi, M. Nishida, M. Hori, T. Kamada, T. Kuzuya, and M. Tada, *J. Pharmacol. Exp. Ther.* **274,** 413 (1995).

[13] S. Pudupakkam, K. A. Harris, W. G. Jamieson, G. DeRose, J. A. Scott, M. W. Carson, M. G. Schlag, P. R. Kvietys, and R. F. Potter, *Am. J. Physiol.* **275,** H94 (1998).

[14] Y. Horie, R. Wolf, and D. N. Granger, *Am. J. Physiol.* **273,** G1007 (1997).

[15] Y. Horie, R. Wolfe, D. C. Anderson, and D. N. Granger, *Am. J. Physiol.* **275,** H520 (1998).

[16] I. Kurose, R. Wolfe, M. B. Grisham, and D. N. Granger, *Circ. Res.* **74,** 376 (1994).

[17] P. Kubes, *Am. J. Physiol.* **265,** H1909 (1993).

[18] T. M. Moore, P. L. Khimenko, P. S. Wilson, and A. E. Taylor, *Am. J. Physiol.* **271,** H1970 (1996).

[19] S. Tanaka, W. Kamiike, H. Kosaka, T. Ito, E. Kumura, T. Shiga, and H. Matsuda, *Am. J. Physiol.* **271,** G405 (1996).

0076-6879/02 $35.00

source has also been identified, both in ischemic and ischemic-reperfused muscle tissue.[20,21] This article describes the direct detection and quantification of NOS-independent NO in ischemic-reperfused skeletal muscle tissue.

How Is NOS-Independent NO Generated?

The generation of NOS-independent NO is known to occur physiologically in the human oral cavity when bacteria reduce dietary nitrate (NO_3^-) to nitrite (NO_2^-) using the enzyme nitrate reductase.[22] NO and other nitrogen oxides are then generated from NO_2^- in the acidic/reducing environment of the stomach.[23,24] A similar type of NO production has been demonstrated on the skin surface.[22]

NOS-independent NO can also be generated via activated neutrophils, which convert NO_2^-, using a myeloperoxidase-dependent pathway, into the inflammatory oxidants nitryl chloride (NO_2Cl) and nitrogen dioxide (NO_2).[25] These nitrogen species serve as a source of NO.

The detection of NOS-independent NO during ischemia–reperfusion injury was first reported by Zweier and colleagues.[20] Oxyhemoglobin, a scavenger of NO regardless of its source, was significantly more effective than an inhibitor of NOS, L-nitro-L-arginine methyl ester (L-NAME), at attenuating the levels of NO detectable during reperfusion in an *ex vivo* cardiac muscle model. The generation of NOS-independent NO from NO_2^- (reaction 1) was shown to be related to the duration of ischemia, a decrease in tissue pH, and to nonenzymatic reducing equivalents, together producing levels of NO that were 4000-fold the maximum possible physiological values generated by NOS.[26] A detailed review of NOS-independent NO generation in ischemic-reperfused cardiac muscle tissue can be found in Zweier *et al.*[27]

$$3NO_2^- + 2H^+ \rightarrow 2NO + NO_3^- + H_2O \qquad \text{(reaction 1)}$$

The generation of significant levels of NOS-independent NO in ischemic-reperfused skeletal muscle was later reported by our laboratory.[21] NO detected 24 hr after ischemia–reperfusion was insensitive to inhibitors of NOS: L-NAME, a nonselective inhibitor of constitutive or inducible NOS, and *S*-methylisothiourea

[20] J. L. Zweier, P. Wang, A. Samouilov, and P. Kuppusmay, *Nature Med.* **1,** 804 (1995).

[21] D. A. Lepore, A. V. Kozlov, A. G. Stewart, J. V. Hurley, W. A. Morrison, and A. Tomasi, *Nitric Oxide Biol. Chem.* **3,** 75 (1999).

[22] E. Weitzberg and J. O. N. Lundberg, *Nitric Oxide Biol. Chem.* **2,** 1 (1998).

[23] J. O. N. Lundberg, E. Weitzberg, J. M. Lundberg, and K. Alving, *Gut* **35,** 1543 (1994).

[24] H. Bartsch, H. Ohshima, and B. Pignatelli, *Mutat. Res.* **202,** 307 (1988).

[25] J. P. Eiserich, M. Hristova, C. E. Cross, D. A. Jones, B. A. Freeman, B. Halliwell, and A. Van Der Vliet, *Nature* **391,** 393 (1998).

[26] A. Samouilov, P. Kuppusamy, and J. L. Zweier, *Arch. Bioch. Biophy.* **357,** 1 (1998).

[27] J. L. Zweier, A. Samouilov, and P. Kuppusamy, *Biochim. Biophys. Acta* **1141,** 250 (1999).

(SMT), a potent inhibitor of inducible NOS.[21] The production of NOS-independent NO was consistent with extensive necrosis found in this model, which is associated with poor perfusion leading to low O_2 tension,[5] increased acidity, and a decrease in nicotinamide adenine dinucleotide phosphate (NADPH). These conditions cause NOS, which is dependent on O_2 and NADPH, to lose activity.[8,21,26,28–30]

Direct Detection of Nitric Oxide Using Hemoglobin or Myoglobin as a Spin Trap

Direct detection of the NO radical is difficult because NO is a short-lived species with a half-life of approximately 1 sec.[31] Heme porphyrin (Hb-Fe^{2+}) is a useful trapping compound for the NO radical due to its high affinity for NO.[32] Similarly, myoglobin (Mb), the single polypeptide subunit of Hb, can also bind NO easily (reaction 2).[32] Whether NO can bind directly to Hb/Mb-Fe^{2+}, forming the nitroso–heme complex (Hb/Mb–Fe^{2+}–NO), depends on the state of oxygenation of the tissue. Hb/Mb–Fe^{2+}–NO is abundant in ischemic-reperfused tissue, which is usually poorly oxygenated.[21] If the tissue is highly oxygenated, ferri–heme complexes (Hb–Fe^{3+}) are formed (reaction 3):

$$\text{Hb/Mb–Fe}^{2+} + \text{NO} \rightarrow \text{Hb/Mb–Fe}^{2+}\text{–NO} \qquad \text{(reaction 2)}$$

$$\downarrow \text{O}_2$$

$$\text{Hb/Mb–Fe}^{2+}\text{–O}_2 + \text{NO} \rightarrow \text{Hb/Mb–Fe}^{3+} + \text{NO}_3{}^{-} \qquad \text{(reaction 3)}$$

Nitric oxide bound directly to either Hb–Fe^{2+} or Mb–Fe^{2+} can be specifically identified and measured using electron paramagnetic resonance (EPR), as it forms stable paramagnetic nitroso–heme complexes with a unique electron paramagnetic signal.[20,33–40] The paramagnetic characteristics of the nitroso–heme complex are

[28] R. R. Giraldez, A. Panda, Y. Xia, S. P. Sanders, and J. L. Zweier, *J. Biol. Chem.* **272,** 21420 (1997).

[29] G. J. Southan and C. Szabo, *Biochem. Pharmacol.* **51,** 383 (1996).

[30] D. A. Lepore, *Nitric Oxide Biol. Chem.* **4,** 541 (2000).

[31] J. S. Beckman and W. H. Koppenol, *Am. J. Physiol.* **271,** C1424 (1996).

[32] A. L. Lehninger, "Biochemistry," Worth Publishers, New York, 1977.

[33] M. C. R. Symons, I. J. Rowland, N. Deighton, K. Shorrock, and K. P. West, *Free Radic. Res. Commun.* **21,** 197 (1994).

[34] D. M. Hall, G. R. Buettner, R. D. Matthes, and C. V. Gisolfi, *J. Appl. Physiol.* **77,** 548 (1994).

[35] H. Kosaka, Y. Sawai, H. Sakaguchi, E. Kumura, N. Harda, M. Watanabe, and T. Shiga, *Am. J. Physiol.* **266,** C1400 (1994).

[36] A. Kozlov, A. Bini, A. Iannone, I. Zini, and A. Tomasi, *Methods Enzymol.* **268A,** 299 (1996).

[37] J. R. Lancaster, J. M. Langrehr, H. A. Bergonia, N. Murase, R. L. Simmons, and R. A. Hoffman, *J. Biol. Chem.* **267,** 10994 (1992).

[38] E. Kumura, T. Yoshimine, K. Iwatsuki, S. Yamanaka, T. Tanaka, T. Hayakawa, T. Shiga, and H. Kosaka, *Am. J. Physiol.* **270,** C748 (1996).

[39] H. Kosaka, M. Watanabe, N. Yoshihara, N. Harada, and T. Shiga, *Biochem. Biophys. Res. Commun.* **184,** 1119 (1992).

[40] U. Westenberger, S. Thanner, H. H. Ruf, R. K. Gersonde, G. Sutter, and O. Trentz, *Free Radic. Res. Commun.* **11,** 167 (1990).

commonly recorded as the first derivative of an absorption spectrum that reveals the unique characteristics of the radical occurring in the portion of the magnetic field ranging from 3000 to 3600 Gauss.[41] NO bound to either Hb–Fe^{2+} or Mb–Fe^{2+} shows a unique hyperfine splitting characteristic of the nitroso–heme complex.[36,42] Figure 1a shows sample EPR spectra obtained for nitroso–heme complexes of NO bound to Hb revealing the characteristic hyperfine splitting of the NO radical.[21]

Quantification of Nitric Oxide Using EPR Spectra of Nitroso–Heme Complexes

Quantification of the NO radical is carried out by a double integration of the area contained by the first derivative of the spectra.[41] The first integration converts the first derivative signal into an absorption signal, and the second integration calculates the area contained by the absorption signal.[41] The double integration method is the preferred method for the quantification of NO because it accounts for the subtleties in the unique shape of the nitroso–heme signal.[20,36,43] The double integration method avoids the confounding factors that are found in the "peak height method" relating to iron–sulfur centers of the mitochondrial respiratory chain or other radicals, which absorb in a portion of the higher end of the magnetic field that overlaps with the nitroso–heme signal.[44]

The molar concentration of NO in an unknown specimen can be determined by direct comparison with a standard curve derived from nitroso-hemoglobin (NOHb) of known molar concentration. The preparation of NOHb standards is described in detail by Kozlov *et al.*[36] Our laboratory has quantified NO levels in specimens of ischemic-reperfused rat skeletal muscle *in vivo* by comparison with NOHb standards.[21] First, EPR spectra recorded for NOHb standards on a Brucker IBM ER 200 spectrometer (Fig. 1A) are quantified by the double integration method using Win-EPR software (Bruker, Germany) and are expressed as the signal intensity per micromolar concentration of the NOHb standard to obtain a standard curve (Fig. 1B). Sample data shown in Fig. 2A represent EPR spectra obtained for gastrocnemius and tibialis muscle specimens from rats treated with either placebo control (saline) or inhibitors of NOS (L-NAME, SMT, or dexamethasone) prior to tourniquet ischemia followed by reperfusion. EPR spectra for sample specimens (Fig. 2A) are quantified using the double integration method and converted into micromolar equivalents of NOHb (Fig. 2B) using the NOHb standard curve. Results show micromolar levels of NO that are found to be insensitive to inhibitors of NOS (L-NAME and SMT), but are partially attenuated by the anti-inflammatory agent

[41] M. Symons, "Electron Spin Resonance Spectroscopy." Van Nostrand Reinhold, New York, 1978.

[42] D. H. O'Keeffe, R. E. Ebel, and J. A. Peterson, *J. Biol. Chem.* **253,** 3509 (1978).

[43] Y. Kotake, *Methods Enzymol.* **268A,** 222 (1996).

[44] J. K. Shergill, R. Cammack, C. E. Cooper, J. M. Cooper, V. M. Mann, and H. V. Schapira, *Biochem. Biophys. Res. Commun.* **228,** 298 (1996).

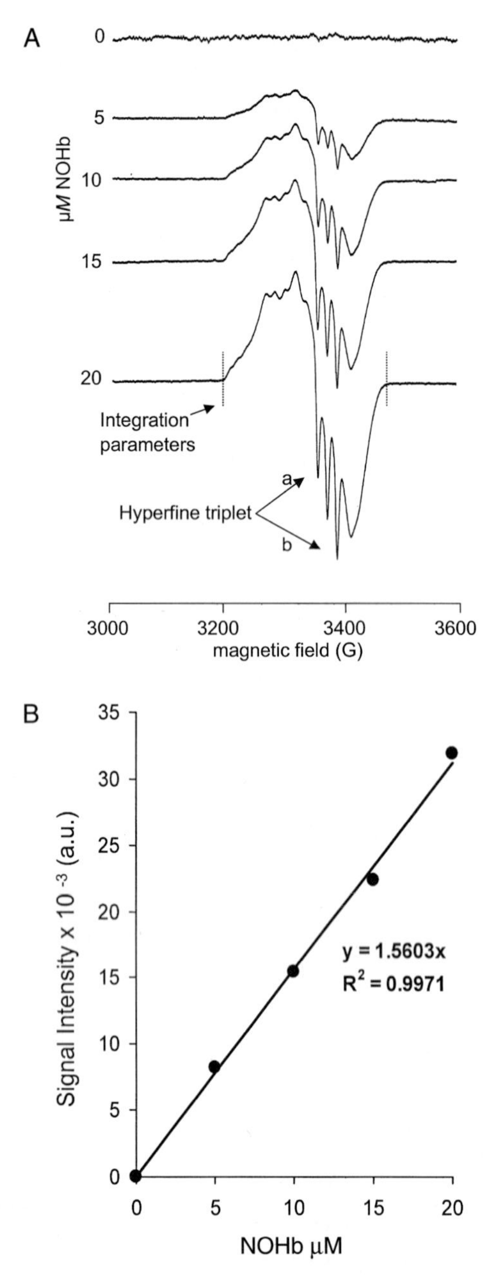

FIG. 1. (A) EPR spectra recorded for increasing concentrations of NOHb standard. The unique splitting characteristic of the nitroso–heme complex known as the hyperfine triplet is indicated by arrows a and b. Vertical dashed lines show the boundaries of the nitroso–heme signal used for integration of the signal. (B) An example of a typical NOHb standard curve. EPR signals recorded for increasing concentrations of NOHb (refer to A) were quantified using the double integration method, and signal intensity was expressed in arbitrary units (a.u.) $\times$ 10^{-3}.

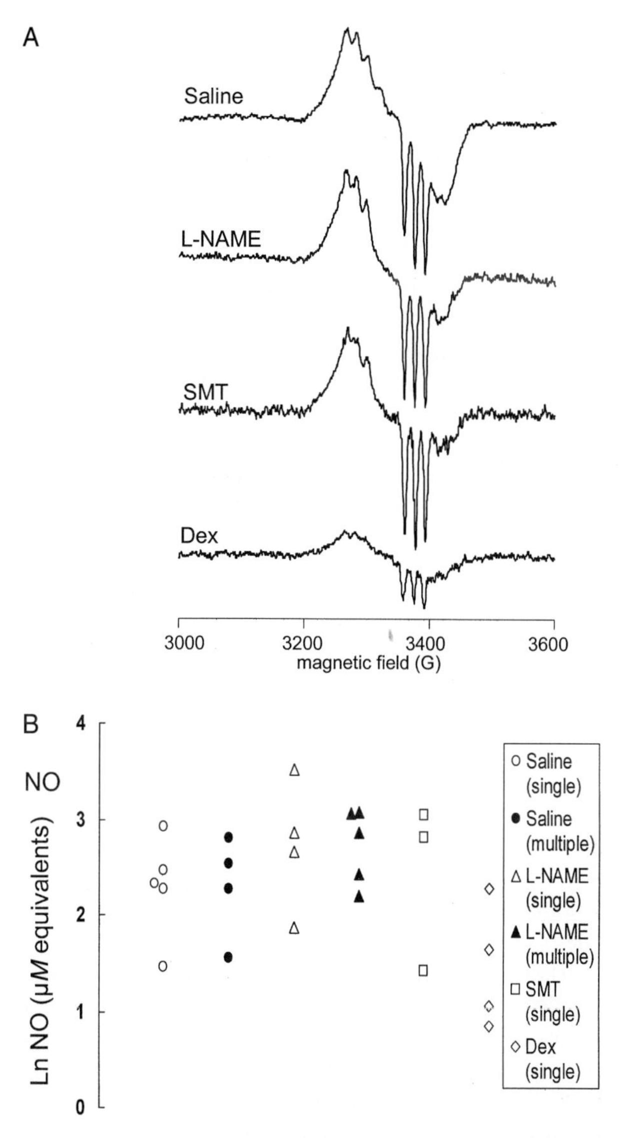

FIG. 2. (A) EPR spectra recorded for rat skeletal muscle 24 hr postischemia–reperfusion. Rats received saline, L-NAME, SMT, or dexamethasone 30 min prior to ischemia. Signals for saline and L-NAME are also representative of those recorded for a second group of rats that received further doses of saline or L-NAME every 4 hr throughout the reperfusion. (B) Muscle nitroso–heme levels were quantified from recorded spectra (A) using the double integration method and were converted into NO (μM equivalents) using a standard curve of NOHb (μM). The NO μM equivalents were then expressed as natural logarithms (Ln). Single or multiple doses of the compounds are indicated on the right-hand side of the figure.

dexamethasone (Fig. 2B). The sensitivity of the method is such that not only are levels of NO detectable in the micromolar concentration range, but differences between treatments can also be detected; e.g., dexamethasone reduces the mean level of nitroso–heme complexes to 66% less than in control-treated animals (one-way analysis of variance, $P = 0.065$, Dunnett's post-hoc test).[21]

Summary

Nitric oxide formation in ischemia–reperfusion injury can be identified and measured directly using EPR of nitroso–heme complexes comprising NO bound to either Mb or Hb–Fe^{2+}. This article described the successful use of this method to detect and quantify the generation of NO formed independently of nitric oxide synthase in ischemia–reperfusion injury to skeletal muscle. The quantification of nitroso-heme complexes using EPR is recommended in ischemia–reperfusion studies of either skeletal or cardiac muscle that aim to characterize the role of nitric oxide.

[17] Nitric Oxide-Dependent Vasodilation in Human Subjects

By JOSEPH A. VITA

Background

There currently is great interest in measuring nitric oxide (NO)-dependent responses in human subjects, particularly responses reflecting the actions of endothelium-derived NO. This interest is based on the growing recognition that the endothelium is a central regular of vascular homeostasis and that impaired biological activity of endothelial factors, such as NO, contributes to the pathogenesis of vascular disease.[1] Endothelium-derived NO has a number important effects in the vasculature, including the ability to inhibit platelet adhesion and aggregation, inhibit leukocyte adhesion, and inhibit growth of vascular smooth muscle cells.[2] However, most *in vivo* studies focus on the vasodilator properties of NO, as they are the most readily measured, particularly in intact human subjects. This article reviews invasive methodology for the assessment of NO-dependent vasodilation in the coronary and peripheral arteries and the more recently developed noninvasive

[1] N. Gokce, J. F. J. Keaney, and J. A. Vita, *in* "Thrombosis and Hemorrhage" (J. Loscalzo and A. I. Shafer, eds.). Williams and Wilkins, 1998.

[2] S. Moncada and A. Higgs, *N. Engl. J. Med.* **329,** 2002 (1993).

0076-6879/02 $35.00

methodology for examination of these responses in the conduit brachial artery. These techniques provide useful information about NO-dependent responses in vascular disease states, the effects of interventions on abnormal responses, and the prognostic importance of such abnormalities in regard to cardiovascular disease risk.

Invasive Studies of Coronary Circulation

NO-dependent vasodilation of coronary arteries may be examined in human subjects undergoing cardiac catheterization. These methods assess changes in coronary artery diameter and/or coronary blood flow during intraarterial agonist infusion using quantitative coronary angiography and intracoronary Doppler, respectively.

Study Protocol

The following protocol is used to examine coronary vasomotor function.[3–8] Patients discontinue vasoactive medications (nitrates, calcium channel blockers, β blockers, angiotensin-converting enzyme inhibitors, etc.) for 24 hr to limit confounding effects on vasomotor tone. Aspirin is not discontinued for ethical reasons. After diagnostic catheterization, a coronary artery suitable for study is identified (diameter >2.0 mm in the portion to be cannulated, supplying noninfarcted myocardium, nonoverlapped segment available for imaging). After giving heparin to make the total dose for the procedure $\geq 10{,}000$ IU, the target vessel is selectively instrumented with a 7- to 8-French (F) guiding catheter, a 3.0-F infusion catheter (e.g., Cook, Inc.), and a 0.014- or 0.018-inch angioplasty wire or Doppler flow wire (Cardiometrics, Inc., Mountain View, CA). Agonists are diluted in normal saline containing heparin (2 IU/ml), and this solution is infused continuously with an infusion pump (Model 55-2222, Harvard Instruments, Needham, MA) to maintain catheter patency. The infusion rate is kept constant at 0.8 ml/min, a rate relatively low compared to the estimated flow in the study vessel (80 ml/min for the left anterior descending or circumflex artery). A list of potential agonists is presented

[3] P. L. Ludmer, A. P. Selwyn, T. L. Shook, R. R. Wayne, G. H. Mudge, R. W. Alexander, and P. Ganz, *N. Engl. J. Med.* **315,** 1046 (1986).

[4] J. A. Vita, C. B. Treasure, E. G. Nabel, J. M. McLenachan, R. D. Fish, A. C. Yeung, V. I. Vekshtein, A. P. Selwyn, and P. Ganz, *Circulation* **81,** 491 (1990).

[5] D. A. Cox, J. A. Vita, C. B. Treasure, R. D. Fish, R. W. Alexander, P. Ganz, and A. P. Selwyn, *Circulation* **80,** 458 (1989).

[6] C. B. Treasure, J. A. Vita, D. A. Cox, R. D. Fish, J. B. Gordon, G. H. Mudge, W. S. Colucci, M. G. Sutton, A. P. Selwyn, R. W. Alexander, and P. Ganz, *Circulation* **81,** 772 (1990).

[7] C. B. Treasure, J. A. Vita, D. A. Cox, R. D. Fish, A. P. Selwyn, R. W. Alexander, and P. Ganz, *Am. J. Cardiol.* **65,** 255 (1990).

[8] J. A. Vita, A. C. Yeung, M. Winniford, J. M. Hodgson, C. B. Treasure, J. L. Klein, S. W. Werns, M. Kern, D. Plotkin, W. J. Shih, Y. Mitchel, and P. Ganz, *Circulation* **102,** 846 (2000).

in Table I. After accounting for dead space, we typically infuse each agonist for 2 min, record five beats of Doppler signal, and then perform quantitative coronary angiography.

Angiography is performed using nonionic contrast and a power injector (Medrad, Inc.) to standardize injection rate and volume (7 to 9 ml/sec for a total of 9 to 12 ml, adjusted to achieve maximal opacity). Currently, images are digitized on line using the Cineangiographic system in our clinical catheterization laboratory (Phillips Medical, Inc.) and then analyzed off line in a blinded manner using customized software (Pie Medical, Inc.). However, many other options are currently available for image acquisition and analysis. The software identifies the lumen borders of the studied segment and then calculates the length and the average diameter, using the coronary guiding catheter for calibration. Images from the same point in the cardiac cycle (typically end diastole) are selected for analysis (three per condition). Fixed anatomic landmarks (side branches) are used to ensure that the same portion of the artery is measured each time.

An important issue is how segments are selected for analysis, as the vasodilator response may vary by segment and by vessel within the same individual. Our approach to this issue has evolved over time, but at present, we measure the entire length of vessel distal to the point of agonist infusion (typically 3–4 mm in length). In the past, when studies involved examination of the effects of an intervention, we selected a relatively short (2 to 5 mm) segment for study that displayed the maximal vasoconstrictor or vasodilator response.[4] However, choice of the most extreme response is subject to errors associated with "regression to the mean." For this reason, we recommend analysis of a longer segment. The resulting changes in coronary diameter are less extreme, but are more representative of the entire vessel and appear to be more reproducible over time.[8]

Safety Issues

Only a trained interventional cardiologist should perform these studies. Patient selection criteria include the following: (1) Clinically stable, without recent rest angina or acute myocardial infarction (at least 72 hr). (2) No significant stenoses in the left main or proximal left anterior descending or circumflex artery. (3) No significant collaterals to the AV nodal artery in the studied vessel (because of the concern about acetylcholine-induced AV block). (4) The target vessel should not supply more than approximately 40% of the viable myocardium because of the possibility of hemodynamic compromise if the artery is occluded during instrumentation or drug infusion. We do not study the right coronary or dominant circumflex coronary artery because of the possibility of inducing complete heart block, however, other investigators have performed such studies with a temporary pacemaker in place.

More than a thousand patients have safely undergone study using these methods. Regarding adverse events, the most common occurrence is severe coronary constriction during acetylcholine infusion. This effect resolves within a few seconds

TABLE I

AGONISTS FOR STUDY OF CORONARY EPICARDIAL FUNCTION

Agonist	Final concentration	Cited source	Rationale[a]	Epicardial response in normal subjects[b]	Epicardial response in CAD patients
Acetylcholine	10^{-9} to 10^{-6} M	Miochol Ciba Vision, Inc.	Stimulates EDNO production and direct VSMC constrictor	Range −53 to 37% Mean −7 ± 21%[4]	Range −100 to 10% Mean −21%[8]
Substance P	5 to 40 pmol/min	Sigma, Inc.	Stimulates EDNO production	12 ± 2%[31]	1 ± 1%[8]
Bradykinin	0.5 to 2.5 μg/min	Sigma, Inc.	Stimulates EDNO production	11 ± 5%[32]	—
Monomethyl-L-arginine	32 to 64 μmol/min	ClinAlfa, Inc.	Inhibits EDNO synthesis	−15 ± 2	−4 ± 1%[33]
Adenosine	0.022 to 2.2 mg/min	Available for clinical use	Endothelium-independent vasodilation	28 ± 5%	29 ± 19%[13]
Nitroglycerin	1 to 20 μg/min	Available for clinical use	NO-dependent, endothelium-independent vasodilation	31 ± 4%	34 ± 4%[3]
Sodium nitroprusside	20 to 40 μg/min	Available for clinical use	NO-dependent, endothelium-independent vasodilation	23 ± 10%	27 ± 19%[13]
L-Arginine	160 μmol/min	Sigma, Inc.	NOS substrate	—	—
Flow	Up to 400% increase	Stimulated by adenosine or papaverine infusion into distal vessel	Stimulates EDNO production	13 ± 1%	2 ± 2%[5]

[a] EDNO, endothelium-derived nitric oxide; VSMC, vascular smooth muscle cell; NOS, nitric oxide synthase.

[b] Depends on the presence of risk factors and reflects population of patients referred for catheterization.

of stopping the infusion and has no consequence if the investigators are vigilant. Use of the Doppler wire to monitor coronary flow continuously is helpful in this regard. We are aware of three major complications worldwide since the mid-1980s, although it is likely that others have occurred and not been reported. Two patients experienced thrombotic coronary occlusion of the study vessel and suffered acute myocardial infarction. In both cases, the patient had received inadequate heparinization (a minimum of 10,000 IU is recommended). The third patient died during intracoronary infusion of acetylcholine when a severe coronary spasm occurred and the patient developed profound hypotension and could not be resuscitated. In this case, a large left anterior descending coronary artery supplying the majority of the left ventricle was the target vessel. Furthermore, the right coronary artery was occluded, and collateral vessels originating from the left anterior descending supplied its territory. When acetylcholine-induced spasm occurred, there was no blood flow to the majority of the left ventricle. All of these complications were avoidable.

Interpretation of Results

Healthy subjects with angiographically normal coronary arteries and few or no coronary risk factors will display arterial dilation during acetylcholine infusion. In contrast, patients with coronary artery disease display vasoconstriction to acetylcholine, but preserved vasodilation to nitroglycerin (Fig. 1). Although very healthy subjects dilate in response to acetylcholine, most subjects who have a clinical indication for cardiac catheterization display a modest constriction in response to this stimulus, even if they have angiographically normal coronary arteries.[4] The degree of constriction relates to the presence of coronary risk factors,[4] but there is substantial variability in the response among individuals, and even within different parts of the coronary circulation.[9] The severity of constriction is believed to reflect the severity of endothelial dysfunction, and reduced constriction in response to therapy is interpreted as an improvement in endothelial vasomotor function (increased endothelium-derived NO and/or decreased endothelium-derived constricting factors, such as endothelin), assuming that the vasodilator response to nonendothelium-dependent vasodilators remain unaffected.[10–12] Similarly, the vasodilator response to other endothelium-dependent vasodilators, such as substance

[9] H. El-Tamimi, M. Mansour, T. J. Wargovich, J. A. Hill, R. A. Kerensky, C. R. Conti, and C. J. Pepine, *Circulation* **89,** 45 (1994).

[10] T. J. Anderson, I. T. Meredith, A. C. Yeung, B. Frei, A. Selwyn, and P. Ganz, *N. Engl. J. Med.* **332,** 488 (1995).

[11] G. B. Mancini, G. C. Henry, C. Macaya, B. J. O'Neill, A. L. Pucillo, R. G. Carere, T. J. Wargovich, H. Mudra, T. F. Luscher, M. I. Klibaner, H. E. Haber, A. C. Uprichard, C. J. Pepine, and B. Pitt, *Circulation* **94,** 258 (1996).

[12] C. B. Treasure, J. L. Klein, W. S. Weintraub, J. D. Talley, M. E. Stillabower, A. S. Kosinski, J. Zhang, S. J. Boccuzzi, J. C. Cedarholm, and R. W. Alexander, *N. Engl. J. Med.* **332,** 481 (1995).

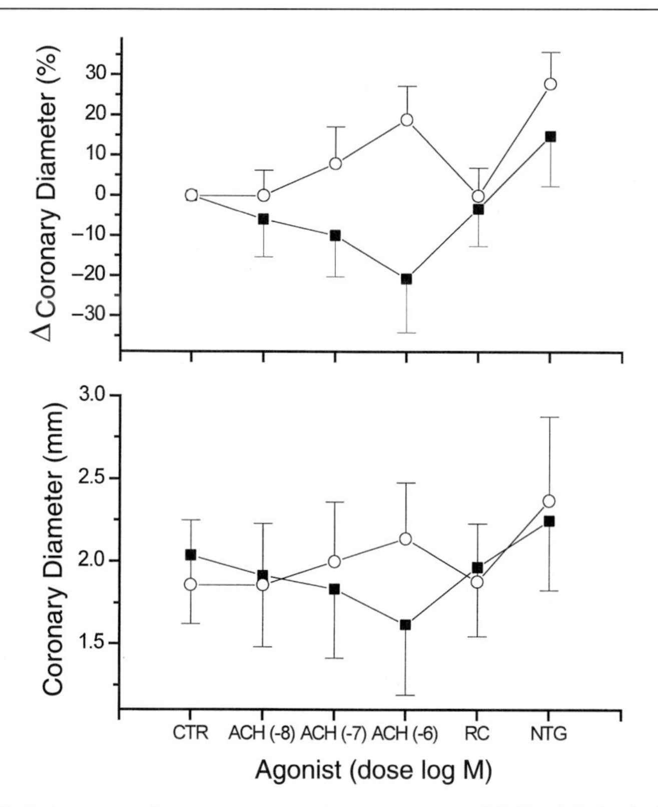

FIG. 1. Typical coronary diameter response to intracoronary acetylcholine infusion in 26 patients with coronary artery disease (■) and 10 subjects with no coronary risk factors (○). (Bottom) Absolute coronary diameter and (top) percentage change in coronary diameter during saline infusion (CTR), during infusion of increasing doses of acetylcholine (10^{-8}, 10^{-7}, 10^{-6} M), during repeat control infusion (RC), and during infusion of 16 μg/min nitroglycerin (NTG). Dose-dependent vasodilation is provoked by acetylcholine in normal subjects, whereas coronary constriction is provoked by acetylcholine in patients with angiographically evident coronary disease. Coronary diameter returns to baseline during the 5-min recontrol infusion, and nitroglycerin produces equivalent coronary vasodilation in both groups (unpublished data). Plotted values are mean $\pm$ SD.

P, bradykinin, and flow, tends to be blunted in the setting of coronary atherosclerosis (Table I). Monomethyl-L-arginine, a selective NOS inhibitor, produces more marked vasoconstriction in healthy subjects compared to patients with coronary disease and/or risk factors, reflecting inhibition of basal nitric oxide synthesis.[13] As with the conduit coronary artery, the vasodilator response of coronary

[13] A. A. Quyyumi, N. Dakak, N. P. Andrews, S. Husain, S. Arora, D. M. Gilligan, J. A. Panza, and R. O. Cannon, *J. Clin. Invest.* **95,** 1747 (1995).

microvessels, assessed as changes in coronary blood flow, is blunted in patients with cardiovascular disease compared to normal subjects.[8] Most notably, there is growing evidence that the severity of the impairment in acetylcholine response or flow-mediated dilation identifies individuals at risk for future cardiovascular disease events.[14,15]

Invasive Studies of Forearm Circulation

Given the limitations associated with study of the coronary circulation, there has been great interest in the study of NO-dependent vasodilation in peripheral arteries. In these studies, agonists are infused into the brachial artery, and vasodilation of forearm microvessels is assessed as increased forearm blood flow using venous occlusion plethysmography.[16] Vasodilation of conduit vessels in the forearm may be assessed during agonist infusion using high-resolution vascular ultrasound.[17]

We use the following protocol to assess forearm microvessels responses.[18,19] All vasoactive medications are withheld for 24–48 hr and patients are asked to fast overnight and, if applicable, refrain from smoking for at least 12 hr. Patients are studied in a temperature-controlled room at 24°. A 20- to 22-gauge arterial catheter (Arrow, Inc.) is inserted in the nondominant brachial artery near the antecubital crease using sterile techniques and local anesthesia (1% lidocaine). A mercury-in-silastic strain gauge, upper arm and wrist cuffs, a plethysmograph (Hokanson, Inc.), and a physiological recorder (Gould Instrument Systems) are used to perform venous occlusion plethysmography. During these studies, the upper arm venous occlusion cuff is inflated to 40 mm Hg (or adjusted to optimize the tracing), and circulation to the hand is excluded by inflation of the wrist cuff to suprasystolic pressure before initiation of flow measurements. At least five measurements are made and averaged for each condition. Arterial pressure is monitored continuously and is used to calculate forearm vascular resistance, when appropriate.

The procedure is extremely safe, although a trained physician should perform the arterial catheter insertion. In study of over 500 patients, the most common adverse event is the development of an ecchymosis at the puncture site. In our

[14] V. Schachinger, M. B. Britten, and A. M. Zeiher, *Circulation* **101,** 1899 (2000).

[15] J. A. Suwaidi, S. Hamasaki, S. T. Higano, R. A. Nishimura, D. R. Holmes, and A. Lerman, *Circulation* **101,** 948 (2000).

[16] M. A. Creager, J. P. Cooke, M. E. Mendelsohn, S. J. Gallagher, S. M. Coleman, J. Loscalzo, and V. J. Dzau, *J. Clin. Invest.* **86,** 228 (1990).

[17] E. H. Lieberman, M. D. Gerhard, A. Uehata, A. P. Selwyn, P. Ganz, A. C. Yeung, and M. A. Creager, *Am. J. Cardiol.* **78,** 1210 (1996).

[18] D. L. Sherman, J. F. Keaney, Jr., E. S. Biegelsen, S. J. Duffy, J. D. Coffman, and J. A. Vita, *Hypertension* **35,** 936 (2000).

[19] S. J. Duffy, E. S. Biegelsen, M. Holbrook, J. D. Russell, N. Gokce, J. F. Keaney, and J. A. Vita, *Circulation* **103,** 2799 (2001).

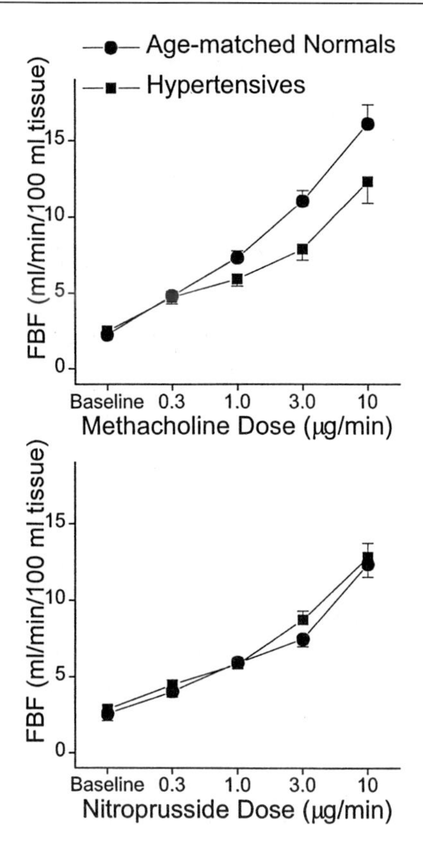

FIG. 2. Typical forearm microvascular response to intraarterial agonist infusion. Forearm blood flow responses to methacholine (top) were examined using venous occlusion plethysmography in 22 patients with hypertension and 20 age-matched controls. The response to methacholine was lower in hypertensive patients ($P < 0.001$ by repeated measures ANOVA). The response to nitroprusside (bottom) was equivalent in 9 patients with hypertension and 9 age-matched controls ($P =$ NS). Reproduced from D. L. Sherman, J. F. Keaney, Jr., E. S. Biegelsen, S. J. Duffy, J. D. Coffman, and J. A. Vita, *Hypertension* **35,** 936 (2000) with permission.

experience, two patients have developed transient hand numbness, presumably due to median nerve trauma. Clearly, the procedure has the potential to produce arterial injury, but no surgical repair has been necessary in our experience.

As in the coronary circulation, forearm blood flow responses during infusion of the muscarinic agonists methacholine or acetylcholine are impaired in patients with coronary artery disease[19] or coronary risk factors such as hypertension (Fig. 2),[18,20] hypercholesterolemia,[16] and diabetes mellitus.[21] The methodology is ideally suited

[20] J. A. Panza, A. A. Quyyumi, J. E. Brush, and S. E. Epstein, *N. Engl. J. Med.* **323,** 22 (1990).

[21] H. H. Ting, F. K. Timimi, K. S. Boles, S. J. Creager, P. Ganz, and M. A. Creager, *J. Clin. Invest.* **97,** 22 (1996).

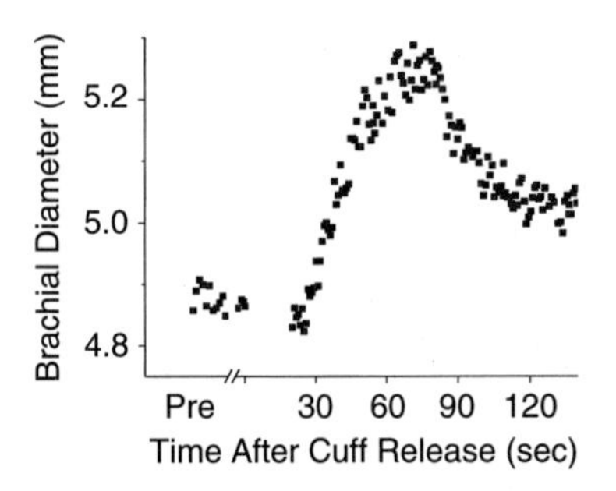

FIG. 3. Typical flow-mediated dilation response in a healthy individual. Flow-mediated dilation was determined with the occlusion cuff on the upper arm as described previously [G. N. Levine, B. Frei, S. N. Koulouris, M. D. Gerhard, J. F. Kearney, Jr., and J. A. Vita, *Circulation* **96,** 1107 (1996)]. Images of the brachial artery were digitized (one image/cardiac cycle on the R wave) at baseline (Pre) and continuously for 2 min beginning 20 sec after cuff release using a commercially available image acquisition system (CVI Acquisition, Information Integrity, Inc., Stow, MA). Brachial artery diameters were measured using an automated edge-detection system (Brachial Tools, Medical Imaging Applications, Iowa City, IA). Reproduced from M. Corretti *et al., J. Am. Coll. Cardiol.* **39,** 257 (2002).

to assess potential mechanisms of vascular dysfunction in human subjects, and a wide and expanding variety of specific inhibitors are available to examine changes in dose response to endothelium-dependent and endothelium-independent agonists. Two studies demonstrated that abnormalities of endothelium-dependent vasodilation in the forearm circulation predict risk for future cardiovascular disease events.[22,23]

Noninvasive Studies in Forearm Circulation

The growing recognition of the importance of endothelial dysfunction has driven the development of noninvasive methods. We use vascular ultrasound to examine flow-mediated dilation in the brachial or femoral arteries as developed by Lieberman and colleagues,[17] Creager and colleagues,[24] and Vita and Keaney.[25] In these studies, reactive hyperemia is induced by 5-min cuff occlusion of the limb. The resultant increase in conduit artery flow increases shear stress at the endothelial surface and provokes NO-dependent vasodilation. The conduit artery is imaged by two-dimensional (2-D) ultrasound, and as shown in Fig. 3, flow-mediated dilation is readily detectable with a peak response approximately 60 sec after cuff release. Studies have demonstrated that this response is largely NO

[22] F. Perticone, R. Ceravolo, A. Pujia, G. Ventura, S. Iacopino, A. Scozzafava, A. Ferraro, M. Chello, P. Mastroroberto, P. Verdecchia, and G. Schillaci, *Circulation* **104,** 191 (2001).

[23] T. Heitzer, T. Schlinzig, K. Krohn, T. Meinertz, and T. Munzel, *Circulation* **104,** 2673 (2001).

[24] D. S. Celermajer, K. E. Sorensen, V. M. Gooch, D. J. Spiegelhalter, O. I. Miller, I. D. Sullivan, J. K. Lloyd, and J. E. Deanfield, *Lancet* **340,** 1111 (1992).

[25] J. A. Vita and J. F. J. Keaney, *in* "Diagnostics of Vascular Diseases: Principals and Technology" (P. Lanzer and M. Lipton, eds.). Springer, Berlin, 1996.

dependent in healthy subjects.[17,26] Anderson and co-workers[27] showed a close correlation between ultrasound-determined, flow-mediated dilation in the brachial artery and the vasomotor response to acetylcholine in coronary arteries of patients undergoing studies with both techniques. Evolution of this noninvasive technique has permitted examination of endothelial vasomotor function in large numbers of individuals and in lower risk populations. It is also ideally suited for repeated studies in the same individual and, thus, is useful for studies of interventions designed to improve endothelial function.

Study Protocol at Boston University School of Medicine

As for our invasive studies, patients withhold vasoactive medications for 24–48 hr and fast overnight prior to study. Any required blood sample collections and blood pressure measurements are done in the nonstudy arm, leaving the study arm (usually the right arm) unperturbed. A narrow-gauge occlusion cuff or pediatric cuff is positioned on the *upper portion* of the study arm, as high up as possible. The patient's arm is positioned with the hand supinated in a comfortable position and cushioned with pillows, as necessary.

The brachial artery ultrasound presets on a Toshiba 140A ultrasound system are as follows: dynamic range, 60 dB; persistence, 4; edge enhancement, 3; post-processing curve, 3; predepth gain compensator, "on," and echo filter "resolution" (7.5 to 9.5 MHz). These are adjusted further for individual patients to optimize the image. After initially scanning the area to identify the brachial artery, the depth and angle are adjusted to position the vessel in the center of the screen. Pan expand or zoom is turned on to maximize the number of pixels per millimeter. The sonographer then spends several minutes obtaining an optimal image of the artery (see Fig. 4). The features of an optimal image are listed in Table II.

We digitize end-diastolic images (R-wave gated) using a workstation connected to the ultrasound system. The acquisition hardware used in our laboratory was custom built by Cardiovascular Engineering, Inc. (Holliston, MA), but a number of other systems are available commercially. Ten baseline 2-D and Doppler images are recorded. Then the sonographer inflates the cuff to 200 mm Hg or 50 mm Hg above the systolic blood pressure for 5 min. Immediately after cuff deflation, the sonographer obtains 10 sec of Doppler signal to capture the peak hyperemic flow signal and then the mode is switched to 2-D and images are recorded up to 120 sec after cuff deflation. A typical response in a healthy subject is shown in Fig. 3.

Following completion of the deflation scan, the subject rests for 15 min to reestablish baseline diameter and flow. Then 2-D images are recorded before and

[26] R. Joannides, W. E. Haefeli, L. Linder, V. Richard, E. H. Bakkali, C. Thuillez, and T. F. Luscher, *Circulation* **91,** 1314 (1995).

[27] T. J. Anderson, A. Uehata, M. D. Gerhard, I. T. Meredith, S. Knab, D. Delagrange, E. Leiberman, P. Ganz, M. A. Creager, A. C. Yeung, and A. P. Selwyn, *J. Am. Coll. Cardiol.* **26,** 1235 (1995).

TABLE II
FEATURES OF AN OPTIMAL TWO-DIMENSIONAL ULTRASOUND IMAGE

The artery lies horizontally on the screen
The artery is centered vertically on the screen
The visualized segment is straight
A clear media–adventia hypoechoic line (M line) is well visualized on both near and far walls over at least 50% of the vessel length
The lumen is sonolucent without shadowing in the center of the lumen
An identifiable landmark is seen

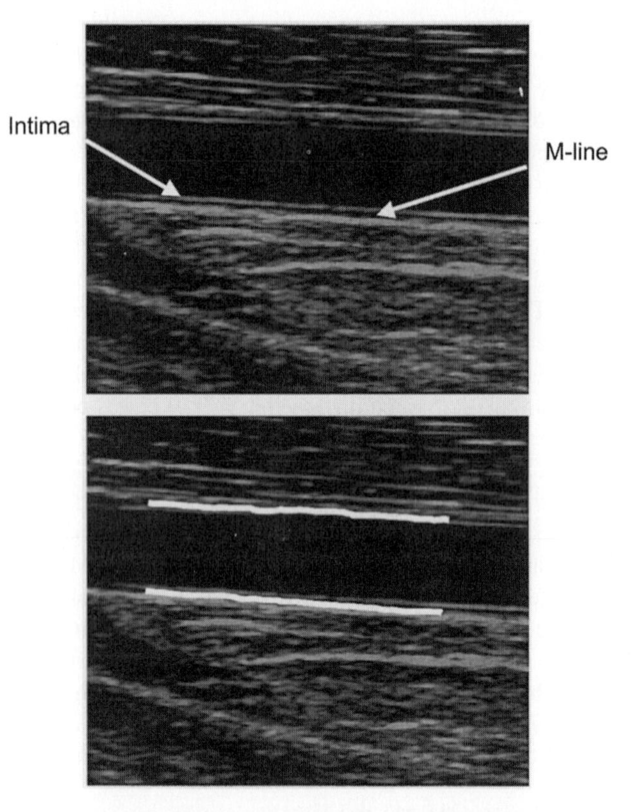

FIG. 4. (Top) A two-dimensional ultrasound image of the brachial artery providing good visualization of the intima and M line (arrows). (Bottom) An example of where the vessel wall would be traced (either manually or using automatic edge detection software). Although this image provides good definition of the intima and has many other features of an optimal image, it could be more horizontal on the screen.

3 min after administration of sublingual nitroglycerin (0.4 mg). The nitroglycerin portion of the study is omitted if systolic blood pressure is low (<100 mm Hg systolic), if the subject has nitrate intolerance, history of migraine headaches or Viagra use, or, in the case of premenopausal female subjects, if a pregnancy test has not been performed to confirm that the subject is not pregnant.

A number of procedures are used to ensure reproducibility when serial studies are performed in the same individual over time. All studies are performed in a temperature-controlled room with the patient in a fasting state at approximately the same time of day. The same ultrasound system presets are maintained for all studies, and the skin site is noted in a log. When the subject undergoes follow-up ultrasound, the sonographer locates and reimages the identical segment of artery using this information and stored images from the initial study. Whenever possible, the same sonographer performs all studies for an individual subject and for a particular study.

We use a commercially available software package for the analysis of brachial artery flow-mediated dilation (Brachial Tools, Medical Imaging Applications, Inc., Iowa City, IA). Briefly, this software uses an automatic edge detection algorithm to detect the M line at the near and far wall (Fig. 4). The distance between lines is then measured and averaged. Flow-mediated dilation is then calculated using the baseline diameter and the vessel diameter 60 sec after cuff deflation and may be expressed as absolute difference or as percentage change from baseline. There currently is some interest in examining the peak response, which does not necessarily occur at 60 sec, but the utility of this approach remains unknown. Three Doppler images are traced to calculate the velocity time integral before and immediately following cuff deflation.

A major controversy in the literature is whether the occlusion cuff should be placed on the upper or lower arm. We use an upper arm cuff position, but many investigators advocate a lower arm position because of concerns that ischemia may stimulate nonendothelium-dependent dilation of the brachial artery. The absolute vasodilator response is substantially lower with the lower arm position, presumably because the extent of reactive hyperemia is less. It is our feeling that the upper arm position is desirable because a larger hyperemic response is induced and there is a consequently greater vasodilator response (both are approximately 50% greater with the upper cuff position).[28] Given that the resolution of vascular ultrasound with a 7.5-MHz probe is approximately 0.1 mm (2.5% of a 4-mm artery), maximizing the "signal-to-noise ratio" is highly desirable. The vasodilator response with an upper arm cuff position has been shown to depend largely on NO production, as approximately 75% of the response is blocked during concomitant infusion of the NOS inhibitor NG-monomethyl-L-arginine.[17] Preliminary studies in our laboratory

[28] T. C. Mannion, J. A. Vita, J. F. J. Keaney, E. J. Benjamin, L. Hunter, and J. Polak, *Vasc. Med.* **3,** 263 (1998).

TABLE III
MEAN VALUES WITH 95% CONFIDENCE INTERVALS (CI) FOR BRACHIAL ARTERY ULTRASOUND STUDIES[a]

	Women			Men		
Age (years):	<30	30 to 49	≥50	<30	30 to 49	≥50
Sample size:	$n = 69$	$n = 62$	$n = 45$	$n = 61$	$n = 91$	$n = 46$
Baseline diameter (mm)	3.00	3.37	3.58	3.97	4.18	4.62
95% CI	2.90–3.10	3.23–3.50	3.37–3.79	3.84–4.11	4.07–4.30	4.44–4.81
Difference (mm)[b]	0.44	0.49	0.42	0.39	0.40	0.39
95% CI	0.40–0.48	0.45–0.53	0.36–0.47	0.35–0.43	0.37–0.44	0.32–0.46
FMD (%)[c]	15.1	15.3	12.4	10.1	9.9	8.7
95% CI	13.4–16.7	13.7–16.9	10.5–13.7	8.8–11.4	8.9–11.0	6.9–10.5

[a] Results shown are for patients without a clinical history of coronary artery disease, peripheral vascular disease, diabetes mellitus, or hypertension.
[b] Absolute value of flow-mediated dilation.
[c] Flow-mediated dilation expressed as percentage change in diameter.

have demonstrated that there is no change in blood pH or lactate concentration during a 5-min period of cuff occlusion.

Studies performed with either cuff position have comparable results in terms of the differences between normal subjects and subjects with various vascular disease states, and in the response to interventions. We present mean values with 95% confidence intervals for men and women without coronary artery disease (Table III). Many interventions that improve flow-mediated dilation reduce cardiovascular disease risk, including lipid-lowering therapy, angiotensin-converting enzyme inhibitors, smoking cessation, and exercise.[1] Table IV provides sample size estimates for intervention studies with two groups and a parallel study design. Finally, there are emerging data that the presence of impaired flow-mediated dilation predicts cardiovascular risk (Fig. 5),[29] and the long-term prognostic value of the methodology is currently the subject of intense investigation. All of these findings support the idea that the methodology will prove to be a useful surrogate marker of cardiovascular disease with utility for identifying novel risk factors and new approaches to therapy.

[29] N. Gokce, J. F. Keaney, Jr., L. Hunter, M. T. Watkins, J. O. Menzoian, and J. A. Vita, *Circulation* **105,** 1567 (2002).
[30] G. N. Levine, B. Frei, S. N. Koulouris, M. D. Gerhard, J. F. Keaney, Jr., and J. A. Vita, *Circulation* **96,** 1107 (1996).
[31] K. Egashira, S. Suzuki, and Y. Hirooka, *Hypertension* **25,** 201 (1995).
[32] M. Kato, N. Shiode, T. Yamagata, H. Matsuura, and G. Kajiyama, *Heart* **78,** 493 (1997).
[33] A. A. Quyyumi, N. Dakak, D. Mulcahy, N. P. Andrews, S. Husain, J. A. Panza, and R. O. Cannon, *J. Am. Coll. Cardiol.* **29,** 308 (1996).
[34] M. Corretti, T. J. Anderson, E. J. Benjamin, D. Celermajer, F. Charbonneau, M. Creager, W. Daley, J. Deanfield, H. Drexler, M. Gerhard, D. Herrington, P. Vallance, J. Vita, and R. Vogel, *J. Am. Coll. Cardiol.* **39,** 257 (2002).

TABLE IV
SAMPLE SIZE ESTIMATES FOR A TWO-GROUP PARALLEL STUDY DESIGN[a]

Change in FMD to be detected	Number of patients per group: Power = 0.80	Number of patients per group: Power = 0.90
2.5	30	40
2.0	46	61
1.5	81	107
1.1	149	198
1.0	179	240

[a] FMD, flow-mediated dilation, two-tailed, $\alpha = 0.05$.

Conclusions

Methodology for the study of NO-dependent vasodilation in intact human subjects has evolved greatly since the mid-1980s. There is great interest in bringing some or all of these techniques into the clinical arena, where they could be used as a tool to estimate cardiovascular risk or to guide risk reduction therapy. However,

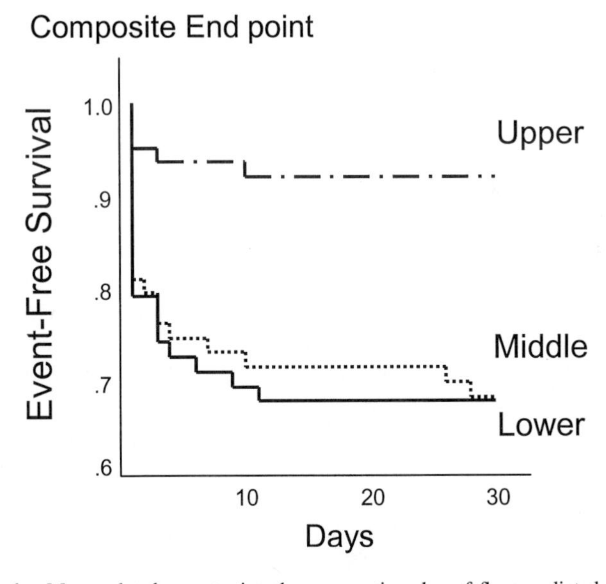

FIG. 5. Kaplan Meyer plot demonstrating the prognostic value of flow-mediated dilation of the brachial artery measured noninvasively. Patients undergoing vascular surgery had ultrasound studies performed preoperatively, and cardiovascular disease events within 30 days of surgery were tabulated. Patients with relatively preserved flow-mediated dilation (upper tertile, >8.1% dilation) had significantly fewer events compared to patients with more impaired flow-mediated dilation (middle and lower tertiles). Adapted from N. Gokce, J. F. Keaney, Jr., L. Hunter, M. T. Watkins, J. O. Menzoian, and J. A. Vita, *Circulation* **105,** 1567 (2002) with permission.

a great deal of work remains to be done to standardize the methodology and establish the ranges of normal and abnormal responses. In the meantime, it is clear that these methods provide much useful information about vascular physiology and pathophysiology in intact human subjects.

Acknowledgments

A Program Project Grant (HL60886) and a Specialized Center of Research Grant (HL55993) from the National Institutes of Health provide support for portions of this work.

[18] Antioxidant and Diffusion Properties of Nitric Oxide in Low-Density Lipoprotein

By HOMERO RUBBO, HORACIO BOTTI, CARLOS BATTHYÁNY, ANDRÉS TROSTCHANSKY, ANA DENICOLA, and RAFAEL RADI

Introduction

Nitric oxide (·NO) is an endogenously synthesized free radical that plays an important role in regulating critical oxidant reactions. Nitric oxide crosses biological membranes readily and can concentrate in hydrophobic compartments, being able to participate in chemical reactions in both aqueous and lipid phases. Although being a free radical, ·NO does not react readily with most organic molecules. In fact, ·NO is neither a strong oxidant nor a strong reductant.[1] However, ·NO reacts rapidly with both carbon- and oxygen-centered radicals to yield a variety of reactive intermediates.[2] Although ·NO-derived metabolites may exert oxidative modifications in low-density lipoprotein (LDL) through peroxynitrite (PN), nitrogen dioxide (·NO_2), and/or the nitrite-myeloperoxidase system,[2–4] ·NO itself inhibits lipid oxidation-dependent processes.[5,6] In fact, ·NO diffusion into LDL can represent a key antioxidant mechanism by acting in the hydrophobic core of LDL where the ratio of oxidizable lipids to endogenous antioxidants is much

[1] R. Radi, *Chem. Res. Toxicol.* **9,** 828 (1996).

[2] R. Radi, A. Denicola, G. Ferrer, B. Alvarez, and H. Rubbo, *in* "Nitric Oxide Biology and Pathobiology" (L. J. Ignarro, ed.), p. 57. San Diego, Academic Press, 2000.

[3] C. Leeuwenburgh, M. Hardy, S. Hazen, P. Wagner, S. Oh-ishi, U. P. Steinbrecher, and J. W. Heinecke, *J. Biol. Chem.* **272,** 1433 (1997).

[4] E. A. Podrez, D. Schmitt, and S. L. Hazen, *J. Clin. Invest.* **103,** 1547 (1999).

[5] H. Rubbo, S. Parthasarathy, B. Kalyanaraman, S. Barnes, M. Kirk, and B. A. Freeman, *Arch. Biochem. Biophys.* **324,** 15 (1995).

[6] S. P. Goss, N. Hogg, and B. Kalyanaraman, *J. Biol. Chem.* **272,** 21647 (1997).

0076-6879/02 $35.00

greater than at the particle surface. Low-density lipoprotein oxidation by different *in vitro* systems, including copper and azo compounds, and the antioxidant role of ·NO have been discussed previously in this series.[7] This article describes the underlying diffusion and antioxidant properties of ·NO during low fluxes of PN-mediated LDL oxidation.

Peroxynitrite-Mediated Oxidative Modifications of LDL

Low-density lipoprotein oxidation represents a critical event in the pathogenesis of atherosclerosis. Although ·NO predominantly elicits antioxidant actions in the vasculature under low concentrations, excess ·NO formation can result in the formation of ·NO-derived oxidants that cause biological damage. In fact, ·NO reacts at diffusion limited rates[8] with superoxide anion ($O_2^{\cdot -}$) to form PN ($ONOO^- + ONOOH$, $k \sim 10^{10}\ M^{-1}\ s^{-1}$), a reactive nitrogen species with strong oxidizing properties.[2]

Peroxynitrite is synthesized as described previously[9–11] from acidified nitrite and hydrogen peroxide and is stored in 1 *N* NaOH ($\varepsilon = 1.7\ mM^{-1}\ cm^{-1}$ at 302 nm). Human LDL is obtained from normolipidemic donors by ultracentrifugation, followed by size-exclusion HPLC purification as described elsewhere.[9,12] Then, PN is added to LDL either as a single bolus or as a continuous flux using a motor-driven syringe.[9] Specifically, PN (20 μl, 6–12 m*M*) is placed into a gas-tight Hamilton syringe attached to the syringe infusion pump (SAGE Instruments, Boston, MA) so that it leads directly into the LDL (2 ml, 0.4 mg ml^{-1}) suspension in 200 m*M* potassium phosphate buffer, pH 7.4, 0.1 m*M* DTPA. The pump is set to infuse PN over 60–500 min at a constant rate of 1 $\mu M\ min^{-1}$. As stock solutions of PN are prepared in 1 *N* NaOH, control samples have the same volume of 1 *N* NaOH infused into LDL. Sample pH is monitored routinely. The influence of the stable decomposition products of PN is tested using PN previously decomposed at pH 7.4 in buffer before addition to LDL. Incubations are performed under continuous stirring at 37°, and aliquots (20–50 μl) are removed at different time points for the determination of conjugated dienes ($\varepsilon = 26\ mM^{-1}\ cm^{-1}$ at 234 nm), cholesteryl ester hydroperoxides, lipophilic antioxidants (α-tocopherol, γ-tocopherol, ubiquinol-10, and carotenoids), free amino

[7] S. P. Goss, B. Kalyanaraman, and N. Hogg, *Methods Enzymol.* **301,** 444 (1999).

[8] R. Kissner, T. Nauser, P. Bugnon, P. Lye, and W. H. Koppenol, *Chem. Res. Toxicol.* **10,** 1285 (1997).

[9] A. Trostchansky, C. Batthyány, H. Botti, R. Radi, A. Denicola, and H. Rubbo, *Arch. Biochem. Biophys.* **395,** 225 (2001).

[10] R. Radi, J. S. Beckman, K. Bush, and B. A. Freeman, *Arch. Biochem. Biophys.* **288,** 481 (1991).

[11] H. Rubbo, R. Radi, M. Trujillo, M. Telleri, B. Kalyanaraman, S. Barnes, M. Kirk, and B. A. Freeman, *J. Biol. Chem.* **269,** 26066 (1994).

[12] C. Batthyány, C. Santos, H. Botti, C. Cerveñansky, R. Radi, O. Augusto, and H. Rubbo, *Arch. Biochem. Biophys.* **384,** 335 (2000).

group content (with trinitrobenzene sulfonic acid), and fluorescence emission spectra analysis ($\lambda_{exc} = 365$ nm).[9] For lipophilic antioxidant determination, LDL samples (10 μl) are mixed with 100 μl ethanol and vortexed for 10 sec following hexane (500 μl) addition. Then samples are centrifuged at 1000*g* for 5 min at 4°, and the hexane phase is removed (400 μl), evaporated under a stream of argon, and resolvated immediately in 50 μl reagent alcohol (methanol/ ethanol/2-propanol, 100/95/5, v/v/v).[13,14] Simultaneous antioxidant analyses are performed on a Supelcosil LC-18 column (25 × 0.46 cm, 5 μm), mobile phase of 31.7 m*M* ammonium formate in methanol/ethanol/2-propanol (765/173/62, v/v/v) at a flow rate of 1 ml min^{-1}. The detection system includes an electrochemical detector (600 mV) connected in series with a fluorometric ($\lambda_{ex} = 295$ nm, $\lambda_{em} = 330$ nm) and a UV/VIS detector ($\lambda_1 = 266$ nm, $\lambda_2 = 450$ nm). Cholesteryl ester hydroperoxides are extracted using methanol/hexane and are analyzed by reverse-phase HPLC on the same LC-18 column.[15] Analyses of oxidation products are performed using acetonitrile/isopropanol/water (44/54/2; v/v/v) with detection at 210 and 234 nm.

Peroxynitrite oxidizes and nitrates LDL with a concomitant depletion of antioxidants and formation of lipid–protein adducts, converting LDL to a proatherogenic form recognized by macrophage scavenger receptors.[9,16–18] The biological half-life of PN can be estimated in the range of 10–100 ms[19]; therefore, at the intravascular/tissue level, it could potentially diffuse some distance before enacting target molecule reactions, e.g., in LDL. Mechanisms for diffusion of PN in lipid environments have been revealed.[19]

Nitric Oxide Inhibition of LDL Oxidation

Nitric oxide plays a critical role in regulating lipid oxidation induced by reactive oxygen and nitrogen species.[11] Nitric oxide donors, including NONOates (1-substituted diazen-1-ium 1,2-diolates), represent an ideal tool for studying the effects of the slow release of ·NO on LDL oxidation due to the ability to release ·NO thermolytically at selected constant rates depending on its particular half-lives.[7] Importantly, the high pH dependence of the rates of ·NO liberation from NONOates allows activation or inactivation when desired. As an example,

[13] T. Menke, P. Niklowitz, S. Adam, M. Weber, B. Schluter, and W. Andler, *Anal. Biochem.* **282,** 209 (2000).

[14] E. Teissier, E. Walters-Laporte, C. Duhem, G. Luc, J. C. Fruchart, and P. Duriez, *Clin. Chem.* **42,** 430 (1996).

[15] L. Kritharides, W. Jessup, J. Gifford, and R. T. Dean, *Anal. Biochem.* **213,** 79 (1993).

[16] A. Graham, N. Hogg, B. Kalyanaraman, V. J. O'Leary, V. Darley-Usmar, and S. Moncada, *FEBS Lett.* **330,** 181 (1993).

[17] N. Hogg, V. Darley-Usmar, M. T. Wilson, and S. Moncada, *FEBS Lett.* **326,** 199 (1993).

[18] S. R. Thomas, M. Davies, and R. Stocker, *Chem. Res. Toxicol.* **11,** 484 (1998).

[19] A. Denicola, J. Souza, and R. Radi, *Proc. Natl. Acad. Sci. U.S.A.* **95,** 3566 (1998).

a stock solution of 1-hydroxy-2-oxo-3-(*N*-ethyl-2-aminoethyl)-3-ethyl-1-triazene (NOC-12, $t_{0.5} = 90$ min) is prepared in degassed 50 m*M* potassium phosphate buffer, pH 9.5, and is added to buffer systems used for LDL oxidation reactions (100 m*M* potassium phosphate, pH 7.4, 20°). Nitric oxide production rates and solution concentrations are measured electrochemically using ·NO-selective electrode (WPI Instruments, Ann Arbor, MI) calibrated by measuring ·NO released from 50 μM KNO_2, 0.1 *M* KI, and 0.1 *M* H_2SO_4, according to the following reaction performed under anaerobic conditions: $2KNO_2 + 2KI + 2H_2SO_4 \rightarrow 2^{\cdot}NO + I_2 + 2H_2O + 2K_2SO_4$.[20]

Nitric oxide causes a prolongation of the lag time and inhibition of the propagation phase of lipid oxidation (Fig. 1) through its chain-breaking activity.[5,6] Moreover, loss of amino groups by oxidants and protein–lipid fluorescent adduct formation are also prevented by ·NO (Fig. 1).[9] Nitric oxide has multiple physicochemical qualities that make it an effective lipid antioxidant, including its ability to react with unsaturated lipid-reactive species to yield nitrogen-containing radical-radical termination products.[5,11] The almost diffusion-limited reaction[21] of ·NO with peroxyl radicals (LOO·, $k = 3 \times 10^9\ M^{-1}\ s^{-1}$) will be more facile than the initiation of secondary peroxidation propagation reactions of LOO· with unsaturated lipids ($k = 30 - 200\ M^{-1}\ s^{-1}$). Moreover, ·NO has a partition coefficient of 6–8 for *n*-octanol-H_2O and concentrates in lipophilic milieu such as the hydrophobic core of LDL (Fig. 2).[22]

Several *in vitro* studies[5,11,23,24] demonstrated the presence of nitrogen-containing products of polyunsaturated fatty acids (LH), including alkylnitrites (LONO), alkylnitrates ($LONO_2$), and nitrolipids (LNO_2). In particular, the product of the LOO·/·NO condensation reaction (LOONO) may be either hydrolyzed to form LOOH and nitrite or cleaved by homolysis to LO· and ·NO_2, with rearrangement of LO· to L(O)·, followed by its recombination with ·NO_2.[23] Nitrated[24] (E. Lima, P. DiMascio, H. Rubbo, and D. Abdalla, *Biochemistry,* in press) lipid formation is now shown to occur both *in vitro* and *in vivo* as potential indicators of the inhibitory role that ·NO plays during lipid oxidation processes.

Nitric Oxide–Antioxidant Interactions during LDL Oxidation

We have observed that by virtue of its high reactivity with lipid radicals, ·NO protects lipophilic antioxidants from oxidation in lipid model systems, exhibiting

[20] H. Rubbo, R. Radi, D. Anselmi, M. Kirk, S. Barnes, J. Butler, J. P. Eiserich, and B. A. Freeman, *J. Biol. Chem.* **275,** 10812 (2000).

[21] S. Padmaja and R. E. Huie, *Biochem. Biophys. Res. Commun.* **195,** 539 (1993).

[22] A. Denicola, C. Batthyány, E. Lissi, B. A. Freeman, H. Rubbo, and R. Radi, *J. Biol. Chem.* **277,** 932 (2002).

[23] V. B. O'Donnell, J. P. Eiserich, P. H. Chumley, M. J. Jablonsky, N. R. Krishna, M. Kirk, S. Barnes, V. M. Darley-Usmar, and B. A. Freeman, *Chem. Res. Toxicol.* **12,** 83 (1999).

[24] M. Balazy, T. Iesaki, J. L. Park, H. Jiang, P. M. Kaminski, and M. S. Wolin, *J. Pharmacol. Exp. Ther.* **299,** 611 (2001).

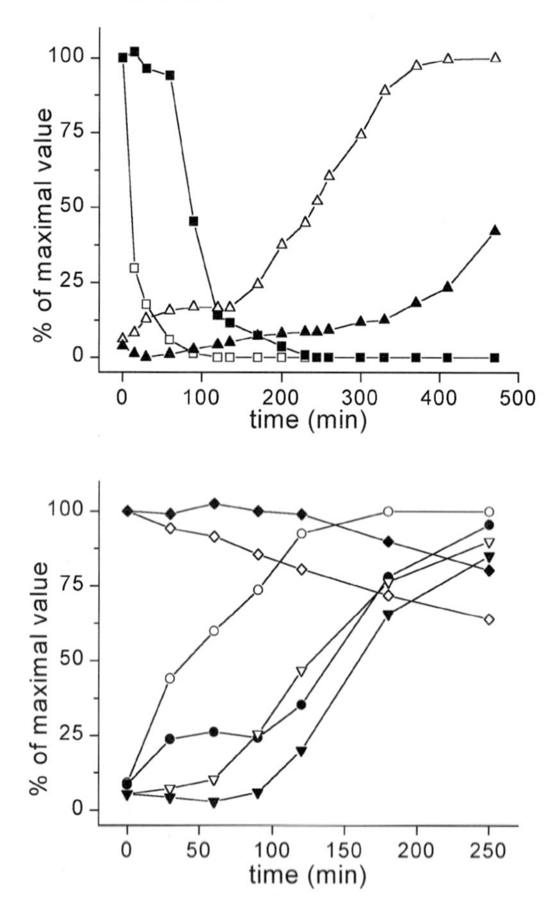

FIG. 1. Oxidative modifications of LDL by PN and the protective role of ·NO. LDL (0.4 mg ml^{-1}) was incubated at 37° in 200 m*M* phosphate buffer, 0.1 m*M* DTPA, pH 7.4, with PN (1 μM min^{-1}) in the absence (open symbols) and presence (closed symbols) of 20 μM NOC-12. (Top) Conjugated dienes (△) and α-tocopherol (□). (Bottom) Free amino groups (◇), cholesteryl linoleate hydroperoxide (◁), and fluorescent adducts (◯).

greater antioxidant capacity if α-tocopherol is present.[20] Human LDL contains a number of antioxidants that inhibit lipid oxidation, with α-tocopherol the most abundant (~six α-tocopherol molecules per LDL particle), with other antioxidants (e.g., carotenoids, ubiquinol-10) present in much lower quantities. α-Tocopherol, localized at the surface of the LDL particle, provides minimal protection to lipid components in the hydrophobic core of LDL. Indeed, the principal oxidizable lipid, cholesteryl linoleate, is localized in the core of the lipoprotein, away from

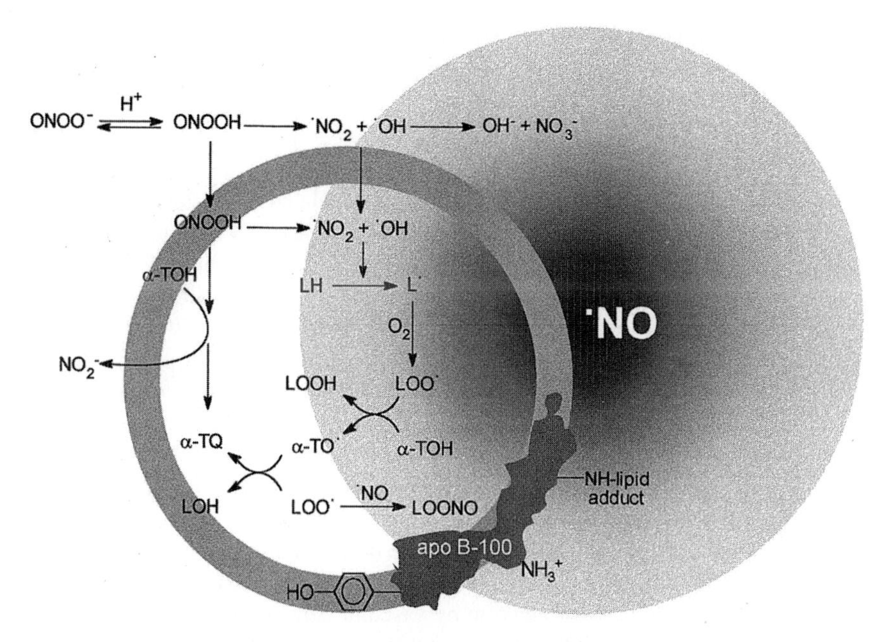

FIG. 2. Proposed mechanism for the antioxidant actions of ·NO in LDL. Lipid hydroperoxides (LOOH) generated in LDL from lipid oxidation by fluxes of PN can react with apo B-100 free amino groups to form fluorescent lipid–protein adducts. In addition, PN causes α-tocopherol (α-TOH) depletion in LDL via one or two electron oxidation mechanisms, yielding α-tocopheryl radical (α-TO·) or α-tocopherylquinone (α-TQ), respectively. Nitric oxide diffuses into the surface, as well as the hydrophobic core of LDL, to inhibit lipid oxidation, lipid–protein adduct formation, and antioxidant depletion through its radical–radical diffusion-limited termination reaction with lipid radicals.

the more polar tocopherols.[25] Because the reaction of LOO· with α-tocopherol ($k = 2.5 \times 10^6\ M^{-1}\ \mathrm{s}^{-1}$) occurs with a rate constant three orders of magnitude less than for the reaction of LOO· with ·NO and because ·NO may access to hydrophobic sites where α-tocopherol is not present, ·NO could act more readily than or in concert with α-tocopherol as an antioxidant defense against oxygen radical-derived oxidized lipid species.

The concentrations of ·NO in the subendothelium of small arterioles have been estimated to be in the range of 250–500 nM (reviewed in Ref. 26), and these concentrations seem to be more than sufficient to produce antioxidant action in LDL. At these physiologically relevant concentrations, ·NO represents a primary source of lipid-soluble antioxidant activity in LDL, protecting endogenous lipophilic antioxidants from low fluxes of PN-mediated depletion. In fact, ·NO at

[25] H. Esterbauer, J. Gebicki, H. Puhl, and G. Jurgens, *Free Radic. Biol. Med.* **13,** 341 (1992).
[26] D. G. Buerk, *Annu. Rev. Biomed. Engin.* **3,** 109 (2001).

initial rates of 300 nM min^{-1} (from 20 μM NOC-12) and added to LDL prior to PN infusion (1 μM min^{-1}) inhibited all PN-dependent oxidative processes (Fig. 1).

Nitric Oxide versus Peroxynitrite Fluxes

Many biological effects of ·NO are critically related to the rate of ·NO formation, and therefore timing of flux of ·NO determines the extent of antioxidant and even its prooxidant actions.[5,11,20] Because lipid oxidation involves a propagation phase by which initial radical reactions are amplified severalfold, an effective ·NO-mediated inhibition requires a threshold ·NO flux and steady-state concentration under which propagation reactions predominate over inhibition.[27] In fact, PN added to LDL as bolus produces fragmentation and nitration of apolipoprotein B-100 and a fast depletion of carotenoids and tocopherols but not extensive lipid oxidation.[3,9,17,18,28] This is probably due to the fact that in the bolus addition condition, PN yields a burst of initiator radicals (i.e., ·OH and $\cdot NO_2$) from the homolysis of ONOOH that rapidly lead the lipid oxidation process to termination by radical-radical reactions.[9,10] In contrast to the bolus addition, a flux of PN generates a low continuous source of lipid radicals, favoring lipid oxidation propagation reactions.[9] Enhanced $O_2^{\cdot -}$ production by vascular cells has been observed under different pathogenic stimuli that may shift the proportion of ·NO yielding PN and decreasing ·NO bioactivity and its antioxidant properties.[29,30]

Nitric Oxide Diffusion into LDL

The diffusion of ·NO into LDL and membranes has been studied, taking advantage of the fact that ·NO is a good collisional quencher of pyrene fluorescence. If collisional quenching occurs then, there is contact between the fluorophore (pyrene) and the quencher (·NO) that provokes deactivation. The decrease in fluorescence intensity is described by the Stern–Volmer equation:

$$\frac{I_o}{I} = 1 + K_{SV}\,[\cdot NO] \quad (1)$$

$$K_{SV} = k_{NO}\tau_o \quad (2)$$

where τ_o is the lifetime of the excited probe in the absence of quencher, k_{NO} is the bimolecular quenching constant, and K_{SV} is the Stern–Volmer constant.[22,31,32]

[27] H. Rubbo and R. Radi, *in* "Handbook in Antioxidants" (E. Cadenas, ed.), p. 689. Dekker, New York, 2001.

[28] O. M. Panasenko, V. S. Sharov, K. Briviba, and H. Sies, *Arch. Biochem. Biophys.* **373,** 302 (2000).

[29] M. E. Pueyo, J. F. Arnal, J. Rami, and J. B. Michel, *Am. J. Physiol.* **274,** C214 (1998).

[30] S. K. Wattanapitayakul, D. M. Weinstein, B. J. Holycross, and J. A. Bauer, *FASEB J.* **14,** 271 (2000).

[31] A. Denicola, J. M. Souza, R. Radi, and E. Lissi, *Arch. Biochem. Biophys.* **328,** 208 (1996).

[32] X. Liu, M. J. Miller, M. S. Joshi, D. Thomas, and J. Lancaster, Jr., *Proc. Natl. Acad. Sci. U.S.A.* **95,** 2175 (1998).

The idea is to select pyrene derivatives that could be incorporated into LDL or membranes at different depths. The probes selected were 1-(pyrenyl) methyltrimethylammonium (PMTMA), 11-(1-pyrenyl)undecyltrimethyl ammonium (PUTMA), and 1-(pyrenyl)methyl 3-(9-octadecenoyloxy) 22,23-bisnor-5-cholenate (PMChO). The pyrene moiety is responsible for the fluorescence (λ_{exc} = 337 nm, λ_{em} = 396 nm), and the substituent determines the location of the probe. Due to its cationic character, PMTMA is located at the surface of LDL or membranes that are rich in negatively charged phospholipids, whereas the undecyl derivative PUTMA is expected to locate deeper in the membrane. The cholesteryl ester derivative PMChO penetrates to the hydrophobic core of LDL, rich in esterified cholesterol.

The pyrenes are incorporated into LDL by adding aliquots of an ethanolic stock solution (final probe concentration 10^{-6} *M*) and incubating overnight at 4°. The excess of fluorescent probe is removed by gel filtration (Sephadex G-25, Pharmacia, equilibrated with 50 m*M* potassium phosphate buffer, pH 7.4). Probe leakage is observed with PMTMA (after 5 hr), but not with PMChO, and must be controlled during long incubations. The differential degree of penetration of the probes is confirmed by different susceptibility to quenching by hydrophilic quenchers such as iodide and tryptophan.[22] The lifetimes of the probes (τ) either in solution or incorporated into the different systems are measured by following the kinetics of fluorescence decay after excitation with a nitronite nitrogen laser in argon.[22,31] Fluorescence quenching experiments are performed by measuring steady-state fluorescence intensities in the absence (I_o) and presence (I) of different ·NO concentrations. The experiments are conducted in an anaerobic fluorimetric cuvette where the buffer (50 m*M* potassium phosphate, pH 7.4) is deoxygenated extensively by bubbling argon (a critical step, as oxygen also quenches pyrene fluorescence) before addition through the rubber septum of the sample suspension with the probe previously incorporated. The mixture is bubbled very gently with argon (extensive bubbling disrupts the LDL particle), and the fluorescence intensity is recorded before (I_o) and after ·NO addition (I) from a freshly prepared stock solution. The concentration of the stock ·NO solution is measured using the oxyhemoglobin method,[33] and the final ·NO concentration in the cuvette is calculated using the dilution factor. The alkaline wash of the ·NO gas is important to get rid of nitrite, which is a moderate quencher. With these data, a Stern–Volmer plot is constructed (Fig. 3), and from the slope (K_{SV}), the apparent second-order quenching constant (k_{NO}) is determined knowing the lifetime of the probe (τ_o).

Using the Einstein–Smoluchowsky equation, the apparent diffusion coefficient of the quencher (D'_{NO}) can be estimated according to Eq. (3):

$$D'_{NO} = \frac{k_{NO} \times 10^3}{4\pi RN} \tag{3}$$

[33] M. E. Murphy and E. Noack, *Methods Enzymol.* **233,** 240 (1994).

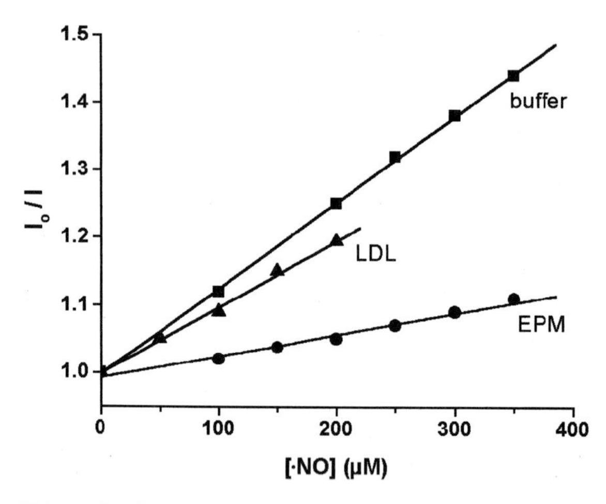

FIG. 3. Stern–Volmer plots for pyrene derivative fluorescence quenching by ·NO. PMTMA fluorescence (λ_{exc} = 337 nm, λ_{em} = 396 nm) in 50 m*M* phosphate buffer, pH 7.4 (■), erythrocyte plasma membranes (EPM, ●), and LDL (▲) quenched by increasing concentrations of ·NO.

where *N* is Avogadro's number and *R* is the sum of the molecular radii of the probe plus ·NO (i.e., 6.9, 7.7, and 9.1×10^{-8} cm for PMTMA, PUTMA, and PMChO, respectively). Results obtained are summarized in Table I. The apparent diffusion coefficient of ·NO in native LDL is approximately 2000 $\mu m^2\ s^{-1}$, only half the value obtained for the probe in solution, which indicates that ·NO can easily get access to the particle, even the hydrophobic core of the LDL particle. It is interesting to note that the D'_{NO} in the erythrocyte plasma membrane is significantly lower (Table I), indicating that the diffusivity of ·NO in LDL exceeds that of biomembranes.

TABLE I
FLUORESCENCE LIFETIME (τ_o), BIMOLECULAR QUENCHING CONSTANTS (k_{NO}), AND DIFFUSION COEFFICIENTS (D_{NO}) FOR NO

System	Probe	τ_o (ns)	k_{NO} ($M^{-1}\ s^{-1}$)	D_{NO} ($\mu m^2\ s^{-1}$)
Buffer	PMTMA	50	25.4×10^9	4500
Liposomes	PMTMA	90	6.9×10^9	1700
Liposomes	PUTMA	170	6.8×10^9	1500
EPM	PMTMA	136	2.3×10^9	500
EPM	PUTMA	230	5.8×10^9	1300
LDL	PMTMA	80	12.0×10^9	2300
LDL	PMChO	200	12.0×10^9	2000

Conclusions

The capacity of ·NO to diffuse and concentrate in the hydrophobic core of LDL (Fig. 2) and its rapid combination reactions with lipid radical species affirm a role of ·NO as a major lipophilic antioxidant in LDL. The recent observations that vascular ·NO production is antiatherogenic, limiting LDL oxidation and foam cell formation, in part implies contributory antioxidant actions of ·NO by mechanisms discussed here.

Acknowledgments

This work was supported by NIH Grants R03 TW00999, TW001493 (B.A.F., A.D., H.R., and R.R.), SAREC (Sweden), ICGEB (Italy), and the Howard Hughes Medical Institute (R.R.), TWAS (Italy), and CSIC (Uruguay) (A.D.), Pfizer-Fundación Manuel Pérez (H.R. and C.B.). C.B. and A.T. were partially supported by a fellowship from PEDECIBA, Uruguay. R.R. is an International Research Scholar of the Howard Hughes Medical Institute.

[19] Measurements of Redox Control of Nitric Oxide Bioavailability

By ANNONG HUANG, SHANE R. THOMAS, and JOHN F. KEANEY, JR.

Introduction

Nitric oxide (NO) is produced in the vascular endothelium and is crucial in the control of vascular tone,[1] arterial pressure,[2–4] smooth muscle cell proliferation,[5] and platelet activity.[6] The bioactivity of endothelium-derived NO is impaired in many vascular diseases,[7,8] and this defect is thought to contribute to the clinical manifestations of vascular disease.[9]

[1] D. C. Lefroy, T. Crake, N. G. Uren, G. J. Davies, and A. Maseri, *Circulation* **88,** 43 (1993).
[2] D. D. Rees, R. M. Palmer, and S. Moncada, *Proc. Natl. Acad. Sci. U.S.A.* **86,** 3375 (1989).
[3] P. L. Huang, Z. Huang, H. Mashimo, K. D. Bloch, M. A. Moskowitz, J. A. Bevan, and M. C. Fishman, *Nature* **377,** 239 (1995).
[4] E. G. Shesely, N. Maeda, H. S. Kim, K. M. Desai, J. H. Krege, V. E. Laubach, P. A. Sherman, W. C. Sessa, and O. Smithies, *Proc. Natl. Acad. Sci. U.S.A.* **93,** 13176 (1996).
[5] U. C. Garg and A. Hassid, *J. Clin. Invest.* **83,** 1774 (1989).
[6] H. Azuma, M. Ishikawa, and S. Sekizaki, *Br. J. Pharmacol.* **88,** 411 (1986).
[7] P. L. Ludmer, A. P. Selwyn, T. L. Shook, R. R. Wayne, G. H. Mudge, R. W. Alexander, and P. Ganz, *N. Engl. J. Med.* **315,** 1046 (1986).
[8] J. A. Vita, C. B. Treasure, E. G. Nabel, J. M. McLenachan, R. D. Fish, A. C. Yeung, V. I. Vekshtein, A. P. Selwyn, and P. Ganz, *Circulation* **81,** 491 (1990).
[9] V. Schachinger, M. B. Britten, and A. M. Zeiher, *Circulation* **101,** 1899 (2000).

0076-6879/02 $35.00

The action of NO is subject to modulation by oxidative stress and cellular antioxidant defense status. For example, both superoxide[10] and the oxidized form of low-density lipoprotein (LDL)[11] may inactivate NO directly, limiting its bioactivity from endothelial cells. Similarly, arteries with reduced SOD activity exhibit defective vasodilation to both endogenous and exogenous sources of NO, presumably through the formation of peroxynitrite.[12,13] Ascorbic acid, an important determinant of intracellular redox state, enhances the production of NO in endothelial cells,[14,15] principally through an increase in the NOS cofactor tetrahydrobiopterin.[15] In platelets, vitamin E status modulates platelet NO bioactivity by altering the balance between agonist-stimulated NO and superoxide production.[16] Thus, cellular redox status and oxidative stress modulate NO bioactivity through a variety of mechanisms.

The purpose of this article is to review methods employed for assessing endothelial cell NO bioactivity in response to modulating oxidative stress or cellular antioxidant status. The methods contained herein allow for the assessment of all facets of NO bioactivity, including NO production, NO inactivation, and activation of guanylyl cyclase.

Overall Assessment of NO Bioactivity

Nitric oxide mediates many of its biological effects by activating soluble guanylyl cyclase, resulting in increased intracellular cGMP concentration.[17] Fortunately, cultured endothelial cells contain soluble guanylyl cyclase, thereby facilitating the use of cellular cGMP accumulation as an overall assessment of NO bioactivity. This assay encompasses all the components of NO action and metabolism, including NO production, its degradation by reaction with superoxide, and stimulation of guanylyl cyclase. We typically assess NO bioactivity as the nitro-L-arginine methyl ester (L-NAME)-sensitive accumulation of cGMP.[15]

The assay starts with confluent monolayers of endothelial cells in six-well plates. We typically use cells with less than 10 passages that have been confluent for 1–2 days. We have found that porcine and human aortic endothelial cells produce a more robust response than either bovine aortic endothelial cells or human umbilical vein endothelial cells. Cells are equilibrated for 30 min in HEPES-buffered

[10] R. J. Gryglewski, R. M. Palmer, and S. Moncada, *Nature* **320,** 454 (1986).

[11] K. Kugiyama, S. A. Kerns, J. D. Morrisett, R. Roberts, and P. D. Henry, *Nature* **344,** 160 (1990).

[12] A. Mugge, J. K. Elwell, T. E. Peterson, and D. G. Harrison, *Am. J. Physiol.* **260,** C219 (1991).

[13] H. A. Omar, P. D. Cherry, M. P. Mortelliti, T. Burke-Wolin, and M. S. Wolin, *Circ. Res.* **69,** 601 (1991).

[14] R. Heller, F. Münscher-Paulig, R. Gräbner, and U. Till, *J. Biol. Chem.* **274,** 8254 (1999).

[15] A. Huang, J. A. Vita, R. C. Venema, and J. F. Keaney, Jr., *J. Biol. Chem.* **275,** 17399 (2000).

[16] J. E. Freedman, L. Li, R. Sauter, and J. F. Keaney, Jr., *FASEB J.* **14,** 2377 (2000).

[17] M. A. Moro, R. J. Russell, S. Cellek, I. Lizasoain, Y. Su, V. M. Darley-Usmar, M. W. Radomski, and S. Moncada, *Proc. Natl. Acad. Sci. U.S.A.* **93,** 1480 (1995).

physiologic salt solution (PSS) containing 22 m*M* *N*-(2-hydroxyethyl)piperazine-*N'*-2-ethanesulfonic acid (HEPES, pH 7.4), 124 m*M* NaCl, 5 m*M* KCl, 1 m*M* $MgCl_2$, 1.5 m*M* $CaCl_2$, 0.16 m*M* HPO_4, 0.4 m*M* H_2PO_4, 5 m*M* $NaHCO_3$, and 5.6 m*M* D-glucose. The HEPES-buffered PSS also contains 200 μM 3-isobutyl-1-methylxanthine (IBMX, to inhibit phosphodiesterases) and 200 μM L-arginine. Equilibrated cells are then treated with the agonist of interest (A23187, bradykinin, serotonin, etc.) for 1–5 min in the absence or presence of 300 μM L-NAME. These time points are chosen because exposures >5 min are associated with loss of intracellular cGMP. After treatment, cells are lysed by the addition of 1 ml 6% ice-cold trichloroacetic acid. Cell lysates are then scraped, and precipitated cellular proteins are cleared by centrifugation at 13,000*g* for 10 min at 4°. Both the cellular protein and supernatant fractions are saved for later analysis, with the latter containing acid-stable cGMP that is stable frozen at −80° for many weeks. We use a commercially available ELISA (Cayman, Ann Arbor, MI) to determine cGMP. In using the ELISA, we do not find a need to acetylate samples derived from porcine cells, but all other samples benefit from the increased sensitivity derived from acetylation. The cGMP content can be normalized to cell protein after solubilizing the protein pellet in 0.1 *N* NaOH and protein determination by the BCA assay (Pierce). In our hands, we find that resting porcine endothelial cells contain ~5.3 ± 0.8 pmol cGMP/mg protein, and cells stimulated with 1 μM A23187 exhibit peak values of 58.2 ± 20.2 pmol cGMP/mg protein.

We have reported that loading cells with ascorbic acid[15] tends to enhance NO bioactivity, whereas thiol oxidation with diamide or thiol modulation with CDNB impairs NO bioactivity.[18] As a first step in sorting out the mechanism(s) for such observations, we needed to determine if this is a consequence of altered NO production or a change in guanylyl cyclase responsiveness to NO.

Guanylyl Cyclase Response to NO

To examine the guanylyl cyclase response, we simply provide an exogenous source of NO. The experimental design is exactly as described earlier, except that endothelial cells are treated with NO derived from diethylamine nonoate (DEANO; 0.01–1 μM). We have employed DEANO because it releases authentic NO rather than NO+, as is the case with sodium nitroprusside.[19] If one observes that some alteration in the intracellular redox state materially changes the cellular response to exogenous NO, two possible explanations include a change in the catalytic activity of guanylyl cyclase or direct inactivation of NO, perhaps through superoxide. To evaluate the former, one can take advantage of the fact that the soluble and particulate forms of guanylyl cyclase share a common catalytic subunit, differing

[18] A. Huang, H. Xiao, J. M. Samii, J. A. Vita, and J. F. Keaney, Jr., *Am. J. Physiol* **281,** C719 (2001).
[19] J. S. Stamler, D. J. Singel, and J. Loscalzo, *Science* **258,** 1898 (1992).

only in the regulatory subunit. Therefore, in order to test the catalytic potential of guanylyl cyclase, one can stimulate the particulate form, and we typically do this with atrial natriuretic peptide (0.01–1 μM) using the experimental setting described earlier for DEANO. If there is no change in the ANP response based on the intracellular redox state, one can conclude that the catalytic capacity of guanylyl cyclase is intact. It is then important to investigate other reasons for reduced responsiveness to NO. One possibility is that NO can become inactivated by superoxide. To investigate this scenario, treatment of cells with a cell-permeable SOD mimic (EUK-8, MnTMPyP, PEG-SOD, etc.) would be expected to normalize the response to DEANO.

In our studies with ascorbic acid, the cellular responses to NO and ANP were independent of redox status.[15] These findings prompted us to examine any effect of ascorbate status on endothelial cell eNOS catalytic activity and NO production.

Determination of Endothelial NO Production

L-[^{3}H]Arginine to L-[^{3}H]Citrulline Conversion Assay

In order to examine endothelial cell NO production, we start with an intact cell assay of eNOS catalytic activity by assessing the conversion of L-[^{3}H]arginine to L-[^{3}H]citrulline under conditions in which eNOS catalytic activity is stimulated. We assure specificity for eNOS by using only that L-[^{3}H]citrulline production that is sensitive to inhibition by L-NAME. The advantage of this approach is that many facets of NO production are involved, including L-arginine uptake and the cellular content of cofactors required for eNOS activity. For this assay, confluent endothelial cell monolayers in six-well plates are washed with HEPES-buffered PSS and incubated in PSS for 30 min. Cells are then treated with 10 μM L-arginine with 3.3 μCi of L-[^{3}H]arginine (NEN, 53.4 Ci/mmol) in the absence or presence of an agonist to stimulate endothelial NO production, such as 1 μM calcium ionophore (A23187 or ionomycin). This experimental design is equally applicable to other agonists, including bradykinin, serotonin, and H_2O_2. After 15 min, cells are washed with HEPES-buffered PSS to remove unincorporated radioactivity and are lysed with 250 μl 100% ethanol (15 min), after which 2 ml of ice-cold stop buffer (20 mM sodium acetate, pH 5.5, 1 mM L-citrulline, 2 mM EDTA, 2 mM EGTA) is added. The content of L-[^{3}H]citrulline is then determined using anion-exchange chromatography. For chromatography, we use 2-ml Dowex AG50W-X8 columns (Bio-Rad, Hercules, CA) preequilibrated with stop buffer. The cell lysate is applied to the column, L-citrulline is eluted with three 2-ml washes of distilled water, and the eluent is assayed by scintillation counting. We report data as L-[^{3}H]citrulline that is sensitive to pretreatment of the PAEC for 30 min with L-NAME (500 μM) and is expressed as dpm per 10^6 endothelial cells. The total cell uptake of L-[^{3}H]arginine can be determined from an aliquot of the total cell lysate just prior to chromatography, as all incorporated ^{3}H must have entered the cell as L-[^{3}H]arginine. Important controls include L-[^{3}H]citrulline production in the absence of eNOS stimulation and cell-free controls to account for L-[^{3}H]arginine that may not be retained by the anion-exchange chromatography.

In this assay, it is important to make sure that the conditions are such that L-[^{3}H]arginine uptake is equivalent, as the quantitative assessment of eNOS catalytic activity is no longer valid. Moreover, we have found that prolonged exposure of the cells to L-[^{3}H]arginine before stimulation of eNOS yields a higher background (L-[^{3}H]citrulline signal in the absence of eNOS stimulation) and less L-NAME-inhibitable L-[^{3}H]citrulline production. The precise reasons for such observations are not clear but may reflect some preference for newly incorporated L-arginine as a substrate for eNOS, consistent with the colocalization of eNOS and the L-arginine (y+) transporter.[20]

We have found previously that certain thiol-modulating agents impair the conversion of L-[^{3}H]arginine to L-[^{3}H]citrulline without modifying the cellular uptake of L-[^{3}H]arginine.[18] In this instance, it is important to determine if this is a direct effect on eNOS itself or is related to some alteration in eNOS cofactors. This point is particularly germane in light of recent studies indicating that tetrahydrobiopterin and NADPH, two NOS cofactors, are subject to redox modulation within the cell.[15,21,18] This can be accomplished easily by testing eNOS catalytic activity in lysed cell preparations, as these assays involve exogenously added cofactors.

In order to perform these assays, we typically use confluent endothelial cells in 100-mm dishes that are suspended in 1 ml PBS with a rubber policeman and washed twice with PBS. Harvested cells are sonicated in 1 ml of lysis buffer consisting of 50 m*M* Tris–HCl, pH 7.5, 100 μM diethylenetriaminepentaacetate (DTPA), 10% glycerol, and protease inhibitors. The homogenate is centrifuged at 65,000 rpm for 60 min using a Beckman Ti70 rotor. The pellet is then resolubilized by sonication (15 pulses) in 100 μl lysis buffer with 0.1% Triton X-100. The protein concentration in solubilized membrane preparations was determined by the bicinchoninic acid (BCA) protein assay (Bio-Rad). Assays (100 μl) contain 150–200 μg of membrane protein, 50 m*M* Tris, pH 7.5, 10 μM tetrahydrobiopterin, 1 m*M* $CaCl_2$, 10 μg/ml calmodulin, 1 μM FAD, 1 μM FMN, 50 μM L-[^{3}H]arginine ($\sim 10^5$ cpm), 0.5 m*M* NADPH, and 100 μM DTPA. Assays are terminated after 15–30 min with 1 ml ice-cold stop buffer containing 20 m*M* sodium acetate, pH 5.5, 1 m*M* L-citrulline, 2 m*M* EDTA, and 2 m*M* EGTA. In our studies, we found that 1-chloro-2,4-dinitrobenzene (CDNB), a thiol-modulating agent,

[20] K. K. McDonald, S. Zharikov, E. R. Block, and M. S. Kilberg, *J. Biol. Chem.* **272,** 31213 (1997).

[21] R. Heller, A. Unbehaun, B. Schellenberg, B. Mayer, G. Werner-Felmayer, and E. R. Werner, *J. Biol. Chem.* **276,** 40 (2001).

impairs eNOS catalytic activity when tested in whole cells, but the activity of eNOS was "rescued" by the addition of exogenous cofactors in lysed cell assays.[18]

Endothelial Production of $NO_2^- + NO_3^-$

In some instances, the cellular uptake of L-arginine may be altered by one of the experimental conditions, thereby invalidating the L-[^{3}H]arginine to L-[^{3}H]citrulline conversion assay outlined earlier. One alternative under these circumstances is to measure the total nitrogen oxide production from cells stimulated to produce NO. We typically use confluent endothelial cells in six-well plates that are washed in phenol red-free Minimum Essential Medium (MEM) (without serum) and incubated in phenol red-free MEM for 30 min. Nitric oxide synthesis is then stimulated with bradykinin, serotonin, or A23187 for 8 hrs and the media $NO_2^- + NO_3^-$ determined using the method of Miles and colleagues.[22] Initially, NO_3^- in the medium (3-ml sample) is reduced by the addition of 40 μM NADPH and 14 mU/ml of nitrate reductase (Sigma) and incubation for 60 min at room temperature. Medium NO_2^- is then derivatized with 2,3-diaminonaphthalene (DAN; 10 μM final concentration) for 10 min in the dark under acidic conditions (0.4 ml 1 *N* HCl added to the 3-ml reaction) and stopped by adding excess NaOH. The DAN is dissolved in dimethylformamide as a 20 m*M* stock solution. Fluorescence of nitrite derivatives is determined at excitation and emission wavelengths of 375 and 415 nm, respectively. Nitrite concentrations are calculated using a standard curve generated from varying concentrations of nitrite derivatized under the same conditions.

Bioassay for Endothelium-Derived Nitric Oxide

Although the assays just outlined provide a firm working knowledge for NO action and metabolism in endothelial cells, they are limited to cells in culture. Because cultured cells may not perfectly mirror events that occur *in vivo,* we recommend confirming cell culture studies with an intact vessel bioassay whenever possible. For this assay, we use a water-jacketed organ chamber system that is available commercially (20 ml, Radnoti Glass, Monrovia, CA). The temperature is maintained at 37° using a water bath with a recirculating pump. We harvest thoracic aortae from New Zealand White rabbits (2.5–3.5 kg) that are first treated with 100 IU/kg heparin and then euthanized with an overdose of pentobarbital (120 mg/kg) via the marginal ear vein. The harvested aorta is placed rapidly in HEPES-buffered PSS and cleaned gently of adventitia using small dissecting forceps and scissors. The aorta is then cut into 3–5 mm size strips and suspended onto stirrups in the organ chambers with one stirrup fixed in space and the other connected to a force transducer (Model FT03, Grass Instruments, Quincy, MA)

[22] A. M. Miles, D. A. Wink, J. C. Cook, and M. B. Grisham, *Methods Enzymol.* **268,** 105 (1996).

and a graphic recorder (Model 7, Grass). The vessel ring is then immersed in Krebs buffer consisting of 118.3 m*M* NaCl, 4.7 m*M* KCl, 1.2 m*M* $MgSO_4 \cdot 7H_2O$, 1.2 m*M* KH_2PO_4, 2.5 m*M* $CaCl_2$, 25 m*M* $NaHCO_3$, 11.1 m*M* glucose, 0.026 m*M* $Na_2EDTA \cdot 2H_2O$, and 10 μM indomethacin. The Krebs buffer is aerated constantly with a gas mixture containing 5% CO_2, 15% O_2, and 80% N_2. We have refrained from using a 95% O_2 mixture due to the hyperoxic environment. The vessel is equilibrated in the organ chamber for 30 min and then passive tension is added in 1 g increments by lengthening the distance between stirrups. After each 1 g increase in resting tension, the rings are contracted actively by substituting Krebs buffer with KCl–Krebs that contains 80 m*M* KCl and 38.3 m*M* NaCl but is otherwise identical in composition to Krebs buffer. This evokes a maximum contraction, and the extent of this contraction is recorded. We plot the KCl-induced contraction as a function of passive tension and choose that point at which further increases in the passive tension no longer enhance KCl-induced contraction. For rabbit aortae, this number is typically ~6 g of passive tension.

Once the optimal passive tension is chosen, the rings are equilibrated in Krebs buffer for 90 min with a complete change of buffer every 30 min. At this point, the rings are ready for assessment of endothelium-dependent (i.e., NO-mediated) arterial relaxation. The rings are treated with a concentration of phenylephrine that induces a 50–60% maximal contraction (defined as the response to 80 m*M* KCl Krebs). After a stable contraction is reached, arterial relaxation is examined in response to increasing doses of acetylcholine or A23187 (both 1 n*M* to 10 μM) to examine receptor-mediated and receptor-independent vasodilation, respectively. We add the agonists for endothelial NO release directly to the organ chambers as a minor volume (1/1000) of a concentration solution (1000×). The extent of relaxation is reported as the percentage reduction in tension evoked by phenylephrine and is typically 85–100% in response to 1 μM acetylcholine or A23187. Relaxation is reversed by washing the rings with fresh Krebs buffer. Rings that are relaxed with acetylcholine can be tested three or four times in other experimental conditions. In contrast, once rings are treated with A23187, the vessels cannot be used again. Direct smooth muscle cell relaxation to NO can be tested by substituting DEANO (1 n*M* to 10 μM) for acetylcholine in the aforementioned protocol.

It is important in these assays to match the extent of arterial contraction among different treatment groups or experimental conditions, as comparisons may not be valid. One can also control for the effect of the endothelium by removing the endothelial lining before mounting the vessel in the organ chamber. This is accomplished readily by placing a soft cotton swab in the lumen of the vessel and rolling it back and forth across a Krebs-soaked paper towel. The lack of functional endothelium can then be confirmed by testing the relaxation to acetylcholine, which should then be absent.

TABLE I
ANTICIPATED CHANGE COMPARED WITH CONTROL IN SPECIFIC ASSAYS OF NO ACTION BASED ON MECHANISM OF REDOX-MEDIATED MODULATION OF NO BIOACTIVITY[a]

Assay		Putative change in NO action or metabolism					
		↑ NO Production	↑ NO Inactivation	↑ GC response to NO	↓ NO Production	↓ NO Inactivation	↓ GC response to NO
Agonist-stimulated cGMP accumulation	EC	↑	↓	↑	↓	↑	↓
DEANO-stimulated cGMP accumulation	EC	↔	↓	↑	↔	↑	↓
L-[^{3}H]Arginine to L-[^{3}H]citrulline conversion		↑	↔	↔	↓	↔	↔
Production of $NO_2^- + NO_3^-$		↑	↔	↔	↓	↔	↔
Vessel bioassay		↑	↓	↑	↓	↑	↓

[a] EC, endothelial cell; DEANO, diethylamine NONOate.

Summary

The assays outlined in this article have been employed successfully by ourselves and many other investigators to assess endothelial NO production under conditions in which the redox status of the cell has been manipulated. Table I contains expected changes in each of the assays (compared to the control state) that one would expect to see for specific perturbations in the NO–cGMP axis in endothelial cells. Used in combination, these assays can provide considerable insight into the effect of cellular redox status on NO bioactivity and provide the investigator with a good starting point for further mechanistic studies.

Acknowledgments

This work was supported by a CJ Martin Postdoctoral Research Fellowship from Australian National Health and Medical Research Council (to S.R.T.) and Grants DK55656 and HL60886 from the National Institutes of Health (to J.F.K.). John F. Keaney, Jr. is an Established Investigator of the American Heart Association.

Section II

Nitrosothiols and Nitric Oxide in Cell Signaling

[20] Reactions of S-Nitrosothiols with L-Ascorbic Acid in Aqueous Solution

By TARA P. DASGUPTA and JAMES N. SMITH

Introduction

The chemistry and biology of *S*-nitrosothiols (RSNOs) have been the subjects of major research interest in recent years[1–4] due to their unique abilities to release the powerful biological messenger nitric oxide (NO) under physiological conditions. RSNOs are essentially NO carriers and play an important role in NO storage, transport, and delivery in the biological system.[5] Hence, nitrosothiols can be used as therapeutic drugs for diseases such as hypertension, cerebrovascular disorders, atherosclerosis, cardiovascular disorders, and penile erectile dysfunction.[6] Specifically *S*-nitrosoglutathione (GSNO), *S*-nitroso-*N*-acetyl-DL-penicillamine (SNAP), and *S*-nitrosocaptopril (CapSNO) have shown great biological diversity[7,8] at the appropriate concentrations. Interestingly, GSNO and SNAP also inhibit growth of the malarial parasite *Plasmodium falciparum*[9], whereas CapSNO acts both as an NO donor and an angiotensin-converting enzyme inhibitor.[10,11]

Cleavage of the S–N bond in RSNOs can occur homolytically or heterolytically depending on the conditions[12] yielding $^{\bullet}NO$, NO^+, or NO^-. The ease of cleavage of the S–N bond determines stability and hence the potency[13,14] of RSNOs. In general, RSNOs are quite inert[15,16] in the presence of metal chelators such as ethylenediaminetetraacetic acid (EDTA) or diethylenetetraminepentaacetic acid

[1] N. Hogg, *Free Radic. Biol. Med.* **28,** 1478 (2000).
[2] K. Szacilowski and Z. Stasicka, *Prog. Reac. Kinet. Mech.* **26,** 1 (2000).
[3] J. Karwin, Jr., J. R. Lancaster, Jr., and P. L. Feldman, *J. Med. Chem.* **38,** 4343 (1995).
[4] Y. Hou, J.-Q. Wang, J. Ramirez, and P. G. Wang, *Methods Enzymol.* **301,** 242 (1999).
[5] A. R. Butler and D. L. H. Williams, *Chem. Rev.* **22,** 233 (1993).
[6] L. J. Ignarro, H. Lippton, J. C. Edwards, W. H. Baricos, A. L. Hyman, P. J. Kadowitz, and C. A. Gruetter, *J. Pharmacol. Exp. Ther.* **218,** 739 (1981).
[7] E. A. Konorev, M. M. Tarpey, J. Joseph, J. E. Baker, and B. Kalyanaraman, *J. Pharmacol. Exp. Ther.* **274,** 200 (1995).
[8] M. W. Radomski, D. D. Rees, A. Dutra, and S. Moncada, *Br. J. Pharmacol.* **107,** 745 (1992).
[9] K. A. Rockett, M. M. Awburn, W. B. Cowden, and I. A. Clark, *Infect. Immun.* **59,** 3280 (1991).
[10] L. Jia and R. C. Blantz, *Eur. J. Pharmacol.* **354,** 33 (1998).
[11] W. R. Mathews and S. W. Kerr, *J. Pharmacol. Exp. Ther.* **267,** 1529 (1993).
[12] D. R. Arnelle and J. S. Stamler, *Arch. Biochem. Biophys.* **318,** 279 (1995).
[13] B. Gaston, J. Reilly, J. M. Drazen, J. Fackler, P. Ramdev, D. Arnelle, M. E. Mullings, D. J. Sugarbaker, C. Chee, D. J. Singel, J. Loscalzo, and J. S. Stamler, *Proc. Natl. Acad. Sci. U.S.A.* **90,** 10957 (1993).
[14] E. A. Kowaluk and H. L. Fung, *J. Pharmacol. Exp. Ther.* **255,** 1265 (1990).
[15] D. L. H. Williams, *Methods Enzymol.* **268,** 299 (1996).
[16] R. J. Singh, N. Hogg, J. Joseph, and B. Kalyanaraman, *J. Biol. Chem.* **271,** 18596 (1996).

METHODS IN ENZYMOLOGY, VOL. 359
0076-6879/02 $35.00

(DTPA). In absence of a metal chelator, the stability in aqueous solution decreases[1] in the order SNOCAP > GSNO > SNAP > SNOCYS, presumably due to the presence of Cu^{2+} as an impurity in water.[15] However, in the presence of many reductants, such as certain thiols[17,18] and ascorbic acid,[19,20] reductive cleavage of the S–N bond of RSNOs produces NO readily.

There is increasing interest in the study of electron transfer processes that are important in biological systems. L-Ascorbic acid (vitamin C) is known to exist in cellular systems at relatively high concentration[16] and is a very important cellular antioxidant.[17] From a physiological point of view, the interaction of nitrovasodilators with cellular antioxidants is really important as it can be one of the decisive steps in their metabolism.

Decomposition of RSNOs by thiols occurs via transnitrosation,[17,18,21] but decomposition by ascorbic acid proceeds either by an outer-sphere one-electron transfer process[20] or by electrophilic nitrosation by RSNO at one of the oxygen sites of the ascorbate ion.[22] The reduction of *S*-nitrosothiols to give NO is strongly pH dependent due to the stepwise acid dissociation of L-ascorbic acid.[23,24] Detailed synthetic methods along with the experimental procedure for studying NO release kinetics in the presence of ascorbate ion are presented here.

Synthesis of RSNOs

The chemical structures of three RSNOs (GSNO, SNAP, and CAPSNO) that have been investigated are shown in Fig. 1.

GSNO

GSNO is prepared[25] by adding sodium nitrite (0.864 g, 12.5 mmol) in one portion to an ice-cold, acidic solution (25 ml, 2 *N* HCl) of glutathione (3.83 g, 12.5 mmol). After about 40 min at 5° the red solution is treated with acetone (25 ml) and stirred for a further 10 min. The resulting fine pale-red precipitate is

[17] A. P. Dicks, H. R. Swift, D. L. H. Williams, A. R. Butlar, H. H. Al-Sa'doni, and B. J. Cox, *J. Chem. Soc. Perkin Trans.* **2,** 481 (1996).

[18] A. P. Dicks, P. H. Beloso, and D. L. H. Williams, *J. Chem. Soc. Perkin Trans.* **2,** 1429 (1997).

[19] D. J. Barnett, A. Rios, and D. L. H. Williams, *J. Chem. Soc. Perkin Trans.* **2,** 1279 (1995).

[20] J. N. Smith and T. P. Dasgupta, *Nitric Oxide Biol. Chem.* **4,** 57 (2000).

[21] K. Wang, Z. Wen, W. Zhang, M. Xian, J.-P. Cheng, and P. G. Wang, *Bioorg. Med. Chem. Lett.* **11,** 433 (2001).

[22] A. J. Holmes and D. L. H. Williams, *J. Chem. Soc. Perkin Trans.* **2,** 1 (2000).

[23] K. Tsukahara and Y. Yamamoto, *Bull. Chem. Soc. Jpn.* **54,** 2642 (1981).

[24] R. A. Rickman, R. L. Sorensen, K. O. Watkins, and G. Davies, *Inorg. Chem.* **16,** 1570 (1977).

[25] T. W. Hart., *Tet. Lett.* **26,** 2013 (1985).

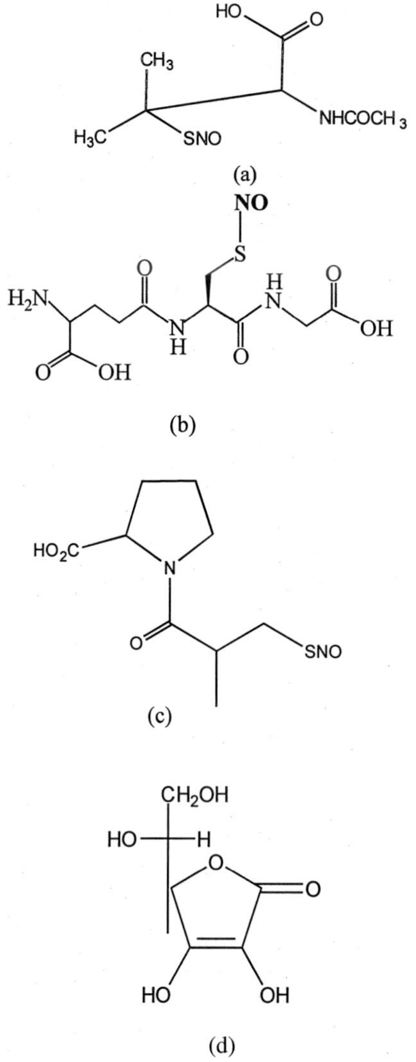

(a)

(b)

(c)

(d)

FIG. 1. Chemical structures of (a) SNAP, (b) GSNO, (c) CapSNO, and (d) L-ascorbic acid.

filtered off and then washed with successive portions of ice-cold water (5×3 ml), acetone (3×25 ml) and diethyl ether (3×25 ml) to afford GSNO. Yield = 75%, (Lit. 76%); $\lambda_{max}(H_2O) = 336$ and 546 nm; $\varepsilon_{336} = 927$; $\varepsilon_{546} = 16$ dm^3 mol^{-1} cm^{-1}; (Lit. $\varepsilon_{335} = 922$; $\varepsilon_{545} = 15.9$ dm^3 mol^{-1} cm^{-1}). The purity is further supported by infrared spectroscopy (NO *str.* 1740, 1675 cm^{-1}).

SNAP

SNAP is prepared[26] by adding $NaNO_2$ (1.38 g or 20 mmol) in 20 ml of deionized water to *N*-acetyl-DL-penicillamine (1.91 g or 10 mmol) dissolved in MeOH–1 *N* HCl (20 ml each) with 2 ml of concentrated H_2SO_4 in 20 min with vigorous stirring at ca, 25°. After an additional 15 min, SNAP is filtered off, washed well with cold deionized water, and air dried to give deep green crystals with red reflections. Yield = 77%, (Lit. 68%); λ_{max} (H_2O) = 340 and 590 nm; $\varepsilon_{340} = 1000$; $\varepsilon_{590} = 13\ dm^3\ mol^{-1}\ cm^{-1}$; (Lit. $\varepsilon_{340} = 815$; $\varepsilon_{590} = 12.4\ dm^3\ mol^{-1}\ cm^{-1}$)

CapSNO

CapSNO is prepared *in situ,*[27] although it has been isolated[28] as red feather-like crystals. It is prepared by reacting equimolar amounts of $NaNO_2$ and captopril (CapSH) under acidic conditions (pH 1.5). After standing for 5 min at ~23°, the red CapSNO solution is neutralized with NaOH.

Repetitive Scanning

All spectroscopic measurements are carried out utilizing UV-VIS spectroscopy with the Diode Array Hewlett-Packard 8452A/8453A spectrophotometer assembled with the Hi-Tech Pneumatic Drive rapid-mixing system. The repetitive scanning for all reactions between RSNOs and L-ascorbic acid show spectral changes (absorbance vs wavelength) with time over the entire wavelength range of interest. There is a gradual disappearance in absorption between the ranges 300–400 nm and 530–630 nm. Scans obtained for each RSNO exhibit close similarities. Figures 2a and 2b represent that of GSNO and CapSNO with L-ascorbic acid, respectively.

The maximum absorption changes occur in the high UV/low visible (~330–350 nm) region with a much smaller change in the higher visible (~530–590 nm) region. The actual wavelength of maximum absorption change (λ_{max}) depends on the nature of the RSNO.[29,30] It is, therefore, not surprising that the loss of NO from RSNOs is naturally accompanied by a decrease in absorbance in the corresponding regions of the spectrum. It should be noted, however, that residual absorption (in the 330- to 350-nm region) at the end of the kinetic reactions is largely due to the presence of dehydroascorbic acid (DHA) and excess unreacted

[26] L. Field, R. V. Dilts, R. Ravichandran, P. G. Lenhert, and G. E. Carnahan, *J. Chem. Soc., Chem. Commun.* 249 (1978).

[27] J. W. Park, *Biochem. Biophys. Res. Commun.* **189,** 206 (1992).

[28] L. Jia, X. Young, and W. Guo, *J. Pharm. Sci.* **88,** 981 (1999).

[29] M. D. Bartberger, K. N. Houk, S. C. Powell, J. D. Mannion, K. Y. Lo, J. S. Stamler, and E. J. Toone, *J. Am. Chem. Soc.* **122,** 5889 (2000).

[30] D. L. H. Williams, *Acc. Chem. Res.* **32,** 869 (1999).

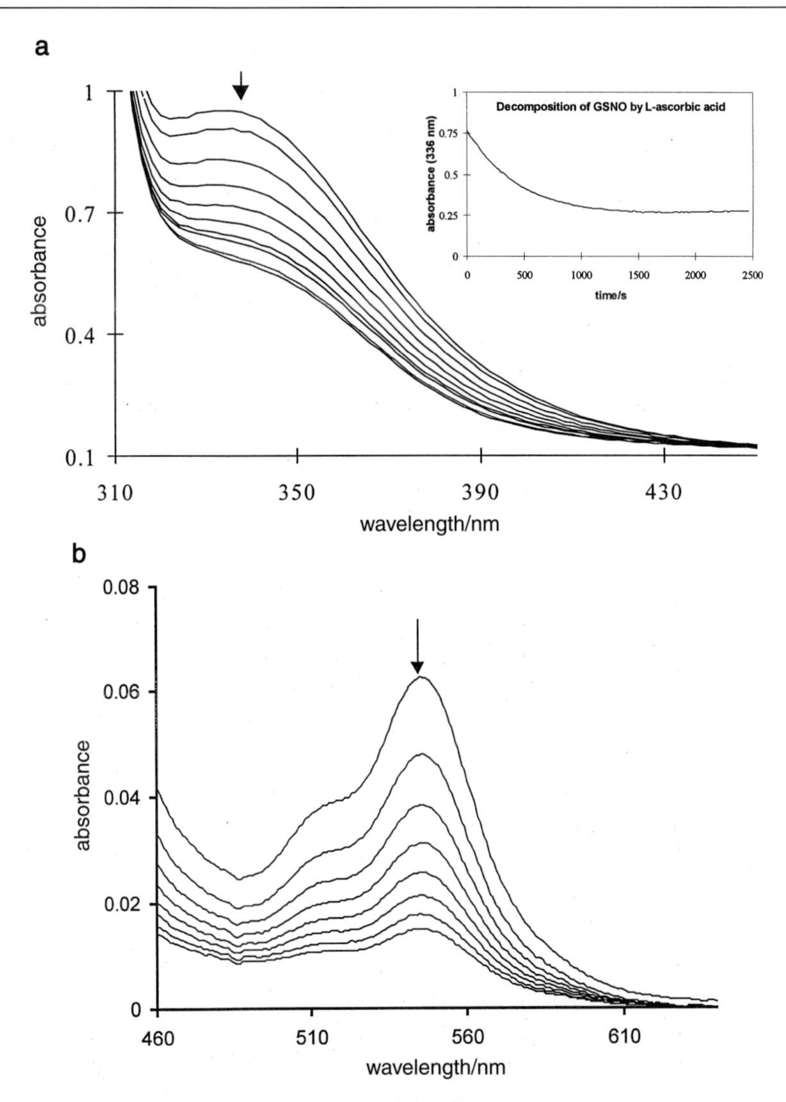

FIG. 2. (a) Change in UV-VIS absorption spectra during the reductive decomposition of GSNO by L-ascorbic acid. Conditions: 1 m*M* GSNO, 16 m*M* $[H_2A]_t$, pH 7.45 (0.1 *M* phosphate buffer), 0.5 m*M* EDTA, $I = 0.5$ *M* (NaCl electrolyte), and at 37°. Spectra were recorded at 1-min intervals. (Inset) Decay curve of GSNO by ascorbic acid at 336 nm. (b) Spectral changes for the reaction of CapSNO with ascorbic acid. Conditions: 5 m*M* CapSNO, 50 m*M* $[H_2A]_t$, pH 7.73, 200 μM EDTA, $I = 0.5$ *M* (NaCl electrolyte), $\theta = 25°$. Spectra were recorded at 12-min intervals.

L-ascorbic acid and is not necessarily due to incomplete decomposition of RSNOs (Fig. 2a, inset).

General Kinetics Measurements

All kinetic measurements are thermostatted at their specific temperatures (±0.1°) between 20° and 37°, and reactions are followed at their respective λ_{max} values. Kinetic experiments are performed under pseudo-first-order conditions, i.e., at least 10-fold excess of ascorbic acid to RSNOs. Pseudo-first-order rate constants (k_{obs}) are obtained by fitting absorbance vs time data generated from the kinetic experiments to the equation: $A_t = (A_o - A_\infty)e^{-kt} + A_\infty$, where A_o, A_t, and A_∞ are the absorbances at $t = 0$, $t = $ t, and $t = \infty$ (final), respectively, and $k = k_{obs}$. For all reactions, very good first-order behavior is observed, and k_{obs} values generated are the average of at least two kinetic runs with less than 5% standard deviation.

Ionic strength is maintained at 0.5 mol dm^{-3} using NaCl as the supporting electrolyte.

Stoichiometric Determination

Various mole ratios of ascorbic acid and RSNOs are mixed together under inert (argon) atmosphere and left for several hours to allow the reaction to be completed. The final absorbance values are then measured at 330, 332, and 336 nm for Cap-SNO, SNAP, and GSNO, respectively, and plotted against the [ascorbate]/[RSNO] ratio in order to determine the reaction stoichiometry. In each case, the break-point of the plot indicates a 1 : 1 mole ratio reaction between ascorbic acid and the RSNO, i.e., ascorbic acid here acts as a one-electron donor.

Determination of Dehydroascorbic Acid

The formation of DHA as the oxidized product of reactions involving ascorbate is determined by standard procedures.[31] DHA is prepared by the complete aerial oxidation of a stock solution of a known concentration of ascorbic acid, which also contains about 20 μM Cu^{2+} as catalyst. This is achieved by bubbling air into the solution at 25° for approximately 8 hr to ensure that all ascorbic is converted to DHA.

Standard solutions in 10.0-ml flasks containing 0.01–0.2 mM DHA are prepared from the just-described final solution and used for the calibration curve. One drop of 10% thiourea solution is then added to 4.0-ml samples of each of the

[31] P. B. Hawk, L. O. Bernard, and W. H. Summerson, "Practical Physiological Chemistry," 12th Ed., p. 1137. Blakiston, New York, 1951.

standard solutions, followed by the addition of 1.0 ml 2,4-dinitrophenylhydrazine (2,4-DNP) reagent (2 g 2,4-DNP dissolved in 100 ml of 4.5 mol dm^{-3} sulfuric acid). The samples are placed in a water bath maintained at 37° for exactly 3 hr. They are later removed and placed in an ice-water bath along with the blank (the blank contains 4.0 ml deionized water and 1 drop thiourea). To each flask in the bath, 5.0 ml of 85% sulfuric acid is added drop wise with constant shaking. Finally, 1.0 ml of 2,4-DNP reagent is added to the blank. The flasks are removed from the bath and allowed to stand for approximately 30 min to ensure full color development. The absorbances of the red solutions formed are measured spectrophotometrically (Diode Array hp 8453) at 525 nm using the blank solution to zero the instrument. A calibration curve is constructed with absorbance plotted against [DHA]. A similar procedure is carried out for our experimental samples. Interpolation of the calibration curve gives the concentrations of DHA present in the experimental samples. It should be noted that although this procedure is the best available, it is not 100% accurate, as other side reactions may interfere with the final yield of DHA.

The oxidation of L-ascorbic acid by RSNOs results in the formation of DHA as its final oxidized product. On analysis of the final reaction solution, an average of about 90% of the expected DHA could be accounted for. The missing percentage could be due to (i) experimental errors associated with this tedious analytical procedure, (ii) any subsequent reaction(s) between $NO^{\bullet}$ and ascorbic acid radical species, and (iii) an incomplete reaction between RSNO and ascorbic acid. The overall results clearly supports a 1 : 1 mole ratio reaction between RSNO and ascorbic acid.

pH Measurements

The pH in all experiments is controlled using a 0.1 *M* (Na_2HPO_4/NaH_2PO_4) buffer system. NaOH is used to adjust the pH of L-ascorbic acid stock solutions to near neutral before usage. All pH measurements are done using Cole-Parmer H^+-sensitive electrodes connected to an Orion Research, expandable ion analyzer EA 920 in the pH mode. The pH meter is calibrated using standard $KHC_8H_4O_4$ (50 m*M*) and KH_2PO_4/Na_2HPO_4 (9 and 30 m*M*) buffer solutions at pH 4.01 and 7.41, respectively.

Nitric Oxide Measurements

Rates and amount of NO released from the reactions are monitored directly using World Precision ISO-NOP electrodes fitted to a WPI ISO-NO mark II meter. The electrode is calibrated by the chemical generation of NO from the reaction with $NaNO_2$, excess H_2SO_4, and KI according to Eq. 1:

$$2\,NaNO_2 + 2\,KI + 2H_2SO_4 \longrightarrow 2\,NO + K_2SO_4 + Na_2SO_4 + I_2 + 2\,H_2O \quad (1)$$

$$
\begin{array}{l}
H_2A \\
\Big\updownarrow K_1 \\
HA^- + RSNO \xrightarrow{k_a} NO + RS^- + A^{\cdot -} + H^+ \\
\Big\updownarrow K_2 \\
A^{2-} + RSNO \xrightarrow{k_b} NO + RS^- + A^{\cdot -}
\end{array}
$$

SCHEME 1

The generation of NO is monitored under anaerobic conditions after saturation of the reactant solutions with argon gas for approximately 10 min prior to the start of each reaction. A continuous flow of argon is maintained at approximately 250 ml/min and vessel pressure about 1 atm.

Decomposition of RSNOs by L-Ascorbic Acid

Decomposition of RSNOs by L-ascorbic acid[20,22,32] to give NO is affected by certain factors, such as light, temperature, pH, and catalysts. We proposed[20] a simplified mechanism (Scheme 1) for the interaction of RSNOs with all three species of ascorbic acid: H_2A, HA^-, and A^{2-}. The reduction of RSNOs by ascorbic acid proceeds in one stage, which involves the reduction of the S–N bond followed by rapid labilization of the NO group. This is clearly indicated by a general decrease in absorption in the electronic spectrum (Fig. 2a). Any intermediate formed as a precursor complex between RSNO and ascorbic acid is transient and is not detected spectrally by the rapid-scanning technique employed. The release of NO via the A^{2-} pathway is by far the most effective, followed by HA^- and H_2A, respectively (Table I). The concentration of A^{2-} increases with pH and, as expected, the potency of ascorbic acid as a reductant increases naturally, which improves the rate of RSNO decay and hence NO production significantly. In the pH range studied, HA^- is the predominant species when compared to A^{2-}, which only exists in trace amounts, and because HA^- is of much lower reactivity, the reactions are still slow. H_2A was found to be totally inactive in all cases studied.

A plot of k_{obs} values versus total ascorbate concentration, $[H_2A]_t$, showed excellent first-order dependence up to approximately 80 m*M* (see Fig. 3). The intercepts observed were very small or totally absent in the case of CapSNO, which clearly indicates that reversed or parallel reactions, thermal, photochemical, and/or catalytic decomposition are negligible under the reaction conditions and

[32] D. Aquart and T. P. Dasgupta, *Nitric Oxide* (in preparation).

TABLE I
SECOND-ORDER RATE CONSTANTS FOR REDUCTIVE DECOMPOSITION OF RSNOs BY L-ASCORBIC ACID AT $I = 0.5$ *M* (NaCl) AND 37°

RSNO	k_a (10^{-2} M^{-1} s^{-1})	k_b (M^{-1} s^{-1})
SNAP	0.98 ± 0.13	662 ± 38
GSNO	1.43 ± 0.33	396 ± 11
CapSNO	2.57 ± 1.29	50 ± 1

that RSNO decay is solely due to the reduction by ascorbic acid. This is further testimony to the stability of these selected RSNOs.

The effect of pH changes on rates was studied in the pH range 5–10 and it was found that there was little reaction and hence NO production below 5.5, but k_{obs} values and NO release increase drastically above pH 7 (Fig. 4 and 5). In light of this, we can safely confirm that the reactivity of ascorbic acid follows the sequence: $A^{2-} \gg HA^- \gg H_2A$.

The rate expression derived from Scheme 1 is shown in Eq. (2)

$$\text{Rate} = (k_a[\text{HA}^-] + k_b[\text{A}^{2-}])[\text{RSNO}] \quad (2)$$

Under pseudo-first-order conditions, Eq. (2) can be written as a function of acid concentration, $[H^+]$, and the first and second acid dissociation constants of

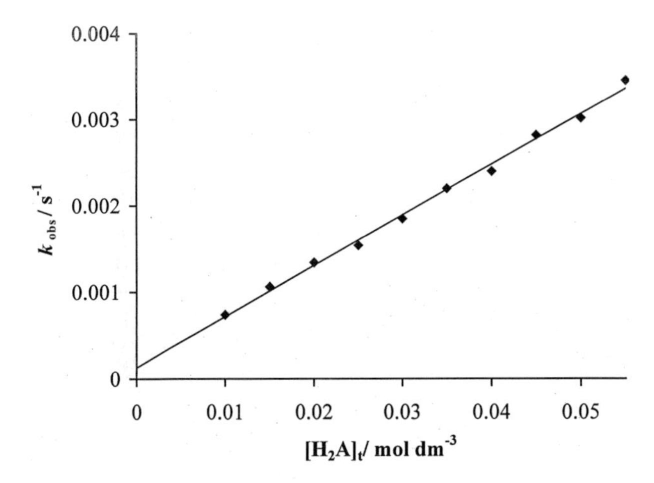

FIG. 3. L-Ascorbic dependent decomposition of GSNO. Conditions: 1 m*M* GSNO, 10–55 m*M* $[H_2A]_t$, pH 7.45 (0.1 *M* phosphate buffer), 0.5 m*M* EDTA, $I = 0.5$ *M* (NaCl electrolyte), $\lambda = 336$ nm, and at 25°.

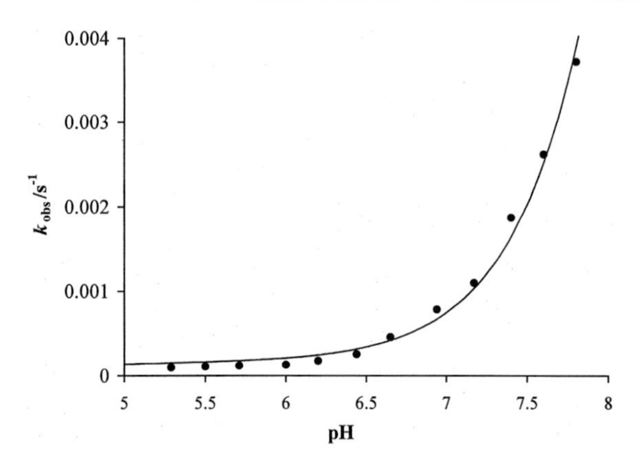

FIG. 4. Observed first-order rates (k_{obs}) versus pH for GSNO decay induced by L-ascorbic acid. Conditions: 1 m*M* GSNO, 30 m*M* $[H_2A]_t$, 0.5 m*M* EDTA, $I = 0.5$ *M* (NaCl), $\lambda = 336$ nm, and at 25°. (●), Experimental points; plotted curve calculated from Eq. (3).

L-ascorbic acid, K_1 and K_2, as shown:

$$k_{obs} = \frac{k_a K_1[H^+] + k_b K_1 K_2}{[H^+]^2 + K_1[H^+] + K_1 K_2}[H_2A]_t \tag{3}$$

The second order rate constants, k_a and k_b, are evaluated by nonlinear regressional analysis for three RSNOs.

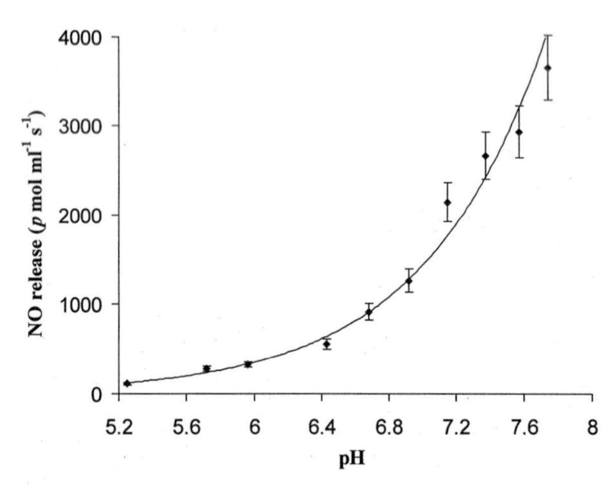

FIG. 5. pH-dependent NO release from the induced breakdown of GSNO by L-ascorbic acid. Conditions: 1 m*M* GSNO, 10 m*M* $[H_2A]_t$, 0.5 m*M* EDTA, $I = 0.5$ *M* (NaCl electrolyte), and at 25°.

Data in Table I show that the reactivity of the A^{2-} species is significantly higher than that of the HA^- species and hence the specific constants, k_b, are much greater than the corresponding k_a values.

Clear evidence can now be drawn from our data and a previous report[22] that the bulkiness (steric hindrance) of the RSNO and, to a lesser degree, the polarity (electronic effect) of the S–N bond are the main factors governing the reactivity with the ascorbate anion. Because the mechanism of attack by ascorbate on the nitrosyl (^+NO) moiety of the RSNO is favored, then higher molecular weight and bulky RSNOs, such as SNAP, GSNO, and CapSNO as expected, react much slower with ascorbate than their lighter and less sterically hindered counterparts, such as *S*-nitrosocysteine (CySNO) and *S*-nitrosopenicillamine (PSNO). Electron-releasing groups such as alkyl and acetyl that are bonded closely to the —S–NO group are strongly deactivating and appear to stabilize the S–N bond, rendering it less susceptible to nucleophilic attack by the ascorbate ion. This results in a decrease in the rate of NO release.

In a previous report[21] between RSNOs and L-ascorbic acid, data within show the reactivity to be PSNO $\gg$ SNAP and CySNO $\gg$ *S*-nitroso-*N*-acetylcysteine. This is further testimony to the role played by the acetyl group as both an electron-withdrawing group and a bulky sterically hindered group. This combined effect will significantly reduce the reactivity of the *S*-nitrosothiol with ascorbate as well as with other biological reductants.

Acknowledgments

The authors extend gratitude to the Department of Chemistry and the Board of Graduate Studies at the University of the West Indies, Mona, for funding this research project. Invaluable contribution from Miss Danielle Aquart in putting the paper together is gratefully acknowledged.

[21] Nitrosothiols and S-Nitrosohemoglobin

By CARLO A. PALMERINI, GIUSEPPE ARIENTI, and ROBERTO PALOMBARI

Introduction

Nitric oxide (NO) is one of the simplest odd electron species. Although it behaves generally as expected for radicals, it also shows a marked tendency to lose the unpaired electron, forming the nitrosonium ion (NO^+).[1] The ability of NO to

[1] Y. Henry, C. Ducrocq, J. C. Drapier, D. Servent, C. Pellat, and A. Giussani, *Eur. Biophys.* **20,** 1 (1991).

0076-6879/02 $35.00

bind to both Fe(II) and Fe(III) of hemoglobin (Hb) is well known, whereas NO^+ is involved in the *S*-nitrosylation of thiols, such as cysteine and glutathione (GSH), to form the nitroso derivatives nitrosocysteine (CysNO) and nitrosoglutathione (GSNO), respectively.

Proteins, such as albumin and Hb, can also be transformed to nitroso derivatives in the presence of low molecular weight *S*-nitrosothiols (SNOs).[2,3] The *S*-nitrosylation of Hb may have special physiological meaning because *S*-nitrosylation/*S*-denitrosylation cycles imply the allosteric transition from the R to the T form of Hb[4]; therefore, *S*-nitrosohemoglobin (SNO-Hb) would act as a reservoir for NO, whose release as the free form or whose transfer to other thiols (GSH) depends on Hb oxigenation.[5,6]

The reaction of NO with O_2 forms reactive nitric oxide species toxic for the cell. The synthesis of SNOs would subtract NO, thus decreasing the possibility of reactive nitric oxide species production.[7]

Several techniques have been used to determine SNOs. The classical Griess reaction[8] and fluorimetric and colorimetric methods[9,10] require a preliminary displacement of the NO^+ group with $HgCl_2$[11] or $CuCl_2$.[9] Additional methods are based on photoluminescence of the NO_2 produced on exposure to ozone[12,13] and on electron paramagnetic resonance[14]; these approaches entail the previous decomposition of SNOs by UV radiation.[15] Electrospray ionization mass spectroscopy[5,16] has been used to identify -SNOs groups in hemoglobin.

This article describes an electrochemical determination of NO bound to high and to low molecular weight SNOs. The method exploits a specific solid-state

[2] A. J. Gow and J. S. Stamler, *Nature* **391,** 169 (1998).

[3] J. S. Stamler, O. Jaraki, J. Osborne, D. I. Simon, J. V. Keaney, J. Vita, D. Singel, C. R. Valeri, and J. Loscalzo, *Proc. Natl. Acad. Sci. U.S.A.* **89,** 7674 (1992).

[4] L. Jia, C. Bonaventura, J. Bonaventura, and J. S. Stamler, *Nature* **380,** 221 (1996).

[5] C. Bonaventura, G. Ferruzzi, S. Tesh, and R. D. Stevans, *J. Biol. Chem.* **274,** 24742 (1999).

[6] R. P. Patel, N. Hogg, N. Y. Spencer, B. Kalyanaraman, S. Matalon, and V. M. Darley-Usmar, *J. Biol. Chem.* **274,** 15487 (1999).

[7] D. A. Wink, M. B. Grisham, A. M. Miles, R. W. Nims, M. C. Krishna, R. Pacelli, D. Teague, C. M. Poore, J. A. Cook, and P. C. Ford, *Methods Enzymol.* **268,** 120 (1996).

[8] L. C. Green, D. A. Wagner, J. Glogowski, P. L. Skipper, J. S. Wishnok, and S. R. Tannenbaum, *Anal. Biochem.* **126,** 131 (1982).

[9] D. A. Wink, S. Kim, D. Coffin, J. C. Cook, Y. Vodovotz, D. Chistodoulou, D. Jourd'heuil, and M. B. Grisham, *Methods Enzymol.* **301,** 201 (1999).

[10] P. Kostka and J. K. Park, *Methods Enzymol.* **301,** 227 (1999).

[11] B. Saville, *Analyst* **83,** 670 (1958).

[12] S. Archer, *FASEB J.* **7,** 349 (1993).

[13] M. T. Gladwin, F. P. Ognibene, L. K. Pannell, J. S. Nichols, M. E. Pease-Fye, J. H. Shelhamer, and A. N. Schechter, *Proc. Natl. Acad. Sci. U.S.A.* **97,** 9943 (2000).

[14] R. Clancy, A. I. Cederbaum, and D. A. Stoyanovsky, *J. Med. Chem.* **44,** 2035 (2001).

[15] V. A. Tyurin, S. X. Liu, Y. Y. Tyurina, N. B. Sussman, C. A. Hubel, J. M. Roberts, R. N. Taylor, and V. E. Kagan, *Circ. Res.* **88,** 1210 (2001).

[16] P. Ferranti, A. Malorni, G. Mamone, N. Sannolo, and G. Marino, *FEBS Lett.* **400,** 19 (1997).

amperometric sensor[17–19] and permits the determination of NO as soon as it is released from SNOs, with a sensitivity of the order of nanomolar concentrations.

Materials and Methods

Materials

Glutathione (GSH), sodium hydrosulfite (dithionite), and cysteine are from Sigma Chemical Co. (St. Louis, MO). Sephadex G-25 is a product of Pharmacia Fine Chemicals (Uppsala, Sweden). Other reagents are from C. Erba, Milan, Italy.

Synthesis of Low Molecular Weight SNOs

SNOs are synthesized from thiols and NO_2^- in acidic medium. GSNO and CysNO are prepared by mixing equimolecular amounts of GSH or cysteine and NO_2^- in 0.1 *M* HCl for about 15 min at room temperature.[20] The formation of GSNO is then assessed by measuring light absorbance at 340 nm.[20]

Synthesis of SNO-Hb through the Reaction with CysNO

Human Hb is obtained from venous blood samples and is purified as described.[21] Briefly, red blood cells are disrupted with hypotonic phosphate buffer, samples are centrifuged to eliminate the ghosts, and the supernatant is purified on a Sephadex G-25 column (60 × 2 cm) equilibrated in 0.1 *M* phosphate, 1 m*M* EDTA, pH 7.4.

Hb (100–200 μM) is then incubated at 37° for 10–60 min in the dark with 2 ml of the same buffer, also containing various concentrations CysNO. The reaction is carried out under a gentle O_2 flux.[22] Hb and SNO-Hb are then purified through a passage on a Sephadex G-25 (60 × 2 cm) column and are eluted with void volume. SNO-Hb is assessed by using the amperometric sensor later. Total Hb is determined by light absorption at 430 nm after treatment with dithionite.[22]

Synthesis of SNO-Hb from metHb through the Reaction with Gaseous NO

Human purified Hb is treated with $K_3Fe(CN)_6$ (10-fold *M* excess for 10 min at room temperature), formed (FeIII) Hb, which is purified on a Sephadex G-25 column (60 × 2 cm), equilibrated in 0.1 *M* phosphate, 1.0 m*M* EDTA, pH 7.4. (FeIII)Hb (150 μM) is then incubated at 37° for 1 hr in the dark with 3 ml of the

[17] G. Alberti, R. Palombari, and F. Pierri, *Solid State Ionics* **97,** 359 (1997).

[18] C. A. Palmerini, G. Arienti, R. Mazzolla, and R. Palombari, *Nitric Oxide* **2,** 375 (1998).

[19] C. A. Palmerini, G. Arienti, and R. Palombari, *Nitric Oxide* **4,** 546 (2000).

[20] D. L. H. Williams, *Chem. Commun.* 1085 (1996).

[21] R. J. Kilbourn, G. Joly, B. Cashon, J. De Angelo, and J. Bonaventura, *Biochem. Biophys. Res. Commun.* **199,** 155 (1994).

[22] T. J. McMahon and J. S. Stamler, *Methods Enzymol.* **301,** 99 (1999).

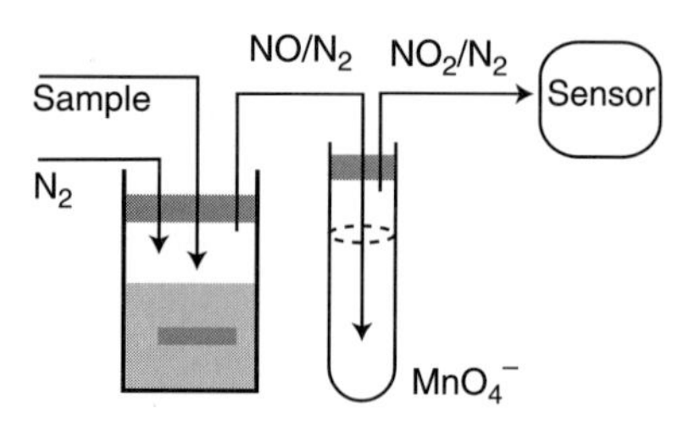

FIG. 1. Schematic representation of the apparatus and the amperometric sensor used for the analysis of free NO, GSNO, SNO-Hb, and NO_2^- in biological samples. Adapted from C. A. Palmerini, G. Arienti, R. Mazzolla, and R. Palombari, *Nitric Oxide* **2,** 375 (1998).

same buffer. The concentration of (FeIII)Hb is determined by light absorbance at 406 nm, and a full spectrum is made to check the complete oxidation to (FeIII)Hb. The reaction mixture is bubbled with N_2 for 30 min and then with gaseous NO (1000 ppm in N_2; 20 ml min^{-1}) for 1 hr. Subsequently, it is again bubbled with N_2 (for 10 min) to remove unreacted NO. Products of the reaction are determined as described next.

Electrochemical Analysis. The apparatus, shown schematically in Fig. 1, consists of a reaction vessel (5 ml) equipped with an injector, maintained at a constant temperature, and supplied with a flow of 10 ml/min of N_2. The NO formed in the vessel is carried to the amperometric sensor described previously.[17–19] This consists of a solid-state cell with a graphite-sensing electrode and two Ag/Ag_2O electrodes acting as a reference and a counter connected by a protonic conductor based on zirconium phosphate (Fig. 2). The sensor is especially designed to obtain high current yield and low noise and could detect, in different ranges of potential, both NO and NO_2, the latter with higher sensitivity. In addition, NO_2 is more stable than NO and therefore NO is transformed into NO_2 using a small trap filled with about 2 ml of an acidic solution of permanganate (50 m*M* $KMnO_4$, 0.5 *M* $HClO_4$). This procedure also avoids errors due to accidental penetration of O_2, which would transform NO into NO_2. The presence of NO in the reaction vessel is recorded vs time as a peak of electric current.

The apparatus is calibrated by injecting known amounts of standard nitrite solution into the reaction vessel that also contain 50 m*M* cysteine in 50 m*M* HCl + 50 m*M* $CuCl_2$ in a final volume of 1 ml. The height of the peaks is linearly related to the amount of added nitrite (correlation coefficient ≥ 0.98).

SNO-Hb, obtained by the reaction with CysNO, is eluted from the Sephadex G-25 and then injected into the reaction vessel containing cysteine and $CuCl_2$, as described earlier.

Samples containing SNO-Hb and NO_2^-, obtained following the reaction of metHb with gaseous NO, are divided into three aliquots: (a) one is injected as such into the reaction vessel containing $CuCl_2$ and cysteine; in parallel experiments, the

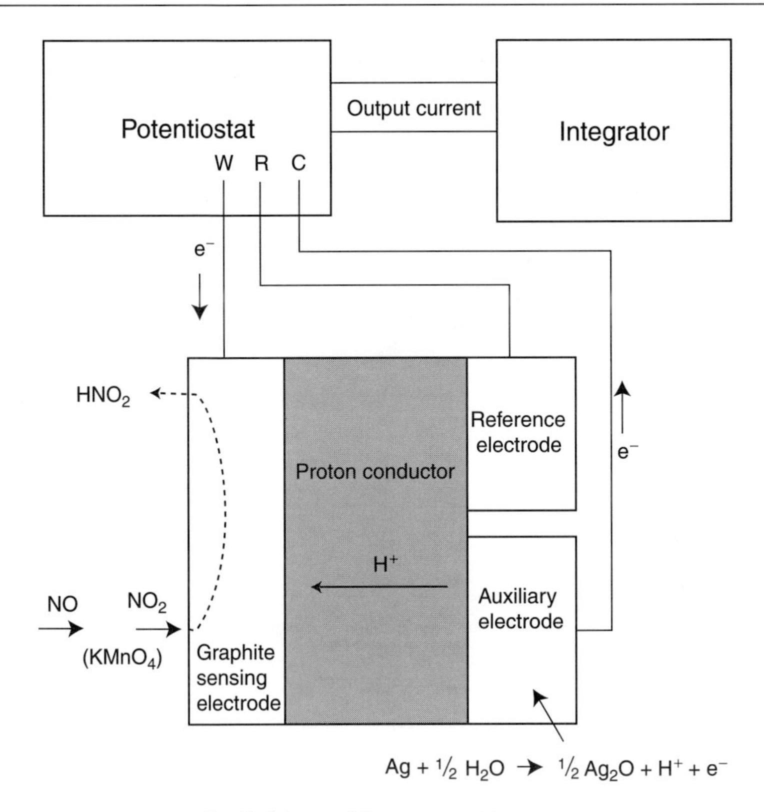

FIG. 2. Scheme of the amperometric sensor.

concentration of heme is determined by light absorbance; (b) before the amperometric assay, an aliquot is reacted with an isovolumetric solution of ammonium sulfamate (10%, w/v) in 0.1 *M* HCl for 5 min at room temperature to remove HNO_2 and to assess SNO-Hb only; and (c) the last aliquot is chromatographed on a Sephadex G-25 column (60 × 2 cm) equilibrated in 0.1 *M* phosphate, pH 7.4, containing 1 m*M* EDTA. SNO-Hb is collected with the void volume. This fraction is used to check SNO-Hb through the amperometric sensor and heme concentration by light absorbance.

Results and Discussion

We describe a new electrochemical technique for the rapid determination of SNOs in biological samples.

The response of the apparatus (described earlier) was first examined using standard solutions of NO_2^-. A typical calibration curve is shown in Fig. 3.

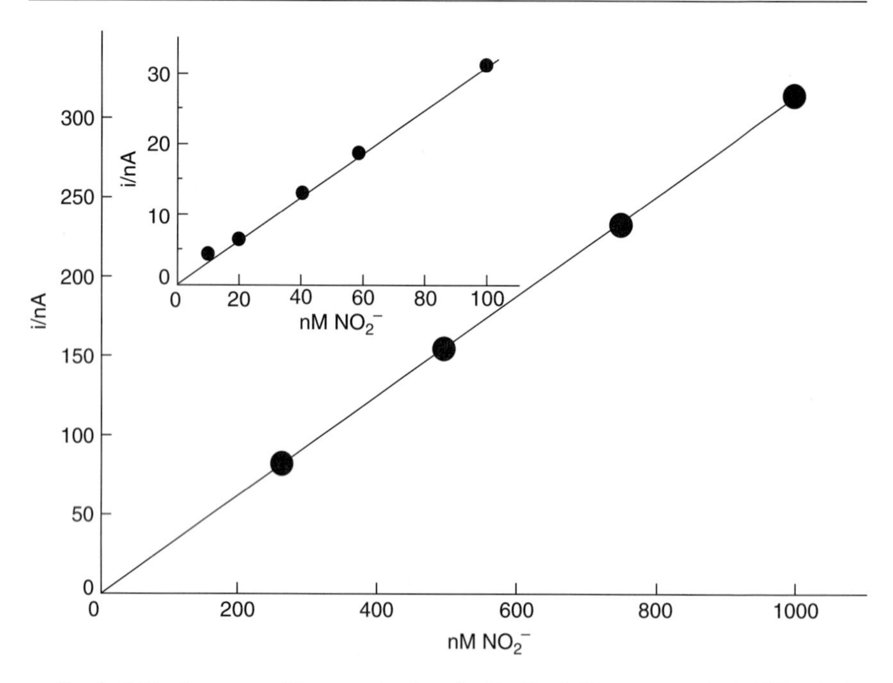

FIG. 3. Calibration curve of the apparatus described in Fig. 1. The concentration of NO_2^- in the reaction vessel (final volume 1 ml) is plotted against the output current peak height. (Inset) Concentrations of NO_2^- below 100 n*M* are shown. Reprinted from C. A. Palmerini, G. Arienti, R. Mazzolla, and R. Palombari, *Nitric Oxide* **2,** 375 (1998).

Calibrations were performed by using cysteine and $CuCl_2$ in acidic medium at a constant temperature (25°). The recorded signal was related linearly to the amount of NO. The method allowed the determination of 10–20 n*M* NO (10–20 pmol in the reaction vessel). Linearity was present from 10 n*M* to 100 μM NO (the part of the graph from 1 to 100 μM was omitted from Fig. 3 for clarity sake). A similar control, with identical results, was performed using standard solutions of GSNO or CysNO, prepared as described earlier. Results obtained with this procedure were in agreement with those achieved by following the method described by Saville[11] in which Hg(II) was used to displace NO^+.

Cysteine and $CuCl_2$ were able to displace NO from SNOs. The mechanism of this reaction entailed the reduction of Cu(II) to Cu(I) and the oxidation of cysteine to cystine:

$$\text{Cu(II)} + \text{cysteine} \rightarrow \text{Cu(I)} + \text{cystine}$$

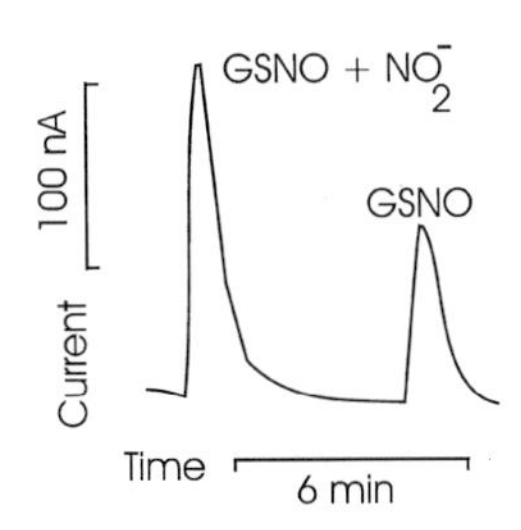

FIG. 4. Response of the apparatus to NO_2^- and GSNO. The peak GSNO + NO_2^- was produced by mixing GSH with an excess of NO_2^- in acidic medium. The contribution of nonreacted NO_2^- was eliminated with ammonium sulfamate (peak GSNO). Reprinted from C. A. Palmerini, G. Arienti, R. Mazzolla, and R. Palombari, *Nitric Oxide* **2,** 375 (1998).

Cu(I) removes NO from SNOs and reduces HNO_2 to NO,[23] following the reactions:

$$\mathrm{Cu(I)} + \mathrm{SNO} \rightarrow \mathrm{S}^- + \mathrm{NO} + \mathrm{Cu(II)}$$
$$\mathrm{Cu(I)} + \mathrm{HNO_2} \rightarrow \mathrm{NO} + \mathrm{Cu(II)}$$

In a previous work,[18] we used a mixture of saturated CuCl in NaCl at pH 4. The use of the $CuCl_2$/cysteine described here increases the availability of Cu(I).

This procedure was tested by preparing standard solutions of GSNO in concentration ranges from 10^{-4} to 10^{-3} *M* and by measuring light absorbance at 340 nm. The molar extinction coefficient was $980 \pm 20\ M^{-1}\ \mathrm{cm}^{-1}$.[20] The solutions were diluted to 10^{-5}–10^{-7} *M* and injected into the reaction vessel. Data obtained by interpolating the height of the peaks in the calibration curve were in agreement with expected values.

The reaction with Cu(I) is not selective for SNOs, but occurs quantitatively with NO_2^- and forms NO in both cases. However, determination of the amount of GSNO was possible, as NO_2^- could be reduced selectively to N_2 by using 10% (w/v) ammonium sulfamate in acidic medium.[10,11]

Figure 4 shows signals due to GSH and NO_2^- in excess (5 μM GSH and about 10 μM NO_2^-). Determination of GSNO was performed by treating the solution with ammonium sulfamate. In parallel experiments, we incubated GSNO and/or NO_2^- (both at known concentrations) with ammonium sulfamate at acidic pH values and found that the determination of GSNO was unaffected by the presence of NO_2^-. Ammonium sulfamate selectively reduced NO_2^-, up to a concentration of 1 m*M*.

SNO-Hb is usually determined as NO_2^- [8] after $HgCl_2$ treatment.[11] We displaced NO from SNO-Hb using Cu(II) and cysteine instead of Hg(II). The

[23] D. L. H. Williams, *Acc. Chem. Res.* **32,** 869 (1999).

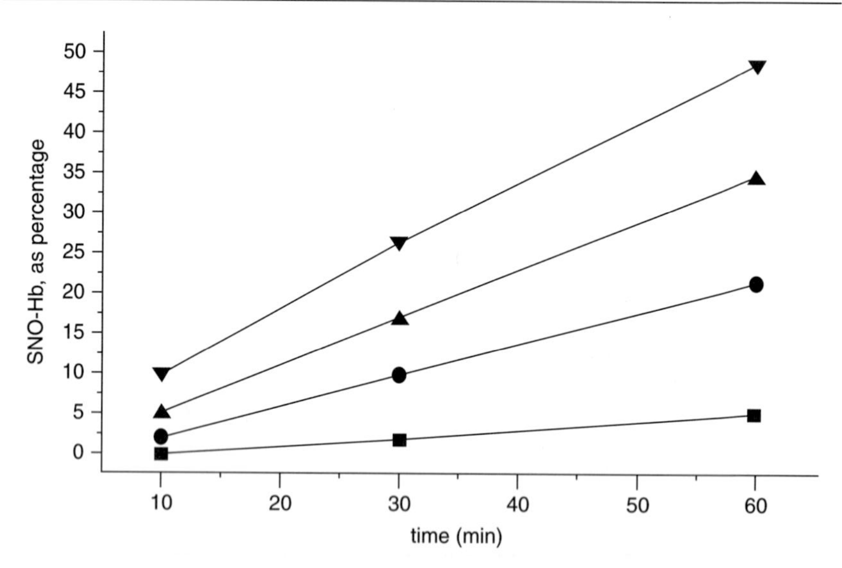

FIG. 5. Production of SNO-Hb from Hb and CysNO at different CysNO/heme Hb molar ratios and for different incubation times at 37°. ■, CysNO/heme ratio 1 : 1; ●, CysNO/heme ratio 2 : 1; ▲, CysNO/heme ratio 5 : 1; ▼, CysNO/heme ratio 10 : 1. Results are expressed as percentage of Hb (heme groups) transformed to SNO-Hb ± SE. Reprinted from C. A. Palmerini, G. Arienti, and R. Palombari, *Nitric Oxide* **4,** 546 (2000).

amperometric determination of released NO permits the assessment of SNO-Hb on varying either the time of exposure to CysNO or the CysNO/hem ratio (Fig. 5).

Moreover, the incubation of (FeIII) Hb with gaseous NO at 37° for 1 hr formed SNO-Hb in the absence of O_2 and at neutral pH values (Fig. 6). The reduction of (FeIII)Hb with gaseous NO forms (FeII)Hb and NO^+ [2,24,25]; NO^+ is necessary for the reaction with thiol groups and for the formation of NO_2^-. Ammonium sulfamate reduces NO_2^- selectively, which allows the determination of SNO-Hb produced (Fig. 6A). This reaction was checked by repeating the determination of SNO-Hb deprived of NO_2^- through a purification step on the Sephadex G-25 column (Fig. 6B). This control was necessary because the NO_2^- formed was about 10 times as much as the SNO-Hb produced.

Conclusions

The technique described in this article allows the determination of SNOs at low concentration and exploits the cleavage of the S—NO bond due to Cu(I). This

[24] M. Hoshino, K. Ozawa, H. Seki, and P. C. Ford, *J. Am. Chem. Soc.* **115,** 9568 (1993).

[25] G. Reichenbach, S. Sabatini, R. Palombari, and C. A. Palmerini, *Nitric Oxide* **5,** 395 (2001).

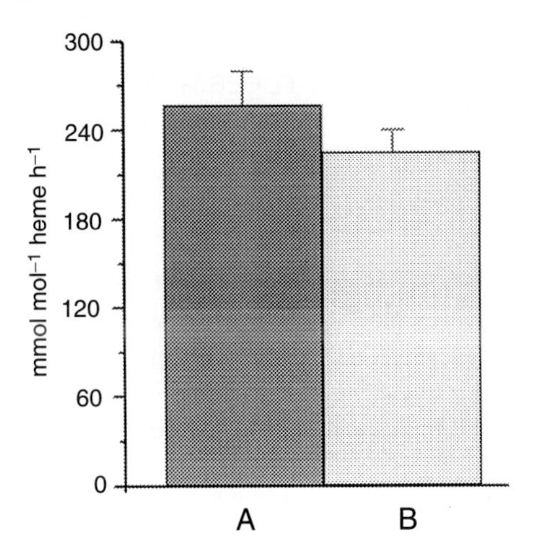

FIG. 6. Recovery of SNO-Hb on bubbling a solution of (FeIII) Hb with NO (1000 ppm, in N_2). The reaction was carried out at 37° for 1 hr. (A) SNO-Hb was determined after reacting with ammonium sulfamate to destroy NO_2^-. (B) SNO-Hb was determined after separation from NO_2^- on Sephadex G-25. Results are expressed as millimoles SNO-Hb mol^{-1} heme hr^{-1} ± SE. NO_2^- that formed in the reaction was 1800 ± 60 mmol mol^{-1} heme hr^{-1}.

procedure is versatile and can be employed not only with the amperometric sensor described here, but also with commercial gas sensors.

We determine SNO-Hb obtained in the reaction between Hb and CysNO by electrochemical assay; however, this method cannot identify the Hb cysteine residues that undergo the *S*-transnitrosylation reaction, although it permits a detailed kinetic study of the NO^+ transfer between Hb and SNOs and between SNO-Hb and glutathione (*in vivo* and *in vitro*). The same applies to the reaction between gaseous NO and metHb.

Colorimetric methods have a low sensitivity, fluorimetric methods have a greater sensitivity, but both measure NO indirectly.[8] Moreover, chemical determination requires the removal of oxidation products from biological systems. Chemiluminescence and electronic paramagnetic resonance-based applications permit the direct determination of NO, but are not suitable for *in vivo* monitoring.[26]

The procedure described here is particularly suitable for routine, real-time measurements in normal and pathological situations, having two interesting features: SNOs may easily be analyzed and distinguished from nitrite and liberated NO can be measured in tissues at nanomolar concentrations without any additional purification steps.

[26] H. Tu, J. Xue, X. Cao, W. Zhang, and L. Jin, *Analyst* **125,** 163 (2000).

[22] Nitrosothiol Processing by Platelets

By JOHN S. HOTHERSALL and ALBERTO A. NORONHA-DUTRA

Introduction

Nitrosothiols (RS-NO), low molecular weight protein thiol compounds containing cysteine-bound nitric oxide (NO), were first implicated as biological effectors by Ignarro *et al.,*[1] in 1980. It is recognized that several RS-NOs are present in biological fluids and tissues[2–4] and are the predominant form of NO in human plasma.[5] RS-NOs have a biological role as NO donors either by direct release[6,7] [reaction (1)] or via transfer of NO to other thiol groups by *S*-transnitrosation[8] [reaction (2)]. *S*-Transnitrosation is important as NO can be efficiently targeted by transfer from one protein or nonprotein to another on the surface of or within cells. Thus release of NO is not obligatory for RS-NO action[9] and NO can remain in a stable and storable form.

$$2\text{RS-NO} \rightarrow \text{RSSR} + 2\text{NO} \quad (1)$$

$$\text{RS-NO} + \text{R}^1\text{SH} \rightarrow \text{R}^1\text{S-NO} + \text{RSH} \quad (2)$$

Platelets represent a major biological target for RS-NOs, and NO-dependent inhibition of activation (adhesion and aggregation) occurs at between 10^{-7} and 10^{-8} *M*,[10,11] compatible with concentrations found in plasma. In a nucleated cell, NO signaling was originally thought to be by targeting to iron present in the heme

[1] L. J. Ignarro, J. C. Edwards, D. Y. Gruetter, B. K. Barry, and C. A. Gruetter, *FEBS Lett.* **110,** 275 (1980).

[2] I. Kluge, U. Gutteck-Amsler, M. Zollinger, and K. Q. Do, *J. Neurochem.* **69,** 2599 (1997).

[3] B. Gaston, J. Reilly, J. M. Drazen, J. Fackler, P. Ramden, D. Arnelle, M. E. Mullins, D. Sugarbaker, C. Chee, D. Singel, J. Loscalzo, and J. S. Stamler, *Proc. Natl. Acad. Sci. U.S.A.* **90,** 10957 (1993).

[4] R. M. Clancy, D. Levartovsky, J. Leszczynska-Piziak, J. Yegudin, and S. B. Abramson, *Proc. Natl. Acad. Sci. U.S.A.* **91,** 3680 (1994).

[5] J. S. Stamler, D. I. Simon, J. A. Osborne, M. E. Mullins, O. Jaraki, T. Michel, D. J. Singel, and J. Loscalzo, *Proc. Natl. Acad. Sci. U.S.A.* **89,** 444 (1992).

[6] L. J. Ignarro, H. Lippton, J. C. Edwards, W. H. Baricos, A. L. Hyman, P. J. Kadowitz, and C. A. Gruetter, *J. Pharmacol. Exp. Ther.* **218,** 739 (1981).

[7] G. Bannenberg, J. Xue, L. Engman, I. Cotgreave, P. Moldeus, and A. J. Ryrfeldt, *J. Pharmacol. Exp. Ther.* **272,** 1238 (1995).

[8] Z. Liu, M. A. Rudd, J. E. Freedman, and J. Loscalzo, *J. Pharmacol. Exp. Ther.* **284,** 526 (1998).

[9] G. Bannenberg, J. Xue, L. Engman, I. Cotgreave, P. Moldeus, and A. Ryrfeldt, *J. Pharmacol. Exp. Ther.* **272,** 1238 (1995).

[10] B. T. Mellion, L. J. Ignarro, C. B. Myers, E. H. Ohlstein, B. A. Ballot, A. Hyman, and P. J. Kadowitz, *Mol. Pharmacol.* **23,** 653 (1983).

[11] M. W. Radomski, D. D. Rees, A. Dutra, and S. Moncada, *Br. J. Pharmacol.* **107,** 745 (1992).

0076-6879/02 $35.00

group of the regulatory protein guanylate cyclase and thus mediated exclusively by its product the second messenger cGMP.[12] However, it is now recognized that RS-NOs can also act through a cGMP-independent mechanism.[13–15] This mechanism probably involves transnitrosation.

Some processing of RS-NOs by cells has been demonstrated[16,17] and is also implied, as biological activity of RS-NOs appears to be stereoselective.[18,19]

The ability of RS-NOs to modulate platelet interactions, be it platelet–platelet, platelet–neutrophil, or platelet–endothelium interaction, is dependent on transnitrosation or NO release. However, compounds that release or transfer NO fastest are not necessarily the most potent.[20] The trigger for release or transfer of NO from RS-NO can be by photochemical or thermal decomposition or by transition metal catalysis.[21,22] Copper I ions (Cu^+) are most active in mediating NO release,[23] and in biological settings the copper must be in the reduced active form. Copper I bound to cysteine is an ideal form as copper I is poorly soluble.

Another route of RS-NO transfer into the cell is via transport of low molecular weight nitrosothiols on their corresponding transporter. As is the case in many peripheral tissues, glutathione can be taken up on a platelet transporter, although glutathione-S-NO (GS-NO) appears to be a poor ligand.[24] Alternatively, GS-NO could be processed by the ectoenzyme γ-glutamyl transpeptidase (γGT) to cysteine-NO (CS-NO) via intracellular dipetidase action on cysteinylglycine S-NO (CGS-NO). In contrast to GS-NO, CS-NO can be taken up on an active transport system,[25] and evidence in neurons indicates that CS-NO is stereoselectively transported.[19] However, we have found no evidence for γGT involvement in GS-NO processing by platelets.[26]

[12] E. H. Lieberman, S. O'Neill, and M. E. Mendelsohn, *Circ. Res.* **68,** 1722 (1991).

[13] M. P. Gordge, J. S. Hothersall, and A. A. Noronha-Dutra, *Br. J. Pharmacol.* **124,** 141 (1998).

[14] D. Tsikas, M. Ikic, K. S. Tewes, M. Raida, and J. C. Frolich, *FEBS Lett.* **442,** 162 (1999).

[15] N. Sogo, K. S. Magid, C. A. Shaw, D. J. Webb, and I. L. Megson, *Biochem. Biophys. Res. Commun.* **279,** 412 (2000).

[16] M. P. Gordge, J. S. Hothersall, G. H. Neild, and A. A. Dutra, *Br. J. Pharmacol.* **119,** 533 (1996).

[17] A. Zai, M. A. Rudd, A. W. Scribner, and J. Loscalzo, *J. Clin. Invest.* **103,** 393 (1999).

[18] R. L. Davisson, M. D. Travis, J. N. Bates, A. K. Johnson, and S. J. Lewis, *Am. J. Physiol.* **272,** H2361 (1997).

[19] A. J. Lipton, M. A. Johnson, T. Macdonald, M. W. Lieberman, D. Gozal, and B. Gaston, *Nature* **413,** 171 (2001).

[20] E. A. Kowaluk and H. L. Fung, *J. Pharmacol. Exp. Ther.* **255,** 1256 (1990).

[21] V. R. Zhelyaskov, K. R. Gee, and D. W. Godwin, *Photochem. Photobiol.* **67,** 282 (1998).

[22] M. P. Gordge, D. J. Meyer, J. Hothersall, G. H. Neild, N. N. Payne, and A Noronha-Dutra, *Br. J. Pharmacol.* **114,** 1083 (1995).

[23] G. Stubauer, A. Giuffre, and P. Sarti, *J. Biol. Chem.* **274,** 28128 (1999).

[24] D. J. Sexton and B. Mutus, *Biochem. Cell Biol.* **73,** 155 (1995).

[25] M. Kashiba, E. Kasahara, K. C. Chien, and M. Inoue, *Arch. Biochem. Biophys.* **363,** 213 (1999).

[26] M. P. Gordge, P. Addis, A. A. Noronha-Dutra, and J. S. Hothersall, *Biochem. Pharmacol.* **5,** 657 (1998).

To examine *S*-transnitrosation-mediated platelet inhibition it is necessary to identify thiol targets, their reversible thiol-disulfide redox status and the processing they undergo during activation. One mechanism by which *S*-transnitrosation can be regulated is through thiolation, as this effectively blocks thiol groups of the target proteins.

Platelet low molecular weight and protein thiol exchange is a complex process and involves thiolation of intra- and extracellular protein thiols. The disulfides cystine (CS-SC) and cystinylglycine (CGS-SGC) appear to be more effective in forming protein-mixed disulfides.[27] They are generated from plasma or platelet glutathione via the platelet ectoenzyme γ-glutamyl transpeptidase (γGT).

The mechanism by which RS-NOs, at concentrations arising under pathophysiological conditions, specifically target platelets rather than the microvascular endothelium[28,29] has important implications for both cardiovascular disease and the clinical use of RS-NOs. Insight into the precise mechanism of RS-NO processing by platelets is vital to the understanding of this phenomenon.

Methods

Preparation of RS-NO

This is based on the principle of HNO_2 formation from acidified (HCl) nitrite reacting with a specific thiol [Reactions (3) and (4)].

$$NO_2^- + H^+ \rightarrow H_2NO_2^- \rightarrow NO^+ + H_2O \quad (3)$$

$$NO^+ + \text{R-SH} \rightarrow \text{RS-NO} + H^+ \quad (4)$$

Using stoichiometric quantities of nitrite and HCl, 20 m*M* standards are freshly prepared and maintained in solution at acidic pH at 4° and protected from light. Immediately prior to use, RS-NOs are diluted in water (chelex treated) to micromolar concentrations and added to platelets. Protein *S*-nitrosothiols (PS-NO) can be prepared from the protein reduced with 1 m*M* dithiothreitol (1 m*M* for 1 hr at room temperature) either with acidified nitrite or, when acid conditions are to be avoided, with butyl nitrite (100 m*M*) as an alternative nitrosating agent. Separation of low molecular weight reactants from PS-NO on Sephadex G-25 is essential. Quantitation of each RS-NO is derived from the absorbance at 334 nm. Care should be taken as extinction coefficients (mM^{-1} cm^{-1}) vary: typical values

[27] D. Giustarini, G. Campoccia, G. Fanetti, R. Rossi, F. Giannerini, L. Lusini, and P. Di Simplicio, *Arch. Biochem. Biophys.* **380,** 1 (2000).

[28] E. J. Langford, A. S. Brown, R. J. Wainwright, A. J. de Belder, M. R. Thomas, R. E. Smith, M. W. Radomski, J. F. Martin, and S. Moncada, *Lancet* **344,** 1458 (1994).

[29] A. J. de Belder, R. MacAllister, M. W. Radomski, S. Moncada, and P. J. Vallance, *Cardiovasc. Res.* **28,** 691 (1994).

are *S*-nitrosocysteine, 0.74; *S*-nitrosohomocysteine, 0.73; *S*-nitroso-*N*-acetyl-DL-penicillamine, 1.00; *S*-nitroso-*N*-acetylcysteine, 0.87, and *S*-nitrosoglutathione, 0.85.[13,30]

Preparation and Treatment of Platelets

Platelet rich plasma (PRP) and washed platelets (WP) are prepared from fresh blood taken in acid citrate dextrose from volunteers who, during the preceding 10 days, had not taken drugs known to affect platelet function. Blood is centrifuged for 20 min at 160*g* to prepare PRP. Washed platelets (WP) are prepared from PRP either with prostacyclin treatment as described[13] or by citric acid adjustment of RPP, pH to 6.5, with washing in HEPES-buffered saline (HBS) containing 3 m*M* EGTA.[31] Platelets are suspended at a count of 200×10^9 in HBS with 1 m*M* calcium chloride and magnesium chloride. Platelet function should be checked for each preparation by measuring aggregation with standard quantities of agonist and sensitivity toward GS-NO checked.

Shear stress-mediated activation of PRP in 500-μl aliquots containing 200–400×10^6/ml platelets is carried out at 37° by stirring at 1000 rpm for 15 min in an aggregometer (equivalent to low shear of 0.15 Nm^{-2}). Alternatively, WP can be activated in the presence of fibrinogen by either shear stress or 20 μg/ml collagen.[32] Exogenous NO may be added at this point. In PRP, RS-NO can be used at between 5 and 20 μM, depending on the rate of NO release or transfer. Other nonthiol NO donors can be employed if contributions from NO_x-mediated *S*-nitrosation rather than direct *S*-transnitrosation are to be investigated.

L-*N*-(1-Iminoethyl) ornithine (50 μM) or other suitable NOS inhibitors can be used to abrogate endogenous platelet NO formation, and SOD or SOD mimetics may be used to prolong the effectiveness of NO in resulting in (trans)nitrosation.

Platelet Extraction (Sample Preparation)

The low levels of RS-NOs associated with platelets make measurement difficult. Distinction between RS-NO and other NO end products such as nitrate and nitrite must be made, thus strong acidic conditions should be avoided. To prevent breakdown of RS-NOs formed *in situ* by endogenous copper, the highest grade reagents and water available should be employed throughout. In many instances, we have found it advantageous to use chelex-treated buffers and extraction media.

To measure RS-NO associated with the platelets, samples after shear activation and treatment with NO donors and/or NOS inhibitors are centrifuged at 8000*g* for

[30] W. R. Mathews and S. W. Kerr, *J. Pharmacol. Exp. Ther.* **267,** 1529 (1993).

[31] C. C. Smith, L. Stanyer, M. B. Cooper, and D. J. Betteridge, *Biochim. Biophys. Acta* **1473,** 286 (1999).

[32] A. Hirayama, A. A. Noronha-Dutra, M. P. Gordge, G. H. Neild, and J. S. Hothersall, *Nitric Oxide* **3,** 95 (1999).

20 sec and the supernatant discarded. The platelet pellet is washed twice in HBSS and is resuspended in chelex-treated water. Extraction is carried out by sonication at 4° or by digitonin treatment (2.5 μM with three freeze thaw cycles or 15 μM for 5 min). Permeabilization with digitonin is particularly advantageous for measuring PS-NO because it can maximize exposure of detectable PS-NO groups and, when employed in the absence of lysis (freeze thaw processing), it allows convenient removal of low molecular weight S-NOs and soluble PS-NOs from membrane-associated PS-NOs.

Preparation of the samples to distinguish between total low molecular weight components and PS-NO can be performed using protein precipitation after harvesting platelets from the PRP or WP incubation sample. Metaphosphoric acid (0.5%) can be used, but care with all controls must be exercised, as artifactual *S*-nitrosation under acid conditions must be accounted for. Strong oxidizing acids such as perchloric acid should be avoided. Extraction in 50% acetonitrile or other organic solvents can be used to overcome acidification problems; however, complete precipitation of proteins is rarely achieved. Alternatively, separation of high and low molecular weight components can be performed with ultrafiltration, however, protein absorption to the various membranes can result in an underestimation of PS-NO.

Homogenization of platelets to investigate cell compartmentalization of targets for *S*-nitrosation has presented difficulties because of their resistance to mechanical shearing forces and osmotic stress. Intact organelle preparation is best done by nitrogen cavitation followed by sucrose gradient centrifugation[33] or, alternatively, loading platelets in a glycerol gradient (0–40%) followed by mechanical lysis[34] (vortexing or forcing through an 18-gauge steel needle) in hypotonic media. However, if a simple distinction between membrane and cytosolic fractions is required, sonication in hypotonic solution followed by centrifugation at 100,000*g* for 15 min is a rapid and reliable method of preparation. To prevent calcium-dependent changes in membrane surface and protein rearrangement, a high concentration of EGTA (10 mM), together with protease inhibitors, is used routinely. In addition, the presence of bathocuproine disulfonic acid to remove any free copper is advisable.

Detection of RS-NO (Release of NO)

The choice of reductant to release NO from RS-NO is important. Photolysis is one possibility but is not widely available and is difficult to use with fluorescence assays (photobleaching), and chemiluminescence measurement is usually restricted to the ozone reactor. The use of vanadium III under acidic conditions may misidentify NO_3^- as RS-NO.[35] The present preferred method of NO release

[33] M. J. Broekman, *Methods Enzymol.* **215,** 21 (1992).

[34] J. T. Harmon, N. J. Greco, and G. A. Jamieson, *Methods Enzymol.* **215,** 32 (1992).

[35] K. Fang, N. V. Ragsdale, R. M. Carey, T. MacDonald, and B. Gaston, *Biochem. Biophys. Res. Commun.* **252,** 535 (1998).

from SNO is via the addition of copper I plus cysteine.[35] CuCl(100 μM) with 1 mM cysteine at 50° is used. Because the release by copper is pH dependent, this should be conducted in buffered solutions ideally at pH 6. Mercuric chloride (1 mM) can be employed but is less sensitive than copper I, which has detection limits in the region of 10 pmol.

Detection of NO Released from RS-NO

Following release of NO from RSNO, detection can be achieved by a variety of assays, the most widely used being fluorescence. Platelets can be preloaded with the fluorescent probe or the probe can be added at the time of detection. Fluorescence is measured in a luminescence spectrometer at an excitation of 501 nm and an emission of 521 nm for dihydrodichlorofluorescein diacetate (H_2DCFDA) and 495–515 nm for diaminofluorescein diacetate (DAF2DA). Platelet RS-NOs are measured following reductive NO release using H_2DCFDA as a probe.[32] For preloading, platelet samples are protected from light and incubated at 37° with 25 μM H_2DCFDA for 30 min. H_2DCFDA is taken up rapidly and deacetylated by cellular esterases and, unlike other redox-sensitive fluorescent probes, is oxidized by NO.[36] Alternatively, DAF2DA (1–5 μM), a new probe purported to be specific for NO,[37] can be used. In our hands, diaminonaphthalene is not sufficiently sensitive to detect platelet RS-NOs. These probes, particularly DAF2DA, in some cells is cytotoxic at low micromolar concentrations and may alter platelet NO and RS-NO homeostasis. When using fluorescent probes to determine RS-NO, to minimize autooxidation of deacetylated H_2DCF and DAF2, all manipulations should be carried out in O_2-free conditions using helium-purged solutions, under argon, and protected from light. Alternatively, the probes may be added at the end of the experiment either as commercially available nonacetylated forms or prepared from the acetylated probes by hydrolysis in a mild base. (0.01 N NaOH for 30 min in the dark).

Reversed-phase HPLC on C_{18}[35] with spectrophotometric (220 or 336 nm) or electrochemical detection is ideal for low molecular weight RS-NO separation and detection and does not rely on release of NO. However, nonspecific nitrosation during sample handling arising because of the acidic pH must be controlled.

Luminol chemiluminescence can be used to measure NO release from RS-NO when a simultaneous source of superoxide is supplied (xanthine plus xanthine oxidase). We have also used Pholasin (Knight Scientific, Plymouth, UK), a 34-kDa glycoprotein from the bioluminescent mollusc, *Pholas dactylus,* which emits intense light oxidation. We have observed inhibition of Pholasin (1 μg/ml)

[36] S. L. Hempel, G. R. Buettner, Y. Q. O'Malley, D. A. Wessels, and D. M. Flaherty, *Free. Radic. Biol. Med.* **27,** 146 (1999).

[37] H. Kojima, N. Nakatsubo, K. Kikuchi, S. Kawahara, Y. Kirino, H. Nagoshi, Y. Hirata, and T. Nagano, *Anal. Chem.* **70,** 2446 (1998).

resting chemiluminescence by NO released from RS-NO.[38] Sources of other oxidants and antioxidants, such as peroxides and thiols, will change Pholasin chemiluminescence. Thus light in the presence of the extract and cysteine (50 μM) is monitored during resting chemiluminescence, and copper I is injected into the sample to release NO from RS-NOs.

Nitrosylation Targets

Platelet Thiol Homeostasis. The *S*-transnitrosation of thiols is regulated by both the ratio of thiol to mixed disulfide of target protein (P-SH:PS-SR) and by the nitrosation flux (NO transfer rate from Rs-NO to PS-NO). This in turn is determined by the availability of redox-active copper and the occurrence of plasma low molecular weight thiols (R-SH) such as glutathione, cysteinylglycine, cysteine, and homocysteine. In platelets, glutathione influx and efflux, together with the action of γ-GT, result in a complex exchange of low molecular weight thiols. How these parameters and specific P-SHs change during platelet activation is relatively unknown.

Separation by reversed-phase HPLC is commonly employed to measure RSH and RS-SR or R^1S-SR, usually as their monobromobimane derivatives. After separation of platelets from their media by centrifugation, it is important to disrupt the platelet pellet fully for cellular contents to be measured. Lysis by sonication, homogenization of digitonin-treated cells, and by freeze thaw cycling can be employed. Inhibition of λGT (10 μM acivicin) and proteases must be employed.

Neutral deproteinized extracts are treated with monobromobimane (1 mM for 15 min in the dark), centrifuged, and thiol derivatives separated on Beckman Ultrasphere IP column phase C18 as described by Newton and Fahey.[39] For disulfide measurements, samples are derivatized with *N*-ethylmaleimide and then reduced with 2 mM DTT. The low molecular weight components of protein mixed disulfides are released by treatment with DTT. Total cell, cytosolic, mitochondrial, and membrane thiols–disulfides can be studied after separation of the relevant cell component. Centrifugation through silicon oil is preferred, as contamination from other compartments is minimized. P-SH are usually measured in extracts that have been depleted of glutathione either by separation of low molecular weight components by ultrafiltration or passing down a G-25 column.

In summary, the relationship between NO transfer from RS-NO to other thiol groups associated with platelets and the redox status of these thiols is important in RS-NO action during platelet activation. The techniques outlined here are in use in many laboratories, and their use in investigations should bring future improvement in our understanding of important biological and clinical events.

[38] J. S. Hothersall and A. A. Dutra, unpublished observations.

[39] G. L. Newton and R. C. Fahey, *Methods Enzymol.* **251,** 148 (1995).

[23] Nitric Oxide-Activated Glutathione Sepharose

By PETER KLATT and SANTIAGO LAMAS

Introduction

Cellular signal transduction relies on the regulation of protein function by posttranslational modifications such as phosphorylation, acetylation, or alkylation. Studies on the mechanisms by which cells transduce the formation of the ubiquitous signaling molecule nitric oxide (NO) into a functional response revealed a number of novel protein modifications with regulatory potential implicated in physiologically relevant NO signaling, as well as in situations of nitrosative stress.[1–5] Undoubtedly, critical cysteine residues are the principal target for NO and other NO-derived signaling molecules, including *S*-nitrosoglutathione (GSNO) and peroxynitrite. Depending on the reactivity and the structural context of the targeted protein thiol, on the one hand, and the structure and chemical reactivity of the NO-derived species, on the other, cysteine groups in proteins may be modified by *S*-nitrosylation and disulfide bridge formation, as well as oxidation to sulfenic, sulfinic, and sulfonic acids. Formation of a mixed disulfide between a protein thiol and the tripeptide glutathione (GSH), a mechanism termed *S*-glutathionylation, has emerged recently as a novel mechanism by which proteins sense the generation of reactive nitrogen species.[6] Although the exact mechanisms by which the glutathione moiety is transferred from GSNO to the protein thiol remain to be established, experimental evidence supports the idea that NO-induced protein *S*-glutathionylation critically depends on the NO-mediated conversion of GSH to GSNO.[6] In addition, molecular modeling of *S*-glutathionylated c-Jun suggests that mixed disulfide formation may be facilitated by specific interactions between the negatively charged glutathionyl moiety and the positively charged lysine and arginine residues surrounding the targeted protein thiol.[7] Such a GSH-binding site-like motif can be found in a number of other transcription factors, such as members of the Fos, ATF/CREB, and Rel/NF-κB families, as well as various cytosolic proteins, including glycogen phosphorylase b, glyceraldehyde-3-phosphate

[1] J. S. Stamler, S. Lamas, and F. C. Fang, *Cell* **106,** 675 (2001).

[2] D. T. Hess, A. Matsumoto, R. Nudelman, and J. S. Stamler, *Nature Cell Biol.* **3,** E46 (2001).

[3] S. R. Jaffrey, H. Erdjument-Bromage, C. D. Ferris, P. Tempst, and S. H. Snyder, *Nature Cell Biol.* **3,** 193 (2001).

[4] J. S. Stamler, E. R. Toone, S. A. Lipton, and N. J. Sucher, *Neuron* **18,** 691 (1997).

[5] J. S. Stamler and A. Hausladen, *Nature Struct. Biol.* **5,** 247 (1998).

[6] P. Klatt and S. Lamas, *Eur. J. Biochem.* **267,** 4928 (2000).

[7] P. Klatt, E. Pineda-Molina, M. G. De Lacoba, C. A. Padilla, E. Martínez Galisteo, J. A. Bárcena, and S. Lamas, *FASEB J.* **13,** 1481 (1999).

METHODS IN ENZYMOLOGY, VOL. 359
0076-6879/02 $35.00

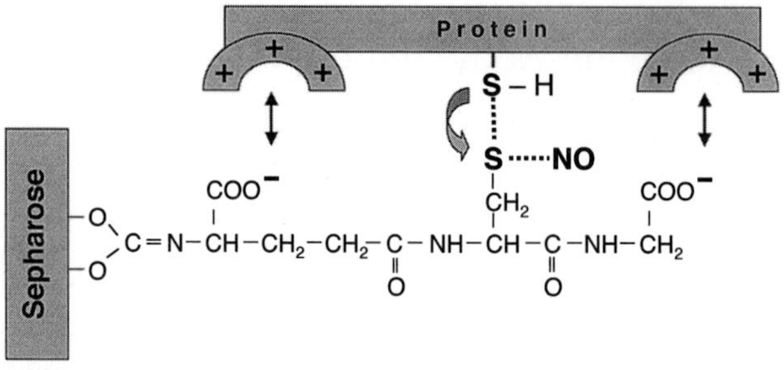

FIG. 1. Proposed scheme for the interaction between proteins and NO-AGS. NO-AGS can be obtained by reacting GSH Sepharose with acidified nitrite as described in the text. Evidence from Ref. 8 suggests that mixed disulfide formation between proteins and NO-AGS may be facilitated by electrostatic interactions that involve positively charged lysine and arginine residues in the proximity to the target thiol of the protein and negatively charged carboxyl groups of the immobilized GSH moiety. These interactions are proposed to position the NO-activated thiol group of the matrix in close vicinity to the targeted protein cysteine, thereby increasing the probability of mixed disulfide formation, which is likely to occur through the nucleophilic attack of the protein thiol on the nitrosothiol group.

dehydrogenase, and creatine kinase. It is tempting to speculate, therefore, that GSNO-mediated *S*-glutathionylation may be a general mechanism by which NO modifies proteins. The identification of proteins that are candidate targets for NO-induced *S*-glutathionylation should be facilitated greatly by a recently developed methodological approach[8] using nitrosylated GSH Sepharose as a mechanism-based affinity matrix that exploits two principal characteristics of site-specific *S*-glutathionylation induced by NO: (i) activation of the thiol moiety of GSH by NO and (ii) specific electrostatic interactions between GSH and the target protein, which are thought to increase the probability of mixed disulfide formation by positioning the protein thiol close to the nitrosothiol of the Sepharose (Fig. 1). The following sections describe the use of NO-activated GSH Sepharose (NO-AGS) as a rapid, simple, and nonradioactive method to screen proteins for their capacity to undergo *S*-glutathionylation, as well as a novel tool for the isolation of targets for GSNO-induced mixed disulfide formation.

Preparation of NO-Activated GSH-Sepharose (NO-AGS)

Starting material for the preparation fo NO-AGS is 2,2′-dipyridyl disulfide-activated GSH Sepharose, better known as "activated thiol Sepharose 4B" (Amersham Pharmacia Biotech, Barcelona, Spain). According to the specifications of the manufacturer, this preparation of Sepharose contains approximately 2.5 μmol activated thiol groups per ml swollen gel. The matrix is suspended in

[8] P. Klatt, E. Pineda-Molina, D. Pérez-Sala, and S. Lamas, *Biochem. J.* **349,** 567 (2000).

10 volumes of water and is incubated at room temperature for about 1 hr under gentle agitation. The swollen gel is washed with 10 volumes of water prior to reduction of the thiol groups by incubation for 1 hr at ambient temperature in 10 volumes of 0.3 *M* $NaHCO_3$ (pH 8.5) containing 1 m*M* EDTA and 5% (v/v) 2-mercaptoethanol. This procedure yields fully reduced GSH Sepharose, which is washed with 10 volumes of water followed by 30 volumes of 10 m*M* HCl. At this stage, the amount of free thiol groups can be determined by a colorimetric assay (a modification of the Ellmann's assay).[9] Reduction should be quantitative and yield 2.5 μmol free thiol groups/ml swollen gel. To avoid oxidation, GSH-Sepharose is immediately processed for *S*-nitrosylation and resuspended in 2 volumes of 10 m*M* HCl. Subsequently, 2 volumes of an aqueous solution of 10 m*M* $NaNO_2$ are added under constant agitation, and the suspension is incubated for a further 15 min at room temperature. Acidification of nitrite results in the release of stoichiometric amounts of NO, which reacts rapidly with available thiol groups of the Sepharose to form *S*-nitrosoglutathione. The obtained matrix, which displays a slightly red color typical of nitrosothiols, is washed with 5 volumes of 10 m*M* HCl, 5 volumes of H_2O, and 30 volumes of a 50 m*M* Tris/HCl buffer (pH 7.4) containing 250 m*M* NaCl and 1 m*M* EDTA. The *S*-nitrosothiol content of the matrix is determined photometrically by a previously described modification of Saville's assay[9] as $HgCl_2$-releasable NO. The conversion of thiol groups into nitrosothiols is virtually quantitative, i.e., it should be about 2.5 μmol/ml swollen gel. NO-activated Sepharose is relatively stable when stored at 4° in the dark under nitrogen or argon. Under these conditions, the loss of *S*-nitrosothiol groups is less than 10% within 1 week.

NO-AGS Binding Assay

Special attention should be paid to the protein preparation used for this assay. Mixed disulfide formation between proteins and activated GSH-Sepharose is completely abolished by millimolar concentrations of 2-mercaptoethanol, GSH, or DTT, which are usually present in protein preparations to prevent oxidation of protein thiols. However, we found that binding of *S*-glutathionylation-sensitive proteins such as c-Jun and p50 to NO-AGS is not significantly affected by up to 100 μM 2-mercaptoethanol. We, therefore, recommend working with concentrated protein preparations (>1 mg/ml) in buffers, which may contain up to 10 m*M* 2-mercaptoethanol as a reducing agent, and to dilute the reduced protein at least 100-fold into the assay mixture. Alternatively, thiols can be removed immediately prior to the binding assay by size-exclusion chromatography. The total amount of purified protein recommended for this binding assay is 10 μg. The protein is added to a suspension of the Sepharose (0.1 ml bed volume) in a final volume of 0.5 ml

[9] D. Gergel and A. I. Cederbaum, *Biochemistry* **35,** 16186 (1996).

of a 50 m*M* Tris/HCl buffer (pH 7.4) containing 250 m*M* NaCl, 1 m*M* EDTA, and 0.01% (v/v) NP-40. Following incubation at ambient temperature for 30 min under constant agitation, the incubation mixture is transferred to 0.5-ml spin filters (CytoSignal spin filters from Affiniti Research Products Ltd, Exeter, Devon, UK), which are centrifuged for 1 min at 300*g*. The resin is washed three times by the addition of 0.5 ml incubation buffer and centrifugation at 300*g* for 1 min. If necessary, nonspecific binding can be reduced further by making the washing conditions more stringent using high-salt (1 *M* NaCl) or detergent (e.g., 5% SDS) containing buffers. Bound protein is eluted by the addition of 0.1 ml of incubation buffer, which additionally contains 1% (v/v) 2-mercaptoethanol, incubation for 10 min at ambient temperature, and centrifugation at 5000*g* for 2 min. Protein binding to the matrix can be monitored by SDS–PAGE or quantified by determining protein concentrations. Of note, the presence of detergent (NP-40) in the eluate may interfere with some methods for the determination of protein concentrations such as the Bradford assay.[10] It is recommended, therefore, to use a detergent-compatible assay such as the DC protein assay kit from Bio-Rad (Madrid, Spain).

Characteristics of Protein Binding to NO-AGS

Figure 2 illustrates the characteristics of covalent protein binding to NO-AGS using the purified DNA-binding domain of c-Jun DNA as an *S*-glutathionylation-sensitive model protein.[7,11] Figure 2 (left) shows that c-Jun binding to GSNO Sepharose follows saturation kinetics. Binding to 0.1 ml NO-AGS is virtually quantitative (>95% of total protein) when the amounts of c-Jun are <0.2 nmol and approaches saturation at >1 nmol protein. c-Jun binding to GSNO Sepharose is not reverted by a high salt concentration (1 *M* NaCl) or detergent (5% SDS) but is undetectable in the presence of millimolar concentrations of thiol-reducing agents, such as 2-mercaptoethanol, GSH, and DTT. However, as outlined earlier, binding is apparently not affected by the presence of up to 100 μM 2-mercaptoethanol.

As shown in Fig. 2 (right), c-Jun binding to NO-AGS shows a linear dependency on the total amount of the matrix in the presence of saturating amounts of protein (2 nmol c-Jun). From the slope of this graph, the c-Jun-binding capacity of NO-AGS is calculated as 12 nmol c-Jun subunit/ml of swollen gel. This value appears rather low taking into account that 1 ml of the matrix contains 2.4 ± 0.1 μmol ($n = 6$) *S*-nitrosothiol groups, i.e., only 0.5% of the available *S*-nitrosothiol groups has reacted with protein sulfhydryls. This low yield may be explained, at least in part, by steric constraints. It has been reported that 1 ml of highly reactive 2,2′-dipyridyl disulfide-activated GSH Sepharose, the starting material for the preparation of NO-AGS, binds a maximum of 10–50 nmol of native proteins, such

[10] M. M. Bradford, *Anal. Biochem.* **72,** 248 (1976).

[11] P. Klatt, E. Pineda-Molina, and S. Lamas, *J. Biol. Chem.* **274,** 15857 (1999).

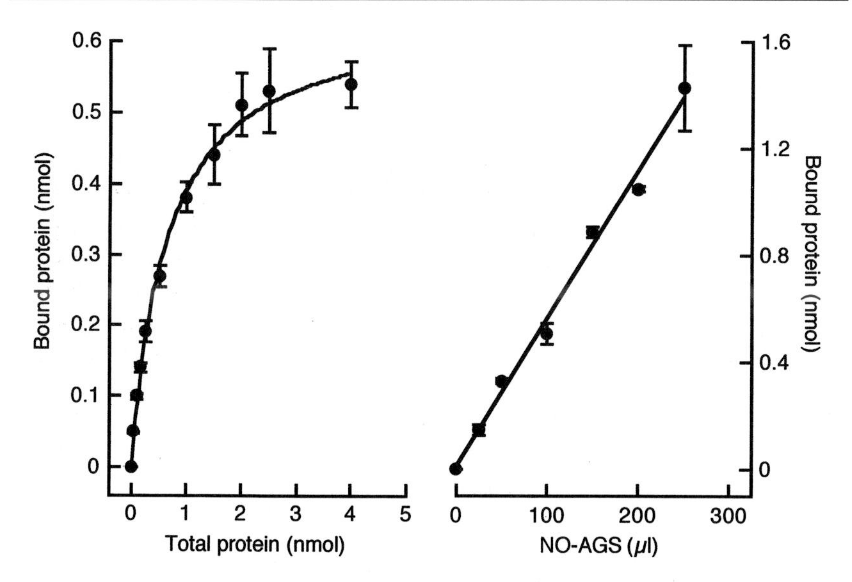

FIG. 2. Basic characteristics of protein binding to NO-AGS. Recombinant c-Jun DNA-binding domains were incubated in a final volume of 0.5 ml with NO-AGS and assayed for covalent binding to this matrix through mixed disulfide formation as described in the text. The dependency of c-Jun binding on the total amount of c-Jun protein (left) was assayed with increasing amounts of c-Jun (0.05–4 nmol) and a fixed quantity (0.1 ml) of GSNO Sepharose. To determine the dependency of c-Jun binding on the amount of NO-AGS (right), a fixed amount of c-Jun (2 nmol) was incubated with increasing amounts of NO-AGS (25–250 μl). Data are mean values $\pm$ SEM of 4–12 different experiments.

as Hb, erythrocyte band III protein, and collagen,[12–14] whereas up to 200 nmol of denatured proteins and peptides is retained by 1 ml of the same matrix.[15] Thus, the binding capacity of NO-AGS is within the range reported for protein binding to activated glutathione Sepharose. The site specificity of NO-AGS is illustrated by a comparison of covalent protein binding to immobilized GSNO, i.e., NO-AGS, with the binding to 2,2′-dipyridl disulfide-activated GSH Sepharose and by correlating these data with protein *S*-glutathionylation induced by free GSNO. Figure 3 shows a typical experiment using the DNA-binding domain of c-Jun and the NF-κB subunit p50 as model proteins. Both proteins contain a reactive cysteine residue that is highly susceptible to *S*-glutathionylation (cysteine 269[7] and cysteine 62,[16] respectively). When mixed disulfide formation is studied in the presence of

[12] A. Kahlenberg, *Anal. Biochem.* **74,** 337 (1976).

[13] S. H. De Bruin, J. J. Joordens, and H. S. Rollema, *Eur. J. Biochem.* **75,** 211 (1977).

[14] B. C. Sykes, *FEBS Lett.* **61,** 180 (1976).

[15] T. A. Egorov, A. Svenson, L. Ryden, and J. Carlsson, *Proc. Natl. Acad. Sci. U.S.A.* **72,** 3029 (1975).

[16] E. Pineda-Molina, P. Klatt, J. Vazquez, A. Marina, M. Garcia de Lacoba, D. Perez-Sala, and S. Lamas, *Biochemistry* **40,** 14134 (2001).

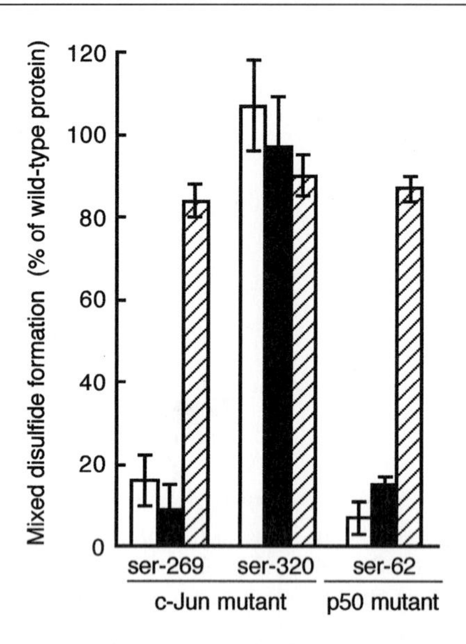

FIG. 3. Site specificity of protein binding to NO-AGS. Wild-type and mutant c-Jun and p50 DNA-binding domains in which either the cysteine located in the DNA-binding site (ser-269) and adjacent leucine zipper (ser-320) of c-Jun or the cysteine in the DNA-binding site of p50 (ser-62) was substituted by serine were assayed for GSNO-induced *S*-glutathionylation (open bars) by the addition of [^{3}H]GSH/GSNO as described elsewhere (see Ref. 17) or precipitated with NO-AGS (closed bars) and 2,2′-dipyridyl disulfide-activated GSH Sepharose (hatched bars) as described in the text. Mixed disulfide formation between mutant proteins and free GSH or Sepharose-bound GSH is expressed as the percent (%) of mixed disulfide formation observed with the corresponding wild-type protein. Data are mean values ± SEM of at least three different experiments.

^{3}H-labeled GSH and GSNO as described elsewhere,[17] wild-type c-Jun and p50 homodimers incorporate 2.0 ± 0.1 ($n = 6$) and 0.9 ± 0.1 ($n = 4$) mol GSH/mol protein, respectively. As shown in Fig. 3 (open bars), mutation of the c-Jun cysteine residue 269 to serine decreases mixed disulfide formation induced by free GSNO to $16 \pm 5\%$ ($n = 5$) of the wild-type protein, whereas a cysteine-to-serine mutation at residue 320 apparently does not affect c-Jun *S*-glutathionylation ($107 \pm 11\%$, $n = 3$). Similarly, mutation of cysteine residue 62 in the DNA-binding domain of p50 almost completely abolishes GSNO-induced *S*-glutathionylation of the NF-κB subunit as compared to the wild-type protein ($7 \pm 4\%$, $n = 4$). This site specificity of *S*-glutathionylation induced by free GSNO is perfectly reflected in the covalent binding of these proteins to a NO-AGS. As shown in Fig. 3 (closed bars), NO-AGS

[17] P. Klatt and S. Lamas, *Methods Enzymol.* **348,** 157 (2002).

precipitates the c-Jun mutants ser-269, ser-320, and the p50 mutant ser-62 with an efficacy of $9 \pm 6\%$ ($n = 5$), $97 \pm 12\%$ ($n = 5$), and $15 \pm 2\%$ ($n = 4$), respectively, as compared with the corresponding wild-type protein. In contrast to these data obtained with NO-AGS, activation of GSH Sepharose with 2,2′-dipyridyl disulfide instead of NO generates a matrix that lacks site specificity and binds both the investigated wild-type protein and mutants to a comparable extent (Fig. 3, hatched bars). This experiment demonstrates that cysteine residues different from cysteine-269 (c-Jun) and cysteine 62 (p50) are fully available for mixed disulfide formation with GSH Sepharose activated with 2,2′-dipyridyl disulfide. However, only a subset of these residues, 1 out of 2 in the case of c-Jun and 1 out of 14 in the case of p50, forms a mixed disulfide with NO-AGS. In conclusion, this suggests that activation of the GSH resin with NO confers site specificity to the matrix and reflects the susceptibility of a given protein to *S*-glutathionylation induced by free GSNO. This idea is further supported by experiments with various purified preparations of thiol-containing proteins as shown in Fig. 4. For these experiments, 10 μg of the indicated protein is incubated with 0.1 ml NO-AGS under the conditions described earlier. Aliquots obtained from the various steps of this assay are analyzed by SDS–PAGE. It can be seen that the c-Jun DNA-binding domains (c-Jun), glycogen phosphorylase (GP-b), glyceraldehyde-3-phosphate dehydrogenase (GAPDH), the caspase-3 subunits p17 and p12 (Caspase-3), creatine kinase (CK), glutaredoxin (Grx), and the NF-κB subunit p50 can be precipitated efficiently by NO-AGS, whereas no appreciable binding is observed with alcohol dehydrogenase (ADH), glycerol-3-phosphate dehydrogenase (GPDH), hemoglobin (Hb), thioredoxin (Trx), carbonic anhydrase (CA), Cu,Zn superoxide dismutase (SOD), and BSA. Here, NO-AGS binding is compared to the susceptibility of these proteins to undergo GSNO-induced *S*-glutathionylation, which was determined in the presence of [^{3}H]GSH as described elsewhere.[17] Such a comparison confirms that proteins susceptible to a GSNO-induced *S*-glutathionylation, i.e., that incorporate $\geq$0.9 mol GSH/mol protein, bind to NO-AGS, whereas proteins that do not readily form mixed disulfides with GSNO, i.e., that incorporate $\leq$0.4 mol GSH/mol protein, are not precipitated by the NO-activated matrix. To exclude that lack of mixed disulfide formation is due to oxidation or inaccessibility of protein cysteines, we recommend control experiments with the thiol-oxidizing agent diamide in the presence of [^{3}H]GSH as described elsewhere.[17] Diamide is supposed to potently direct mixed disulfide formation in a nonspecific manner to any protein thiol that is accessible to GSH. Diamide-induced *S*-glutathionylation, thus, is taken as a measure for the maximal number of cysteine residues in a protein accessible to mixed disulfide formation. As shown in Fig. 4, some of the investigated proteins, namely ADH, GPDH, Hb, and BSA, do not form mixed disulfides with free GSNO or NO-AGS, although they possess one or more cysteines that become readily thiolated in the presence of diamide ($\geq$0.9 mol GSH/mol protein).

Protein	Binding to NO-AGS						SH	RSSG	
	Input	Unbound	1st wash	2nd wash	3rd wash	Eluate		GSNO	Diamide
c-Jun							4	2.0	3.8
GP-b							18	2.0	14.3
GAPDH							16	1.3	4.2
Caspase-3							14	1.2	n.d.
CK							8	1.0	3.1
Grx							4	0.9	1.0
p50							14	0.9	8.1
ADH							32	0.4	16.9
GPDH							22	0.4	14.5
Hb							2	0.2	1.8
Trx							3	0.2	0.6
CA							1	0.1	0.4
SOD							6	0.2	0.3
BSA							35	0.1	0.9

FIG. 4. Binding of purified proteins to NO-AGS compared with their susceptibility to GSNO- and diamide-induced *S*-glutathionylation. A suspension of 0.1 ml NO-AGS was incubated under the conditions described in the text with 10 μg of one of the indicated purified proteins, including recombinant human c-Jun DNA-binding domains (c-Jun), glycogen phosphorylase purified from rabbit muscle (GP-b), glyceraldehyde-3-phosphate dehydrogenase purified from rabbit muscle (GAPDH), the recombinant human caspase-3 subunits p17 and p12 (caspase-3), creatine kinase purified from rabbit muscle (CK), recombinant human glutaredoxin (Grx), the human recombinant NF-κB subunit p50 (p50), baker's yeast alcohol dehydrogenase (ADH), glycerol-3-phosphate dehydrogenase from rabbit muscle (GPDH), bovine erythrocyte hemoglobin (Hb), human recombinant thioredoxin (Trx), bovine erythrocyte carbonic anhydrase (CA), bovine liver Cu,Zn superoxide dismutase (SOD), and BSA. Aliquots (25 μl) of the incubation mixture (input), NO-AGS flow through (unbound), various washing steps (first, second, and third wash), and NO-AGS eluate (eluate) were analyzed by SDS–PAGE on 13% polyacrylamide gels. Gels were stained for protein with Coomassie Blue. The gels shown are representative for at least three similar experiments. The total number of cysteines per protein (SH) was taken from the published amino acid sequences and is given as mole cysteine residues per mole protein. GSNO- and diamide-induced *S*-glutathionylation (RSSG) of the indicated purified proteins (2–10 μM) was determined in the presence of [^{3}H]GSH (3 mM) and 1 mM GSNO (GSNO) or 10 mM diamide (diamide), respectively, as described elsewhere.[17] Data are given as mole mixed disulfide per mole protein. Molar ratios refer to the established quaternary structure of the protein. Data are mean values of at least three independent experiments. n.d., not determined.

Isolation of Target Proteins for GSNO-Induced *S*-Glutathionylation by NO-AGS Chromatography

As outlined earlier, the binding characteristics of NO-AGS support the idea that this matrix might be a useful tool for the isolation and identification of cellular proteins that are potential targets for GSNO-induced *S*-glutathionylation. The identification of such proteins might be performed by functional assays or proteomic methods. This section describes the application of NO-AGS for the precipitation of nuclear transcription factors that are potential targets for GSNO-induced *S*-glutathionylation from HeLa cell extracts.

For this experiment, pooled concentrated nuclear extracts [10 mg total protein in a 20 m*M* HEPES buffer adjusted to pH 7.9 containing 500 m*M* NaCl, 1 m*M* EDTA, 10% (v/v) glycerol, 10 m*M* 2-mercaptoethanol and 2 m*M* Pefabloc] from exponentially growing HeLa cells are diluted 1 : 100 (v/v) into 50 ml of a 50 m*M* Tris/HCl buffer (pH 7.4) containing 250 m*M* NaCl, 1 m*M* EDTA, 0.01% (v/v) NP-40, 1 m*M* of the protease inhibitor Pefabloc, and 5 ml of thoroughly suspended NO-AGS. The suspension is incubated for 30 min at ambient temperature with occasional shaking prior to filtration over a fritted column (ID = 1 cm). Subsequently, the column is washed with 30 volumes of incubation buffer and eluted with 3 volumes of the same buffer, which additionally contains 1% (v/v) 2-mercaptoethanol. The eluate, which usually contains 50–100 μg protein, is concentrated to $\leq$1 ml on Vivapore 20 concentrators (molecular weight cut-off: 7500; Vivascience, Binbrook, Lincoln, UK), and aliquots are stored at $-80°$ for further analysis. To determine the specific enrichment of proteins that are targets for GSNO-induced *S*-glutathionylation, the column input (total extract), unbound protein (flow through), and bound protein (eluate) are assayed for the total amount of cysteine residues available for mixed disulfide formation with GSH and for the amount of cysteines that undergo GSNO-induced mixed disulfide formation in the presence of diamide/[^{3}H]GSH and GSNO/[^{3}H]GSH, respectively, as described elsewhere.[17] As shown in Table I, NO-AGS chromatography of HeLa cell extracts results in an about 20-fold enrichment of GSNO-reactive proteins, whereas the number of diamide-reactive thiols increases only about 2-fold.

DNA-binding assays may be used to identify nuclear transcription factors isolated by NO-AGS chromatography. Figure 5 shows results from electrophoretic mobility shift assays (EMSA) for the detection of activator protein-1 (AP-1), nuclear factor-*k*B (NF-κB), cAMP-response-element-binding protein (CREB), and CCAAT-binding protein-1 (CP-1). For this assay, 1 μg of total protein from the NO-AGS flow through (unbound) and eluate (bound) are incubated for 30 min on ice in a 20 m*M* Tris/HCl buffer (pH 7.5) containing 50 m*M* NaCl, 5 m*M* $MgCl_2$, 1 m*M* EDTA, 5% (v/v) glycerol, 0.01% (v/v) NP-40, 0.2 mg/ml bovine serum albumin, 0.1 mg/ml poly(dL-dC), and one of the following: ^{32}P-radiolabeled double-stranded oligonucleotides containing the DNA-binding sites for the transcription factors AP-1 (5′-GGG CTT GAT GAG TCA GCC GGA CC-3′), NF-κB (5′-GGA GAG GGG ATT CCC TGC G-3′), CREB (5′-AGA GAT TGC CTG ACG TCA GAG AGC TAG-3′), and CP-1 (5′-CCA CAA ACC AGC CAA TGA GTA ACT GCT GG-3′). Incubations are performed in the absence or presence of a 50-fold excess of an unlabeled competitor oligonucleotide identical to the radiolabeled probe

TABLE I
ISOLATION OF NUCLEAR HeLa CELL PROTEINS SUSCEPTIBLE TO GSNO-INDUCED *S*-GLUTATHIONYLATION BY NO-AGS CHROMATOGRAPHY[a]

	Total extract	Flow through	Eluate
Total protein (mg)	12 ± 3	11 ± 3	0.05 ± 0.01
GSNO-induced *S*-glutathionylation (nmol [^{3}H]GSH/mg protein)	2 ± 2	3 ± 1	37 ± 14
Purification of GSNO-reactive sulfhydryls (-fold)	1	1	19
Diamide-induced *S*-glutathionylation (nmol [^{3}H]GSH/mg protein)	53 ± 12	67 ± 19	122 ± 12
Purification of diamide-reactive sulfhydryls (-fold)	1	1	2

[a] HeLa cell nuclear extracts were subjected to NO-AGS chromatography as outlined in the text. The total nuclear extract, column flow through, and eluate were analyzed for protein content by the Bradford assay [M. M. Bradford, *Anal. Biochem.* **72,** 248 (1976)] and *S*-glutathionylation, which was induced either by the addition of [^{3}H]GSH/GSNO or [^{3}H]GSH/diamide as described elsewhere [P. Klatt and S. Lamas, *Methods Enzymol.* **348,** 157 (2002)]. Data are mean values ± SEM of three different preparations.

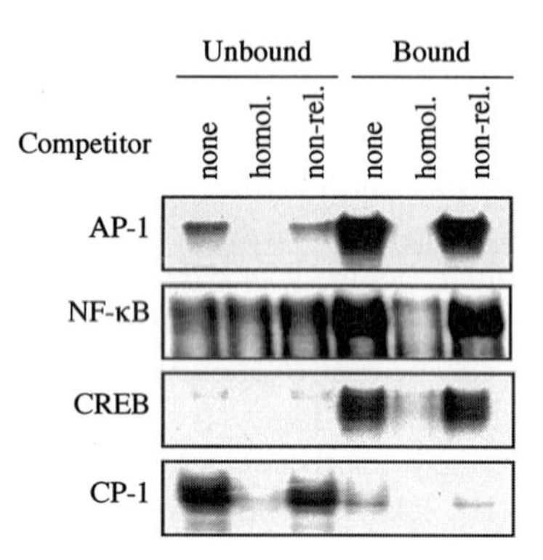

FIG. 5. Isolation of AP-1, NF-κB, CREB, and CP-1 DNA-binding activities from nuclear HeLa cell extracts by NO-AGS chromatography. HeLa cell nuclear extracts were subjected to NO-AGS chromatography, and the column flow through (unbound) and eluate (bound) were analyzed for AP-1, NF-κB, CREB, and CP-1 DNA-binding activities by EMSA as described in the text. Specificity of protein–DNA complex formation was confirmed by homologous competition with a 50-fold excess of an unlabeled competitor oligonucleotide identical to the radiolabeled probe (homol.) or an unrelated oligonucleotide (non-rel.). The unlabeled double-stranded AP-1 oligonucleotide was used as a nonrelated competitor for NF-κB binding, and the NF-κB oligonucleotide was used as a nonrelated competitor for AP-1, CREB, and CP-1 binding. Gels shown are representative of four similar experiments.

or an unrelated oligonucleotide. The unlabeled double-stranded AP-1 oligonucleotide described earlier can be used as a nonrelated competitor for NF-κB binding, and the unlabeled NF-κB oligonucleotide as a nonrelated competitor for AP-1, CREB, and CP-1 binding. Subsequent to the binding reaction, samples are subjected to electrophoresis at 180 V on preelectrophoresed 6% nondenaturing polyacrylamide gels with 22 m*M* Tris–borate/0.5 m*M* EDTA as running buffer. Gels are dried and visualized by autoradiography. Consistent with NO-AGS binding data obtained with the purified AP-1 and NF-κB subunits c-Jun and p50 (see earlier discussion), the DNA-binding activities of these transcription factors are almost exclusively detected in the NO-AGS eluate (Fig. 5). The specificity of the nitrosothiol matrix is further confirmed by the observation that the DNA-binding activity of CREB, which in terms of its DNA-binding site can be considered as a structural homologue of c-Jun, is retained by the matrix, whereas the transcription factor CP-1, despite the presence of reactive cysteine residues in its DNA-binding domain, does not bind to NO-AGS.

In conclusion, we believe that NO-AGS may serve as a probe of site-specific *S*-glutathionylation and as a useful tool to isolate proteins that are candidate targets for GSNO-induced mixed disulfide formation.

Acknowledgments

This report is based on experiments developed in the laboratory of S. Lamas at the Centro de Investigaciones Biológicas, CSIC, Madrid, Spain, with collaboration of E. Pineda-Molina, D. Pérez-Sala, and M. García de Lacoba. The authors thank J. Rey (Centro de Investigaciones Biológicas, CSIC, Madrid, Spain), J. Alcamí (Instituto de Salud Carlos III, Madrid, Spain), J. A. Bárcena (Departamento de Bioquímica y Biología Molecular, Facultad de Veterinaria, Universidad de Córdoba, Córdoba, Spain), and B. Brüne (Department of Medicine IV, Experimental Division, Faculty of Medicine, University of Erlangen-Nürnberg, Erlangen, Germany) for their generous gifts of expression plasmids and purified recombinant proteins, and also thank P. Ramos (Facultad de Ciencias Experimentales de la Salud, Universidad San Pablo CEU, Boadilla del Monte, Madrid, Spain) for critical reading of the manuscript. Financial support of the Plan Nacional I+D+I (SAF 97-0035 and SAF 2000-0149), the Comunidad Autónoma de Madrid (CAM 08.4/0032/1998), and the European Community (BMH4-CT98-5052) is gratefully acknowledged.

[24] Measurement of Protein Nitration and S-Nitrosothiol Formation in Biology and Medicine

By KEVIN P. MOORE and ALI R. MANI

Introduction

Reactive nitrogen species (RNS) are involved in cell signaling and cellular injury, being formed in a variety of physiological and pathological processes. Most RNS are short-lived species and are difficult to measure directly. However, RNS may react with tyrosine residues in proteins to form nitrotyrosine (a nitration reaction) or with cysteine residues to form *S*-nitrosylated proteins (*S*-nitrosothiols, RSNOs), an *S*-nitrosation reaction. RNS may also cause the nitration of other amino acids such as tryptophan,[1] lipids such as γ-tocopherol,[2] and unsaturated fatty acids,[3] or carbohydrates such as glucose.[3] RNS may also cause *N*-nitrosation reactions to form nitrosamines, which have carcinogenic potential. However, this article focuses on the measurement of nitrotyrosine and RSNOs in biological samples.

In proteins, most nitration reactions (addition of $-NO_2$) occur on a tyrosine residue to form 3-nitrotyrosine, which is chemically stable but which ultimately undergoes metabolism in the liver to form 3-nitro-4-hydroxyphenylacetic acid, which is excreted in urine.[4] The nitration of critical tyrosine residues may have important effects on protein function such as cell signaling, as it may prevent tyrosine phosphorylation. However, at present it is not clear to what extent nitration affects protein function *in vivo* or whether the formation of nitrotyrosine merely represents a footprint for the formation of RNS *in vivo*. Initially it was thought that the formation of nitrotyrosine was an index for peroxynitrite formation, but we now know that a variety of other RNS, such as dinitrogen trioxide (N_2O_3) and nitryl chloride (NO_2–Cl), are important or maybe more important in the formation of nitrotyrosine *in vivo*.[5,6]

[1] B. Alvarez, H. Rubbo, M. Kirk, S. Barnes, B. A. Freeman, and R. Radi, *Chem. Res. Toxicol.* **9,** 390 (1996).

[2] K. Hensly, K. S. Williamson, and R. A. Floyd, *Free Radic. Biol. Med.* **24,** 520 (2000).

[3] V. B. O'Donnell, J. P. Eiserich, A. Bloodsworth, P. H. Chumley, M. Kirk, S. Barnes, V. M. Darley-Usmar, and B. A. Freeman, *Methods Enzymol.* **301,** 454 (1999).

[4] H. Ohshima, M. Friesen, I. Brouet, and H. Bartsch, *Food. Chem. Toxicol.* **28,** 647 (1990).

[5] S. Baldus, J. P. Eiserich, A. Mani, L. Castro, M. Figueroa, P. Chumley, W. Ma, A. Tousson, C. R. White, D. C. Bullard, M. L. Brennan, A. J. Lusis, K. P. Moore, and B. A. Freeman, *J. Clin. Invest.* **108,** 1759 (2001).

[6] S. Pfeiffer, A. Lass, K. Schmidt, and B. Mayer, *FASEB J.* **15,** 2355 (2001).

METHODS IN ENZYMOLOGY, VOL. 359
0076-6879/02 $35.00

From the work of Lascalzo's group[7] in the early 1990s it became increasingly evident that *S*-nitrosation of the sulydryl moiety of cysteine in proteins or peptides to form RSNOs was a potentially important pathway to regulate protein function. Initially, the function of RSNOs was described in terms of their vascular effects as vasodilators and then as modulators of platelet function. More recently, the *S*-nitrosation of hemoglobin has been described as a regulator of hemoglobin function,[8] although this function is currently contentious.[9] However, the demonstration that *S*-nitrosation of proteins is an important cell signaling mechanism controlling protein or enzyme function seems to be an important emerging mechanism in biology and medicine.

Identification of Protein Targets to Nitration or Nitrosation

Based on *in vitro* studies, about 100 proteins have been shown to undergo regulation by *S*-nitrosation but only a handful of proteins have been identified as targets for *S*-nitrosation *in vivo*.[10] Loss or rearrangement of the S–NO group during sample preparation has been the most important technical limitation, but recent advances, using thiol-blocking agents to stabilize RSNOs and the use of the biotin switch method, have provided new tools for identifying S–NO-modified proteins formed *in vivo*.[10,11] With the exception of *S*-nitrosoalbumin, which represents $<0.01\%$ of total albumin,[11] we do not know what proportion of proteins regulated in this way are *S*-nitrosated at any one time, and this is similar to that observed for nitrotyrosine, which is also present at $<0.01\%$ of total tyrosine residues in tissue proteins. The nitrosation and nitration of proteins are relatively selective processes, as specific cysteine or tyrosine residues in proteins undergo modification by RNS. Site-directed mutagenesis methods have been used to determine the specific residue responsible for altered protein function nitration or nitrosation.[12–14] Liquid chromatography linked to electrospray ionization tandem

[7] J. S. Stamler, O. Jaraki, J. Osborne, D. I. Simon, J. Keaney, J. Vita, D. Single, C. R. Valeri, and J. Loscalzo, *Proc. Natl. Acad. Sci. U.S.A.* **89,** 7674 (1992).

[8] A. J. Gow and J. S. Stamler, *Nature (Lond.)* **391,** 169 (1998).

[9] M. Wolzt, R. J. MacAllister, D. Davis, M. Feelisch, S. Moncada, P. Vallance, and A. J. Hobbs, *J. Biol. Chem.* **274,** 28983 (1999).

[10] S. R. Jaffrey, H. Erdjument-Bromage, C. D. Ferris, P. Tempst, and S. H. Snyder, *Nature Cell Biol.* **3,** 193 (2001).

[11] R. Marley, R. P. Patel, N. Orie, E. Ceaser, V. Darley-Usmar, and K. Moore, *Free Radic. Biol. Med.* **31,** 688 (2001).

[12] T. Shimokawa, R. J. Kulmacz, D. L. DeWitt, and W. L. Smith, *J. Biol. Chem.* **265,** 20073 (1990).

[13] I. Perez-Mato, C. Castro, F. A. Ruiz, F. J. Corrales, and J. M. Mato, *J. Biol. Chem.* **274,** 17075 (1999).

[14] J. Sun, C. Xin, J. P. Eu, J. S. Stamler, and G. Meissner, *Proc. Natl. Acad. Sci. U.S.A.* **98,** 11158 (2001).

mass spectrometry (LC-MS/MS) is an alternative method for identification of a selective site(s) of nitration or nitrosation in relatively pure proteins.[15,16]

Nitrotyrosine can be localized in tissues by immunohistochemistry and quantitated using mass spectrometry. Mass spectrometry is the most sensitive and specific method to quantitate nitrotyrosine levels, providing care is taken to use sample preparation procedures that avoid the artifactual nitration of tyrosine. Western blotting and confocal microscopy using double labeling with antibodies directed at specific proteins suspected of being surrogate targets for nitration have been used to identify nitrated proteins.[17] More recently, the combination of two-dimensional gel electrophoresis and matrix-assisted laser desorption ionization/time-of-flight (MALDI-TOF) mass spectrometric analysis has identified several hepatic proteins nitrated during endotoxemia.[18] This article discusses methods used for the measurement of *S*-nitrosated and nitrated proteins.

Measurement of RSNOs

The first thing to understand when handling RSNOs is that they are unstable under certain conditions. In buffer solution they can be broken down by ultraviolet (UV) light or by trace amounts of copper ions. Therefore, solutions should be kept in the dark, and trace metal ions chelated by diethylenetriaminepentaacetic acid (DTPA, $\sim$50–100 μM). They are most stable under acidic conditions (low molecular weight RSNOs are even stable in strong acid). The second important thing to understand is that thiols in biological samples, particularly low molecular weight thiols, will readily undergo transnitrosation reactions with RSNOs to form less stable intermediates such as *S*-nitrosocysteine or *S*-nitrosoglutathione, which decompose readily. This can be prevented by adding *N*-ethylmaleimide (NEM; final concentration of 5 mM) to samples during collection. For example, we found that whereas 40% of added *S*-nitrosoalbumin (1 μM) decomposes within 2 hr,[19] it was stable for up to 24 hr at room temperature when NEM was present. NEM alkylates thiol groups rapidly and prevents thiol-dependent decomposition of RSNOs. The decomposition of RSNOs may also be enhanced by ascorbate or ceruloplasmin,[20] although this is not a major problem in plasma, providing free thiol groups are blocked. All investigators that measure RSNOs

[15] C. Ducrocq, M. Dendane, O. Lapreote, L. Serani, B. C. Das, N. Bouchemal-Chibani, B. T. Doan, B. Gillet, A. Karim, A. Carayon, and D. Payen, *Eur. J. Biochem.* **253,** 146 (1998).

[16] A. S. Petersson, H. Steen, D. E. Kalume, K. Caidahl, and P. Roepstorff, *J. Mass Spectrom.* **36,** 616 (2001).

[17] M. H. Zou, M. Leist, and V. Ullrich, *Am. J. Pathol.* **154,** 1359 (1999).

[18] K. S. Aulak, M. Miyagi, L. Yan, K. A. West, D. Massillon, J. W. Crabb, and D. J. Stuehr, *Proc. Natl. Acad. Sci. U.S.A.* **98,** 12056 (2001).

[19] R. Marley, M. Feelisch, S. Holt, and K. Moore, *Free Radic. Res.* **32,** 1 (2000).

[20] A. P. Dicks and D. L. H. Williams, *Chem. Biol.* **3,** 655 (1996).

should assess their stability in fresh tissue homogenate or plasma for themselves and confirm that their sample handling methods ensure stability. When carrying out stability studies, it is important to use a near-physiological concentration of added RSNO, as the decomposition of RSNOs is retarded if high concentrations are used, as these presumably accelerate the consumption of low molecular weight thiols.

Many methods have been described for the measurement of RSNOs in the literature, but only a few have been validated and applied successfully to their quantitative measurement in biological samples. To validate a method, one should be able to add the analyte to the biological fluid in question and demonstrate that the assay quantitatively measures an increment of the analyte concentration equal to the amount added. This validation still does not exclude the possibility that other compounds endogenously present in biological samples may cause overestimation. This is a significant problem in the measurement of RSNOs, as nitrite, present in all biological samples, is usually detected by the methods used to measure RSNOs. Because background nitrite concentrations are ~300 n*M* in plasma, it is necessary to remove nitrite. Some groups have used gel filtration methods,[21] but these are time-consuming and do not remove nitrite completely. We discovered that nitrite ions can be removed effectively by reaction with acidified sulfanilamide, and this has no impact on the nitric oxide (NO) signal released by authentic RSNOs added to plasma.[19]

The method that we developed was based on the release of NO by reaction by a mixture of Cu^+, iodide in acid and release of free iodine, and quantification by ozone-dependent chemiluminescence.[19] Chemiluminescence-based assays can detect plasma concentrations of RSNOs as low as ~5 n*M*.[19,21,22] This method is similar to, and gives identical results to, that developed by Gladwin *et al.*[21] Because the peaks of NO released from RSNOs are somewhat broader than that released from nitrite alone, we find it hard to reliably detect concentrations much lower than 5 n*M*, but the quantitation may be enhanced by using Origen software to integrate the peaks (M. Gladwin and R. Patel, personal communication). With this method the concentration of RSNOs in venous plasma of healthy human volunteers was estimated as 28 ± 7 n*M*, which is considerably lower than previous reports (0.2–7 μM).[7,23] Many investigators were initially confused by their own findings, as it was originally reported by Stamler *et al.*[7] that plasma RSNOs (predominantly *S*-nitrosoalbumin) were present at a concentration of 7 μM. This measurement was based on photolytic cleavage of the S–NO bond. These levels are now widely acknowledged to be too high, with most recent estimates

[21] M. T. Gladwin, F. P. Ognibene, L. K. Pannell, J. S. Nichols, M. E. Pease-Fye, J. H. Shelhamer, and A. N. Schechter, *Proc. Natl. Acad. Sci. U.S.A.* **97,** 9943 (2000).

[22] A. Samouilov and J. L. Zweier, *Anal. Biochem.* **258,** 322 (1998).

[23] R. K. Goldman, A. A. Vlessis, and D. D. Trunkey, *Anal. Biochem.* **259,** 98 (1998).

of plasma RSNO concentrations as being in the low nanomolar range.[19,24,25] The specificity of chemiluminescence-based analysis of NO-related compounds is high because the majority of other molecules potentially able to give chemiluminescence with ozone are not volatile or do not occur in biological systems. Two exceptions are ammonia and sulfur gases, which can produce chemiluminescence with ozone, and a number of methods for the removal of such interferences have been described.[26,27] We have also applied the chemiluminescence-based method for the measurement of total RSNO content in solid tissue homogenates (unpublished).

To detect specific RSNOs, such as *S*-nitrosoglutathione or *S*-nitrosoalbumin, requires a separation system such as high-performance liquid chromatography (HPLC). Akaike *et al.*[28] developed a novel and sophisticated HPLC method to separate nitrite and RSNOs with postcolumn copper-based decomposition of RSNOs and derivatization by the Griess reaction. However, this method was not sensitive enough to detect basal concentrations of RSNOs in plasma. Liquid chromatography–mass spectrometry (LC-MS) is also reported to have a good sensitivity in measurement of *S*-nitrosoglutathione in tissue homogenates.[29] However, low molecular weight nitrosothiols (e.g., *S*-nitrosoglutathione or *S*-nitrosocysteine) can be quantitated by chemiluminescence following the precipitation of proteins, which is described next.

RSNO Methods

Synthesis of S-Nitrosothiols

S-Nitrosocysteine. Prepare by reacting equal volumes of sodium nitrite (10 m*M*) with L-cysteine hydrochloride (10 m*M*) at pH 2.

S-Nitrosoglutathione. Prepare by dissolving 614 mg (2 mmol) of reduced glutathione in 3 ml 0.67 *M* HCl and cooled on ice. Add 138 mg sodium nitrite (20 mmol) as powder and stir for 40 min on ice. A thick pink precipitate is formed. Filter under a stream of nitrogen and wash with 2 × 10 ml of ice-cold water, followed by 2 × 10 ml ice-cold acetone, and then 2 × 10 ml ice-cold ether. Dry in a vacuum desecrator in the dark. The yield of *S*-nitrosoglutathione is high, and only trace amounts of glutathione are evident on LC-MS analysis (unpublished).

[24] T. Akaike, *Free Radic. Res.* **35,** 461 (2000).
[25] R. Rossi, D. Giustarini, A. Milzani, R. Colombo, I. Dalle-Donne, and P. Di Simplicio, *Circ. Res.* **89,** E47 (2001).
[26] S. J. Chung and H. L. Fung, *Anal. Lett.* **25,** 2021 (1992).
[27] A. R. Butler and P. Rhodes, *Anal. Biochem.* **249,** 1 (1997).
[28] T. Akaike, K. Inoue, T. Okamoto, H. Nishino, M. Otagiri, S. Fujii, and H. Maeda, *J. Biochem.* (*Tokyo*) **122,** 459 (1997).
[29] I. Kluge, U. Gutteck-Amsler, M. Zollinger, and K. Q. Do, *J. Neurochem.* **69,** 2599 (1997).

S-Nitrosoalbumin. Human albumin (20 mg/ml) is initially treated with dithiothreitol (2 m*M*) in phosphate-buffered saline (PBS) to reduce the Cys-34 thiol group and is then dialyzed for 48 hr against four changes of 3 liters of PBS containing DTPA (100 μM). The albumin solution is then diluted with water containing DTPA (100 μM) to give a final concentration of ~10 mg/ml (~140 μM). This is then incubated with equal volume of *S*-nitrosocysteine (prepared as described earlier) at room temperature for 30 min in the dark to form *S*-nitrosoalbumin with a yield of >80% (with respect to reactive thiols). Any remaining unreacted thiol groups are then alkylated with NEM (1 m*M*) at room temperature, followed by dialysis at 4° against 4 × 3 liters PBS supplemented with DTPA (100 μM) for 48 hr. *S*-Nitrosoalbumin can be stored at −20°, and the concentration of the stock solution (~145 μM) is determined immediately prior to use using the Saville reaction (see later).

UV Measurement of RSNOs

The concentration of low molecular weight RSNOs in buffer can be estimated by measuring their absorbance at 335 nm where *S*-nitrosocysteine, *S*-nitrosoglutathione, and *S*-nitrosoalbumin have a molar extinction coefficient of 503, 589, and 3869 $M^{-1}cm^{-1}$, respectively.

Saville Reaction

A more accurate method of determining the concentration of the stock solution is to use the Saville reaction.[30] This method is based on the measurement of nitrite present in the sample before and after the addition of Hg^{2+}, which catalyzes the release of NO^+ from the *S*-nitrosated thiol, which then reacts with the aromatic amine, sulfanilamide, to form a diazonium salt, followed by coupling to another aromatic amine, *N*-(1-naphthyl)ethylenediamine HCl, and measuring its absorbance at 540 nm. This is then calibrated against a nitrite standard curve. The method we use is as follows: 50 μl of sample is added to 1950 μl of either solution A (sulfanilamide 0.5% dissolved in 0.25 *M* HCl) or solution B (sulfanilamide 0.5% dissolved in 0.25 *M* HCl containing 0.2% $HgCl_2$). The two samples are incubated at room temperature for 5 min to allow formation of the diazonium salt, and at the end of this time, add 1 ml of *N*-(1-naphthyl)ethylenediamine solution (0.02% dissolved in 0.5 *M* HCl). Color formation, indicative of the azo dye product, is complete by 5 min. The absorbance is measured at 540 nm and is then compared to the nitrite standard curve. The reaction in solution A gives background nitrite concentrations, and that with solution B gives background nitrite *plus* RSNO. Although the Saville reaction is the gold standard for quantifying RSNOs in

[30] B. Saville, *Analyst* **83,** 670 (1958).

aqueous solution, it cannot be used in biological milieu where many side reactions will yield erroneous results.

Method Protocol for the Measurement of Plasma RSNOs

At present we do not know if plasma samples can be frozen and analyzed later, and therefore we carry all our analyses on fresh samples. The main problem facing investigators is how to measure RSNOs in the presence of nitrite. Following the stabilization of RSNOs with NEM, the addition of acidified sulfanilamide, which reacts with nitrite, can remove background nitrite up to at least 10 μM nitrite and has no effect on the release of NO in the purge vessel. This method is reliable, simple, and works. It is important to make sure that there is enough acid present to decrease the pH adequately to enable the reaction of sulfanilamide and nitrite.

Blood Collection. Blood should be collected into ethylenediaminetetraacetic acid (EDTA) tubes containing NEM to achieve a final concentration of ~5 m*M* and then centrifuged. Use a high concentration of NEM added (e.g., 200 m*M*) to minimize dilution of plasma. We assume that much of this NEM is consumed by erythrocytic glutathione.

Plasma Handling. A known volume of plasma is removed, e.g., 0.5 ml, and a further 25 μl of 100 m*M* NEM in PBS is added to give a final concentration of ~5 m*M* (this ignores NEM already added to whole blood). Plasma concentrations of RSNOs can then be measured later. We always leave samples for at least 5 min before proceeding.

Removal of Nitrite. To the 500 μl of plasma treated with NEM, add 100 μl of 1 *M* HCl containing 2.5% sulfanilamide and vortex mixed.

Purge Vessel Conditions. The purge vessel we use is ~8 times as large as the conventional Sievers purge vessel (custom made). We maintain a temperature of 65°. It contains 8 ml glacial acetic acid and 2 ml of potassium iodide (50 mg/ml) and 400 μl of copper sulfate (200 m*M*). The addition of copper sulfate makes the solution brown due to release of iodine.

Injection of Plasma Samples. Once the baseline has settled, the sample is injected and the release of NO is monitored. The signal from normal human plasma is small, and typically only goes a few millivolts above baseline. Once the signal has returned to baseline, stop recording and change the solutions. This is essential or there will be frothing with subsequent samples. We now avoid antifroth reagent. Changing the solutions is time-consuming but essential.

Standard Curves and Structural Confirmation. At the beginning and end of each run we run a known standard curves of *S*-nitrosoalbumin ranging from 10 n*M* to 1 μM. To confirm that the species is an RSNO, one should run the same sample before and after the addition of mercury (Hg^{2+}) (0.2% $HgCl_2$).

What are the limitations of this method? This method measures total RSNOs and does not distinguish between low molecular weight and high molecular weight nitrosothiols such as *S*-nitrosoglutathione and *S*-nitrosoalbumin. Furthermore, nitrosylhemoglobin, which is the product of heme nitrosylation, will release NO in this system.[21] However, Gladwin *et al.*[21] has described a maneuver to determine the contribution from nitrosyl hemoglobin. To distinguish between low and high molecular weight *S*-nitrosothiols, one can precipitate proteins with 10% trichloroacetic acid following the addition of NEM. Following centrifugation, the supernatant contains low molecular weight RSNOs such as *S*-nitrosoglutathione. It is important to treat samples with NEM first, as acid and nitrite may cause artifactual nitrosation of thiol groups. Using this method, we can add *S*-nitrosoglutathione to plasma (prestabilized with NEM) and recover 85% of added *S*-nitrosoglutathione up to 2 hr later. The loss of 15% *S*-nitrosoglutathione is probably due to the incomplete recovery of plasma.

Identification of S-Nitrosylated Proteins

To identify tissue proteins that are uniquely sensitive to *S*-nitrosation *in vivo,* a biotin switch-based method has been described,[10] which identifies among a mixture of proteins those that contain NO bonds to cysteine residues. This method is indirect, involving the substitution of a biotin group at all cysteines that had been modified by *S*-nitrosation. Proteins in a tissue extract are initially denatured and free SH groups are chemically blocked by treatment with methanethiosulfonate. Subsequently, the S–NO bond in RSNOs is reduced selectively to a free SH group by ascorbate, and this is then biotinylated with a sulfhydryl-specific reagent, *N*-[6-(biotinamido)hexyl]-3′-(2′-pyridyldithion)propionamide. The biotin serves as a placeholder that identifies thiol residues in proteins that had been *S*-nitrosated and provides an efficient strategy to concentrate these proteins by affinity chromatography using immobilized streptavidin. The biotinylated proteins can then be resolved by SDS–PAGE and identified by immunoblotting with protein-specific antibodies or mass spectrometry-based fingerprint analysis. This method does assume that all of the methanethiosulfonate is removed after the first step, as it will prevent subsequent biotinylation of free SH groups.

Measurement of Protein Nitration

The measurement of protein nitration presents several problems. Most investigators have used immunoassays. While immunoassays imply specificity by using specific antibodies, it should be realized that it is easy to generate nonspecific staining if high concentrations of antibodies are used, and great care is needed to use this technique. Second, immunoassays are limited by the fact that we do not know what is measured. It is based on a faith that immunoassays are specific, which for proteins holds true, but when using a small immunogen, such as nitrotyrosine, many difficulties may occur. The antibodies raised recognize nitrotyrosine within certain epitopes. Therefore, negative staining may miss important increases in tissue nitrotyrosine. The ideal method should be sensitive, specific, and quantitiative. Therefore, we have developed a mass spectrometric assay for nitrotyrosine in biological fluids or tissues.[31] A second approach is to measure the urinary excretion of 3-nitro-4-hydroxyphenylacetic, the major urinary metabolite of nitrotyrosine. The advantage of this method is that it gives a time-integrated approach to the measurement of total *in vivo* nitrotyrosine formation. We have developed a GC-MS-based assay for this metabolite, which will be submitted for publication in the near future.

The main problem with measuring nitrotyrosine again is nitrite, which causes artifactual nitration of tyrosine if acidic conditions ($pH < 2$) are employed during sample preparation.[32] To measure protein-bound nitrotyrosine by gas chromatography mass spectrometry (GC-MS) or HPLC, investigators have usually employed acid hydrolysis, which causes artifactual nitration of tyrosine or enzymic hydrolysis, which may be incomplete or release nitrotyrosine from autohydrolysis of the enzymes employed. Acid hydrolysis may also cause conversion of nitrotyrosine to 3-aminotyrosine during preparation.[33] To circumvent these problems, we developed a method in which alkaline hydrolysis of proteins is used to release nitrotyrosine and tyrosine followed by GC-MS analysis. By using deuterated tyrosine ([2,3,5,6-^{2}H]tyrosine) as an internal standard, one can assess artifactual nitration by measuring the formation of [2,5,6-^{2}H]nitrotyrosine, and, in our hands, this does not occur.[31]

HPLC is the most widely used chromatographic technique for separation and analysis of nitrotyrosine, however, the limit of sensitivity of HPLC coupled with UV detector is 0.2–0.6 μM.[34,35] Sensitivity can be improved using electrochemical detection, but it is recognized that there may be interference from coeluting species that have a similar reduction potential.[36] HPLC fluorescence detection has also been used for the detection of nitrotyrosine. Although nitrotyrosine is not fluorescent, it can be detected using a fluorescence detector after reduction to 3-aminotyrosine or derivatization with 7-fluoro-4-nitrobenzo-2-oxa-1,3-diazole.[33,37] Reduction of nitrotyrosine to 3-aminotyrosine has also been used in the HPLC-coupled electrochemical detection of nitrotyrosine.[38,39] Although

[31] M. T. Frost, B. Halliwell, and K. P. Moore, *Biochem. J.* **345,** 453 (2000).

[32] M. E. Knowles, D. J. McWeeny, L. Couchman, and M. Thorogood, *Nature* (*Lond.*) **247,** 288 (1974).

[33] C. Herce-Pagliai, S. Kotecha, and D. E. G. Shuker, *Nitric Oxide* **2,** 324 (1998).

[34] H. Kaur and B. Halliwell, *FEBS Lett.* **350,** 9 (1994).

[35] N. Fukuyama, Y. Takebayeshi, M. Hida, H. Ishida, K. Ichimori, and H. Nakazawa, *Free Radic. Biol. Med.* **22,** 771 (1997).

[36] H. Kaur, L. Lyras, P. Jenner, and B. Halliwell, *J. Neurochem.* **70,** 2220 (1998).

[37] Y. Kamisaki, K. Wada, K. Nakamoto, Y. Kishimoto, M. Kitani, and T. Itoh, *J. Chromatogr. B.* **685,** 343 (1996).

[38] M. K. Shigenaga, H. H. Lee, B. C. Blount, S. Christen, E. T. Shigeno, H. Yip, and B. N. Ames, *Proc. Natl. Acad. Sci. U.S.A.* **94,** 3211 (1997).

[39] H. Ohshima, I. Celan, L. Chazotte, B. Pignatelli, and H. F. Mower, *Nitric Oxide* **3,** 132 (1999).

these electrochemical-based assays are highly sensitive (100 f*M*), the major disadvantage is that 3-aminotyrosine is produced endogenously *in vivo* via tyrosine amination,[40] so endogenous 3-aminotyrosine might cause interference with the sample of interest.

Several GC-MS-based assays have been reported, employing stable isotopic dilution methods with electron capture negative-ion chemical ionization. These methods have been used to measure nitrotyrosine in low-density lipoprotein,[41] platelets,[42] and rat heart.[43] However, any method that employs acidic hydrolysis or acidic derivatization conditions will predominantly measure artifactual nitration of tyrosine, and the results are uninterpretable. Using alkaline hydrolysis of proteins, Frost *et al.*[31] have developed a sensitive GC-MS assay for the measurement of nitrotyrosine and can detect nitrotyrosine at the 1-pg level. Based on this method, the free and protein-bound nitrotyrosine contents of normal human volunteers were determined as 1.2 ± 0.3 ng/ml and 1.3 ± 0.1 ng/mg, respectively. LC-MS, particularly with MS/MS detection, is an alternative approach that avoids artifactual nitration of tyrosine.[44] Advances using the LC-MS/MS assay have shown a direct and specific method 100 times more sensitive than electrochemical detection.[45]

Nitrotyrosine Methods

Isotope Dilution GC-MS Analysis of Protein-Bound and Free Nitrotyrosine in Plasma

Internal Standards. [$^{13}C_9$]Nitrotyrosine and [2,3,5,6-^{2}H]tyrosine are used as internal standards for the measurement of nitrotyrosine and tyrosine, respectively. Deuterated and ^{13}C-labeled tyrosine are available from Cambridge Isotopes. [$^{13}C_9$]nitrotyrosine is synthesized by reacting 1 mg [$^{13}C_9$]tyrosine with 25 μl of a 1 : 4 (v/v) dilution of tetranitromethane/ethanol in 1 ml sodium carbonate buffer 0.5 *M* (pH 9.0) for 30 min. The yellow solution is aspirated, avoiding any unreacted tetranitromethane, and is extracted on a reverse phase LC_{18} column (Sigma), as described for the extraction of free nitrotyrosine later, and then lyophilized under vacuum. To remove any unreacted tyrosine and by-products, the sample can be purified by HPLC on a Techsphere C_{18} column (25×4.6 mm). This employs a

[40] R. S. Sodum and E. S. Fiala, *Chem. Res. Toxicol.* **10,** 1420 (1997).

[41] C. Leeuwenburgh, M. M. Hardy, S. T. Hazen, P. Wagner, S. Oh-ishi, U. P. Steinbrecher, and J. W. Heinecke, *J. Biol. Chem.* **279,** 1433 (1997).

[42] H. Jiang and M. Balazy, *Nitric Oxide* **2,** 350 (1998).

[43] J. R. Crowley, K. Yarasheski, C. Leeuwenburgh, J. Turk, and J. W. Heinecke, *Anal. Biochem.* **259,** 127 (1998).

[44] D. Yi, B. A. Ingelse, M. W. Duncan, and G. A. Smythe, *J. Am. Soc. Mass Spectrom.* **11,** 578 (2000).

[45] J. S. Althaus, K. R. Schmidt, S. T. Fountain, M. T. Tseng, R. T. Carrol, P. Galatsis, and E. D. Hall, *Free Radic. Biol. Med.* **29,** 1085 (2000).

gradient of water containing 0.1% (v/v) trifluoroacetic acid (TFA) (solution A) and 0.1% TFA/acetonitrile (solution B). Initial conditions are 100% A, changing to 90% A/10% B over 15 min and then to 50% A/50% B from 15 to 30 min. Fractions containing [$^{13}C_9$]nitrotyrosine are identified by their retention times (~28 min) and characteristic UV spectra (λ_{max} at 216, 276, and 354 nm). Fractions containing ^{13}C-labeled nitrotyrosine are pooled, and the concentration is determined against unlabeled nitrotyrosine by GC-MS.

Hydrolysis of Plasma Proteins. The plasma proteins are precipitated from 1 ml plasma by adding 2 ml water and 12 ml of ice-cold 2 : 1 (v/v) chloroform/methanol. The mixture is vortex mixed and centrifuged at 9000*g*. This method is mild and has the advantage of removing water (upper layer) and fat-soluble components (lower layer), with protein precipitated at the interface. It is best to remove the upper layer first, followed by the lower layer. The protein slurry is freeze-dried under vacuum, and 1–2 mg of dried protein is weighed into polypropylene tubes (Sarstedt Ltd, UK; heat-resistant tubes with screw tops) and 1 ml of 4 *M* NaOH is added. Internal standards (20 ng of [$^{13}C_9$(nitrotyrosine and 10 μg [2,3,5,6-^{2}H] tyrosine) are added to the samples, purged with argon to create an inert atmosphere, and sealed with heat-resistant poly(tetrafluoroethylene) tape (Fisher Scientific, UK) and screw caps. Samples are heated to 120° for 20 hr. The hydrolysate (which contains free nitrotyrosine) is adjusted to pH 5.0 by adding 0.5 ml citrate buffer (0.5 *M*, pH 5.0), followed by titration with hydrochloric acid.

Sample Preparation for Measurement of Free Nitrotyrosine. To measure free nitrotyrosine and tyrosine, add 20 ng [$^{13}C_9$]nitrotyrosine and 10 μg [2,3,5,6-^{2}H] tyrosine to 1 ml of plasma. The plasma is filtered by centrifugation at 9000*g* in a 30-kDa molecular mass cut-off centrifugal membrane microfuge tube (Millipore). The resulting filtrate contains free nitrotyrosine and low molecular weight peptides. Ideally, this should undergo a further filtration step using a 5000 molecular weight cut-off.

Extraction of Nitrotyrosine and Tyrosine. The resulting hydrolyzed mixture (pH adjusted to 5.0) or filtered sample is applied to an LC_{18} reverse phase column (Sigma) that has been prewashed with 2 ml of methanol, 2 ml of water, and 6 ml 0.1% (v/v) TFA/water (adjusted to pH 5.0 with ammonia solution). The column is washed with 2 ml water and eluted with 4 ml of 25% (v/v) methanol in water. For measurement of tyrosine, 100 μl of 4 ml eluate is derivatized after being dried under nitrogen. The remaining 3.9-ml eluate is freeze-dried under vacuum, resuspended in 400 μl 1% (v/v) TFA/water (adjusted to pH 4.0 with ammonia solution), and then loaded onto an ENV^+ SPE cartridge (Jones Chromatography, UK). The ENV^+ column is preconditioned with 2 ml of methanol followed by 2 ml of water (pH 4.0) containing 1% TFA. The fraction containing nitrotyrosine is washed with 1 ml of water and is eluted with 1 ml of 50% (v/v) methanol/water, dried under nitrogen, and derivatized as described later.

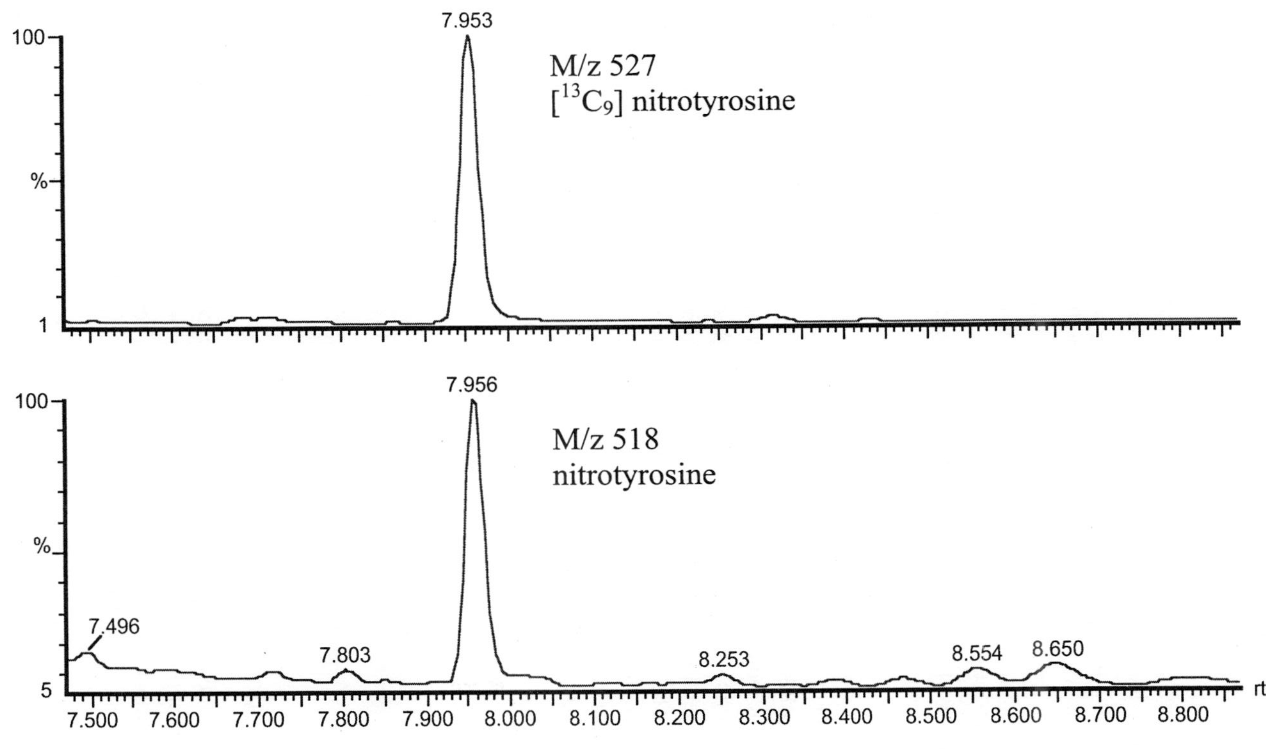

FIG. 1. Typical chromatogram showing peak shape and resolution for nitrotrosine in 1 mg hydrolyzed plasma protein after the addition of 20 ng [$^{13}C_9$]nitrotyrosine.

Derivatization of Nitrotyrosine and Tyrosine. After extraction, nitrotyrosine and tyrosine are converted into the respective amides by the addition of 200 μl anhydrous dimethylformamide and 20 μl of diisopropyl ethylamine. Samples are left for at least 10 min on ice, and then 40 μl ethyl heptafluorobutrate is added. After being left for a further 30 min at room temperature, samples are sonicated for 2 hr on ice. Unreacted and excess reagents are evaporated under nitrogen. The amino acid amides are converted into the *t*-butyldimethylsilyl ester or ether by the addition of 40 μl *N*-(*t*-butyldimethylsilyl)-*N*-methyltrifluoroacetamide (TBDMS) at room temperature for 30 min. Derivatized samples are dried under nitrogen and redissolved in 20 μl of undecane containing 5% (v/v) TBDMS.

GC-MS Analysis of Derivatized Samples. Samples are analyzed on a GC equipped with a 15-m DB-1701 (J&W Scientific) capillary column (0.25 mm internal diameter, 0.25 μm film thickness) interfaced with a mass spectrometer. The ion source and interface temperature are set at 200 and 300°, respectively. Samples are analyzed in negative-ion chemical ionization mode with ammonia as the reagent gas. For measurement of nitrotyrosine, derivatized samples are injected with on-column injection (manual injection of samples directly into the column from a fine-bore microliter syringe), which avoids problems that can arise from the interactions between the analyte and the glass injection liner. The initial column temperature is maintained at 150° for 1 min and is then increased to 300° at 20°/min. The derivatized amino acids are quantified by isotope dilution GC-MS, and ions are monitored at 518 and 527 mass units for nitrotyrosine (Fig. 1) and at 407 and 411 mass units for tyrosine with single ion monitoring. The concentrations are calculated from the known $^{13}C_9$ or [2,3,5,6-^{2}H] internal standards, which, respectively, are heavier 9 and 4 mass units for nitrotyrosine and tyrosine.

[25] S-Glutathionylation of NF-κB Subunit p50

By ESTELA PINEDA-MOLINA and SANTIAGO LAMAS

Introduction

The cellular redox status can modulate the function of multiple molecules such as transcription factors.[1] These proteins may respond to oxidative changes through their postranslational modification, which can represent a reversible mechanism for the regulation of protein function.[2] In this sense, it is known that thiols, by

[1] R. G. Allen and M. Tresini, *Free Radic. Biol. Med.* **28,** 463 (2000).
[2] Y. Morel and R. Barouki, *Biochem. J.* **342,** 481 (1999).

0076-6879/02 $35.00

virtue of their ability to be oxidized reversibly to sulfenic acid or to the formation of inter, intra, or mixed disulfides, are recognized as key components involved in the maintenance of redox balance. Accordingly, the oxidation of highly reactive cysteine residues or the formation of disulfides with other thiols may be the most feasible mechanisms by which the redox state may modulate the function of transcription factors.

The transcription factor NF-κB constitutes a classical model for the study of redox regulation of transcription factors. Its DNA binding depends on the oxidative state of a cysteine located in the basic region of its DNA-binding domain.[3,4] Accordingly, it has been demonstrated that NF-κB needs reduced conditions to bind to the DNA.[3,4] However, oxidative conditions may induce its activation in the cytoplasm.[5–7] The mechanisms underlying the modulation of NF-κB by the redox state remain unclear. Despite that, redox regulation of the DNA binding of p50 (structural component of NF-κB) has been shown to be modulated through the *S*-glutathionylation of a highly conserved cysteine residue (Cys-62) *in vitro*.[8] Furthermore, the oxidation to sulfenic acid also contributes to the DNA-binding inhibition observed in such conditions. Both modifications, *S*-glutathionylation and sulfenic acid formation, occur when changes in the redox pair GSH/GSSG are applied. Although both reactions have different stoichiometry, their total effect produces significant inhibition on the p50 DNA-binding capacity that can be reverted easily by thiol-reducing agents such as dithiothreitol (DTT).

Detection of *S*-glutathionylation of p50, as well as of sulfenic acid formation, is described in detail. For this purpose, the expression and purification of p50 DNA-binding domain and the methodology used for the study of DNA-binding are also comprehensively explained.

Methods

Expression of p50 DNA-Binding Domains

All transcription factors susceptible to redox regulation are characterized by the existence of reactive cysteine residues in their DNA-binding domain. Usually, these cysteines are surrounded by basic amino acids that increase their reactivity.

[3] J. R. Matthews, N. Wakasugi, J.-L. Virelizier, J. Yodoi, and R. T. Hay, *Nucleic Acids Res.* **20,** 3821 (1992).
[4] T. Hayashi, Y. Ueno, and T. Okamoto, *J. Biol. Chem.* **268,** 11380 (1993).
[5] M. Meyer, R. Schreck, and P. A. Baeuerle, *EMBO J.* **12,** 2005 (1993).
[6] J. Satriano and D. Schlondorff, *J. Clin. Invest.* **94,** 1629 (1994).
[7] R. Schreck and P. A. Baeuerle, *Methods Enzymol.* **234,** 151 (1994).
[8] E. Pineda-Molina, P. Klatt, J. Vázquez, A. Marina, M. García de Lacoba, D. Pérez-Sala, and S. Lamas, *Biochemistry* **40** (47), 14134 (2001).

To assay modulation of DNA binding of the NF-κB subunit p50 by changes in oxidative conditions, the expression and purification of its DNA-binding domain were performed as described in the following protocol. This domain contains a highly conserved cysteine (Cys-62). Because this residue has been postulated to be the redox sensor of p50, the purification of a Ser mutant in position 62 is also described.

Procedure. The fragment corresponding to residues 36–385 of the protein KBF1 from the DNA-binding domain of the p50 subunit of NF-κB (EMBL accession number M55643) is obtained by polymerase chain reaction (PCR). As a template, a clone with the full-length human p50 is used. The primers are sense 5′-CGG GGA TCC GCA CTG CCA ACA GCA GAT-3′ and antisense 5′-CTC CCT AAG CTT CCA GCT CCG GCA CCA CTA-3′. PCR products are digested with *Bam*HI and *Hin*dIII and ligated into the expression vector pQE-30 (containing six histidine-tagged residues, Qiagen).

The cysteine of this fragment (Cys-62) is mutated to serine by PCR mutagenesis using the wild-type fragment as a template. The construction is made in two separate steps. First, the 5′ half-site of the mutant (an insert of 120 bp) is amplified using the same upper primer described earlier and a second primer with the sequence 5′-CTC AAG CTT GTA CCC ATG GGA TGG GCC TTC CGA-3′ (named "mutation primer"). This fragment, flanked by *Bam*HI/*Hin*dIII sites, is gel purified. Subsequently, it is digested with *Bam*HI/*Hin*dIII and subcloned into the cloning vector pUC-18. The second fragment (the 3′ half-site) is generated using the "mutation primer" as the upper primer, being the same antisense primer as described for the native fragment. After digestion, the resulting insert (1100 bp) contains *Bam*HI/*Hin*dIII sites. Once amplification is performed, both inserts (5′ and 3′) have a unique and new *Nco*I site. Constructs are digested with *Nco*I/*Hin*dIII or *Nco*I, respectively, and are ligated to obtain an insert of 1220 bp into the pUC-18 vector. The insert is then digested with *Bam*HI and *Hin*dIII and cloned into the expression vector pQE-30. The mutation is confirmed by automatic sequencing.

Purification of p50 DNA-Binding Domains

Procedure. For the expression of fragments, competent *Escherichia coli* M15 [pREP4] cells are transformed by heat shock at 42° for 2 min as described (QIAexpress Protein Purification System, Qiagen). A volume of 10 ml of transformed bacteria is grown in LB medium containing 25 g Bacto-tryptone, 15 g Bacto-yeast extract, 5 g NaCl, 125 mg ampicillin, and 25 mg kanamycine. After an overnight incubation, 7.5 ml is inoculated in 1 liter of LB medium. The culture is grown at 37° with agitation for 2 hr to reach an optical density of 0.6–0.8 (600 nm). Subsequently, bacteria are induced with 1 m*M* isopropyl-β,D-thiogalactopyranosid

(IPTG) and incubated for 4 hr. Then, the suspension is centrifugated for 10 min at 10,000*g*. The supernatant is discarded, and the pellet is resuspended in 20 ml of a 50 m*M* sodium phosphate buffer (pH 8.0) containing 350 m*M* NaCl and 0.1% (v/v) 2-mercaptoethanol (buffer A). The resuspended pellet is lysed in this buffer by sonication. The lysed suspension is then centrifugated for 30 min at 20,000*g*. The supernatant is incubated with agitation for 1 hr on ice with 2–3 ml (bed volume) of a nickel–chelate resin (Ni-NTA, Qiagen), equilibrated previously with buffer A with a supplement of 10 m*M* imidazole. For the purication procedure, the mixture is poured into a chromatography column (inner diameter 1.5 cm), washed with 15 bed volumes of buffer A (10 m*M* imidazole), followed by 5 bed volumes of buffer A with 50 m*M* imidazole. Finally, the protein is eluted with 5 volumes of buffer A with 250 m*M* imidazole. The eluate is then concentrated in Vivapore 20 concentrators (molecular weight cut-off 7500; Vivascience, Binbrook, Lincoln, UK) and preserved with a buffer containing 0.01% NP-40 and 5% glycerol at −80°. The purity of the obtained protein preparations is estimated to be >95% as judged by Coomassie-stained SDS gels. Concentrations of the purified protein, determined by amino acid analysis, Bradford, and absorbance at 280 nm, are in the range of 0.4–0.5 m*M*.

Precautions

a. It is desirable to optimize the sonication of the cell suspension because different yields of soluble protein can be obtained.
b. 2-Mercaptoethanol is essential to maintain the reduced state of the protein so it must be added before each purification step. Because it may react with nickel, facilitating its reduction, it is important to wash the column exhaustively before the elution is done.
c. The preparation may be stored at 4° but some amount of degraded protein can be expected.

Study of p50 DNA-Binding Inhibition by Electrophoresis Mobility Shift Assay (EMSA)

It is well known that the DNA-binding activity of p50 can be inhibited by changes in the GSH/GSSG redox pair promoted by oxidative conditions. The following protocol describes the methodology used to confirm this observation. Hence, it is necessary to preincubate the protein with different GSH/GSSG ratios and subsequently assay its DNA-binding activity.

Procedure. To test the induced inhibition in the DNA-binding activity of p50 by oxidative changes, the DNA-binding domains of wild-type or C62S mutants (10 μM) are preincubated in a 20 m*M* Tris/HCl buffer (pH 7.5) containing 50 m*M* NaCl, 5 m*M* $MgCl_2$, 1 m*M* EDTA, 5% glycerol, and 0.01% NP-40 (buffer B) in the presence of different GSH/GSSG ratios, ranging from 100 to 0.1. After an incubation for 1 hr at 37°, 2 μl of the preincubation mixture is diluted with 15 μl of buffer B plus 0.2 mg/ml bovine serum albumin, 0.1 mg/ml poly(dI-dC), and GSH/GSSG at the same ratio established during preincubation. After that, 3 μl of the ^{32}P-radiolabeled double-stranded NF-κB oligonucleotide 5′-GGA GAG GGG ATT CCC TGC G-3′ from the cyclooxygenase-2 promoter, comprising bases -452 to -433 from the transcriptional start site in the reported sequence,[9] is added. The resulting sample is incubated for 30 min at room temperature to allow DNA binding prior to electrophoresis.[10] At the same time, a 6% nondenaturing gel is prerunning at 200 V for 30 min. When the binding reaction is completed, the samples are loaded into the gel with an additional well containing bromphenol blue to follow the run.

The electrophoresis runs for approximately 1 hr and 15 min. Next, the gel is vacuum dried and visualized by autoradiography. The relative amount of each binding complex is quantified by densitometric analysis.

Following this protocol, it has been possible to observe that the DNA-binding activity of p50 is inhibited when the GSSG concentration is increased (Fig. 1A, upper panel). This inhibition value is obtained with high prooxidative conditions (GSH/GSSG of 0.1). Moreover, this inhibition is almost absent when the binding reaction is done using the C62S mutant protein (Fig. 1B, upper panel). This indicates that the detected inhibition occurs when an oxidized Cys-62 residue, which abrogates DNA binding, is generated. All these results confirm previous ideas about the redox sensitivity of Cys-62 within the p50 DNA-binding domain.

Precautions

a. Because glutathione may undergo a degree of autoxidation during the preincubation reaction, the GSH/GSSG ratios may change slightly. That possibility may be checked calculating both GSH and GSSG amounts using methods described previously.[11,12]
b. It is absolutely necessary to adjust the amount of protein and the specific oligonucleotide. As a general rule the oligonucleotide must be in excess over the total protein.
c. The protein solution used for the incubations must be freshly prepared because oxidized forms can be generated easily. In the latter case, a subestimation of the binding may occur due to less reduced protein availability. To avoid this problem, it is important to be certain that oxidation of the protein in the starting material is prevented by incubating it with a reducing agent such as DTT.

[9] S. B. Appleby, A. Ristimaki, K. Neilson, K. Narko, and T. Hla, *Biochem. J.* **302,** 723 (1994).
[10] P. Klatt, E. Pineda-Molina, M. García de Lacoba, C. A. Padilla, E. Martinez-Galesteo, J. A. Barcena, and S. Lamas, *FASEB J.* **13,** 1481 (1999).
[11] H. Sies and K.-H. Summer, *Eur. J. Biochem.* **57,** 503 (1975).
[12] G. L. Ellman, *Arch. Biochem. Biophys.* **82,** 70 (1959).

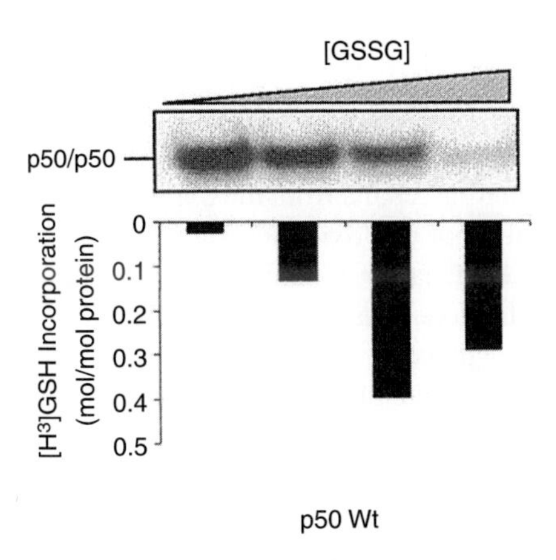

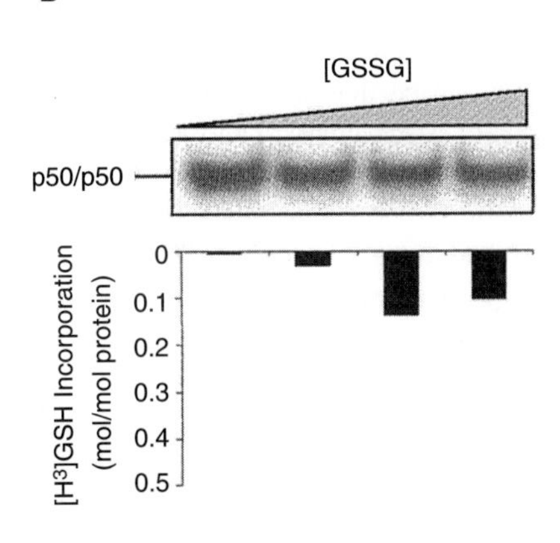

FIG. 1. Oxidative inhibition of p50 DNA binding. (A) The study of p50 DNA-binding activity by EMSA experiments demonstrates that an increase in oxidative conditions (low GSH/GSSG values) produces a decrease in the DNA-binding capacity. The experiment was performed in the presence of different ratios of the redox pair (GSH/GSSG from 100 to 0.1), maintaining the total number of glutathione equivalents. The maximal inhibition (around 70%) is reached at a GSH/GSSG ratio of 0.1 (GSSG 100 times in excess over the GSH amount). Under the same conditions, the inhibition is related to the glutathione incorporation (40% maximum), but there is not a perfect correlation at maximal oxidative conditions (compare the upper panel with the last column of the glutathione incorporation graphed). (B) The mutant protein in Cys-62 was assayed for DNA binding after its incubation with increasing amounts of GSSG as described previously. The maximal glutathione incorporation (14%) is also related to a decrease in DNA binding, but is not in good correlation with the maximal inhibition value (19%).

d. It is essential to test the DNA-binding activity of proteins beforehand in order to avoid low-affinity conditions in DNA binding responsible for any type of inhibition.

Detection of S-Glutathionylated p50 by Mass Spectrometry

DNA-binding inhibition results from the oxidation of Cys-62 in the p50 molecule. However, the nature of this oxidation state is not detected by EMSA experiments. One possible explanation for this inhibition is the formation of a mixed disulfide between glutathione and the thiol group of Cys-62. The fact that preincubation is done in the presence of glutathione supports the possibility of this reaction. In most cases, *S*-glutathionylation has been demonstrated by qualitative experiments that do not permit concluding how susceptible a protein is to undergo this reaction. Qualitative approximations include mass spectrometry experiments that can be done with intact or trypsinized protein. The use of mass spectrometry technology (MALDI-TOF and nano ES QIT) has allowed the identification of many targets for *S*-glutathionylation and, in several cases, it has made it possible to precisely map the specific thiol involved in the glutathione adduct. However, these techniques are not useful for quantifying the degree of glutathione incorporation because they are based on properties such as the ionization capacity of a molecule by laser (MALDI-TOF) or a gas phase (nano ES QIT). Furthermore, it requires an exhaustive and optimal sample preparation because a wide variety of reagents (detergents, salts, etc.) interfere with the mass spectrometry assays. Nevertheless, it has been possible to detect the *S*-glutathionylation of p50 after treatment with GSH/GSSG and to assign the mixed disulfide formation to Cys 62 (Fig. 2), as described next.

Procedure. The DNA-binding domain of p50 (wild-type and mutant) is incubated with 1.5 m*M* GSSG or 1 m*M* DTT for 30 min at 37° and is subsequently diluted in 20 m*M* sodium phosphate buffer, pH 7.5, in the presence of 4 m*M* urea and *N*-ethylmaleimide 10 times in excess over the total number of thiols. With this reaction it is possible to induce the *S*-glutathionylation of p50 and to block all nonreactive thiols. After an incubation period of 30 min at 37°, trypsin (2.3 mg/ml) is added to get a final protein/trypsin ratio of 1/20. Then the mixture is incubated overnight at 37°. The tryptic fragments are separated by reversed-phase HPLC on a Vydac 218TP5415 column (4.6 × 150 mm) using an AKTÄ prime system (Pharmacia Biotech.). The initial mobile-phase composition is held in 100% water containing 0.1% trifluoroacetic acid (TFA) for 10 min, followed by a gradient elution (0.1% TFA to 0.1% TFA–90% acetonitrile) for 100 min. The fractions are collected, and aliquots of 0.5 μl are applied onto the target and dried. Then 0.5 μl of saturated α-ciano-4-hydroxycinnamic acid matrix in water : acetonitrile (1 : 1) containing 0.1% TFA is added and dried. Calibration is performed externally with the use of a set of synthetic peptides.

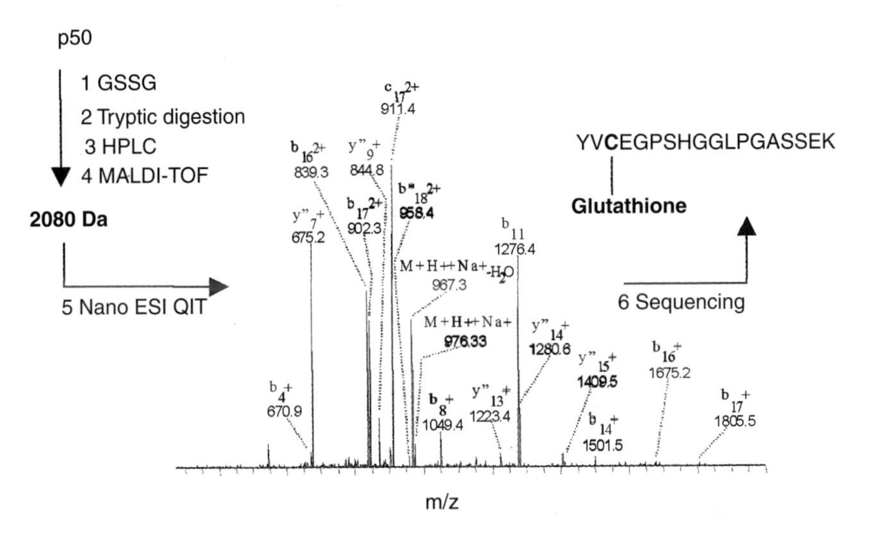

FIG. 2. The *S*-glutathionylation of p50 can be detected by mass spectrometry assays. The sample was incubated with 1.5 m*M* GSSG and after tryptic digestion and HPLC purification was analyzed by MALDI-TOF mass spectrometry. Study of the fractions allowed identification of a 2080-Da peptide whose molecular weight was in agreement with the formation of a mixed disulfide in the cysteine residue of a 1775-Da peptide. Further sequencing of the sample after fragmentation with a nano ESI QIT mass spectrometer corroborated this observation. The procedure and the fragmentation spectrum of the 2080-Da peptide are shown.

Analysis by MALDI-TOF mass spectrometry is performed using a Kompact Probe instrument (Kratos-Shimazdu, Manchester, UK) equipped with an extended flight tube of 1.7 m and delayed extraction, operating in linear mode.

Analysis by nanospray ion trap (nESI-IT) mass spectrometry is performed using an ion trap mass spectrometer (Model LCQ, Finnigan, ThermoQuest, San Jose, CA) equipped with a nanospray interface, exactly as described previously.[13] As mentioned before, the glutathione incorporation into Cys-62 is detected easily with these mass spectrometry assays. Accordingly, a peptide of 2080 Da corresponding to a reduced peptide of 1775 Da (YVCEGPSHGGLPGASSEK) bearing a glutathione adduct (plus 305 Da) in the cysteine residue is observed (Fig. 2). To confirm adduct formation, the peptide can be analyzed by nano-ESI mass spectrometry. The fragmentation series that emerges after the analysis correlates with the calculated theoretical series for a sample with glutathione adducts. Also, the *S*-glutathionylation of another residue not involved in the DNA binding (Cys-135) is also detected.

[13] A. Marina, M. A. García, J. P. Albar, J. Yague, J. A. Lopez de Castro, and J. Vazquez, *J. Mass Spectrom.* **34,** 17 (1999).

Precautions

a. The amount of protein to be analyzed by mass spectrometry assays may be of 10 pmol at least, although lower values can sometimes also be applied.
b. The fractions are first analyzed by MALDI-TOF mass spectrometry to determine the molecular weight of the purified tryptic peptides. The comparison between treated (plus GSSG) and control samples (plus DTT) reveals for the peptides that must be analyzed by nano-ESI mass spectrometry any interesting modification. All possibilities must be tested, including the modification of other residues, such as methionines.
c. It is necessary to clearly establish the molecular weight of the most interesting peptides before nano-ESI mass spectrometry analysis is performed. This is because the sample has a limited half-life in the mass spectrometer, which is sometimes not long enough to allow a correct analysis of the peptide mixture.

Quantitative Determination of p50 S-Glutathionylation

Procedure. The *S*-glutathionylation detected by mass spectrometry assays may represent the mechanism by which the DNA-binding activity of p50 is inhibited.

To test this hypothesis, it is essential to quantify glutathione incorporation. For this purpose, the DNA-binding domains of the wild-type and mutant proteins (10 μM) are incubated for 1 hr at 37° in buffer B in the presence of different ratios of ^{3}H-GSH and ^{3}H-GSSG. The total concentration of GSH equivalents is maintained at 3 m*M* in the incubation mix.

The tritium-labeled glutathione (^{3}H-GSH, 45–50 Ci/mmol, ~0.02 m*M*) is adjusted to a final concentration of 10 m*M* with the addition of 10 volumes of an 11 m*M* solution of unlabeled GSH. GSSG contamination in this preparation may be analyzed by HPLC analysis, but is around 2% according to the manufacturer's instructions. ^{3}H-GSSG is prepared by oxidation of ^{3}H-GSH as described.[14]

After incubation, the protein (with or without incorporated glutathione) is precipitated by adding 1 ml of 10% ice-cold trichloroacetic acid.[14] Finally, incorporation of ^{3}H-GSH to the protein is quantified by liquid scintillation counting. Using this methodology, the *S*-glutathionylation of p50 is around 40% at maximal DNA-binding inhibition values (GSH/GSSG equal to 0.1; Fig. 1A, lower panel). Although we do not observe significant DNA-binding inhibition with the mutant protein, a low amount of glutathione incorporation is obtained (Fig. 1B, lower panel).

[14] M. A. Pajares, C. Duran, F. Corrales, M. M. Pliego, and J. M. Mato, *J. Biol. Chem.* **267,** 17598 (1992).

Precautions

a. To quantify the stoichiometry of the reaction correctly, it is essential to adjust the protein and glutathione concentration in the sample tube.
b. Glutathione incorporation data must be corrected with the precipitation yield value. In this case the protein recovery after precipitation is around 98%.
c. Incorporation data must be corrected by the values obtained from a control sample treated first with a GSH/GSSG ratio of 0.1 and DTT. Thus, the blank value is determined as the non-DTT-releasable radiolabel by treating the radiolabeled proteins with 10 m*M* DTT prior to the precipitation.
d. The amount of bound glutathione depends on different events, such as steric impediments, other oxidative modifications that could compete with *S*-glutathionylation, or the potential presence of some p50 molecules that could keep it in its reduced state.
e. Another possible explanation for the low glutathione incorporation is offered by molecular modeling studies.[15] These methods allow analysis of the possible facilitated electrostatic interactions of the glutathione moiety with the cysteines surrounded by a basic environment within the molecule.

Detection of Sulfenic Acid in Cys-62

The maximal DNA-binding inhibition value is around 70%, and *S*-glutathionylation represents 40% of the total oxidative modifications that contribute to this negative effect. Hence, there is no perfect correlation between the degree of *S*-glutathionylation and the observed inhibition (Fig. 1). *S*-Glutathionylation is the most abundant modification in p50 under oxidative conditions. However, it is conceivable that other oxidative modifications aside from the formation of a mixed disulfide may occur in those conditions. In agreement with this, generation of a reversible sulfenic acid has been suggested as an alternative modification.[16–18] Next we propose a practical approach to check if a sulfenate is generated during an oxidative treatment.

Procedure. p50 domains (wild-type and mutant) are incubated in the presence of 1.5 m*M* GSSG and 1 m*M* dimedone for 1 hr at 37°, and subsequently diluted

[15] S. J. Weiner, P. A. Kollman, D. T. Nguyen, and D. A. Case, *J. Comput. Chem.* **7,** 230 (1986).

[16] H. R. Ellis and L. B. Poole, *Biochemistry* **36,** 15013 (1997).

[17] J. I. Yeh, A. Claiborne, and W. G. Hol, *Biochemistry* **35,** 9951 (1996).

[18] S. Boschi-Muller, S. Azza, S. Sanglier-Cianferani, F. Talfournier, A. Van Dorsselear, and G. Branlant, *J. Biol. Chem.* **275,** 35908 (2000).

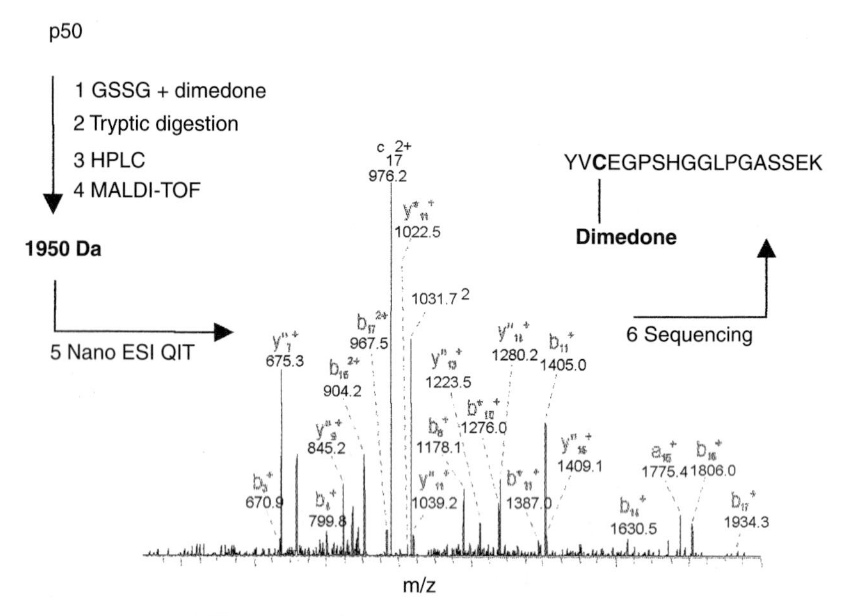

FIG. 3. Detection of a sulfenic acid in Cys-62. p50 was incubated as described in the presence of 1.5 m*M* GSSG and 1 m*M* dimedone. Subsequently, the sample was digested with trypsin and purified by HPLC. The fractions were collected and analyzed by MALDI-TOF mass spectrometry. A peptide of 1950 Da was identified as containing a dimedone adduct and was subjected to sequencing by nano ESI QIT. After analysis of the obtained fragmentation series, it was confirmed that the 1950-Da peptide contained a dimedone adduct at Cys-62. The procedure and the fragmentation spectrum of the 1950-Da peptide with the corresponding series named over the molecular weight of the detected fragments are shown.

in 20 m*M* sodium phosphate buffer, pH 7.5, in the presence of 4 m*M* urea and *N*-ethylmaleimide 10 times in excess over the total number of thiols. The reagent dimedone is used in multiple studies as a sulfenic acid probe. It can form irreversible covalent adducts with the sulfur atom implicated in the sulfenic acid formation.[19] After the incubation period, p50 is subjected to tryptic digestion, HPLC, and mass spectrometry analysis, as described previously.

Formation of a sulfenic acid in the Cys-62 of p50 is detected by mass spectrometry assays[8] (Fig. 3). The presence of a 1950-Da peptide corresponding to a 1775-Da peptide with a mass increment of 176 Da is explained adequately by assuming sulfur oxidation of the dimedone-derivatized Cys, followed by the

[19] L. V. Benitez and W. S. Allison, *J. Biol. Chem.* **249,** 6234 (1974).

formation of a sodium adduct (theoretical monoisotopic mass: 1950.8 Da). This modification can also be detected after treatment of the p50 molecule with H_2O_2.

Precautions

a. In general, sulfenic acid modification is a very unstable reaction that derives easily to irreversible oxidative states, such as sulfinic or sulfonic acids. Furthermore, in the presence of neighboring thiols, it may lead to the formation of mixed disulfides.[20]
b. The detection of sulfenic acid has been approached by other groups using the sulfhydryl/sulfenate probe 7-chloro-4-nitro-2-oxa-1,3-diazole (NBD-Cl). Although this probe can detect an intermediary compound with a maximum absorbance at 347 nm (the absorbance of a sulfenate–NBD–Cl complex), the impossibility of eliminating completely the free probe, which absorbs at 343 nm,[21] can lead to significantly overestimating the amount of sulfenate formed.
c. Because dimedone does not react with free thiols, the experiments can be done assuming no interference from glutathione or DTT.
d. The formation of sodium or potasium adducts is a frequent phenomenon in HPLC-purified proteins or peptides. Hence, the existence of these complexes should be taken into account when the mixture is analyzed.

Acknowledgments

This work was done in collaboration with Dr. Peter Klatt, Dr. Jesús Vázquez, Dr. Anabel Marina, and Dr. Dolores Pérez-Sala. The experiments described have been supported by grants from the Plan Nacional de I+D+I SAF 97-0035 and SAF 2000-0149 and from the Comunidad de Madrid 08.4/0032/98. Estela Pineda-Molina is the recipient of a training grant from the Spanish Ministry of Science and Technology. We also thank Carlos Fernández Tornero (Centro de Investigaciones Biológicas, C.S.I.C., Madrid, Spain) for helpful discussions and comments on the manuscript.

[20] A. Claiborne, J. I. Yeh, T. C. Mallett, J. Luba, E. J. Crane, V. Charrier, and D. Parsonage, *Biochemistry* **38,** 15407 (1999).

[21] A. J. Boulton, P. B. Ghosh, and A. R. Katritzky, *J. Chem. Soc. Ser. B,* 1004 (1966).

[26] Intercellular Signaling Mediated by Nitric Oxide in Human Glioblastoma Cells

By HIDEKI MATSUMOTO, SACHIKO HAYASHI, ZHAO-HUI JIN, MASANORI HATASHITA, TOSHIO OHTSUBO, TAKEO OHNISHI, and EIICHI KANO

Introduction

Nitric oxide (NO) is generated endogenously from L-arginine by NO synthase (NOS) isoenzymes.[1] Inducible NOS (iNOS or NOS2) is expressed in various species of mammalian cells after exposure to many types of inducers, such as cytokines, bacterial lipopolysaccharide (LPS), heat shock, and hypoxia.[2–4] The p53 protein is accumulated in response to genotoxic and nongenotoxic stressors, such as DNA-damaging agents and heat, respectively, and the protein has been shown to be a key protein acting as a cell cycle checkpoint.[5,6]

NO has been suggested to be involved in a p53-dependent response to many kinds of stress, such as heat shock and changes in cellular metabolism. It has also been reported that NO can induce wild-type p53 (wtp53) protein.[7] This article introduces methods for the analysis of intercellular signal transduction mediated by NO using human glioblastoma cells.

Cells

A human glioblastoma cell line, A-172, is from JCRB Cell Bank (Setagaya, Tokyo, Japan). The two human glioblastoma cell lines, A-172 and A-172/m*p53*, are cultured in Dulbecco's modified Eagle's medium containing 10% fetal bovine serum, penicillin (50 U/ml), streptomycin (50 μg/ml), and kanamycin (50 μg/ml) (DMEM-10). A-172 cells have the wild-type *p53* gene (wt*p53*), whereas

[1] C. Nathan, *FASEB J.* **6,** 3051 (1992).

[2] K. L. MacNaul and N. I. Hutchinson, *Biochem. Biophys. Res. Commun.* **196,** 1330 (1993).

[3] I. Y. Malyshev, E. B. Manukhina, V. D. Mikoyan, L. N. Kubrina, and A. F. Vanin, *FEBS Lett.* **370,** 159 (1995).

[4] G. Melillo, T. Musso, A. Sica, L. S. Taylor, G. W. Cox, and L. A. Varesio, *J. Exp. Med.* **182,** 1683 (1995).

[5] A. Murray, *Curr. Opin. Cell Biol.* **6,** 872 (1994).

[6] H. Matsumoto, A. Takahashi, X. Wang, K. Ohnishi, and T. Ohnishi, *Int. J. Radiat. Oncol. Biol. Phys.* **38,** 1089 (1997).

[7] K. Forrester, S. Ambs, S. E. Lupold, R. B. Kapust, E. A. Spillare, W. C. Weinberg, E. Felley-Bosco, X. W. Wang, D. A. Geller, E. Tzeng, T. R. Billiar, and C. C. Harris, *Proc. Natl. Acad. Sci. U.S.A.* **93,** 2442 (1996).

0076-6879/02 $35.00

A-172/m*p53* cells have a mutated *p53* gene (m*p53*).[8,9] The doubling times of A-172 and A-172/m*p53* cells are approximately 24 and 22 hr, respectively. Plating efficiencies of both cell lines are approximately 70%.

Heat Shock

Twenty hours before heat treatment, exponentially growing cells are seeded at $\sim 10^6$ cells per dish in 9-cm dishes or $\sim 10^5$ cells per flask in 25-cm^2 flasks containing DMEM-10 without irradiated feeder cells. Cells are washed once with DMEM-10. Subsequently, cells either in dishes wrapped with parafilm or in screw-capped flasks are immersed into a water bath (Model EPS-47, Toyo Seisakusho Co., Tokyo, Japan) preset at $44 \pm 0.1°$. After the treatment, cells are incubated at 37° in a conventional humidified CO_2 incubator.

Irradiation with X-Rays

Twenty hours before irradiation, exponentially growing cells are seeded at $\sim 10^6$ cells per dish in 9-cm dishes or $\sim 10^5$ cells per flask in 25-cm^2 flasks containing DMEM-10 without irradiated feeder cells. Cells are washed once with DMEM-10. Subsequently, cells are irradiated with X-rays (1.0–10.0 Gy at 1.0 Gy/min) in DMEM-10 using a High Technical System X-ray (Model HW-150, Hitex Co., Tokyo, Japan). After irradiation with X-rays, cells are incubated at 37° in a conventional humidified CO_2 incubator.

Preparation of Crude Extract for Western Blotting of iNOS, p53, and hsp70

Cell Harvest

Cells are treated with PBS(−) containing 0.005% trypsin and 0.02% EDTA at 37° for 5 min, and then DMEM-10 is added to quench trypson. Cells are recovered into 15-ml conical tubes and centrifuged at 800*g* for 3 min. After discarding the supernatant, pelleted cells are resuspended in 1 ml of PBS(−), transfered into a 1.5-ml microtube, and pelleted by centrifugation at 1000*g* for 3 min.

Protein Extraction

Cells are suspended in 100 μl of RIPA buffer (50 m*M* Tris, pH 7.2, 150 m*M* NaCl, 1% NP-40, 1% sodium deoxycholate, and 0.05% SDS) and are then frozen

[8] X. Wang, A. Takahashi, K. Ohnishi, H. Matsumoto, K. Suda, and T. Ohnishi, *Exp. Cell Res.* **237,** 186 (1997).

[9] K. Ohnishi, X. Wang, A. Takahashi, and T. Ohnishi, *Exp. Cell Res.* **238,** 399 (1998).

and thawed three times. After centrifugation at 18,500*g* for 20 min, the supernatant is recovered into a new 1.5-ml microtube as the total protein fraction.

Protein Assay

Concentrations of the total protein fractions are quantified using a commercial protein assay reagent, e.g., Protein Assay CBB Solution (5x) (Code 294-49, Nacalai Tesque, Kyoto, Japan) or Bio-Rad protein assay kit (Bio-Rad Laboratories, Richmond, CA). One microliter of samples is add to 0.8 ml of distilled water, and then 0.2 ml of the reagent is added to the mixture. After mixing by vortexing, the mixtures are left at room temperature for 15 min. At the same time, the bovine serum albumin (BSA) solution as the standard solution is quantified with the reagent in the range of 1–10 mg/ml to plot the standard curve. After 15 min, absorbance at 595 nm (A_{595}) of samples and BSA solution is measured using a photometer. For Western blot analysis, concentrations of the total protein fractions are suitable at 3–5 mg protein/ml. To adjust the concentrations of samples in this range, the samples are diluted with RIPA buffer.

Western Blot Analysis

SDS–PAGE and Blotting

An aliquot (20, 20, and 5 μg) of protein is used for Western blotting analysis of iNOS, p53, and hsp70, respectively. After electrophoresis on 10% polyacrylamide gels containing 0.1% SDS,[10] proteins in the gels are transferred electrophoretically to Immobilon-P transfer membranes (Millipore Corp., Bedford, MA).

Staining with Antibodies and Amplification System

Proteins on the membrane are incubated with anti-iNOS polyclonal antibody (N-20, Santa Cruz Biotechnology, Santa Cruz, CA), anti-p53, or anti-hsp72 monoclonal antibody (PAb 1801; Oncogene Science, Inc., Uniondale, NY, or C92F3A-5; StressGen Biotechnologies Corp., Victoria, B.C., Canada). For visualization of the band, we use the horseradish peroxidase-conjugated anti-mouse IgG antibody (Zymed Laboratories, Inc., San Francisco, CA) and a BLAST (Chromogenic) blotting amplification system (NEN Life Science Products, Boston, MA).

Estimation of Protein Contents

Relative amounts of iNOS, hsp72, and p53 are calculated from the scanning profiles using a personal computer with the public domain NIH Image program (version 1.67).

[10] U. K. Laemmli, *Nature* **227,** 680 (1970).

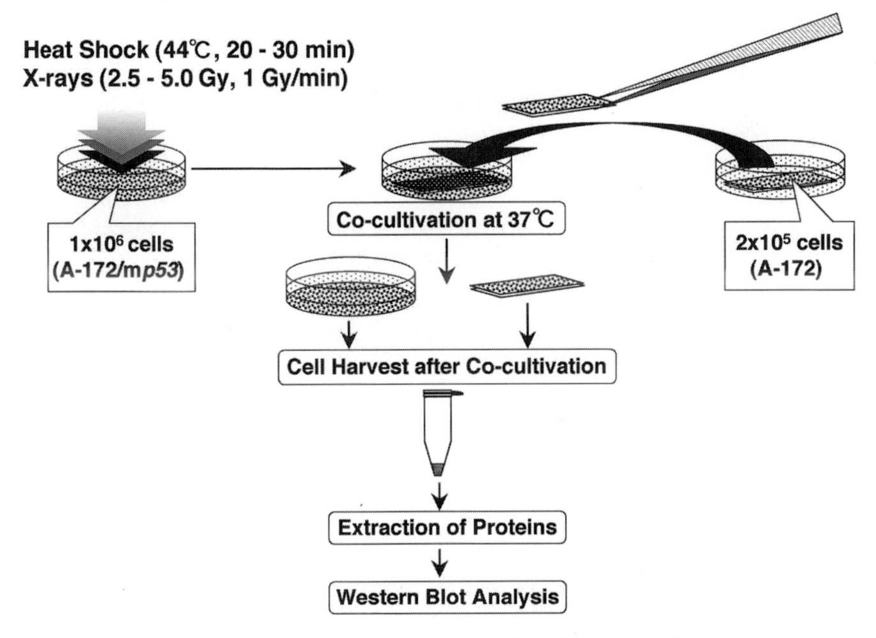

FIG. 1. Cocultivation of nonstressed cells with stressed cells.

Cocultivation of Nonstressed Cells with Stressed Cells

Twenty hours before irradiation with heat shock or X-rays, exponentially growing cells are seeded at $\sim 10^6$ cells per dish in 9-cm dishes containing DMEM-10 without irradiated feeder cells. At the same time, cells are seeded on slide glasses at 2×10^5 cells per slide glass (26×76 mm). Cells in the dishes are heated or irradiated with X-rays. Slide glasses containing nonirradiated cells are then transferred into dishes containing the irradiated cells, and the cocultures are incubated at 37° for up to 10 hr. At various times, cells are harvested from slide glasses and dishes, and total proteins are extracted and subjected to Western blotting (Fig. 1).

Treatment with Conditioned Medium

Twenty hours before treatment, exponentially growing cells are seeded at $\sim 10^6$ cells per flask in 75-cm^2 flasks containing DMEM-10 without irradiated feeder cells. Cells are washed twice with DMEM-10. Subsequently, cells in flasks are heated at $44 \pm 0.1°$ for 15 min. After the heat shock, cells are incubated at 37° in a conventional humidified CO_2 incubator. Then, the conditioned medium is recovered 10 hr after incubation by centrifugation at 800g for 10 min. Twenty hours before treatment, exponentially growing cells are seeded at $\sim 10^5$ cells per

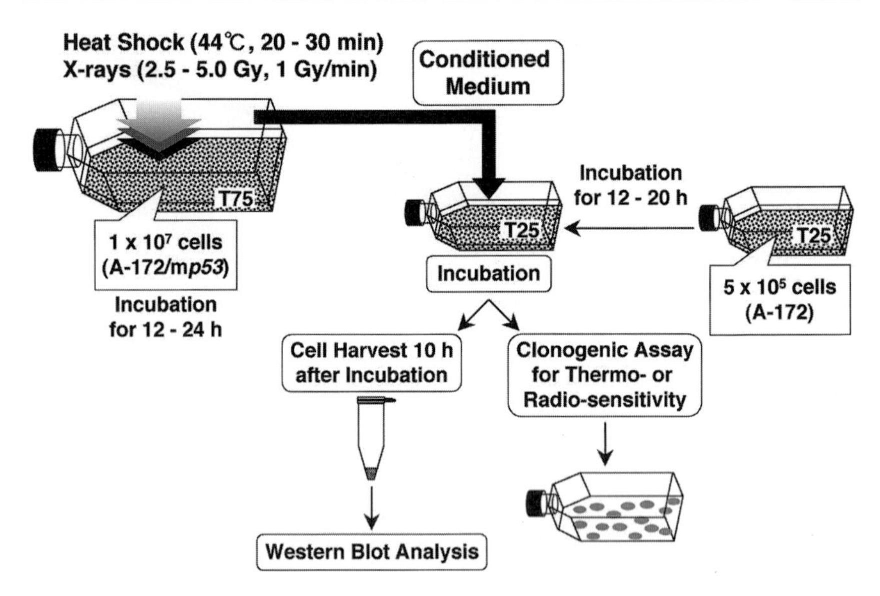

FIG. 2. Treatment of nonstressed cells with conditioned medium from stressed cells.

flask in 25-cm^2 flasks containing DMEM-10 without irradiated feeder cells. Cells are washed twice with DMEM-10 and are then exposed to the conditioned medium for 10–24 hr at 37°. After the treatment, cells are harvested and used for Western blot analysis (Fig. 2).

Measurement of Nitrite in the Conditioned Medium

The nitrite concentration in the medium is measured according to the method of Saltzman[11] with some modifications. One hundred microliters of medium is mixed well with 150 μl of the reagent containing 0.5% sulfanilic acid (analytical grade), 0.002% *N*-1-naphthylethylenediamine dihydrochloride (analytical grade), and 14% acetic acid (analytical grade). After standing at room temperature for 15 min, the absorbance of samples at 550 nm (A_{550}) is measured. The solution of sodium nitrite dissolved in medium is used as the standard solution. In this system, nitrite concentrations can be measured in the range of 0 to 10 μM. In addition, this method is applicable not only to the culture medium not containing the nitrate/nitrite (Eagle's MEM), but also to that containing the nitrate/nitrite (Dulbecco's MEM and RPMI 1640) (Fig. 3).

[11] B. E. Saltzman, *Anal. Chem.* **26,** 1949 (1954).

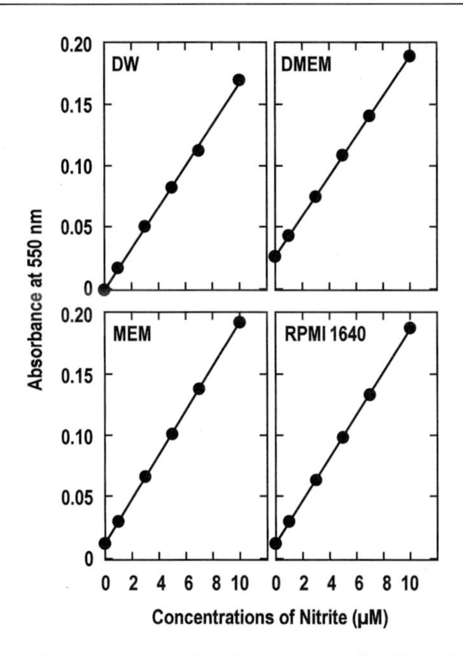

FIG. 3. Measurement of nitrite concentrations in various media. The solutions of sodium nitrite dissolved in these media were used as standard solutions. This assay system is applicable to not only Eagle's MEM not containing nitrate/nitrite, but also to Dulbecco's modified Eagle's medium and RPMI 1640 containing nitrate/nitrite.

Measurement of Cell Survival

The surviving cell fractions after heat shock or X-ray irradiations are determined as colony-forming units and are corrected by the plating efficiency of the nontreated cells as a control. Two replicate flasks are used per experiment, and two or more independent experiments are repeated for each survival point. Colonies obtained are fixed with methanol and stained with 2% Giemsa solution. Visible colonies composed of more than 50 cells after 10 days are counted as having grown from surviving cells.

The T_0 or D_0 value represents the treatment period for the heat shock or dose of irradiation, respectively, required to reduce the survival to 1/e with the exponentially regressing portion of the survival curves. The T_q or D_q (quasi-threshold treatment period or dose) is defined as the treatment period for heat shock or dose of irradiation, respectively, at which the straight portion of the survival curve, extrapolated backward, cuts the treatment period axis drawn through a survival fraction of unity.[12]

[12] E. J. Hall, "Radiobiology for the Radiologist," 5th Ed. Lippincott, New York, 2000.

Concluding Remarks

The effects of NO generated endogenously from stressed cells on nonstressed cells can be examined using the experimental system presented here. We have shown that NO generated endogenously from stressed human glioblastoma cells induces thermo- or radio resistance in nonstressed glioblastoma cells, i.e., the "NO-mediated bystander effect."[13,14] These experimental systems may prove useful in the development of not only the biology of NO, but also in radiation biology and radiation oncology.

Acknowledgments

This work was supported by a Grant-in-Aid for Scientific Research on Priority Areas (No. 09255217), Exploratory Research (No. 09877170), Scientific Research (A) (No. 09307015), and Scientific Research (C) (No. 10670840) from the Ministry of Education, Science, Sports and Culture, Japan.

[13] H. Matsumoto, S. Hayashi, M. Hatashita, K. Ohnishi, T. Ohtsubo, R. Kitai, H. Shioura, T. Ohnishi, and E. Kano, *Cancer Res.* **59,** 3239 (1999).

[14] H. Matsumoto, S. Hayashi, M. Hatashita, K. Ohnishi, H. Shioura, T. Ohtsubo, R. Kitai, T. Ohnishi, and E. Kano, *Radiat. Res.* **155,** 387 (2001).

[27] L-Arginine Metabolism in *Trypanosoma cruzi* in the Regulation of Programmed Cell Death

By LUCÍA PIACENZA, GONZALO PELUFFO, and RAFAEL RADI

Introduction

The unicellular eukaryote *Trypanosoma cruzi* belongs to the Kinetoplastidae family and is the causative agent of Chagas disease, affecting several million people in Central and South America. This protozoan parasite has a complex life cycle and exists in three morphological distinct forms, nonproliferative infective (metacyclic and blood trypomastigote), proliferative insect-borne (epimastigote), and intracellular proliferative (amastigote forms), which grow and replicate intracellularly in a variety of host mammalian cells, especially in macrophages and cardiac muscle cells.[1] The vertebrate host limits and controls the infection through the activated macrophage-dependent generation of reactive oxygen and nitrogen

[1] Z. Brener, *Annu. Rev. Microbiol.* **27,** 347 (1973).

0076-6879/02 $35.00

species. Indeed, maximal trypanocidal activity depends on lymphokines released by T cells responsible for the upregulation of the inducible form of nitric oxide synthase (iNOS) in macrophages.[2] Despite the role of L-arginine metabolism in the control of parasite infection by macrophage-derived ·NO, this amino acid may also promote invasion and infectivity through the formation of polyamines.[3,4] Moreover, two processes previously believed to be exclusive of multicellular organisms and which are potentially linked such as nitric oxide (·NO) production[5] and programmed cell death (PCD)[6] have been recognized in *T. cruzi.*

This article describes methods adapted to the study of L-arginine metabolism and PCD in the unicellular organism *T. cruzi* and shows that the application of these methods in combination with different pharmacological treatment can serve to unravel new aspects of *T. cruzi*–host cell interactions.

L-Arginine Metabolism in *T. cruzi*

In trypanosomatides, as in multicellular organisms, L-arginine serves as precursor for the synthesis of nitric oxide and polyamines (Scheme 1). Nitric oxide is a widespread intra- and intercellular messenger involved in a broad range of functions, such as the regulation of vascular tone, neuronal signaling, and immune response to infection.[7] ·NO is formed by nitric oxide synthases (NOS: EC 1.14.13.39) present in three distinct isoforms in mammalian and plant tissue and is also reported to be present in bacteria of *Nocardia* species[8] and trypanosomatides.[5,9,10] Constitutively expressed isoforms of NOS depend on Ca^{2+} influx for calmodulin binding and maximal activity. In epimastigote forms of *T. cruzi,* a constitutive isoform of NOS has been partially characterized.[5] The enzyme depends on the stimulation of NMDA membrane receptor channels triggered by L-glutamate and other excitatory amino acids such as serine and glycine. Stimulation of the NMDA receptor channel leads to a rise in cGMP, attributed to ·NO-dependent activation of soluble guanylate cyclase.[10] Moreover, *Leishmania donovani* and *T. cruzi* NOS were

[2] S. G. Reed, *J. Immunol.* **140,** 4342 (1988).
[3] F. Kierszenbaum, J. J. Wirth, P. P. McCann, and A. Sjoerdsma, *Proc. Natl. Acad. Sci. U.S.A.* **84,** 4278 (1987).
[4] M. A. Yakubu, B. Basso, and F. Kierszenbaum, *J. Parasitol.* **78,** 414 (1992).
[5] C. Paveto, C. Pereira, J. Espinosa, A. E. Montagna, M. Farber, M. Esteva, M. M. Flawia, and H. N. Torres, *J. Biol. Chem.* **270,** 16576 (1995).
[6] J. C. Ameisen, T. Idzioerk, O. Billaut-Mulot, J. P. Tissier, A. Potentier, and A. Ouaissi, *Cell Death Differ.* **2,** 285 (1995).
[7] R. G. Knowles and S. Moncada, *Biochem. J.* **298,** 249 (1994).
[8] Y. Chen and J. P. Rosazza, *Biochem. Biophys. Res. Commun.* **203,** 1251 (1994).
[9] N. K. Basu, L. Kole, A. Ghosh, and P. K. Das, *FEMS Microbiol. Lett.* **156,** 43 (1997).
[10] C. Pereira, C. Paveto, J. Espinosa, G. Alonso, M. M. Flawia, and H. N. Torres, *J. Eukaryot. Microbiol.* **44,** 155 (1997).

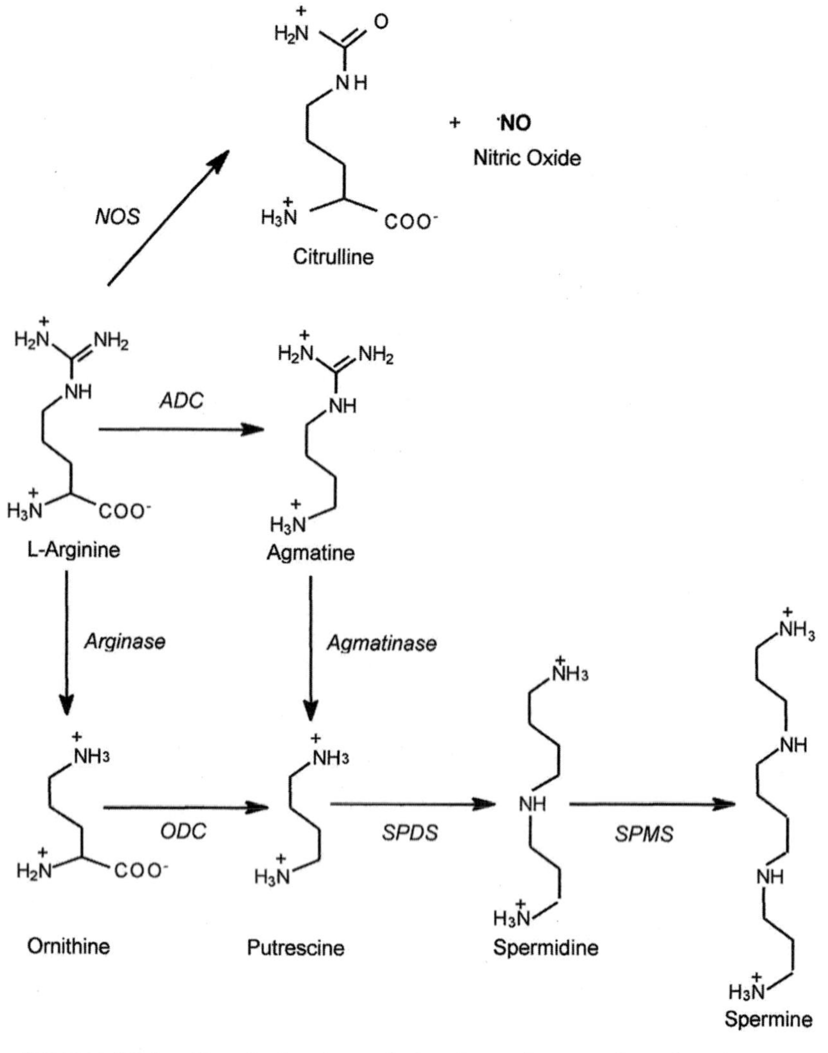

SCHEME 1. Nitric oxide and polyamine synthesis pathways. In mammalian cells, polyamines are produced via ornithine decarboxylase (ODC), whereas in *T. cruzi,* this pathway is carried out probably via arginine decarboxylase (ADC).

immunolocalized using antisera raised against the mammalian neuronal isoform (nNOS).[9,11]

Within the cell, L-arginine is also the precursor for the synthesis of polyamines (putrescine, spermidine, and spermine), polycations involved in processes such as

[11] J. Goldstein, C. Paveto, J. J. Lopez-Costa, C. Pereira, G. Alonso, H. N. Torres, and M. M. Flawia, *Biocell* **24,** 217 (2000).

proliferation, differentiation, and gene expression. With the exception of plants and bacteria, polyamines are synthesized from L-arginine by the combined action of arginase and ornithine decarboxylase (ODC) with the production of putrescine. Putrescine is then converted to spermidine and spermine in reactions catalyzed by spermidine synthase (SPDS) and spermine synthase (SPMS), respectively, by the sequential addition of aminopropyl groups derived from decarboxylated adenosylmethionine. In plants,[12] bacteria,[13] and some parasitic protozoa, such as *Cryptosporidium parvum*[14] and *T. cruzi,*[15,16] polyamines are synthesized by arginine decarboxylase (ADC) through the production of agmatine.

Thus, *de novo* polyamine synthesis in *T. cruzi* is probably carried out through the ADC pathway with the generation of agmatine, which can be metabolized further to putrescine by the action of agmatinase, an enzyme not yet characterized in this parasite. Putrescine is then metabolized as in the rest of the organisms (Scheme I). ODC and ADC activity is efficiently inhibited by their fluorinated derivatives difluoromethyl ornithine (DFMO) and difluoromethyl arginine (DFMA). It has been shown that treatment of epimastigotes with DFMA inhibits not only cell proliferation, but also cell invasion in mammalian cells, whereas DFMO has no effect. These observations establish the importance of ADC in two fundamental processes for the life cycle of this parasite.[3,4]

Programmed Cell Death in Trypanosomatides

Programmed cell death is a widespread phenomenon in multicellular organisms and is characterized by typical biochemical and morphological features, including DNA condensation and fragmentation, cell shrinkage, and ultimately the formation of apoptotic bodies that are eliminated rapidly *in vivo* by phagocytes. Biochemical pathways leading to PCD depend on cellular lineage information.[17] Studies in trypanosomatides revealed the capacity of these unicellular eukaryotes to die by a process of PCD with features of apoptosis (DNA condensation and fragmentation, cell shrinkage, and cytoplasm vacuolization) in response to various extracellular stimuli, including the presence of human serum, culture conditions, and the presence of the antibiotic Geneticin (G418).[6] In *T. cruzi,* the control of cell proliferation is crucial for the maintenance and dissemination of infection during the life cycle of the parasite. As a unicellular parasite, the levels of intracellular polyamines

[12] R. Slocum, *in* "Biochemistry and Physiology of Plants" (F. H. Slocum, ed.). CRC Press, Boca Raton, FL, 1991.

[13] C. W. Tabor and H. Tabor, *Microbiol. Rev.* **49,** 81 (1985).

[14] J. S. Keithly, G. Zhu, S. J. Upton, K. M. Woods, M. P. Martinez, and N. Yarlett, *Mol. Biochem. Parasitol.* **88,** 35 (1997).

[15] S. Majumder, J. J. Wirth, A. J. Bitonti, P. P. McCann, and F. Kierszenbaum, *J. Parasitol.* **78,** 371 (1992).

[16] S. Hernandez and S. Schwarcz de Tarlovsky, *Cell Mol. Biol.* (*Noisy-le-grand*) **45,** 383 (1999).

[17] D. L. Vaux, G. Haecker, and A. Strasser, *Cell* **76,** 777 (1994).

control cell proliferation. It has been shown that polyamine levels in mammalian cells regulate cell cycle progression and that a deficit in these polications leads to cell arrest and apoptosis.[18]

In multicellular organisms, the production of $^{\bullet}NO$ by nitric oxide synthases is associated with modulation of PCD. Induction of apoptosis is generally associated with high levels of $^{\bullet}NO$ (micromolar range, generally produced by iNOS), whereas inhibition of the process is achieved by low levels of $^{\bullet}NO$ (nanomolar range, constitutive isoforms of NOS). Inhibition of apoptosis by low levels of $^{\bullet}NO$ is mediated by pathways independent of and dependent on cGMP production. The independent pathways are associated with *S*-nitrosylation of critical thiols in caspases,[19] inhibition of the activation of transcription factors (NF-κB),[20] control of cell respiration by interactions with cytochrome *c* oxidase, and control of the permeability transition pore at the mitochondrial membrane level.[21,22] However, cGMP-dependent inhibition of apoptosis is mediated by the upregulation of anti-apoptotic proteins such as members belonging to the *Bcl-2* family.[23,24]

Studies carried out in our laboratory demonstrate the importance of *T. cruzi* L-arginine metabolism in the modulation of parasite-induced PCD.[25]

In the context of a parasite infection, investigations of *T. cruzi* L-arginine metabolism may provide new insights into the processes involved in *T. cruzi* survival and death inside mammalian host cells.

Description of Methods

Evaluation of Nitric Oxide Synthase Activity in Epimastigotes of T. cruzi

Nitric oxide production by cultured epimastigotes can be evaluated by measuring nitrite (NO_2^-) accumulation in the culture medium following, e.g., a 10-hr incubation period at 28° in different experimental conditions. The incubation mixture consists of 3×10^8 cells/ml in Krebs–Henseleit buffer (15 m*M* $NaHCO_3$, 5 m*M* KCl, 120 m*M* NaCl, 0.7 m*M* Na_2HPO_4, 1.5 m*M* NaH_2PO_4, pH 7.2) containing 10 m*M* glucose in the presence of the appropriate stimuli, substrate,

[18] J. M. Dypbukt, M. Ankarcrona, M. Burkitt, A. Sjoholm, K. Strom, S. Orrenius, and P. Nicotera, *J. Biol. Chem.* **269,** 30553 (1994).

[19] S. Dimmeler, J. Haendeler, M. Nehls, and A. M. Zeiher, *J. Exp. Med.* **185,** 601 (1997).

[20] H. B. Peng, P. Libby, and J. K. Liao, *J. Biol. Chem.* **270,** 14214 (1995).

[21] B. Beltran, A. Mathur, M. R. Duchen, J. D. Erusalimsky, and S. Moncada, *Proc. Natl. Acad. Sci. U.S.A.* **97,** 14602 (2000).

[22] E. Clementi, G. C. Brown, M. Feelisch, and S. Moncada, *Proc. Natl. Acad. Sci. U.S.A.* **95,** 7631 (1998).

[23] U. K. Messmer, U. K. Reed, and B. Brune, *J. Biol. Chem.* **271,** 20192 (1996).

[24] A. M. Genaro, S. Hortelano, A. Alvarez, C. Martinez, and L. Bosca, *J. Clin. Invest.* **95,** 1884 (1995).

[25] L. Piacenza, G. Peluffo, and R. Radi, *Proc. Natl. Acad. Sci. U.S.A.* **98,** 7301 (2001).

TABLE I
PRODUCTION OF NITRIC OXIDE BY *T. cruzi*[a]

Compound (m*M*)	NO_2^- (pmol min^{-1} mg^{-1})	Change (%)
	In Krebs–Henseleit	
None	$1.94 \pm 0.19^{(a)}$	0
+ L-Arginine (1)	$2.29 \pm 0.28^{(b)}$	18
+ L-Glutamate (0.01)	$2.70 \pm 0.15^{(b)}$	39
+ L-Glutamate (0.1)	$3.01 \pm 0.1^{(b)}$	55
+ L-Glutamate (1)	$3.10 \pm 0.05^{(b)}$	60
+ L-NAME (4)	1.84 ± 0.15	−5
+ MK-801 (0.1)	1.87 ± 0.1	−4
+ D-Arg (1) + L-Glu (1)	1.69 ± 0.14	−13
	In the presence of 10% FHS	
10% FHS	$2.36 \pm 0.18^{(c)}$	0
+ L-Arginine (10)	$3.98 \pm 0.37^{(d)}$	69
+ L-Glutamate (1)	$3.92 \pm 0.3^{(d)}$	66
+ L-NAME (10)	2.10 ± 0.12	−10

[a] $^{\bullet}$NO production is measured as NO_2^- accumulation over a 10-hr period. Results are expressed as means ± standard deviation of at least three independent assays. (b) and (d) represent statistical differences ($p < 0.05$) with respect to (a) and (c), respectively. FHS is fresh human serum. Adapted with permission from L. Piacenza, G. Peluffo, and R. Radi, *Proc. Natl. Acad. Sci. U.S.A.* **98,** 7301 (2001). Copyright © 2001 National Academy of Sciences, U.S.A.

and/or inhibitors. Following the incubation period, cells are lysed by three freeze–thaw cycles, and the supernatant is collected by centrifugation at 13,000*g* for 15 min at room temperature. The NO_2^- concentration is measured using the Griess reagent with a standard curve of $NaNO_2$ (0–10 μM).[26] Briefly, culture supernatants (1 ml) are mixed with 1 ml of the Griess reagent [equal volumes of 0.2% (w/v) *N*-(1-naphthyl)ethylenediamine and 2% (w/v) sulfanilamide in 5% (v/v) phosphoric acid)], incubated for 10 min at room temperature, and absorbance determined at 543 nm. Glutamate-dependent activation of NOS activity is evaluated in the presence of different L-glutamate concentrations (0–1 m*M*) added to the medium in the presence or absence of 1 m*M* L-arginine, and the EC_{50} for L-glutamate is calculated ($EC_{50} \approx 10\ \mu M$; Table I). Inhibition studies are performed by preincubating (20 min) epimastigotes with the channel ion inhibitor MK-801 (0.1 m*M*) and the NOS inhibitor nitroarginine methyl ester (L-NAME 0–4 m*M*) prior to the addition of L-arginine and L-glutamate. Long incubation periods (~10 hr) are needed to accumulate sufficient NO_2^- in the culture medium in order to be measured by the

[26] M. B. Grisham, G. G. Johnson, and J. R. Lancaster, Jr., *Methods Enzymol.* **268,** 237 (1996).

Griess reagent (Table I). The more sensitive fluorescent method for NO_2^- detection based on 2,3-diaminonaphthalene[27] could not be utilized due to interference with parasite-released products.

Alternatively, NOS activity can be measured by following the conversion of L-[^{3}H]arginine to L-[^{3}H]citrulline in epimastigotes or cell extracts.[5,25] For these experiments, 1×10^7 cell/ml are incubated in the presence of 1 mM L-glutamate in Krebs–Henseleit buffer, pH 7.2, containing 1 μCi L-[^{3}H]arginine (40 Ci/mmol) at 28° for 45 min. The presence of the different inhibitors, such as L-NAME, aminoguanidine, L-valine (50 mM, used to inhibit arginase activity), and MK-801 (premixed 20 min after the addition of L-[^{3}H]arginine), are assayed as described previously. Following the incubation period, parasites are lysed as described earlier and formed L-[^{3}H]citrulline is purified from the supernatant (1 ml) by passage through a 4-ml cation-exchange resin [Dowex 50W ion-exchange resin converted to the sodium form by washing four times in 1 M NaOH (until pH reaches 12) and rinsing in distilled water until the pH drops to 8.0]. L-[^{3}H]Citrulline is eluted from the column with 8 ml of 50 mM HEPES buffer, pH 5.0, containing 5 mM EDTA and mixed with 15 ml of liquid scintillation cocktail (LSC). *T. cruzi* NOS activity is estimated to be 1.68 ± 0.05 and 1.12 ± 0.04 pmol L-[^{3}H]citrulline formed min^{-1} mg^{-1} of protein in the absence and presence of 4 mM L-NAME, respectively, several times lower than values reported previously.[5]

Induction and Analysis of Apoptosis in T. cruzi

Induction of apoptosis in epimastigotes can be achieved by incubating parasites in Krebs–Henseleit buffer, pH 7.2, at 28° in the presence of fresh human serum (FHS).[6,25] For all experiments, 10% of FHS is used as death stimuli (as no differences in [^{3}H]thymidine incorporation are observed with higher concentrations of human serum), and epimastigotes death (necrotic or apoptotic) is evaluated using different experimental approaches, including detection of DNA fragmentation on 2% agarose gels stained with ethidium bromide, quantitation of the DNA fragmentation by the diphenylamine method, membrane permeability assays with incorporated [^{3}H]uridine, *in situ* staining with the TUNEL technique, and evaluating changes in parasite proliferating rates by DNA [^{3}H]thymidine incorporation assays.

DNA Fragmentation Analysis

Agarose Gels. Analysis of parasite DNA fragmentation is evaluated in epimastigotes (3×10^8 cells/ml) incubated in Krebs–Henseleit buffer, pH 7.2, in the presence of death stimuli (10% FHS) after an 18- to 20-hr incubation period at 28°. Long incubations are needed to obtain sufficient fragmented DNA to be visualized in agarose gels. After incubation, parasites are pelleted by centrifugation at 700g

[27] A. M. Miles, D. A. Wink, J. C. Cook, and M. B. Grisham, *Methods Enzymol.* **268,** 105 (1996).

for 10 min and are resuspended in 500 μl of lysis buffer TET(10 mM Tris, pH 8.0, 1 mM EDTA, 0.5% Triton X-100) and allowed to incubate on ice for 20 min. Intact chromatin (pellet) is separated from DNA fragments (supernatant) by centrifugation at 13,000g for 45 min at 4°. DNA is precipitated in the presence of 100 μl of 5 M NaCl, 5 μl of 1 M $MgCl_2$, and 1 ml of absolute cold ethanol at −20° overnight. Precipitated DNA is pelleted by centrifugation at 13,000g for 1 hr at 4°, washed three times in cold ethanol, and dried and resuspended in 50 μl of buffer TE (10 mM Tris, pH 8.0, 1 mM EDTA). Further phenol chloroform extraction steps are not necessary, avoiding losses in parasite material. Samples are then treated with RNase A (1 hr at 37°) and proteinase K (1 hr at 60°) and are mixed with loading solution (0.25% bromophenol blue and 30% glycerol) for gel analysis. Fifty microliters of each sample is loaded on a 2% agarose gel, separated electrophoretically at 80 V for 1.5 hr, and analyzed under UV light (Fig. 1A). Quantification of the fragmented DNA in each condition can be measured by the DPA reagent method as described previously (Fig. 1B). In order to screen for caspase-like activity in this trypanosomatide, the peptide inhibitor Ac-Asp-Val-Glu-Asp-aldehyde (Ac-DEVD-CHO) can be used at different concentrations (0–250 μM) in the presence of death stimuli (Figs. 1A and 1B).

TUNEL Assay. Fragmented DNA exposed 3′-OH ends as a result of DNase activity can be detected using the TUNEL assay technique. Parasites are treated using different experimental conditions, spotted on precoated poly-L-lysine slides, dried for 1–2 hr, and fixed with cold acetone at −20° for 1 hr. Assays are performed following the manufacturer's instructions (Apoptosis Detection System, Fluorescein, Promega Co.).[25]

Changes in Parasite Proliferation Rates

In order to evaluate the changes in parasite proliferation rates exposed to death stimuli and/or different experimental conditions, [^{3}H]thymidine incorporation assays are performed, incubating 1×10^6 cells/ml (exponential growth phase) in complete BHI medium (200 μl) containing 1 μCi of [^{3}H]thymidine for 18 hr at 28° in 96 microwell plates. Following incubation, parasite DNA is recovered on nylon filters using a semiautomatic cell harvester and counted for radioactivity as described previously. Results can be expressed as percentages of [^{3}H]thymidine incorporation, taking as 100% the control (10% heat-inactivated human serum) condition (Fig. 1C).

Membrane Permeability Assays

One of the typical features of apoptosis is the maintenance of membrane integrity throughout the process. *In vivo,* apoptotic cells are eliminated rapidly by professional phagocytes. *In vitro* experiments with long incubation periods can switch apoptotic to necrotic death. Therefore, it is necessary to evaluate membrane integrity using suitable assays such as [^{3}H]uridine release from preloaded cells.

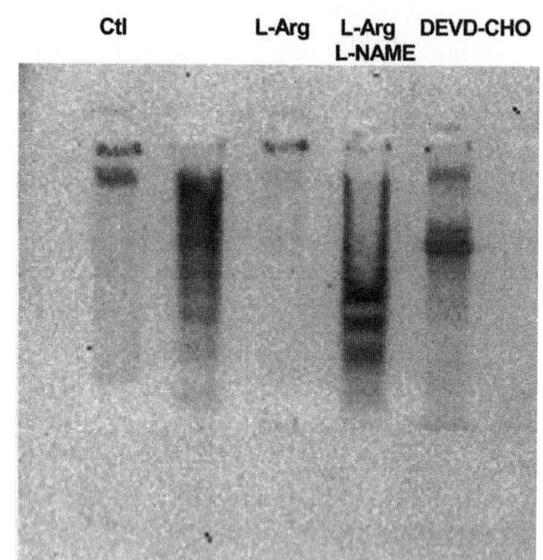
A
10% FHS
Ctl
L-Arg
L-Arg
L-NAME
DEVD-CHO

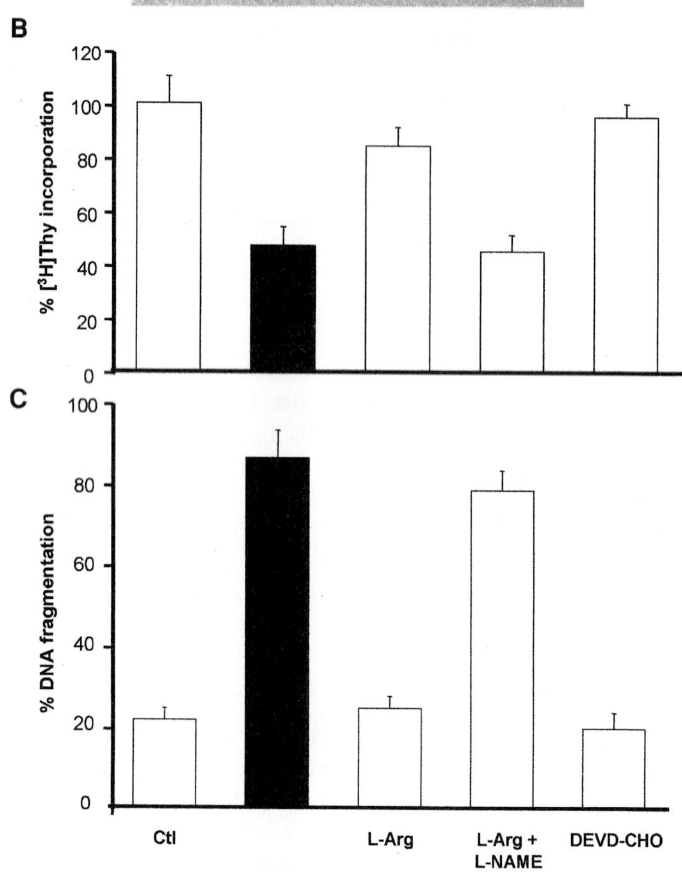
B
120
100
80
60
40
20
0
% [³H]Thy incorporation
C
100
80
60
40
20
0
% DNA fragmentation
Ctl
L-Arg
L-Arg +
L-NAME
DEVD-CHO

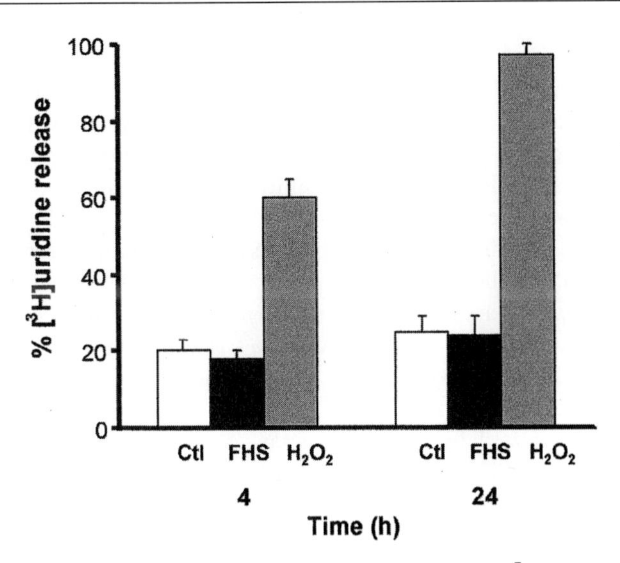

FIG. 2. Membrane permeability assays. Labeled epimastigotes (1×10^7 cells/ml) were incubated in different experimental conditions (Ctl, 10% FHS, and 50 m*M* H_2O_2) for 24 hr at 28°. Aliquots of 1 ml were taken periodically, and parasites were collected by centrifugation. Supernatants and pellets were assayed for radioactivity. Results are expressed as the percentage of [^{3}H]uridine released with respect to the total incorporated in each condition (supernatant + pellet).

Parasites are incubated with [^{3}H]uridine (1 μCi/ml) for 18 hr in BHI complete medium, collected by centrifugation, and washed three times in Krebs–Henseleit buffer in order to eliminate all nonincorporated [^{3}H]uridine. Cells (1×10^7 cells/ml in Krebs–Henseleit buffer) are incubated in different experimental conditions, and aliquots of 1 ml are collected at required time points. Parasites are collected by centrifugation at 700*g*, and the supernatant (released [^{3}H]uridine) and pellet (non-released [^{3}H]uridine) are mixed with 5 ml LSC and counted for total radioactivity. [^{3}H]Uridine released is calculated as follows: [^{3}H]uridine % released = cpm present in the supernatant $\times$ 100/total CPM (Fig. 2).

FIG. 1. Analysis of *T. cruzi* PCD and its modulation by L-arginine. (A) DNA fragmentation analysis. Cells (3×10^8/ml) were incubated for 18 hr in serum-free conditions (Ctl) in the presence of 10% FHS containing 10 m*M* L-arginine, 10 m*M* L-arginine plus 10 m*M* L-NAME, or 200 μM DEVD-CHO. DNA was extracted and analyzed on ethidium bromide-stained 2% agarose gels. (B) [^{3}H]Thymidine incorporation. Parasites at a density of 1×10^6 cells/ml were incubated for 18 hr at 28° in BHI complete medium containing 1 μCi of radiolabeled thymidine under the same conditions as in (A). Following the incubation period, DNA was harvested onto fiber glass filters and counted for radioactivity. (C) Parasites (3×10^8 cells/ml) were incubated, using the different experimental conditions, in Krebs–Henseleit buffer for the quantification of DNA fragments by the DPA method. Adapted with permission from L. Piacenza, G. Peluffo, and R. Radi, *Proc. Natl. Acad. Sci. U.S.A.* **98,** 7301 (2001).

Proteolytic Activities Involved in Apoptosis

In mammalian cells, apoptosis is generally triggered by the activation of a family of cysteine proteases named caspases. Peptide aldehydes with a sequence resembling that of the natural substrate inhibit capase activity in many cellular models. Caspases are also inhibited by *S*-nitrosylation processes involving $^{\bullet}$NO.[28] A search of caspase-like activity in *T. cruzi* is accomplished by incubating parasites (3×10^8 cells/ml) for different time exposures (0–6 hr) in the presence of death stimuli (10% FHS). Following incubation, parasites are pelleted by centrifugation at 700*g* for 10 min and are resuspended in 500 μl of 20 m*M* PIPES buffer, pH 7.2, containing 100 m*M* NaCl, 1 m*M* EDTA, 0.1% CHAPS, 10% sucrose, and 5 m*M* dithiothreitol (DTT). Following a three freeze–thaw cycle at -80°, cell debris is removed by centrifugation at 13,000*g* for 30 min at 4°, and the supernatant is stored at -20° until used. For enzymatic assays, protein extracts (2 mg) are preincubated for 15 min at 37° in the presence of freshly prepared 5 m*M* DTT and/or DEVD-CHO (250 μM). The reaction is initiated by the addition of 400 μM of the chromogenic substrate DEVD coupled to *p*-nitroaniline (pNA), and the increment in absorbance at 380 nm is followed at 37° for 2 hr. Enzyme activity is expressed as picomoles of pNA released per minute per milligram of protein at 37° (pNA $\varepsilon_{380} = 5000\ M^{-1}\ \text{cm}^{-1}$).[25] Because enzyme activity is assayed in the presence of large quantities of substrate and high protein concentration, the participation of other parasite cysteine proteinase activities (such as *T. cruzi* cruzipain[29]) must be ruled out. Cysteine proteinase (cruzipain) activity can be measured fluorometrically using peptides coupled to 7-amino-4-methylcoumarin (NHMec) as substrates.[30] Cell extracts (20 μg) are prepared as described earlier, and assays (final volume 200 μl) are carried out using the peptide z-Phe-Arg-NHMec at a final concentration of 10 μM in in 20 m*M* PIPES, buffer, pH 7.2, containing 100 m*M* NaCl, 1 m*M* EDTA, 0.1% CHAPS, 10% sucrose, and 1 m*M* DTT. The release of NHMec is followed at 37° for 1 hr using a Fluostar Galaxy (BMG Lab Technologies) microplate reader with excitation at 380 nm and emission at 460 nm. Enzyme activity is expressed as micromoles NMHmec released per minute per milligram of protein at 37° using a standard curve of NHMec (0–30 μM). Inhibition studies are performed as described earlier using the general cystein protease inhibitor E64 [*trans*-epoxysuccinyl-L-leucylamido-(4-guanidino)butane, *N*-[*N*-(L-3-transcarboxyrane-2-carbonyl)-L-leucyl]agmatine, final concentration of 10 μM]. Studies in mammalian cells show the involvement of lysosomal cathepsins in the

[28] J. Li, C. A. Bombeck, S. Yang, Y. M. Kim, and T. R. Billiar, *J. Biol. Chem.* **274,** 17325 (1999).

[29] P. Schotte, W. Declercq, S. Van Huffel, P. Vandenabeele, and R. Beyaert, *FEBS Lett.* **442,** 117 (1999).

[30] J. J. Cazzulo, M. C. Cazzulo Franke, J. Martinez, and B. M. Franke de Cazzulo, *Biochim. Biophys. Acta* **1037,** 186 (1990).

TABLE II
PROTEOLYTIC ACTIVITIES INVOLVED IN *T. cruzi* PROGRAMMED CELL DEATH[a]

Compound (mM)	DEVD-pNA (pmol min^{-1} mg^{-1})	Change (%)	Phe-Arg-NHMec (μmol min^{-1} mg^{-1})	Change (%)
Control	$0.072 \pm 0.01^{(a)}$	0	$0.4 \pm 0.02^{(c)}$	0
10% FHS	$0.205 \pm 0.01^{(b)}$	185	$0.95 \pm 0.03^{(d)}$	140
+ DEVD-CHO (200 μM)	$0.09 \pm 0.007^{(b)}$	25	—	—
+ E64 (30 μM)	—	—	$0.025 \pm 0.01^{(d)}$	−94

[a] Proteolytic activities were measured after a 6-hr incubation period in the presence of 10% FHS in parasite extracts using the chromogenic substrate DEVD-pNA and the fluorogenic substrate z-Phe-Arg-NHMec for caspase-like and cysteine protease, activity, respectively. Enzyme activity is expressed as picomole pNA released at 37° per minute per milligram or micromole NHMec released at 37° per minute per milligram or protein extract. Inhibitors (DEVD-CHO and E64) were added to the assay mixture 30 min prior to the addition of the substrate. (b) and (d) represent statistical differences ($p < 0.05$) with respect to (a) and (c), respectively.

induction of PCD.[31] Thus, in *T. cruzi,* further analyses are needed to confirm the real participation of caspase activity (Table II).

Modulation of PCD by L-*Arginine and Derived Metabolites*

Evaluation of parasite NOS activity in the presence of death stimuli is performed as described earlier with the exception that the serum is first removed by centrifugation at 700*g* for 10 min to avoid protein interference. Pelleted parasites are resuspended in 500 μl of Krebs–Henseleit buffer, pH 7.2, and lysed by three freeze–thaw cycles. Cell debris is removed by centrifugation, and the supernatant is used for the quantitation of NO_2^- by the Griess reagent as described previously (Table I).[26] Under conditions of maximal activity (i.e., 10 mM L-arginine + 10% FHS), the intracellular production of $^{\bullet}$NO by a single epimastigote is estimated to be 0.22 μM $^{\bullet}$NO min^{-1} (considering the volume of a single epimastigote as 3×10^{-14} liter, Table I).[25] In these conditions, DNA fragmentation is inhibited and parasite proliferation rates are recovered (Figs. 1B and 1C, respectively). In order to confirm the action of parasite-derived $^{\bullet}$NO, assays with $^{\bullet}$NO donors can be performed. Different $^{\bullet}$NO donors are available commercially, differing in stability and half-life decomposition kinetics. Taking into account the long incubation periods of our experiments and the low $^{\bullet}$NO fluxes derived from parasite NOS activity

[31] V. Stoka, B. Turk, S. L. Schendel, T. H. Kim, T. Cirman, S. J. Snipas, L. M. Ellerby, D. Bredesen, H. Freeze, M. Abrahamson, D. Bromme, S. Krajewski, J. C. Reed, X. M. Yin, V. Turk, and G. S. Salvesen, *J. Biol. Chem.* **276,** 3149 (2001).

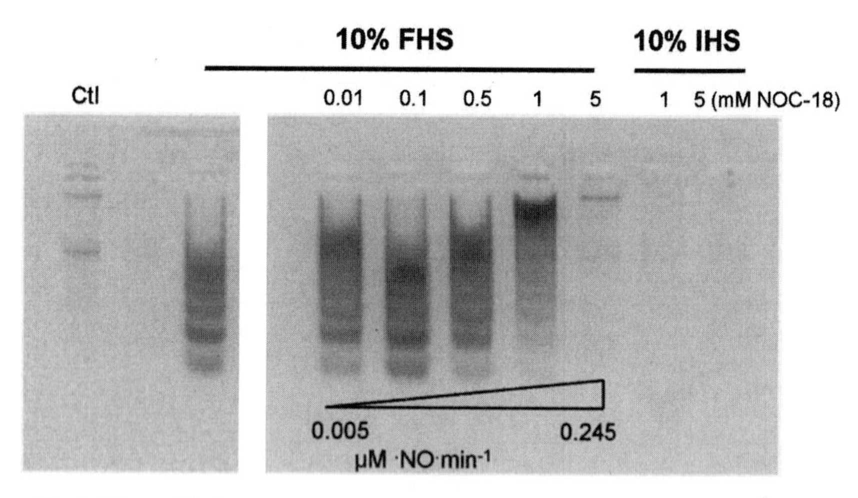

FIG. 3. Effects of $^{\bullet}$NO donors on PCD. DNA fragmentation analysis. Parasites (3×10^8 cells/ml) were incubated for 18 hr at 28° in serum-free conditions (Ctl) with 10% FHS or 10% heat-inactivated human serum (IHS) in the presence of different NOC-18 concentrations (0.01–5 m*M*). Following exposure, DNA was collected as described in Fig. 1, and DNA fragments were visualized in a 2% agarose gel. The concentration of NOC-18 (m*M*) corresponds to the following $^{\bullet}$NO fluxes: 0.01 (0.05 μM min^{-1}), 0.1 (0.018 μM min^{-1}), 0.5 (0.082 μM min^{-1}), 1 (0.145 μM min^{-1}), and 5 (0.245 μM min^{-1}). Adapted with permission from L. Piacenza, G. Peluffo, and R. Radi, *Proc. Natl. Acad. Sci. U.S.A.* **98,** 7301 (2001). Copyright © 2001 National Academy of Sciences, U.S.A.

(0.22 μM $^{\bullet}$NO min^{-1}), we selected two $^{\bullet}$NO donors (1-hydroxy-2-oxo-3,3-bis(2-aminoethyl)-1-triazene NOC-18 and spermine NONOate) with half-lives between 18 and 24 hr, ensuring the release of $^{\bullet}$NO at low levels during the incubation period. The rate of release of $^{\bullet}$NO from donors can be measured using the oxyhemoglobin oxidation assay in Krebs–Henseleit buffer, pH 7.2, containing 10% FHS, following the decrease in absorbance at 577 nm due to $^{\bullet}$NO-dependent oxidation of oxyhemoglobin as described previously.[32]

Appropriate concentrations (0–5 m*M*) of $^{\bullet}$NO donors are resuspended in Krebs–Henseleit buffer, pH 7.2, immediately before use. Figure 3 shows results obtained after an 18-hr incubation of parasites in the presence of death stimuli and different $^{\bullet}$NO donor concentrations on DNA fragmentation. Control experiments are carried out using 1 and 5 m*M* of NOC-18 in the presence of 10% heat-inactivated human serum. Concentrations of NOC-18 leading to $^{\bullet}$NO fluxes of 0.145–0.245 μM $^{\bullet}$NO min^{-1} are able to inhibit parasite DNA fragmentation completely (Fig. 3). In these conditions, parasite proliferation rates are not restored unlike that with the L-arginine (not shown). Importantly, the inability of [^{3}H]thymidine recovery by $^{\bullet}$NO is not due to $^{\bullet}$NO-dependent cytostasis, as

[32] C. C. Winterbourn, *Methods Enzymol.* **186,** 265 (1990).

TABLE III
T. cruzi L-[^{3}H]ARGININE UTILIZATION IN THE PRESENCE OF 10% FHS[a]

Condition	% of radiolabeled products		
	L-Arginine	L-Citrulline	Agmatine + polyamines
Control	58 ± 3	2.5 ± 1	7 ± 3
10% FHS	28 ± 2	9.6 ± 0.8	47 ± 5

[a] Epimastigotes (3×10^8 cells/ml) were incubated in the presence or absence of 10% FHS for a 20-hr period at 28° in Krebs–Henseleit buffer, pH 7.2, containing 1 μCi L-[^{3}H]arginine. Following incubation, parasites were collected and lysed, and supernatants were analyzed by TLC. Identified products were scraped and counted for radioactivity. Results are expressed as the percentage of radiolabeled products with respect to the total incorporated in each condition.

proliferation in the presence of the $^{\bullet}$NO donor remains unchanged in the presence of 10% heat-inactivated human serum (IHS).

L-*Arginine Metabolism by ADC*

$^{\bullet}$NO does not restore parasite proliferation rates (as does 10 m*M* L-arginine; Fig. 1C), suggesting that other L-arginine-derived metabolites may be involved in this process. Proliferation processes are dependent on polyamine production or, alternatively, uptake from the extracellular medium. To explore the possibility that L-arginine is being used as a substrate for ADC in the presence of death stimuli, assays for measuring polyamines using thin-layer chromatography (TLC) are used.[33] Parasites (3×10^8 cells/ml) are incubated in Krebs–Henseleit buffer containing 1 μCi of L-[^{3}H]arginine (40 Ci/mmol) under different experimental conditions for a period of 20 hr at 28°. Following incubation, parasites are collected by centrifugation at 700*g* and washed three times in Krebs–Henseleit buffer to eliminate all nonincorporated L-[^{3}H]arginine. Parasites are resuspended and lysed with 200 μl of buffer TET containing 10 m*M* Tris, pH 8.0, 1 m*M* EDTA, and 0.2% of Triton X-100. Samples are pelleted at 13,000*g* at 4° for 30 min, and supernatants are analyzed by TLC. Fifty microliters of each sample is spotted onto silica gel plates, and chromatography is run with the solvent system chloroform : methanol : ammonium hydroxide : water (1 : 4 : 2 : 1, v/v). Following chromatography, plates are dried at 100° for 20 min and developed with a solution of 0.25% ninhydrin in acetone. Products of interest are identified by comigration with authentic standards (L-arginine, agmatine, putrescine, spermidine, ornithine, and

[33] W. Durante, L. Liao, K. J. Peyton, and A. I. Schafer, *J. Biol. Chem.* **272,** 30154 (1997).

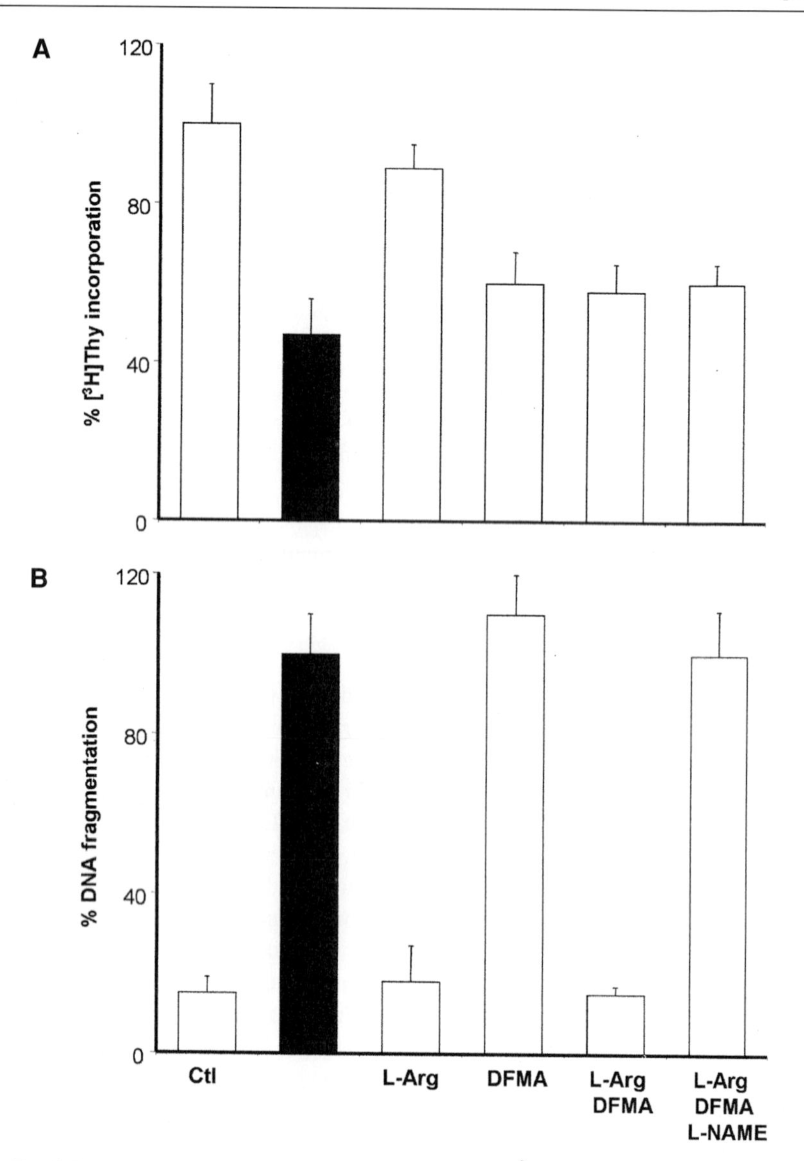

FIG. 4. Involvement of ADC in the modulation of PCD. (A) [^{3}H]Thymidine incorporation assays. Parasites (3×10^8 cells/ml) were incubated in different experimental conditions for a period of 18 hr at 28° and assayed for thymidine incorporation as described in Fig. 1. Concentrations used are L-arginine (10 m*M*), DFMA (10 m*M*), and L-NAME (10 m*M*). Thymidine incorporation is expressed relative to nontreated parasites (control condition) taken as 100%. (B) DNA analysis. Using the same conditions as in (A), parasites were assayed for DNA fragmentation in agarose gels. The percentage of DNA fragmentation was estimated by digital analysis of the electrophoretic bands, taking as 100% the fragmentation observed with 10% FHS.

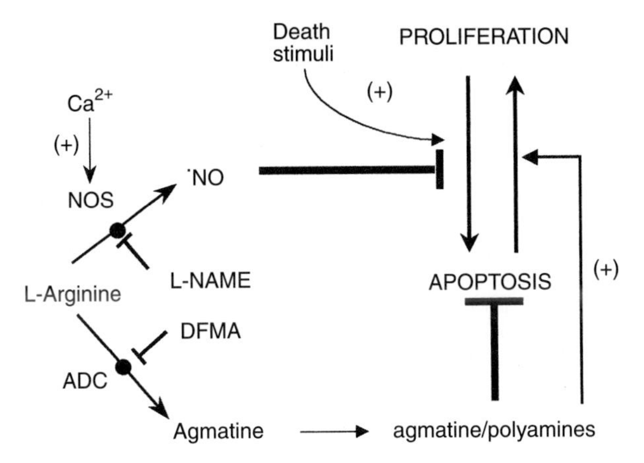

SCHEME 2. Proposed metabolic pathways for L-arginine in the modulation of *T. cruzi* programmed cell death. Modified with permission from L. Piacenza, G. Peluffo, and R. Radi, *Proc. Natl. Acad. Sci. U.S.A.* **98,** 7301 (2001). Copyright © 2001 National Academy of Sciences, U.S.A.

L-citrulline), scraped, and extracted with 100 μl ethyl acetate for 1 hr. The radioactivity of each sample is evaluated by mixing with 5 ml of liquid scintillation cocktail (LSC), and the percentage of labeled products is calculated as follows: cpm of the product $\times$ 100 total cpm present in the sample. This solvent system allows separation of L-arginine, L-citrulline, and agmatine/polyamines from one another. Results are shown in Table III, demonstrating that in the presence of death stimuli, L-arginine is used for the synthesis of agmatine and probably the different polyamines.

Involvement of ADC in L-*Arginine Protection*

ADC activity is blocked effectively by the presence of its fluorinated derivative DFMA ($K_i = 20$ mM in *T. cruzi*[16]). The presence of DFMA (10 mM) but not DFMO (10 mM) prevents L-arginine (10 mM)–[^{3}H]thymidine incorporation but has no effect on DNA fragmentation (Figs. 4A and 4B, respectively). Moreover, exogenous supplementation with agmatine (3–10 mM), putrescine (3–10 mM), spermidine (3–10 mM), but not L-ornithine (10 mM) is able to inhibit parasite PCD (data not shown).

These results suggest the presence of two distinct but complementary L-arginine-dependent metabolic pathways that contribute to the inhibition of PCD induced by the presence of 10% FHS in epimastigotes of *T. cruzi* (Scheme II).

Conclusions and Perspectives

We have discussed methodology for evaluating the metabolism of L-arginine and the occurrence of PCD in the unicellular eukaryotic parasite *T. cruzi*. Future work is needed to evaluate the participation, relevance, and interrelationship of both processes in *T. cruzi* survival and death in mammalian host cells.

Acknowledgments

This work was supported by grants from the Howard Hughes Medical Institute (United States), International Centre for Genetic Engineering and Biotechnology (Italy), Swedish Agency for Research Cooperation (Sweden), and Universidad de la República (Uruguay) to R.R. L.P. and G.P. were partially supported by fellowships from the Programa de Desarrollo de Clencias Básicas (Uruguay) and Programa para la Investigación Biomédica, Facultad de Medicina (Uruguay), respectively. R.R. is an International Research Scholar of the Howard Hughes Medical Institute.

Section III

Nitric Oxide and Mitochondrial Functions

[28] Measurement of Mitochondrial Respiratory Thresholds and the Control of Respiration by Nitric Oxide

By PAUL S. BROOKES, SRUTI SHIVA, RAKESH P. PATEL, and VICTOR M. DARLEY-USMAR

Introduction

The nitric oxide (NO•)–cytochrome *c* oxidase signaling pathway is now emerging as one of the most sensitive physiological mechanisms for the regulation of mitochondrial respiration.[1,2] This is mediated through binding of NO• to the binuclear $Cu_B/heme_{a3}$ center in cytochrome *c* oxidase and has been characterized extensively.[3,4] However, implications for this binding in the control of mitochondrial respiration, ATP synthesis, and reactive oxygen species (ROS) generation have received less attention.[5] It is important to understand the mechanisms underlying these phenomena, as we and others have shown that the degree of control by NO• over mitochondrial function is perturbed in pathological situations.[6]

Basic Concepts of Metabolic Control

Like other complexes of the mitochondrial respiratory chain, cytochrome *c* oxidase (complex IV) is present in excess of the amount required to meet normal metabolic demand.[7] One result of this property is that submaximal inhibition of the enzyme does not affect the rate of the overall metabolic pathway (i.e., respiration) until a threshold level of inhibition is reached. Beyond this threshold, respiratory inhibition occurs. This is illustrated in Fig. 1, which shows a "threshold curve" for cytochrome *c* oxidase of rat heart mitochondria. Implicit in the existence of such thresholds is the degree of control an enzyme, in this case cytochrome *c* oxidase, has over respiration. Although a full discussion of metabolic control

[1] G. C. Brown, *Acta Physiol Scand.* **168,** 667 (2000).

[2] M. W. Cleeter, J. M. Cooper, V. M. Darley-Usmar, S. Moncada, and A. H. Schapira, *FEBS Lett.* **345,** 50 (1994).

[3] A. Giuffre, M. C. Barone, D. Mastronicola, E. D'Itri, P. Sarti, and M. Brunori, *Biochemistry* **39,** 15446 (2000).

[4] J. Torres, V. Darley-Usmar, and M. T. Wilson, *Biochem. J.* **312,** 169 (1995).

[5] P. S. Brookes and V. M. Darley-Usmar, *Free Radic. Biol. Med.* **32,** 370 (2002).

[6] P. S. Brookes, J. Zhang, L. Dai, F. Zhou, D. A. Parks, V. M. Darley-Usmar, and P. G. Anderson, *J. Mol. Cell Cardiol.* **33,** 69 (2001).

[7] J. P. Mazat, T. Letellier, F. Bedes, M. Malgat, B. Korzeniewski, L. S. Jouaville, and R. Morkuniene, *Mol. Cell Biochem.* **174,** 143 (1997).

0076-6879/02 $35.00

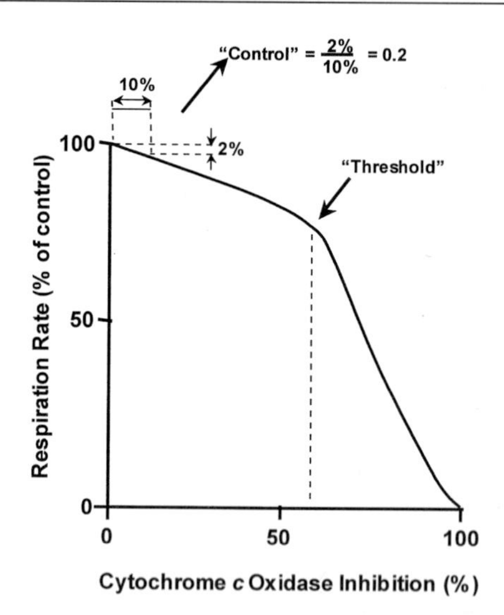

FIG. 1. Mitochondrial complex IV respiratory threshold. A representative trace for isolated rat heart mitochondria is shown, in which respiration rate and complex IV activity were inhibited with cyanide. The specific definitions of the terms "Threshold" and "Control" in the context of this article are indicated.

theory is beyond the scope of this manuscript (see Mazat *et al.*[7] for a more detailed discussion), it should be emphasized that in this context the word control has a very specific meaning, i.e., the "flux control coefficient." This represents the relationship between the fractional change in the metabolic pathway flux caused by a fractional change in the flux of the enzyme in question.[8] In other words, how does inhibition of a specific component in a metabolic pathway affect the overall activity of that pathway? In minimal terms, the control coefficient can be approximated from the slope of the initial portion of the threshold curve, as shown in Fig. 1.

Experimentally, respiratory thresholds and control coefficients are determined by measuring the effects of a specific inhibitor (e.g., cyanide for cytochrome *c* oxidase), both on the isolated activity of a respiratory complex and on the overall respiration rate.[7] By combining the *y* axes of these two graphs at shared concentrations of inhibitor, a threshold curve is constructed showing respiration as a function of the respiratory complex activity. This is shown theoretically in Fig. 2. For clarity, the *x* axis of the threshold curve is often inverted to show respiration as a function of complex inhibition rather than complex activity. This article

[8] A. K. Groen, R. J. Wanders, H. V. Westerhoff, M. R. Van der Meer, and J. M. Tager, *J. Biol. Chem.* **257,** 2754 (1982).

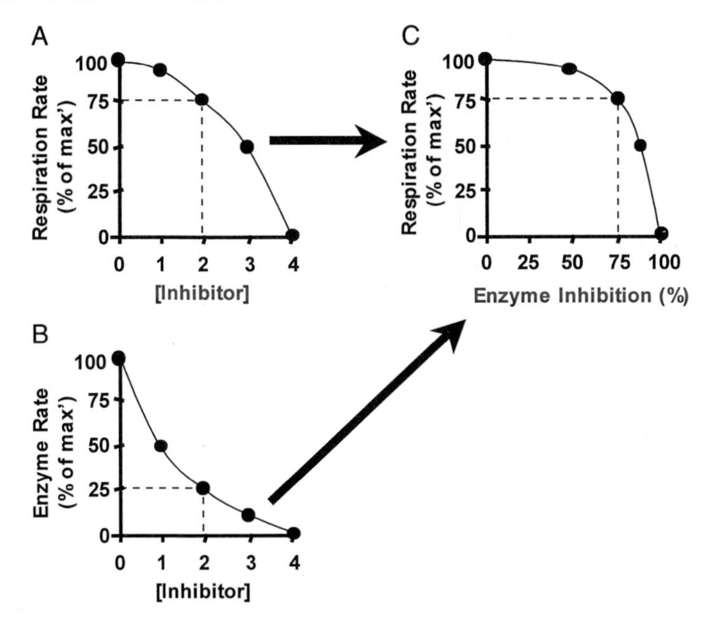

FIG. 2. Construction of threshold curves from inhibitor titration curves. Responses of respiration rate (A) and enzyme rate (B) to the inhibitor are determined. The *y* axes of these graphs (i.e., enzyme rate and respiration rate) are then coplotted at corresponding values of inhibitor concentration to yield the threshold curve (C). Calculation of an example point on the curve is shown by the dotted lines: 2 units of inhibitor results in a 25% inhibition of respiration rate (A) and a 75% inhibition of the enzyme (B). Thus, 75% enzyme inhibition causes a 25% respiration inhibition (C).

outlines developments to this methodology to measure the control over mitochondrial function by NO$^{\bullet}$. These methods are discussed in the context of several important experimental considerations that arise when working with and measuring NO$^{\bullet}$.

Isolation of Mitochondria

Several reliable methods exist for mitochondrial isolation. While a brief method for rat liver mitochondria is included here for completeness, we recommend choosing an isolation procedure specific for the tissue of interest (e.g. *Methods in Enzymology,* Vol. 55). All steps are performed at 4°. Liver tissue from a single animal (male, 250 g body mass) is chopped finely with scissors in 50 ml of isolation buffer (see Table I) and is then rinsed several times to remove blood. Chopped tissue is homogenized in 40 ml of buffer, with 12 strokes of the loose pestle in a glass:glass Dounce homogenizer (e.g., Model 885300-0040 from Kontes Glass,

TABLE I
COMPOSITION OF BUFFERS USED IN THE PREPARATION OF MITOCHONDRIA AND MEASUREMENT OF NO$^{\bullet}$ RESPIRATORY THRESHOLDS[a]

Solution	Composition, notes
Mitochondrial isolation buffer	Liver: 250 m*M* sucrose, 10 m*M* Tris–HCl, 1 m*M* EGTA Heart: 300 m*M* sucrose, 20 m*M* Tris–HCl, 2 m*M* EGTA Both pH 7.4 at 4°. Store at 4° for up to 3 days
Mitochondrial respiration buffer	100 m*M* KCl, 25 m*M* sucrose, 10 m*M* HEPES, 5 m*M* $MgCl_2$, 5 m*M* KH_2PO_4, 1 m*M* EGTA, pH 7.3, at 37°. Store at −20°
Respiratory substrates and ADP	All 0.5 *M* (except TMPD, 1 m*M*) in respiration buffer. Heat gently to dissolve, pH 7. Store at −20°
NONOate NO$^{\bullet}$ donors	100–500 m*M* in 10 m*M* NaOH. Store at −20°
FCCP	1 m*M* stock in 100% ethanol. Store at −20°
DTPA	1 m*M* stock solution in water. Store at 4°. Add to incubation buffers at 10–100 μM final concentration

[a] Storage at −20° for no more than 1 month.

Vineland, NJ). The homogenate is diluted to 80 ml with buffer, split into 2 × 40-ml tubes, and centrifuged for 10 min at 3000*g*. Supernatants (including the salmon-pink layer above the pellet) are then centrifuged for 10 min at 10,000*g*. Supernatants from this spin are discarded, and the pellets are resuspended in 80 ml buffer and centrifuged at 10,000*g* for 10 min. The resulting pellets have an upper light brown layer (broken mitochondria plus microsomes), with a dark brown core. A pestle and 1–2 ml of buffer are used to gently separate the upper layers, which are discarded. The dark brown pellets are then combined in a single tube, resuspended in 80 ml of buffer, and centrifuged at 10,000*g* as described earlier. The resulting pellets are finally resuspended in ~2 ml buffer and retained on ice until use (within 3–4 hr). Heart mitochondria can be isolated using a similar protocol, but with a modified isolation buffer (Table I) and the use of a high-speed mechanical disruption device (e.g., Polytron) to homogenize the tissue. The mitochondrial protein concentration is determined using the Folin-phenol reagent against a standard curve constructed using bovine serum albumin.[9]

Equipment Considerations

The essential apparatus required to measure NO$^{\bullet}$ effects on mitochondrial function is a combined O_2 electrode and NO$^{\bullet}$ electrode chamber (Fig. 3). A cylindrical electrode chamber made of Perspex holds a volume of 1 ml and is water jacketed to maintain the sample at 37° via a circulating water bath. The contents are

[9] O. H. Lowry, N. J. Rosebrough, A. L. Farr, and R. J. Randall, *J. Biol. Chem.* **193,** 265 (1951).

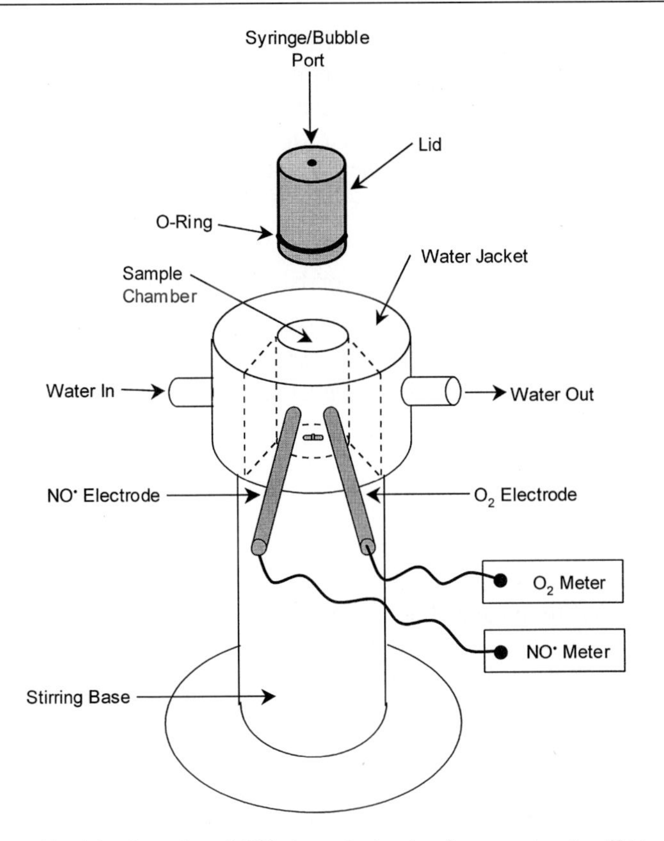

FIG. 3. Combined O_2 electrode and NO$^{\bullet}$ electrode chamber for measuring the effects of NO$^{\bullet}$ on mitochondrial and cellular respiration. For full description, see text.

stirred by a magnetic or electrostatic stirring device with a glass or Teflon-coated flea (Rank Brothers, Bottisham, Cambridge, UK). While commercial apparatus are available, (e.g., WPI, Sarasota, FL), custom construction can also be used to accommodate specific experimental requirements.

A segment of the water jacket is solid Perspex, with a hole drilled from the outside to the center of the chamber, to house a Clark-type O_2 electrode (Instech, Plymouth Meeting, PA). An adjacent hole is also drilled to house a Clark-type NO$^{\bullet}$ electrode. Those made by both WPI (Sarasota, FL) and Harvard Apparatus (Holliston, MA) offer a good sensitivity and signal : noise ratio. The electrodes are positioned sufficiently high in the chamber to avoid contact with the magnetic stirring flea. The NO$^{\bullet}$ and O_2 sensors lead to their respective meters, and data are recorded by a digital data-recording device (Dataq, Akron, OH) connected to a PC. The chamber is capped by a Perspex lid containing an O-ring seal, which slides inside the chamber to meet the sample, leaving no headspace. The importance of this is discussed later. The lid contains a port for the exit of air bubbles and sample injection via Hamilton syringes. The underside of the lid is beveled to direct air bubbles upward toward the hole (not shown). If ambient electrical noise is high, the apparatus (minus the electrode meters and data acquisition device) should be housed in a well-earthed Faraday cage.

Preparing NO• Stock Solutions for Electrode Calibration

When measuring any parameter affected by NO•, it is important to calibrate the NO• electrode accurately using a stock solution of pure NO• in water. Such a solution is prepared by bubbling commercial NO• gas through a solid NaOH trap, followed by a 10 *M* NaOH trap (to remove nitrite), and then finally through a glass bulb containing the deionized water. The entire apparatus is purged with argon for 20 min prior to bubbling with NO•. Typically, 30 min of NO• bubbling is required to saturate the water, at the end of which the bulb is sealed and stored in the dark. NO•-saturated water (~1.5–2 m*M*) is withdrawn via a rubber septum, and its concentration is monitored spectrophotometrically using oxyhemoglobin. In this case, NO• from the bulb is added to oxyHb (30 μM in heme) in a cuvette, and the spectrum between 500 and 700 nm is measured after each addition. OxyHb is oxidized rapidly to metHb by NO• in a 1 : 1 ratio (NO• : heme). The concentration of NO• is determined by calculating the loss of oxyHb by measuring the absorbance at 577 nm and converting to oxyHb concentration using $\varepsilon_{577\ \mathrm{nm}} = 14.6\ \mathrm{m}M^{-1}\mathrm{cm}^{-1}$. A fresh bulb is made when the concentration of NO• decays below 90% of its initial level (~1–2 weeks). Readers are recommended to consult Part A of Nitric Oxide (Vol. 268 of *Methods in Enzymology*) for comprehensive methods on NO• handling and measurement.

NO• Electrode Calibration

To calibrate the NO• electrode, the chamber is filled with the mitochondrial respiration buffer (Table I), and N_2 gas is blown over the chamber to decrease the O_2 concentration to <10% saturation. The lid is then fitted and air bubbles are eliminated. Once a stable baseline reading is obtained from the NO• electrode, bolus additions of saturated NO• solution are injected into the chamber using a gas-tight syringe. Three to five additions of NO• are made over a final calculated concentration range of 0.5–5 μM. Typical NO• decay traces for bolus additions are shown in Fig. 4A (solid trace).

When bolus NO• is added to the chamber, a rapid increase in the electrode signal occurs. The magnitude and slope of this increase are dependent on a combination of the mixing speed within the chamber, the experimental conditions (e.g., buffer composition and temperature), and the age and condition of the electrode itself.

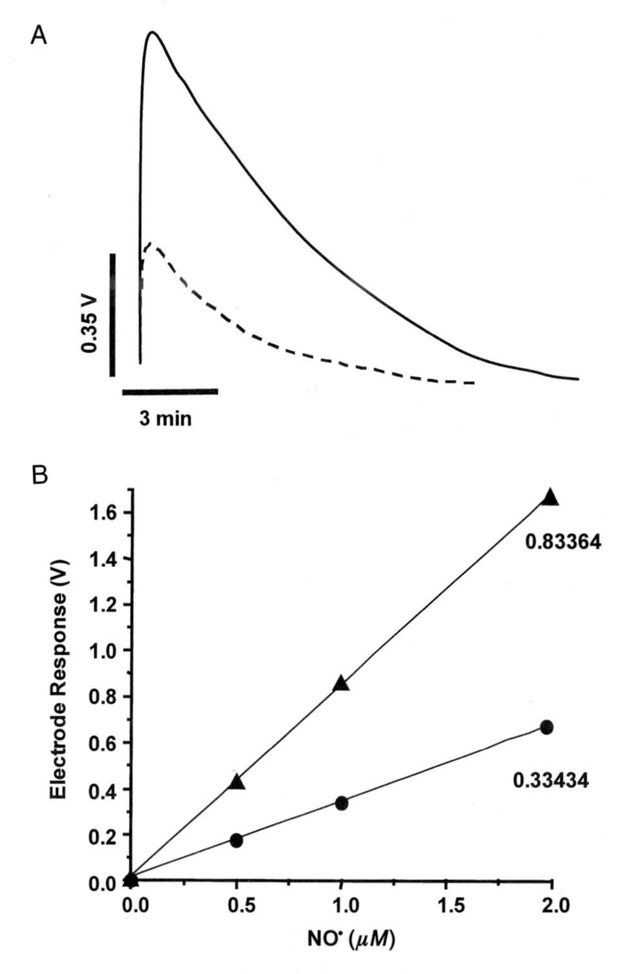

FIG. 4. Calibration of the NO˙ electrode and effects of O_2. (A) Typical NO˙ decay traces recorded in the NO˙ electrode chamber containing mitochondrial respiration buffer (Table I) following the bolus addition of 1 μM NO˙ from a stock solution (see methods), with the lid closed. Solid line: at ~20% O_2 saturation. Dotted line: at ~80% O_2 saturation. The *y* axis is the NO˙ electrode response in volts. (B) The electrode response (volts) as a function of added NO˙ concentration at nominally high (80%, circles) and low (20%, triangles) O_2 tension. Numbers alongside the lines are slopes, in volts/μmol NO˙. Traces and data are representative of at least four identical experiments.

The signal increase is then followed by a much slower decrease in the signal, signifying NO˙ decay in the chamber due to the reaction with oxygen dissolved in the buffers. The maximal value of the electrode signal before decay occurs is taken as the response of the electrode. Knowing the actual amount of NO˙ added from the bulb, a response factor is calculated in volts/μmol NO˙.

Effects of O_2 on $NO^{\cdot}$ Electrode Calibration

The calibration described earlier is performed at nominally "low O_2" to avoid the rapid reaction between $NO^{\cdot}$ and O_2.[10–12] In air-saturated buffer at 37°, O_2 is present at 200–240 μM, depending on buffer composition. Hence at nominally "high O_2" (air-saturated buffer), a given concentration of added $NO^{\cdot}$ produces a much smaller electrode signal than the same amount of $NO^{\cdot}$ added at low O_2 ($<25\ \mu M$). This difference in electrode response at different O_2 tensions (Fig. 4A, dotted vs solid trace) is highly reproducible, such that a "correction factor" can be calculated to compare signals at two different O_2 tensions. For example, a signal produced by $NO^{\cdot}$ at 20% O_2 saturation is 2.5 times greater than the signal from the same amount of $NO^{\cdot}$ added at 80% O_2. This is seen clearly in Fig. 4B, showing the electrode response (volts/μmol $NO^{\cdot}$) at two different O_2 tensions.

The importance of a tight-fitting lid for the $NO^{\cdot}$ electrode chamber is emphasized by data in Fig. 5A, which shows the decay of $NO^{\cdot}$ in mitochondrial respiration buffer with and without the lid on the chamber. The loss of $NO^{\cdot}$ to the headspace is thus a significant source of apparent $NO^{\cdot}$ consumption, and care should be taken to eliminate air bubbles from the chamber, as even a small bubble (~10 μl) can represent a significant $NO^{\cdot}$ sink, as well as a source of O_2 (which is important if O_2 consumption is also being measured).

Thus, by calibrating the electrode at low O_2, to gain a true value of volts/μmol $NO^{\cdot}$, and then again at high O_2 to calculate the difference factor, it is possible to determine the true concentration of $NO^{\cdot}$ that is present in the chamber, even when measurements are made at high O_2 tensions. Finally, it is important to calibrate the electrode frequently in this manner, as this relationship may change from experiment to experiment, depending on the age and condition of the electrode.

$NO^{\cdot}$ Release from $NO^{\cdot}$ Donor Compounds

One way to measure the effect of various concentrations of $NO^{\cdot}$ on both mitochondria and cells is the use of $NO^{\cdot}$ donor compounds. An example is the NONOates, a series of compounds that are stable at alkaline pH but degrade at neutral pH to release $NO^{\cdot}$, with half-lives ranging from 2 min [(*Z*)-1-(*N,N*-diethylamino) diazen-1-ium-1,2-diolate (DEA-NONOate)] to several hours [(*Z*)-1-*N*-(2-aminoethyl)-*N*-(2-aminoethyl)amino] diazen-1-ium-1,2-diolate (DETA-NONOate)].[13] An important property of these compounds is that they release $NO^{\cdot}$

[10] S. Shiva, P. S. Brookes, R. P. Patel, P. G. Anderson, and V. M. Darley-Usmar, *Proc. Natl. Acad. Sci. U.S.A.* **98,** 7212 (2001).

[11] D. D. Thomas, X. Liu, S. P. Kantrow, and J. R. Lancaster, Jr., *Proc. Natl. Acad. Sci. U.S.A.* **98,** 355 (2001).

[12] X. Liu, M. J. S. Miller, M. S. Joshi, D. D. Thomas, and J. R. Lancaster, Jr., *Proc. Natl. Acad. Sci. U.S.A.* **95,** 2175 (1998).

[13] L. K. Keefer, R. W. Nims, K. M. Davies, and D. A. Wink, *Methods Enzymol.* **268,** 281 (1996).

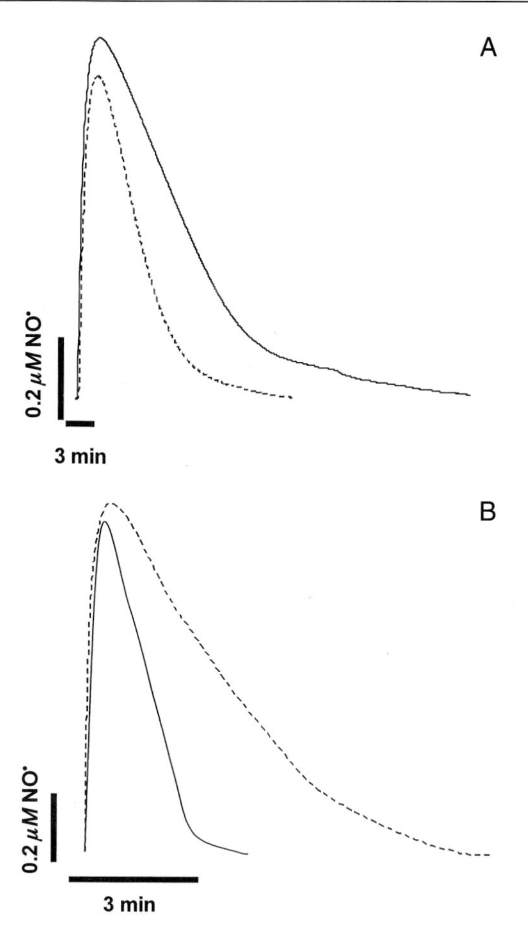

FIG. 5. Effects of chamber lid closure and light/dark on NO• decay in buffer. (A) Typical NO• decay traces recorded in the NO• electrode chamber containing mitochondrial respiration buffer (Table I) following the bolus addition of 1 μM NO• from a stock solution (see methods), with the lid closed (solid line) or open to the atmosphere (dotted line). (B) Decay traces for 1 μM NO•, recorded as just described, in the NO• electrode chamber containing DMEM with the lid on, at ambient laboratory light levels (solid line) or in darkness (dotted line). Traces are representative of at least four identical experiments.

independently of the biological system being studied. This is not the case for other NO• donors, e.g., the *S*-nitrosothiols, which often require the presence of specific biological factors for NO• release to occur.[14] The concentration and exact type of NONOate used should be varied to produce an NO• release profile tailored to

[14] R. P. Patel, J. McAndrew, H. Sellak, C. R. White, H. Jo, B. A. Freeman, and V. M. Darley-Usmar, *Biochim. Biophys. Acta* **1411,** 385 (1999).

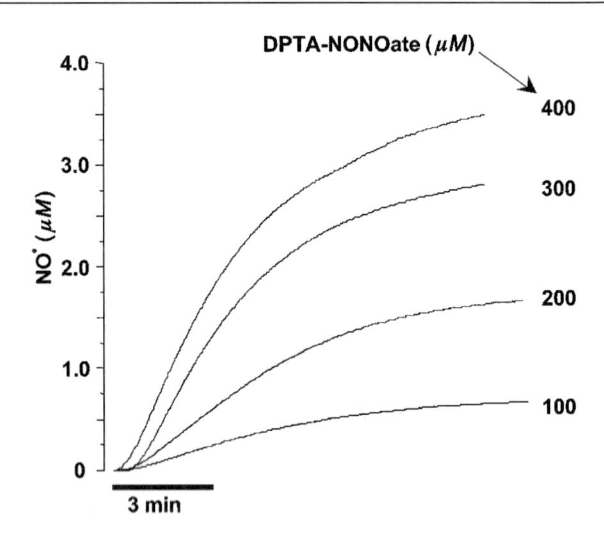

FIG. 6. Release of NO˙ from DPTA-NONOate in DMEM. Traces show NO˙ released from the indicated amounts of DPTA-NONOate, added from a 100 m*M* stock solution prepared in 10 m*M* NaOH. Recordings were made at 100% O_2 saturation, and the electrode was calibrated using authentic NO˙ as detailed in the text. Traces are representative of at least four identical experiments.

meet the experimental requirements. A typical NO˙ release profile from (*Z*)-1-[*N*-3-aminopropyl)-*N*-(3-aminopropyl) amino] diazen-1-ium-1,2-diolate (DTPA-NONOate) is shown in Fig. 6 and consists of an initial positive slope, followed by a flattening out to a steady-state plateau when the rate of NO˙ consumption equals the rate of NO˙ production. Both the initial rate of NO˙ release and the steady-state level of NO˙ are dependent on the type and concentration of NONOate used. A final advantage of NONOates is the ability to monitor their decomposition spectrophotometrically at 251 nm to obtain information on half-life and yields of NO˙ ($\varepsilon = 7680\ M^{-1}\text{cm}^{-1}$ for DETA-NONOate and $7640\ M^{-1}\text{cm}^{-1}$ DPTA-NONOate).

NO˙ Threshold Measurement

As detailed in the Introduction, the experimental derivation of mitochondrial respiratory thresholds consists essentially of two measurements: the effects of the inhibitor on an individual respiratory complex and the concomitant effects of the inhibitor on respiration. Procedures for measuring respiratory thresholds using commonly available pharmacological inhibitors are described in detail elsewhere.[7] These include rotenone for complex I, thenoyl-trifluoroacetone for complex II, myxothiazol or antimycin for complex III, and cyanide for complex IV.

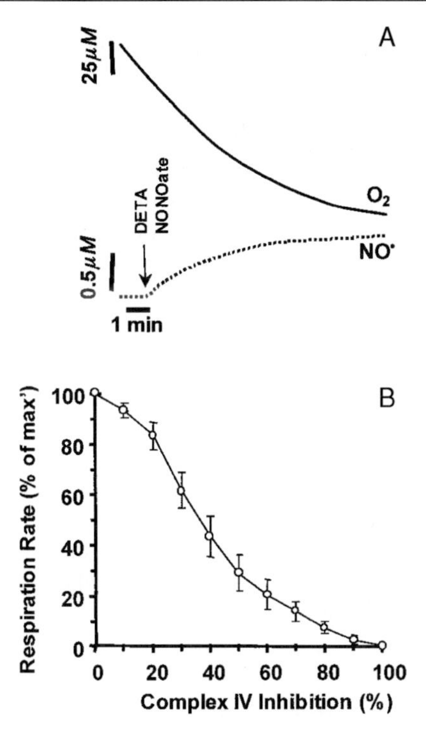

FIG. 7. Complex IV respiratory threshold measured using NO• in rat heart mitochondria. (A) Rat heart mitochondria were incubated in the electrode chamber in respiration buffer at 1 mg/ml in the presence of glutamate, malate, and ADP (state 3 respiration, see text). At the indicated time, DETA-NONOate (0.2 m*M*) was added, and both NO• release and O_2 consumption were monitored. Representative NO• and O_2 traces are shown. The experiment was then repeated in the presence of ascorbate/TMPD/FCCP (to measure isolated complex IV activity), and the threshold curve (B) was constructed from the combined NO• titration curves of respiration and enzyme activity. Data are means ± SEM ($N = 6$).

To measure the NO• threshold of cytochrome *c* oxidase, mitochondria are incubated in the electrode chamber in respiration buffer (Table I) at 1 mg protein/ml. Complex I-linked substrates (10 m*M* glutamate plus 2.5 m*M* malate) and ADP (200 μ*M*) are added to initiate state 3 respiration. After a stable respiration rate is obtained and before O_2 saturation drops below 80% saturation, DETA-NONOate (0.3–1 m*M*) is added, resulting in the release of NO• and a progressive inhibition of respiration. A typical combined O_2/NO• electrode trace is shown in Fig. 7A, with the addition of the NO• donor indicated by the arrow. The O_2 consumption rate (nmol O_2/min/mg protein) is determined at 1-min intervals along the trace. At these same time points, a line is dropped to the NO• trace, and NO• concentration is determined. In this way, a plot of NO• concentration vs respiration can be constructed.

Because cytochrome *c* oxidase is the O_2-consuming enzyme of mitochondria, its activity can be assayed independently of the respiratory chain. This is achieved by replacing the respiratory substrates and ADP in the medium with ascorbate (1 m*M*), tetramethyl phenylenediamine (TMPD, 50 μM), and carbonyl cyanide 4-trifluoromethoxyphenylhydrazine (FCCP, 1 μM). Ascorbate plus TMPD is a reducing system that directly feed electrons into cytochrome *c* oxidase (via cytochrome *c*), and FCCP is an uncoupler that dissipates the mitochondrial proton gradient, so allowing the H^+ pump of cytochrome *c* oxidase to function at maximal activity by removing feedback inhibition by the H^+ gradient. DETA-NONOate is then added to the chamber as before, and both $NO^{\bullet}$ concentration and O_2 consumption rate (cytochrome *c* oxidase activity) are monitored. In the same manner as for the respiration measurement, a plot of $NO^{\bullet}$ concentration vs cytochrome *c* oxidase activity can then be constructed. Finally, the threshold curve is constructed by combining the *y* axes of the two $NO^{\bullet}$ titration curves at corresponding concentrations of $NO^{\bullet}$, yielding a plot of cytochrome *c* oxidase activity vs respiration rate (Fig. 7B). The power of this technique is in the ability to construct a complete threshold curve from only two separate mitochondrial incubations, as a separate incubation is not required for each concentration of inhibitor (as would be the case, e.g., if using cyanide as the inhibitor).

Measurement of $NO^{\bullet}$ Thresholds in Cells

The inhibition of respiration by $NO^{\bullet}$ can also be measured in intact cells using the combined $NO^{\bullet}$ electrode/O_2 electrode/$NO^{\bullet}$ donor approach outlined earlier. While this is not a true respiratory complex threshold, the importance of measuring the dose response of cell respiration to $NO^{\bullet}$ should be stressed, as it can vary greatly under both physiological and pathological conditions.[11] A physiological example is the respiratory state of mitochondria inside cells. In isolated mitochondria, it has been shown that $NO^{\bullet}$ is a more potent inhibitor of respiration in state 3 (ATP generating) than in state 4 (without ADP).[15] However, respiratory states do not strictly exist within cells, and mitochondria are thought to exist in "state 3.5." Thus, cellular activity and ATP consumption may have significant bearing on the control of respiration by $NO^{\bullet}$. In addition, variations in the number of mitochondria per cell, the amount and location of $NO^{\bullet}$ synthase isoforms, and the presence of species reactive toward $NO^{\bullet}$ (e.g., superoxide, myoglobin) may all affect $NO^{\bullet}$ delivery to the mitochondrion and thus affect the control of respiration by $NO^{\bullet}$.[10] From a pathological perspective, we have shown that the mitochondrial $NO^{\bullet}$)–cytochrome *c* oxidase signaling pathway is perturbed in cardiac

[15] V. Borutaite and G. C. Brown, *Biochem. J.* **315,** 295 (1996).

hypertrophy, with hypertrophied myocytes exhibiting a greater sensitivity to both exogenously and endogenously generated NO•.[6,16]

When measuring NO• effects in cells, several complications need to be considered. One example is the consumption of NO• by components of the cell culture media. Such consumption would decrease NO• bioavailability to the cells, and therefore calibration of the NO• electrode should be performed in both the cell culture medium and in a non-NO•-consuming buffer such as phosphate-buffered saline, to quantify and correct for NO• consumption by the medium.

One mechanism by which cell culture components may contribute to NO• consumption is via the generation of ROS. It has been shown that this can be mediated by riboflavin (a common constituent in cell culture media) in the presence of light.[17] Therefore, it is important to conduct experiments under subdued light, as emphasized by data in Fig. 5B showing that the decay of NO• is much steeper in ambient light vs dark conditions. This phenomenon is especially prevalent in, but not limited to, media containing albumin or proteins capable of forming nitrosothiols, as NO• in its reactions with O_2 can form nitrosating agents, leading to *S*-nitrosothiol formation, which then decay depending on experimental conditions. Because some of the most common mechanisms of NO• consumption and *S*-nitrosothiol decay are dependent on metal ions, it is also recommended that a metal chelator such as diethylaminepentaacetic acid (DTPA) be included in the medium to limit such reactions.[14]

The reaction between NO• and O_2 also limits the ability to measure the effect of NO• on cellular respiration.[11,12] While physiological O_2 tension is estimated at 5–20 μM, most tissue culture systems utilize atmospheric O_2 levels. It is not feasible to measure respiration at physiologically low O_2 tensions, as actively respiring cells would consume all of the O_2 in the respiration chamber within a short time period. One solution is to measure respiration in an "Oxystat" type of apparatus, in which a low O_2 tension is maintained by gas replenishment using feedback-regulated gas flow controllers.[18]

Another solution to this problem is to vary the level of NONOate donor added to obtain inhibition of respiration at different O_2 tensions. As shown in Fig. 8, the addition of 2 m*M* DETA-NONOate at 90% O_2 saturation results in a high rate of NO• release (trace iii), and an NO• response curve that is right shifted (Fig. 8B, open diamonds), as inhibition takes place almost entirely at high O_2 tension, where O_2 competes for NO• binding. In contrast, the addition of 0.6 m*M* DETA-NONOate at the same 90% O_2 saturation level (trace i) results in a much slower rate of NO• release, such that the onset of inhibition of respiration is delayed until significant

[16] L. Dai, P. S. Brookes, V. M. Darley-Usmar, and P. G. Anderson, *Am. J. Physiol. Heart Circ. Physiol.* **281,** H2261 (2001).

[17] A. Grzelak, B. Rychlik, and G. Bartosz, *Free Radic. Biol. Med.* **30,** 1418 (2001).

[18] R. P. Cole, P. C. Sukanek, J. B. Wittenberg, and B. A. Wittenberg, *J. Appl. Physiol.* **53,** 1116 (1982).

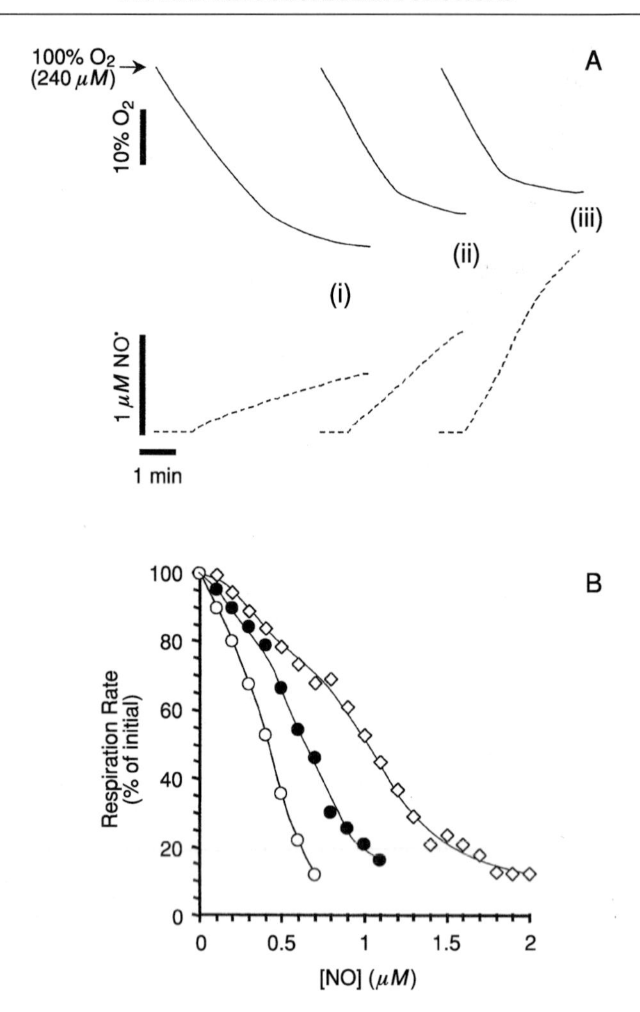

FIG. 8. Response of cell respiration to NO• and the effects of added NONOate level. Isolated adult rat ventricular cardiac myocytes were incubated in the combined electrode chamber in Krebs–Henseleit buffer plus 1 μM FCCP and 1 mM Ca^{2+}. (A) Representative O_2 consumption traces (solid lines) and NO• release traces (dotted lines) for (i) 0.6 mM, (ii) 1 mM, and (iii) 2 mM added DETA-NONOate. (B) Corresponding NO• titration curves calculated from data in (A). ○, 0.6 mM; ●, 1 mM; ◇, 2 mM DETA-NONOate. Traces and data are representative of at least five identical experiments.

O_2 consumption has occurred. This results in an NO• response curve that is left shifted, as the inhibition takes place at a lower O_2 saturation, whereupon NO• is a more potent inhibitor of respiration (Fig. 8B, open circles).[1,2] These results serve to highlight the effects of O_2 tension on NO• inhibition of mitochondrial respiration.

Summary and Conclusions

The methods described serve to illustrate not only the importance of measuring NO• thresholds in mitochondria and cells, but also highlight many of the considerations that must be taken when making such measurements. These include the presence of O_2, the source of NO•, the precise design of the apparatus, and experimental conditions, such as buffer composition, light level, and temperature. When all these factors are considered, accurate interpretations of the role of NO• in controlling mitochondrial function can be made. Such experiments can provide insight into the intricate nature of the interaction between NO• and mitochondria and the variations in this system that can occur under both physiological and pathological conditions.

Acknowledgments

P.S.B. is funded by the American Heart Association's southeast affiliate (Grant-in-Aid 0160174B) and by UAB's Clinical Nutrition Center. S.S. is in receipt of an NIH postgraduate training fellowship. R.P.P. is funded by the Parker B. Francis Foundation. V.D.U. is funded by the National Institutes of Health (AA12613, HL58031, ES10167).

[29] Determination of Nitric Oxide-Induced Effects on Tissue Levels of Glutathione and Mitochondrial Membrane Potential

By TERESA L. WRIGHT, CHUN-QI LI, LAURA J. TRUDEL, GERALD N. WOGAN, and STEVEN R. TANNENBAUM

Introduction

Nitric oxide (NO•) plays a major role in physiology through its regulation of physiological processes, such as vasodilation and neurotransmission, and also as a cytostatic/cytotoxic agent.[1,2] Three major processes govern its fate in biological systems: diffusion, autooxidation to form nitrous anhydride (N_2O_3), and reaction with superoxide to form peroxynitrite.[3] Given the complex chemistry involved, a multitude of subsequent reactions are possible, which are driven by local concentrations of nitric oxide, molecular oxygen, and superoxide. NO• diffuses freely

[1] D. A. Wink and J. B. Mitchell, *Free Radic. Biol. Med.* **25,** 434 (1998).
[2] J. S. Beckman and W. H. Koppenol, *Am. J. Physiol.* **271,** C1424 (1996).
[3] B. Chen, M. Keshive, and W. M. Deen, *Biophys. J.* **75,** 745 (1998).

0076-6879/02 $35.00

into cells and reacts directly with Fe–S protein centers, tyrosyl radicals, and thiols of proteins. N_2O_3 is a powerful nitrosating agent that causes DNA deamination and also nitrosates sulfhydryl groups to form *S*-nitroso compounds or amines to form *N*-nitrosamines. Nitrosation of secondary amines can produce mutagenic or carcinogenic alkylating agents, and nitrosation of primary amines leads to the formation of diazonium ions, which are hydrolyzed readily.[4] Peroxynitrite is capable of oxidation and nitration of biological macromolecules, including DNA and proteins.[5]

Tissue Glutathione Levels Associated with $NO^{\bullet}$ Overproduction in SJL Mice

$NO^{\bullet}$-induced macromolecular damage causes multiple adverse cellular responses, including cell death or genotoxicity, depending on dose and exposure conditions.[6] The nature of $NO^{\bullet}$-induced cytotoxicity reflects the complexity of its chemistry, involving changes in DNA synthesis, mitochondrial function, cellular membrane integrity, cell cycle progression, and induction of necrotic and apoptotic cell death. Among many different modulating factors that can influence the cytotoxic effects of $NO^{\bullet}$, intracellular glutathione and other antioxidants are of primary importance. Glutathione (γ-glutamylcysteinylglycine, GSH) is a tripeptide thiol occurring at intracellular concentrations ranging from 0.5 to 10 m*M*.[7] GSH creates a reducing milieu in the cell and is involved in enzymatic and nonenzymatic protection against reactive species that react directly with reduced glutathione (GSH) to produce oxidized glutathione (GSSG). GSSG is recycled to GSH by glutathione reductase, coupled with the oxidation of NADPH. Nitric oxide and its reactive products undergo a variety of reactions with glutathione, resulting in the formation of GSSG and *S*-nitrosoglutathione (GSNO), as well as other stable and unstable products.[8] GSNO can also be formed by the reaction of N_2O_3 with GSH or by the reaction of $NO^{\bullet}$ with glutathione radical ($GS^{\bullet}$). GSNO can react with GSH to release $NO^{\bullet}$, thus providing a transport mechanism for $NO^{\bullet}$.[9] Peroxynitrite can oxidize GSH to form GSSG.[10]

[4] S. R. Tannenbaum, S. Tamir, T. Derojas-Walker, and J. S. Wishnok, *in* "Nitrosamines and Related N-Nitroso Compounds" (R. N. Loeppky and C. J. Michejda, eds.), Vol. 553, p. 120. American Chemical Society, Washington, DC, 1994.

[5] W. A. Pryor and G. L. Squadrito, *Am. J. Physiol.* **268,** L699 (1995).

[6] S. Tamir, R. S. Lewis, T. de Rojas Walker, W. M. Deen, J. S. Wishnok, and S. R. Tannenbaum, *Chem. Res. Toxicol.* **6,** 895 (1993).

[7] M. E. Anderson, *Chem. Biol. Interact.* **111-112,** 1 (1998).

[8] S. P. Singh, J. S. Wishnok, M. Keshive, W. M. Deen, and S. R. Tannenbaum, *Proc. Natl. Acad. Sci. U.S.A.* **93,** 14428 (1996).

[9] N. Hogg, R. J. Singh, and B. Kalyanaraman, *FEBS Lett.* **382,** 223 (1996).

[10] R. Radi, J. S. Beckman, K. M. Bush, and B. A. Freeman, *Arch. Biochem. Biophys.* **288,** 481 (1991).

Glutathione is important in the detoxification of NO• *in vitro* and *in vivo*.[11] We previously related NO•-induced oxidative stress to glutathione metabolism in rodent and human cells in culture,[12] and describe in the following section alterations in mitochondrial function in the same human cells. We have also extended our investigation into intact animals, using an SJL mouse model shown previously to produce NO• at high levels *in vivo*.[13] NO• overproduction is induced in young SJL mice by growth of transplantable pre-B-cell lymphoma cells of the RcsX cell line.[14,15] Intraperitoneal injection of RcsX cells into SJL mice is followed by rapid growth of the tumor cells, activation of macrophages in the spleen, lymph nodes, and liver, and a 50-fold increase in NO• production over a period of 15 days. Major sites of NO• production are spleen and lymph nodes, and elevated levels of iNOS, apoptosis, 3-nitrotyrosine formation (a marker of peroxynitrite production), increased level of ethenoadenine, and increased mutation frequency have been demonstrated in these tissues.[13,15–17] We report here results of experiments to examine the effects of NO• overproduction on tissue levels of glutathione in SJL mice.

Experimental Design

Male SJL mice (Jackson Laboratories, Bar Harbor, ME) are maintained on a 12-hr light/dark cycle and housed in microisolator cages with free access to feed and water. Fourteen mice are fed a low nitrate control diet (AIN-76A, Bioserve, Frenchtown, NJ) to minimize the background rate of nitrate excretion. After 1 week (day 0), the drinking water of five mice is replaced with a 30 m*M* solution of *N*-methylarginine (NMA), a competitive inhibitor of iNOS,[17] and the drinking water of 9 mice is replaced with a 30 m*M* solution of ammonium acetate (AA). Two days later, five mice receiving NMA and five receiving AA are injected intraperitoneally with 1×10^7 cells of the RcsX transplantable lymphoma. To verify NO• overproduction in tumor-bearing animals, urine is collected from each group (SJL + AA, SJL/RcsX + NMA, SJL/RcsX + AA) on days 2 and 17 into tubes containing NaOH to inhibit bacterial growth. Urinary nitrate excretion is analyzed.[18] SJL mice bearing the RcsX tumor are found to excrete an increased level of nitrate

[11] M. W. Walker, M. T. Kinter, R. J. Roberts, and D. R. Spitz, *Pediatr. Res.* **37,** 41 (1995).

[12] S. Luperchio, S. Tamir, and S. R. Tannenbaum, *Free Radic. Biol. Med.* **21,** 513 (1996).

[13] A. Gal, S. Tamir, S. R. Tannenbaum, and G. N. Wogan, *Proc. Natl. Acad. Sci. U.S.A.* **93,** 11499 (1996).

[14] J. L. Lasky, N. M. Ponzio, and G. J. Thorbecke, *J. Immunol.* **140,** 679 (1988).

[15] A. Gal, S. Tamir, L. J. Kennedy, S. R. Tannenbaum, and G. N. Wogan, *Cancer Res.* **57,** 1823 (1997).

[16] J. Nair, A. Gal, S. Tamir, S. R. Tannenbaum, G. N. Wogan, and H. Bartsch, *Carcinogenesis* **19,** 2081 (1998).

[17] S. Tamir, T. deRojas-Walker, A. Gal, A. H. Weller, X. Li, J. G. Fox, G. N. Wogan, and S. R. Tannenbaum, *Cancer Res.* **55,** 4391 (1995).

[18] L. C. Green, D. A. Wagner, J. Glogowski, P. L. Skipper, J. S. Wishnok, and S. R. Tannenbaum, *Anal. Biochem.* **126,** 131 (1982).

in the urine, reflecting NO$^{\bullet}$ overproduction, and administration of NMA blocks this increase (data not shown). Animals are killed by CO_2 asphyxiation 15 days after injection of RcsX cells, and peripheral lymph nodes, spleen, and liver are removed for analysis.

Glutathione Analysis

The method of Anderson,[19] with modifications, is used to prepare tissue samples.[12] Briefly, tissues are removed and snap frozen rapidly. Prior to analysis, they are thawed, weighed, rinsed with water, blotted dry, and homogenized in 5% 5-sulfosalycilic acid (SSA, 10 volume/g wet tissue weight). Homogenates are centrifuged for 10 min at 20,000g at 4°, and then supernatants are extracted and diluted for analysis. Tissue concentrations (μmol/g wet tissue weight) of total glutathione (GSH + GSSG), reduced glutathione (GSH), and oxidized glutathione (GSSG) are determined according to a modification of the DTNB-GSSG reductase recycling assay,[12,20] as adapted for use in 96-well microtiter plates.[21] For glutathione measurement, 4 μl of 2-vinylpyridine (97% pure) and triethanolamine (to a final pH of 6 or 7) is added sequentially, with vortexing, to a 200-μl aliquot of supernatant for analysis of the GSSG content. Vinyl pyridine is added to derivatize GSH and prevent its participation in the recycling assay. (*Note:* 2-vinylpyridine is toxic and should be handled, following recommended safety procedures, in a fume hood.) The assay is carried out in 96-well plates in a reaction mixture containing 75 μl NADPH (387 μM in 0.1 M sodium phosphate/5 mM EDTA buffer, pH 7.5), 50 μl 5,5′-dithiobis(2-nitrobenzoic) acid (DTNB) (2.4 mM in 0.1 M sodium phosphate/ 5 mM EDTA buffer, pH 7.5), and a 25-μl sample or standard prepared by serial dilution to contain 1.95 to 125 ng GSSG/well. The reaction is initiated by the addition of 50 μl of a glutathione reductase solution (1.28 units/ml in 0.1 M sodium phosphate/5 mM EDTA buffer, pH 7.5) to each well. All samples and standards are analyzed in triplicate. The change in absorbance at 405 nm is monitored for 10 min with a Ceres 900 96-well microplate reader (Bio-Tek Instruments, Winooski, VT). A standard curve is constructed by plotting dA/dT vs ng GSH. The equation describing the line obtained is used to calculate the total glutathione content of the samples, and the amount of GSH is determined by subtracting GSSG from the total glutathione content.

Results

Urinary nitrate excretion verifies that NO$^{\bullet}$ production is strongly elevated in RcsX tumor-bearing mice and that the increase is abolished by the administration

[19] M. E. Anderson, *Methods Enzymol.* **113,** 548 (1985).
[20] F. Tietze, *Anal. Biochem.* **27,** 502 (1969).
[21] M. A. Baker, G. J. Cerniglia, and A. Zaman, *Anal. Biochem.* **190,** 360 (1990).

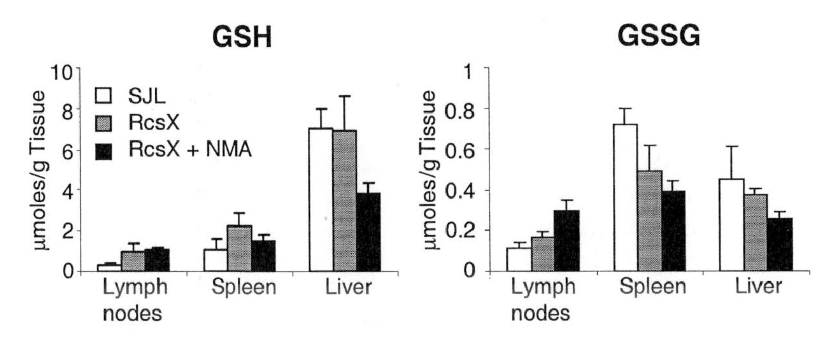

FIG. 1. Levels of reduced (GSH) and oxidized (GSSG) glutathione (mean ± SD) in lymph nodes, spleen, and liver of SJL mice given ammonium acetate (SJL), RcsX-bearing SJL mice given ammonium acetate (RcsX), and RcsX-bearing mice given *N*-methyl arginine (RcsX + NMA).

of NMA in the drinking water, as expected (data not shown). Tissue levels of reduced and oxidized glutathione in lymph nodes, spleen, and liver of animals in all treatment groups are summarized in Fig. 1. NO$^\bullet$ overproduction results in significantly ($P < 0.05$) elevated GSH levels in lymph nodes and spleen, but not in liver. Administration of NMA abolishes the increase in spleen and significantly reduces GSH levels in liver. GSSG levels are significantly lower in spleen and liver in tumor-bearing animals and still lower in tumor-bearing animals treated with NMA.

Increased GSH levels can be a result of *de novo* synthesis or an increased reduction of GSSG to GSH by glutathione reductase as a protective mechanism against oxidative stress. The increase in GSH in this *in vivo* model of NO$^\bullet$ production is probably due to *de novo* synthesis of GSH because the total level of glutathione is increased. This was confirmed by an assay of γ-GCS activity (data not shown). The RcsX cells may themselves have had a direct effect on glutathione metabolism. It can be seen in Fig. 1 that the administration of NMA did not cause glutathione levels in the lymph nodes to decrease, possibly due to an altered uptake of glutathione or an increase in glutathione reductase activity. The latter possibility is supported by the decline in GSSG levels in spleen and liver of tumor-bearing animals. Further enhancement of the decline in GSSG levels by NMA suggests a possible direct effect on glutathione reductase activity. Thus, glutathione appears to play a significant role in NO$^\bullet$-induced toxicity in the intact animal in a similar manner, as observed previously in cultured cells.

Mitochondrial Membrane Depolarization Induced by Nitric Oxide

Extensive evidence indicates that NO$^\bullet$ activates apoptosis in many cell types, whereas in others, including hepatocytes, it inhibits the induction of apoptosis by drugs or growth factor withdrawal.[22] The dose and manner in which NO$^{\bullet}$ is delivered can affect the apoptotic response significantly, and in many of the aforementioned studies, cells were exposed via donor drugs to total NO$^{\bullet}$ doses orders of magnitude higher than those estimated to occur *in vivo*.[23] Additionally, contributions of the drugs themselves or their decomposition products to apoptosis induction cannot be discounted, as their reactivity may be quite distinct from that of NO$^{\bullet}$.[24] In this investigation, we further characterize responses of cells exposed to NO$^{\bullet}$ gas through a delivery system in which delivery rate and total dose are quantified and controlled in order to circumvent these issues. In earlier studies, human lymphoblastoid TK6 cells were shown to respond differently and with greater sensitivity to NO$^{\bullet}$ than Chinese hamster ovary CHO-AA8 cells.[25] These differences were shown subsequently to be related to glutathione metabolism[12] in that TK6 cells exhibited lower capacity than CHO-AA8 cells to synthesize GSH *de novo*. We have extended these observations by characterizing apoptotic signaling pathways induced by NO$^{\bullet}$ in TK6 (wild-type p53) and WTK-1 (mutant p53) cells, two human lymphoblastoid lines used extensively in previous studies of mutagenesis and apoptosis.[26–28] NO$^{\bullet}$ is delivered into the medium of well-stirred cell suspensions by diffusion through Silastic tubing at constant rates comparable to those estimated to occur in inflamed tissues *in vivo*.[6]

Biochemical and cellular mechanisms through which NO$^{\bullet}$ induces apoptosis in these cells are incompletely understood, but involve a reduction in mitochondrial membrane potential (MMP), an event that follows the onset of the mitochondrial permeability transition. Evidence indicates that nuclear apoptosis is preceded by mitochondrial membrane depolarization and that mitochondria are important mediators of apoptosis.[29]

Cell Culture and NO$^{\bullet}$ Exposure

Human lymphoblastoid cells of the TK6 and WTK-1 lines are maintained in exponentially growing suspension culture at 37° in a humidified, 5% CO_2 atmosphere in RPMI 1640 medium supplemented with 10% heat-inactivated calf serum, 100 units/ml penicillin, 100 μg/ml streptomycin, and 2 m*M* L-glutamine. Stock cells are subcultured and maintained at a cell density not greater than

[22] B. Brune, A. von Knethen, and K. B. Sandau, *Eur. J. Pharmacol.* **351,** 261 (1998).

[23] R. P. Patel, J. McAndrew, H. Sellak, C. R. White, H. Jo, B. A. Freeman, and V. M. Darley-Usmar, *Biochim. Biophys. Acta* **1411,** 385 (1999).

[24] C. W. Tabor and H. Tabor, *Microbiol. Rev.* **49,** 81 (1985).

[25] S. Burney, S. Tamir, A. Gal, and S. R. Tannenbaum, *Nitric Oxide* **1,** 130 (1997).

[26] J. C. Zhuang, T. L. Wright, T. deRojas-Walker, S. R. Tannenbaum, and G. N. Wogan, *Environ. Mol. Mutagen.* **35,** 39 (2000).

[27] F. Xia, X. Wang, Y. H. Wang, N. M. Tsang, D. W. Yandell, K. T. Kelsey, and H. L. Liber, *Cancer Res.* **55,** 12 (1995).

[28] Y. Yu and J. B. Little, *Cancer Res.* **58,** 4277 (1998).

[29] Y. Shi, *Nature Struct. Biol.* **8,** 394 (2001).

1×10^6 cells/ml in 150-mm dishes during experiments. All cell culture reagents are purchased from BioWhittaker (Walkersville, MD).

Cells at a density of 4×10^5 cells/ml in 100 ml of medium without calf serum are exposed to $NO^{\bullet}$ through Silastic tubing (0.025 inch ID, 0.047-in OD, from Dow Corning, Midland, MI) in the delivery system described previously.[6] The polymer is freely permeable to $NO^{\bullet}$, which is delivered into stirred cell suspensions for 2 hr at rates controlled by tubing lengths of 5 to 30 cm. Cells exposed to argon gas through 30-cm tubing for 2 hr serve as negative controls. Total amounts of $NO^{\bullet}$ delivered under these conditions, expressed as nitrite plus nitrate concentrations, are quantified by automated analysis using the Griess reagent [*N*-(1-naphthy)ethylenediamine and sulfanilic acid] as described previously.[18] At the end of treatment, cells are collected by centrifugation at 1200 rpm for 6 min, washed once in PBS (pH 7.4), resuspended in fresh RPMI 1640 medium with 10% calf serum, and incubated at 37°.

Qualitative Detection of Mitochondrial Membrane Depolarization

For MMP analysis by fluorescence microscopy, the fluorescent probe 5,5′,6,6′-tetrachloro-1,1′,3,3′-tetraethylbenzimidazolylcarbocyanine iodide (JC-1, from Molecular Probes, Inc., Eugene, OR) is used. JC-1 is a lipophylic cationic dye that exhibits potential-dependent accumulation in mitochondria, indicated by a fluorescence emission shift from green (~530 nm) to red (~585 nm).[30] Consequently, mitochondrial depolarization is accompanied by a decrease in the red/green fluorescence intensity ratio. The potential-sensitive color shift is due to concentration-dependent formation of red fluorescent J aggregates. The ratio of green to red fluorescence is dependent only on the membrane potential and not on other factors, such as mitochondrial size, shape, and density. MMP changes can be detected qualitatively by fluorescence microscopy following JC-1 staining. A 2.5 m*M* stock solution of JC-1 is prepared in dimethyl sulfoxide (DMSO, Sigma) and stored in light-shielded vials at −20°. One-milliliter aliquots of TK6 and WTK-1 cell suspensions collected 8, 24, and 48 hr after $NO^{\bullet}$ treatment are placed into Eppendorf tubes and incubated with 10 μM JC-1 dye (adding 4 μl of 2.5 m*M* JC-1 stock solution in 1 ml of cell suspension) for 15 min at room temperature in the dark. Cells are washed three times in cold phosphate-buffered saline (PBS, GIBCO) and resuspended in 100 μl PBS. JC-1-stained cell suspensions are placed onto microscope slides, and MMP changes are qualitatively estimated immediately with a Nikon Eclipse E 600 photomicroscope equipped with epifluorescence optics and a 100-W mercury bulb. Objective lenses include Plan Fluor 10×/0.3 and Plan Fluor 40×/0.75. For visualizing the green fluorescence of the JC-1 monomer, any filter combinations (such as Zeiss' barrier 515-565, dichroic mirror FT 510 nm, and exciter 450–490 nm) used for fluorescein dye are adequate. Likewise, any rhodamine

[30] S. Salvioli, A. Ardizzoni, C. Franceschi, and A. Cossarizza, *FEBS Lett.* **411,** 77 (1997).

filter sets (such as Zeiss' barrier LP 590 nm, dichroic mirror FT 580 nm, and exciter BP 546/12 nm) are suitable for detecting the red fluorescence of J aggregates. However, to visualize green and red fluorescence simultaneously, a long-pass filter system (such as Zeiss' barrier LP 520 nm, dichroic mirror FT 510 nm, and exciter 450–490 nm) is desirable and is used in this study.

Quantitative Analysis of Mitochondrial Membrane Depolarization

Changes in MMP are quantified dynamically by flow cytometry after JC-1 staining based on the color shift from red to green caused by disaggregation accompanying membrane depolarization. Both colors are detected using filters mounted in the flow cytometer so that the green emission is analyzed in fluorescence channel 1 (FL-1) and the red emission in channel 2 (FL-2). Cell staining is performed as follows. Thirty minutes prior to cytometric analysis, JC-1 is added to the 1.5-ml aliquot of $NO^{\bullet}$-treated TK6 or WTK-1 cells in 5-ml polystyrene round-bottom tubes (Becton Dickinson) to a final concentration of 10 μM and incubated at 37°, 5% CO_2 atmosphere for 30 min. At the end of the incubation period, cells are centrifuged and washed three times in prechilled PBS and resuspended in 1 ml PBS. Five microliters of a propidium iodide solution (3 mg/ml in H_2O_2) is added to the cell suspension at a final concentration of 10 μg/ml 5 min prior to flow cytometry analysis. At designated times, fluorescence emission of 10,000 cells in each sample is quantified on an FL-1 (~530 nm) versus FL-2 (~585 nm) dot plot on a FACScan flow cytometer (Becton Dickinson) equipped with a single 488-nm argon laser. Data are converted to density plots and histogram plots using CellQuest software, and results are expressed as either the mean monomer fluorescence (green) alone or as the ratio of aggregate/monomer (red/green). TK6 and WTK-1 cells treated with argon gas or with 2.5 μM etoposide for 6 hr are used as negative and positive controls, respectively.

Results

When TK6 and WTK-1 cells are exposed to $NO^{\bullet}$ for 2 hr, the cumulative $NO^{\bullet}$ exposure doses and dose rates at varying tubing length are 390 μmol (533 n*M*/sec), 210 μmol (300 n*M*/sec), and 70 μmol (100 n*M*/sec) in the medium, respectively; no nitrite or nitrate is detectable in the medium of cells treated with argon gas. Qualitative evidence of mitochondrial depolarization is observed in TK6 and WTK-1 cells exposed to $NO^{\bullet}$ treatment. The loss of mitochondrial membrane potential in $NO^{\bullet}$-treated TK6 and WTK-1 cells is time dependent. JC-1 aggregates in normal cells with a high mitochondrial membrane potential appear red, whereas JC-1 monomers in damaged cells with the loss of mitochondrial membrane potential emit green in the fluorescence microscope (not shown). The loss of mitochondrial membrane potential induced by $NO^{\bullet}$ is quantified by flow cytometry following JC-1 staining. High FL-2 fluorescence (585 nm) corresponds to the aggregated form of JC-1 and is proportional to the intact mitochondrial membrane

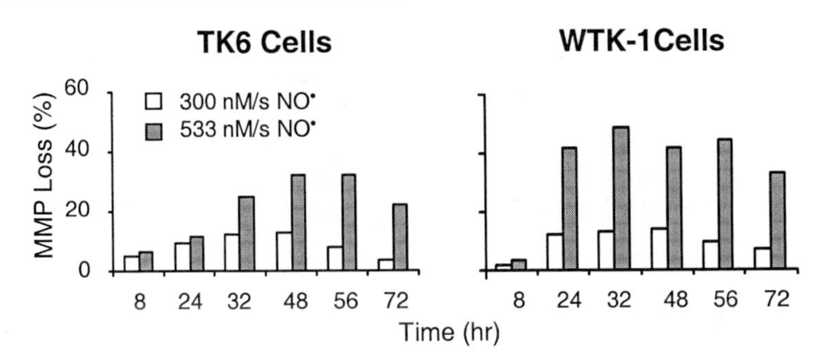

FIG. 2. Mitochondrial membrane depolarization in TK6 and WTK-1 cells treated with varying rates of NO• for 2 hr. Data are normalized to values from control cells treated with argon gas and represent mean values from two experiments, each done in duplicate. Standard deviations are 6 to 16% (not shown).

potential. NO•-induced loss of mitochondrial membrane potential leads to a reduction in FL-2 and a concomitant gain of cells that exhibit high FL-1 fluorescence (530 nm). As shown in Fig. 2, treating TK6 and WTK-1 cells with 300 n*M*/sec NO• for 2 hr (210 μmol total dose) results in slight MMP loss in both cell lines, whereas a dose-dependent loss is induced in cells treated with 533 n*M*/sec NO• for 2 hr (390 μmol NO•). MMP loss is detectable as early as 8 hr and becomes more evident at 48 hr in TK6 cells, with a maximum of 34%, and 32 hr in WTK-1 cells with a maximum of 54%. The magnitude of MMP loss is 1.3- to 3.2-fold higher in WTK-1 cells than in TK6 cells at corresponding time points between 24 and 72 hr.

The loss of mitochondrial membrane potential has been shown to occur in a variety of apoptotic model systems and has been suggested to result in the release of proapoptotic factors, such as apoptosis-inducing factor and cytochrome *c*.[22,29] Evidence showing the presence of mitochondrial NO• synthase and NO•[31,32] further suggests that NO• may modulate mitochondrial permeability transition and thus regulate apoptosis. Our results show that nanomolar/second exposure to NO• induces mitochondrial depolarization in a dose-dependent manner in both TK6 and WTK-1 cells, but of greater magnitude in WTK-1 cells. In contrast, MMP loss is associated with apoptosis in TK6 cells but not in WTK-1 cells (data not shown), suggesting that p53-independent apoptosis induced by NO• in WTK-1 cells is not mediated by mitochondria. Furthermore, whether the association of MMP loss with apoptosis in TK6 cells reflects a cause–effect relationship or independent events is uncertain. These findings are supported by a previous report,[33] which suggests that the release of cytochrome *c*, but not loss of MMP, is a required

[31] P. Ghafourifar, U. Schenk, S. D. Klein, and C. Richter, *J. Biol. Chem.* **274,** 31185 (1999).

[32] M. O. Lopez-Figueroa, C. Caamano, M. I. Morano, L. C. Ronn, H. Akil, and S. J. Watson, *Biochem. Biophys. Res. Commun.* **272,** 129 (2000).

[33] R. M. Kluck, E. Bossy-Wetzel, D. R. Green, and D. D. Newmeyer, *Science* **275,** 1132 (1997).

step for initiation of the cell death program, implying that these events may be cell type and inducer-specific features of apoptosis. It has been reported[34] that a physiological concentration of NO$^•$ (11 nM/sec) reversibly inhibits mitochondrial permeability transition pore opening, but supraphysiological release rates of NO$^•$ (>2 μM/sec) accelerate the process. In this context, the rates of NO$^•$ exposure (100–533 nM/sec) used in this study are in the range estimated to be produced *in vivo* in pathological conditions.

Acknowledgments

This publication was supported by Grant 5 P01 CA26731 from the National Cancer Institute. The contents are solely the responsibility of the authors and do not necessarily represent the official views of the National Cancer Institute.

[34] P. S. Brookes, E. P. Salinas, K. Darley-Usmar, J. P. Eiserich, B. A. Freeman, V. M. Darley-Usmar, and P. G. Anderson, *J. Biol. Chem.* **275,** 20474 (2000).

[30] Pharmacological Regulation of Mitochondrial Nitric Oxide Synthase

By ALBERTO BOVERIS, SILVIA LORES ARNAIZ, JUANITA BUSTAMANTE, SILVIA ALVAREZ, LAURA VALDEZ, ALEJANDRO D. BOVERIS, and ANA NAVARRO

Mitochondrial Production of Nitric Oxide

The production of nitric oxide (NO) was originally described by Ghafourifar and Richter[1] and by Giulivi *et al.*[2] in rat liver mitochondria. Later, NO production was reported in brain,[3] thymus,[4,5] and heart mitochondria.[6] The enzyme responsible for NO production in mitochondria is biochemically similar to other nitric oxide synthases (NOS): NADPH and arginine are utilized to produce NADP, citrulline, and NO. The mitochondrial enzyme, with the attributed physiological function of producing NO to regulate cytochrome oxidase activity and, consequently,

[1] P. Ghafourifar and C. Richter, *FEBS Lett.* **418,** 291 (1997).
[2] C. Giulivi, J. J. Poderoso, and A. Boveris, *J. Biol. Chem.* **273,** 11038 (1998).
[3] A. S. Lores, M. F. Coronel, and A. Boveris, *Nitric Oxide* **3,** 235 (1999).
[4] J. Bustamante, G. Bersier, M. Romero, R. A. Badin, and A. Boveris, *Arch. Biochem. Biophys.* **376,** 239 (2000).
[5] J. Bustamante, G. Bersier, R. Aron Badin, C. Cymeryng, A. Parodi, and A. Boveris, *Nitric Oxide* (in press).
[6] S. French, C. Giulivi, and R. S. Balaban, *Am. J. Physiol. Heart Circ. Physiol.* **280,** H2863 (2001).

0076-6879/02 $35.00

TABLE I
DIFFERENTIAL PROPERTIES OF MITOCHONDRIAL NITRIC OXIDE SYNTHASES AND CLASSICAL NITRIC OXIDE SYNTHASES

NOS type	Organ	MW	Expression	Ca^{2+} requirement	Reacts with antibody
mtNOS-1	Brain	145,000	Regulable	+	Anti-nNOS (C-terminal) Anti-nNOS (N-terminal)
mtNOS-2	Liver, thymus, kidney	130,000	Regulable	+	Anti-iNOS
nNOS (NOS-1)	Nerve tissue	161,000	Constitutive	+	Anti-nNOS (C-terminal) Anti-nNOS (N-terminal)
iNOS (NOS-2)	Macrophages	131,000	Inducible	−	Anti-iNOS
eNOS (NOS-3)	Endothelial cells	133,000	Constitutive	+	Anti-eNOS

mitochondrial and cellular respiration,[7–10] is termed mitochondrial nitric oxide synthase (mtNOS).[11] At present, two different proteins with the same enzymatic activity and physiological function have been recognized: liver mtNOS (130 kDa), which reacts with antibodies anti-iNOS,[12–14] and thymus and kidney mtNOS that show similar molecular mass and antibody reactivity. We call them mtNOS-2 for similarity with iNOS or NOS-2 in terms of antibody reactivity. At variance, brain mtNOS or mtNOS-1, for similarity with nNOS or NOS-1 again in terms of antibody reactivity, is an enzyme of 145 kDa that reacts with antibodies anti-nNOS directed against the amino acids of the C and of the N ends of the protein (Table I).

It was recognized that mtNOS activity is under hormonal regulation by thyroxine in the liver,[13] and that it can be modulated pharmacologically in brain, liver, kidney, and heart.[15]

[7] C. Giulivi, *Biochem. J.* **332,** 673 (1998).
[8] A. Boveris, L. E. Costa, J. J. Poderoso, M. C. Carreras, and E. Cadenas, *Ann. N.Y. Acad. Sci.* **899,** 121 (2000).
[9] P. Sarti, E. Lendaro, R. Ippoliti, A. Bellelli, P. A. Benedetti, and M. Brunori, *FASEB J.* **13,** 191 (1999).
[10] G. C. Brown, *FEBS Lett.* **369,** 136 (1995).
[11] T. E. Bates, A. Loesch, G. Burnstock, and J. B. Clark, *Biochem. Biophys. Res. Commun.* **213,** 869 (1995) and **218,** 40 (1996).
[12] A. Tatoyan and C. Giulivi, *J. Biol. Chem.* **273,** 11044 (1998).
[13] M. C. Carreras, J. G. Peralta, D. P. Converso, P. V. Finochietto, I. Rebagliati, A. A. Zaminovich, and J. J. Poderoso, *Am. J. Physiol. Heart Circ. Physiol.* **281,** H000 (2001).
[14] M. F. Lopez, B. S. Kristal, E. Chernokalskaya, A. Lazarev, A. I. Shestopalov, A. Bogdanova, and M. Robinson, *Electrophoresis* **21,** 3427 (2000).
[15] A. Boveris, A. S. Lores, S. Alvarez, L. E. Costa, and L. B. Valdez, *in* "Free Radicals in Chemistry, Biology and Medicine" (T. Yoshikawa, S. Toyokuni, Y. Yamamoto, and Y. Naito, eds.), p. 256. OICA International, London, 2000.

The main properties of the two mtNOS, and for comparative purposes of the three classical NOS (nNOS, iNOS, and eNOS), are given in Table I. For the three classical NOS enzymes, amino acid sequences have been described, and commercially available antibodies are used routinely for Western blots analysis and for immunohistochemical and electron microscopy detection.[16] The denomination constitutive, as different from inducible, is not fully appropriate, as there are numerous reports on the regulation of constitutive NOS activities.[17] It is worth noting that mtNOS-1 (145 kDa) coexists in neurons with the nNOS (161 kDa) of synaptosomes and cytosol. A similar situation occurs in liver and thymus. In both organs, mtNOS-2 (130 kDa) coexists with NOS isoenzymes in the endoplasmic reticulum or cytosol (133 kDa), which react with the antibody against eNOS (NOS-3).[5,13]

Measurement of mtNOS Activity

Both mtNOS-1 and mtNOS-2 are determined by the same assays, as they have the same biochemical properties. Both mitochondria and submitochondrial particles can be used. Reaction media are based either on 50 m*M* phosphate buffer or 0.23 *M* mannitol, 0.07 *M* sucrose, 30 m*M* Tris–MOPS (pH 7.4). Absolute enzyme requirements are 100–200 μM NADPH ($K_m = 15\ \mu M$),[16] 0.2–1.0 m*M* arginine ($K_m = 6$–$12\ \mu M$),[12] and 0.2–1 m*M* Ca^{2+}. Other components often added to the reaction medium are 1–4 μM Cu,Zn-superoxide dismutase, to prevent O_2^- reaction with NO; 0.1–1 μM catalase, to prevent H_2O_2 interference; 2–10 μM tetrahydrobiopterine, to increase activity up to 20%; 4 μM FAD and 0.1 μM FMN, to increase reaction rates up to 10%[12]; 0.1–0.3 μM calmodulin, with effects of up to 15% depending on [Ca^{2+}]; and 5–20 μM dithiothreitol, observed to have both positive and negative effects. Mitochondrial protein is used in the range of 0.5–4 mg protein/ml.

A series of assays has been used to determine mtNOS activity, and a comparison of sensitivities is given in Table II.

Mitochondrial Preparations for Determination of NO Production

Rat liver, kidney, and heart mitochondria and mouse brain mitochondria used to determine NO production are prepared by a classical procedure using 0.23 *M* mannitol, 0.07 *M* sucrose, 1 m*M* EDTA, and 10 m*M* Tris–HCl (pH 7.4) as homogenization medium and isolating the mitochondrial fraction by precipitation at 8000*g* of a 700-g homogenate supernatant.[18] Submitochondrial particles are

[16] L. J. Ignarro, *in* "Nitric Oxide: Biology and Pathology" (L. J. Ignarro, ed.), p. 3. Academic Press, New York, 2000.

[17] H. Kleinert, J. P. Boissel, P. M. Schwartz, and U. Fösterman, *in* "Nitric Oxide: Biology and Pathobiology" (L. J. Ignarro, ed.), p. 105. Academic Press, San Diego, 2000.

[18] A. Boveris, N. Oshino, and B. Chance, *Biochem. J.* **128,** 617 (1972).

TABLE II
SENSITIVITY OF VARIOUS ASSAYS UTILIZED TO DETERMINE mtNOS ACTIVITY

Assay	NO (μM)	NO production rate (μM/min)
Spectrophotometric		
Double wavelength (581–591 nm)	0.05	0.02
Single wavelength (580 nm)	0.5	0.1
Radioactive ([^{4}C]citrulline)	0.03	0.01
EPR (MGD-Fe^{2+})	2	0.5
NO electrode	0.1	0.05
Nitrate/nitrite determination (Griess reagent)	1	1
Western blot	10-ng mtNOS/band	

either (a) inside-out mitochondrial vesicles prepared by the sonication of heart mitochondria or (b) fragments obtained from mitochondria of other organs by twice freezing and thawing and homogenization by passage through a 29-gauge hypodermic needle and syringe. Further mitochondrial purification can be performed by Ficoll gradient.[19] For Western blot analysis, mitochondrial pellets are washed with isolation buffer and dialyzed against it at 4° for 4 hr.

Spectrophotometric Determination of NO Production

The determination of myoglobin (MbO_2) or hemoglobin (HbO_2) oxidation to Met derivatives gives the best results. The method is based on the original assay developed for perfused organs in which the hemoglobin γ band[20] is used to follow NO production and on its adaptation to the hemoprotein α band for use with neutrophils[21] and lymphocytes.[22] The α band allows the use of 20–40 μM heme groups for effective NO trapping and adapts better to conditions of high light scattering of cellular and mitochondrial suspensions. Mitochondrial suspensions of rat liver and brain mitochondria at 1 mg/ml show absorbances at 581 nm of about 1.5 and 1.7, respectively. Similarly, if mitochondrial fragments are used, absorbances are about 0.5 and 0.8. In such conditions, a highly sensitive instrument is needed to detect the specific absorbance changes (about 0.005–0.010 A/min) over a background of high light absorption. Such requirements are provided by a double-beam and double-wavelength instrument in which the active wavelength is set at 581 nm and the reference wavelength at the isosbestic point at 591 nm

[19] J. B. Clark and W. J. Nicklas, *J. Biol. Chem.* **245,** 4724 (1970).
[20] M. E. Murphy and E. Noack, *Methods Enzymol.* **233,** 240 (1994).
[21] M. C. Carreras, J. J. Poderoso, E. Cadenas, and A. Boveris, *Methods Enzymol.* **269,** 65 (1996).
[22] L. B. Valdez, S. Alvarez, S. L. Arnaiz, F. Schopfer, M. C. Carreras, J. J. Poderoso, and A. Boveris, *Free Radic. Biol. Med.* **29,** 349 (2000).

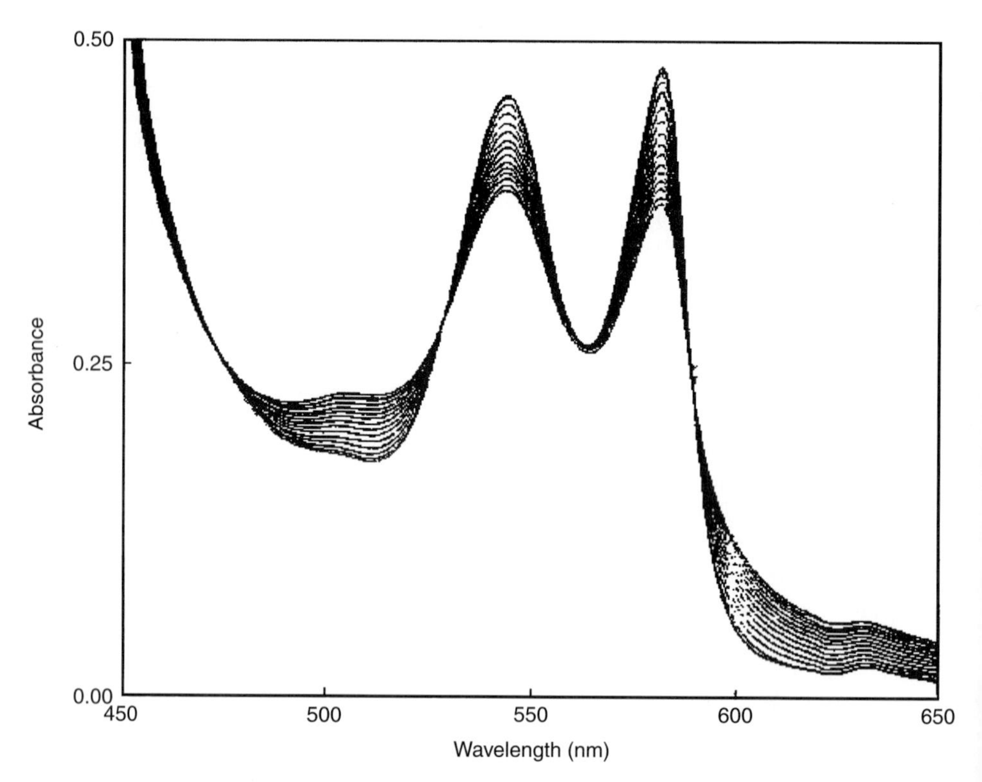

FIG. 1. Oxyhemoglobin spectrum with maxima at 545 and 581 nm, two isobestic points at 530 and 591 nm, and spectral changes due to its oxidation to methemoglobin by 200 μM nitrosoglutathione in the presence of 40 μM dithiothreitol. Spectra were recorded every 30 sec.

($E_{581-591} = 11.2\ mM^{-1}\ cm^{-1}$) (Fig. 1). Cuvettes of either 1.0 or 3.0 ml final volume and a 1-cm optical path are used. Care should be taken with double-beam double-wavelength spectrophotometers that usually have light beams that do not exactly fit the optical path of 1 ml cuvettes. Figure 2 illustrates the determination of NO production by mitochondrial preparations. The determination is initiated by recording a baseline for a couple of minutes to assure optical and electronic stability, and the reaction is initiated by the addition of NADPH and arginine. The upward deflection of the traces indicates a decrease at 581 nm, which identifies metHb formation and NO generation [reaction(1)].

$$HbO_2 + NO \rightarrow metHb + products \quad (1)$$

It is worth noting that the reaction rate decreases after 3 or 4 min, it is then essential to consider initial rates to determine maximal activities. Preincubation for 1–3 min of the mitochondrial preparation with adequate NOS inhibitors produces an almost complete suppression of NO generation. The effects of NOS inhibitors, L-NMMA

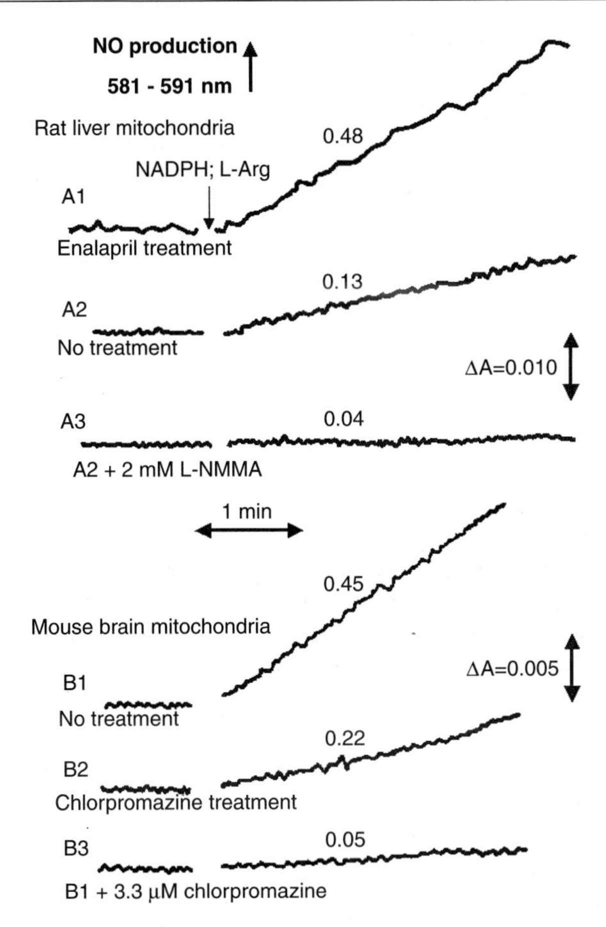

FIG. 2. Spectrophotometric determination of mtNOS at 581–591 nm in the presence of 30 μM HbO_2 at 30°. PerkinElmer 356 double-beam double-wavelength spectrophotometer. Reaction medium: phosphate buffer 50 mM (pH 7.4), 200 μM NADPH, 1.0 mM arginine, 1 mM $CaCl_2$, 4 μM Cu,Zn-superoxide dismutase, and 0.1 μM catalase. Numbers near traces indicate nmol NO/min/mg protein. Rat liver mitochondria: (A1) rat treated with enalapril (30 mg/kg/day ip for 15 days), 1.7 mg protein /ml; (A2) untreated (control) rat, 2.8 mg protein/ml; (A3) idem A2 preincubated 2 min with 2 mM L-NMMA. Mouse brain mitochondria: (B1) untreated (control) mouse, 2.3 mg protein /ml; (B2) mouse treated with chlorpromazine (10 mg/kg, single dose, ip, 1 hr before killing the animal), 1.5 mg protein /ml; (B3) idem preincubated 2 min with chlorpromazine.

and chlorpromazine, are shown in Fig. 2. Then, the rates of NO production can be estimated reliably from the rates of metHb formation sensitive to NOS. Sensitive spectrophotometers with a single light beam at about 577–580 nm or with multiple wavelengths set at 581–592 nm have been used to determine mitocondrial NO production,[1–3] but they have a tendency to produce overestimation of the rates with values in the range of 1–5 μM NO/min/mg protein, probably due to baseline drifts. Both HbO_2 and MbO_2 at equal heme concentrations behave similarly as NO traps and indicators.

Pharmacological Regulation of Liver and Kidney mtNOS-2 and of Heart mtNOS Activities by Enalapril (Spectrophometric Determination)

Figure 3 shows both (a) the dependence of the rates of NO production on the amount of protein and (b) the effect of *in vivo* enalapril treatment on the mtNOS of liver, kidney, and heart. Concerning the first point, the determined rates are related linearly to protein content in the range of 0.5–4.0 mg protein/ml. With respect to the second point, acute treatment with enalapril (30 mg/kg rat, ip) increased NO production by 50, 70, and 50% after 7 days and by 4, 9, and 4 times after 15 days in liver, kidney, and heart mitochondria, respectively. Liver and kidney show an mtNOS-2 that reacts with anti-iNOS antibodies, whereas heart mtNOS did not react with antibodies against nNOS, iNOS, and eNOS. Pioneer immunocytochemical reports on mtNOS, using a monoclonal antibody against eNOS, indicated the presence of a protein, eNOS or a protein with a significant homology to eNOS, in liver, brain, heart, kidney, and skeletal muscle mitochondria.[11] The effect of *in vivo* treatment with enalapril and chlorpromazine that increase and decrease, respectively, the NO production of rat liver and mouse brain mitochondria is also shown in Fig. 2.

Pharmacological Regulation of Liver mtNOS-2 Activity by Enalapril (Determined by EPR)

The effect of enalapril on liver mtNOS-2 activity was assayed by determining NO production by EPR in the presence of the probe MGD-Fe^{2+} (Fig. 4), as described previously by Giulivi *et al.*[2] The NO signal is clearly evident in liver mitochondria from enalapril-treated rats but it is barely seen in liver mitochondria from untreated (control) rats. Supplementation with the NOS inhibitor L-NMMA abolished the NO signal. The pattern of NO production by liver mitochondria from enalapril-treated and untreated rats as detected by EPR is similar to the one observed in the spectrophotometric measurement of mtNOS-2 activity (Figs. 2 and 3).

Pharmacological Regulation of Brain mtNOS-1 by Chlorpromazine (Oxygen Electrode Determination)

Supplementation of respiring mitochondria with arginine or with NOS inhibitors significantly (about 15 to 40%) decreases or increases, respectively, the respiration rate both in metabolic states 4 and 3 with malate-glutamate or succinate as substrates.[2] Figure 5 illustrates use of the oxygen electrode to determine the functional activity of mtNOS-1 in intact respiring mitochondria. The addition of

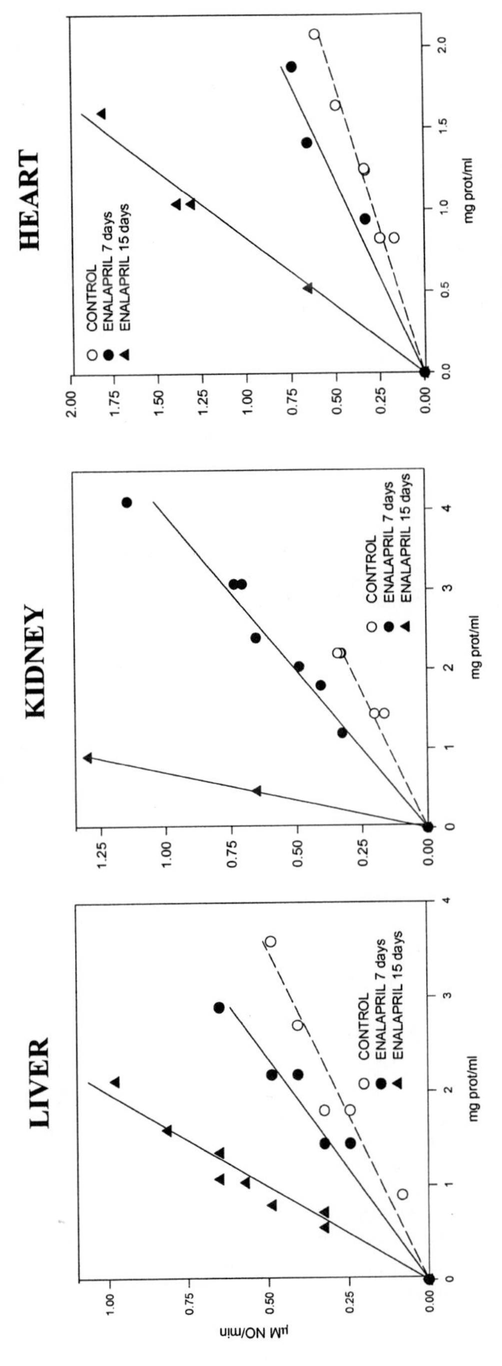

FIG. 3. Production of NO in liver, kidney, and heart mitochondria in rats treated 7 and 15 days with enalapril (30 mg/kg/day ip). Experimental conditions as in Fig. 2.

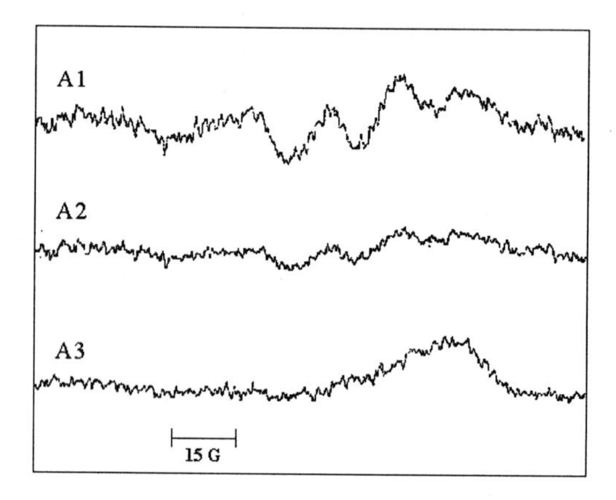

FIG. 4. EPR detection of NO production by liver mtNOS-2. Reaction medium as in Fig. 2 plus 2 m*M* $FeSO_4$ and 20 m*M* MGD (*N*-methy-D-glucamine dithiocarbamate; Oklahoma Medical Research Foundation, Oklahoma City, OK). Liver mitochondria fragments (2.1 mg protein/ml) were incubated for 5–10 min at room (25°) temperature, transferred with Pasteur pipettes, and spectra recorded in a Bruker ECS 106 ESR with a ERY 1025T cavity. (A1) Enalapril-treated rat, (A2) untreated (control) rat, and (A3) A1 + 2 m*M* L-NMMA. EPR settings were microwave power, 20 mW; modulation amplitude, 6.175 G; time constant, 164 ms; modulation frequency, 50 kHz; receiver gain, 2×10^4; sweep with 150 G; conversion time, 163 ms; and microwave frequency, 9.81 Ghz.

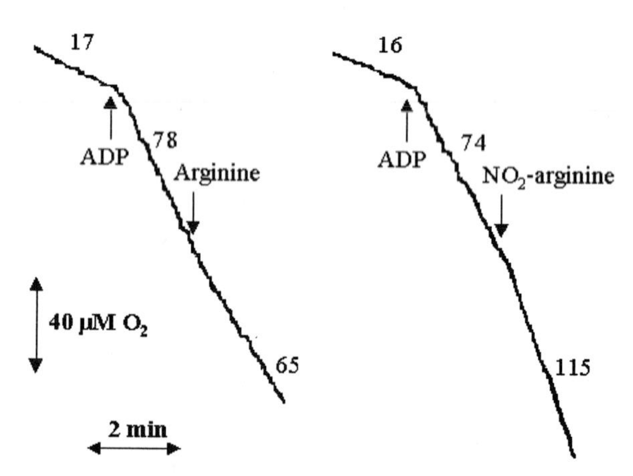

FIG. 5. Determination of functional activity of mtNOS-1 in mouse brain mitochondria by measuring oxygen uptake. Mitochondria (1.17 mg protein/ml) were suspended in 0.23 *M* mannitol, 0.07 *M* sucrose, 20 m*M* Tris–HCl) (pH 7.4), 6 m*M* malate, 6 m*M* glutamate, 5 m*M* phosphate, 0.5 m*M* ADP, and 3 m*M* $MgCl_2$ at 37° (0.3 m*M* arginine and 2 m*M* *Nω*-nitro-L-arginine). Numbers near traces indicate oxygen uptake in nanograms at O/min/mg protein.

TABLE III

EFFECTS OF CHLORPROMAZINE TREATMENT (ONE DOSE 10 mg/kg ip) ON mtNOS ACTIVITY AND RESPIRATORY RATES OF BRAIN MOUSE MITOCHONDRIA

		Respiratory rate	
Substrate/condition	mtNOS activity (nmol NO/min/mg protein)	State 4 (ng at O/min/mg protein)	State 3
Malate glutamate			
No treatment	0.48 ± 0.05	15 ± 2	67 ± 4
Chlorpromazine	0.25 ± 0.06	21 ± 3	94 ± 5
Succinate			
No treatment	0.48 ± 0.05	22 ± 3	76 ± 5
Chlorpromazine	0.25 ± 0.06	29 ± 3	98 ± 4

arginine to brain mitochondria in state 3 with malate-glutamate as the substrate decreased the rate of oxygen uptake by 13%. Conversely, supplementation with the NOS inhibitor NO_2 arginine increased the respiratory rate by 55% in the same experimental conditions. The phenomenon is interpreted in terms of the NO inhibition of cytochrome oxidase and of a continuous generation of NO. Arginine addition increases the intramitochondrial pool of arginine available to mtNOS with consequent increases in the rate of NO production and in the NO intramitochondrial steady-state concentration. A higher level of intramitochondrial NO leads to inhibition of cytochrome oxidase and to a lower respiratory rate. Conversely, it is understood that the NOS inhibitor decreases the rate of mitochondrial NO production and cytochrome oxidase inhibition, resulting in a higher respiratory rate. *In vivo* treatment with chlorpromazine increased respiration in state 4 and in state 3 (Table III). Both (a) the direct increase in state 3 respiration (Table III) or (b) the respiratory difference between the conditions (i) with arginine and (ii) with the NOS inhibitor (Fig. 5) can be taken as an indirect measurement of functional mtNOS activity. A similar effect of thyroxine plasma levels, considering hypothyroidism, euthyroidism, and hyperthyroidism, increasing the rate of oxygen uptake, and inhibiting the activity of liver mtNOS-2, has been reported.[13] Simple respiratory determinations with the oxygen electrode can be used to assay the toxic or pharmacological effects of different *in vivo* treatment on functional mtNOS activity.

Pharmacological Regulation of mtNOS Activity as Determined by Western Blot Analysis

The effects of enalapril and chlorpromazine treatment on the mtNOS of kidney, liver, and brain mitochondria were analyzed by Western blots (Fig. 6). Kidney and

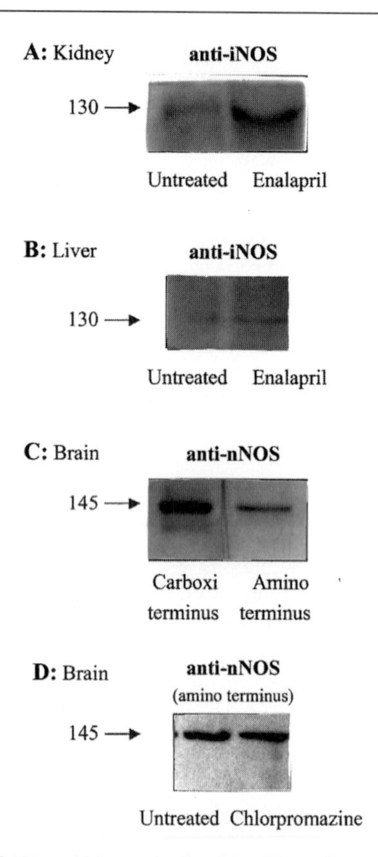

FIG. 6. Western blots of (A) rat kidney mitochondria, (B) rat liver mitochondria, and (C and D) mouse brain mitochondria. Enalapril and chlorpromazine treatment as in Fig. 2. Molecular mass is indicated at the left (in kDa). Submitochondrial particles (150 μg protein) supplemented with the protease inhibitors pepstatin, leupeptin, and aprotinin (at 1 μg/ml) and 0.4 m*M* phenylmethylsulfonyl fluoride were separated by SDS–PAGE (7.5%), blotted onto a nitrocellulose membrane (Bio-Rad, München, Germany), and probed with rabbit polyclonal antibodies (dilution 1 : 500) for nitric oxide synthases. Antibodies were anti-nNOS, epitopes of amino acids 2–300 at the amino terminus [Santa Cruz NOS1 (H-299): sc8309]; anti-nNOS, epitopes of amino acids 1409–1429 at the carboxy terminus [Santa Cruz NOS1 (R-20): sc 648]; anti-1NOS, epitopes at the carboxy terminus of NOS II (Santa Cruz C-19); and anti eNOS, epitopes at the amino terminus of NOS III (N-20). All are from Santa Cruz Laboratories (Santa Cruz, CA). Nitrocellulose membranes were incubated with a secondary goat anti-rabbit antibody conjugated with horseradish peroxidase (dilution 1 : 5000), followed by chemiluminescence development with ECL reagent for 2–4 min.

liver mitochondria show a protein of about 130 kDa (mtNOS-2) reacting with an anti-iNOS antibody that exhibits a marked increased expression after enalapril treatment. The increases in band intensities after enalapril treatment agree quantitatively with the measured enzymatic activities (Figs. 2 and 3). Brain mitochondria show a protein of about 145 kDa (mtNOS-1) reacting with the antibodies

TABLE IV
PHARMACOLOGICAL REGULATION OF MITOCHONDRIAL NITRIC OXIDE SYNTHASE ACTIVITY

Drug	Daily dose/treatment	Organ	Effect on mtNOS (% of control activity)
Haloperidol	2 mg/kg mouse, 1 day	Brain	−54
Chlorpromazine	10 mg/kg mouse, 1 hr	Brain	−48
Enalapril	30 mg/kg rat, 15 days	Liver, heart	313–330
Enalapril	30 mg/kg rat, 15 days	Kidney	890
Enalapril	10 mg/kg rat, 6 months	Heart, liver	105–120
Losartan	10 mg/kg rat, 6 months	Heart	95
Thyroxine (T_4)	0.2 mg/kg rat, 2 weeks	Liver	−45
Methylprednisolone	3 μg/ml medium, 30 min	Thymocytes	450

anti-nNOS directed against the amino and carboxy terminus of the polypeptide chain. Chlorpromazine treatment has no effect on the band intensity of the protein reacting with the anti-nNOS (amino terminus) antibody, despite the measured decrease in enzymatic activity (Fig. 2 and Table III) indicating a direct inhibitory effect of chlorpromazine on the enzymatic protein.

Pharmacological Regulation of mtNOS-1 and mtNOS-2

Table IV lists a series of *in vivo* treatments involving the use of five pharmacological substances used commonly that affect both mtNOS-1 and mtNOS-2 activity. Both increases and decreases of mtNOS activity are described. The *in vitro* effect of methylprednisolone in thymocytes is included, considering its pharmacological relevance.

[31] Characterization of Mitochondrial Nitric Oxide Synthase

By PEDRAM GHAFOURIFAR

The diverse roles of nitric oxide (NO) in biology have been studied broadly in the last two decades. NO reacts with hemoproteins, thiols, and ions, such as superoxide anion (O_2^-). Many of the actions of NO are due to its ability to react with soluble guanylate cyclase and subsequently increase cyclic GMP levels. However, a considerable portion of the biological functions of NO are cyclic GMP independent. Mitochondria, the lipoprotein cytoplasmic particles of eukaryotic cells,

0076-6879/02 $35.00

possess many hemoproteins and thiols and are the main cellular sources of O_2^-. Thus, mitochondria play an important role in mediating cGMP-independent effects of NO.

Mitochondria are heavily compartmentalized organelles. The mitochondrial matrix, inner membrane, and intermembrane space differ in a number of ways, such as in electrochemistry, redox state, and pH, as well as radical and protein content. Thus NO and its metabolites interact with these different mitochondrial components in distinct manners. For example, cytochrome oxidase, an abundant hemoprotein embedded in the mitochondrial inner membrane, interacts with NO in a transient, reversible, and O_2 concentration-dependent manner.

Physiologically relevant concentrations of NO inhibit cytochrome oxidase in a manner resembling a pharmacological competitive antagonism with oxygen.[1] NO can also react with reduced thiols to produce nitrosothiols. This reaction is also reversible, however, it is redox and pH sensitive. Although *S*-nitrosylation is generally fast and *S*-nitrosylated proteins are relatively stable *in vitro,* their formation and stabilization are not favored in certain biological compartments. For example, inorganic phosphate, which is abundant in the mitochondria matrix, inhibits the *S*-nitrosylation reaction.[2] Also, some enzymes present in the matrix, such as glutathione peroxidase[3] or thioredoxin reductase,[4] can decompose *S*-nitrosothiols. In contrast to the matrix, the mitochondrial intermembrane space seems the preferred site for *S*-nitrosylation. To date, *S*-nitrosylation of a mitochondrial intermembrane space protein caspase-3 has been demonstrated.[5] Caspase-3 is *S*-nitrosylated as long as it is within the organelle, rendering it apoptotically silent. This is likely a mechanism for protecting mitochondria against the protease activity of the caspase.

Nitric oxide can also react with O_2^- to produce the powerful oxidizing agent, peroxynitrite ($ONOO^-$). This reaction is stoichiometric, diffusion limited, and requires relatively high pH, which is provided in the mitochondrial matrix. In intact, tightly coupled succinate-energized mitochondria, the pH is in the range of 7.5 to 7.8 in the matrix[6] and 6.9 to 7.0 in the intermembrane space.[7] Thus, the mitochondrial matrix environment can favor the formation and reactions of $ONOO^-$. In contrast to the reversible reactions of NO with mitochondrial proteins, the ones of $ONOO^-$ are mostly irreversible. A direct measurement of $ONOO^-$ in biological samples is difficult, however, biomarkers such as nitrotyrosine or lipid peroxides are good indicators. Tyrosine nitration of mitochondrial matrix

[1] P. Ghafourifar, U. Bringold, S. D. Klein, and C. Richter, *Biol. Sign. Recep.* **10,** 57 (2000).
[2] E. G. DeMaster, B. J. Quast, and R. A. Mitchell, *Biochem. Pharmacol.* **53,** 581 (1997).
[3] Y. Hou, Z. Guo, J. Li, and P. G. Wang, *Biochem. Biophys. Res. Commun.* **228,** 88 (1996).
[4] D. Nikitovic and A. Holmgren, *J. Biol. Chem.* **271,** 19180 (1996).
[5] J. B. Mannick, C. Schonhoff, N. Papeta, P. Ghafourifar, M. Szibor, K. Fang, and B. Gaston, *J. Cell Biol.* **154,** 1111 (2001).
[6] P. Bernardi, *J. Biol. Chem.* **267,** 8834 (1992).
[7] J. D. Cortese, A. L. Voglino, and C. R. Hackenbrock, *Biochim. Biophys. Acta* **1100,** 189 (1992).

proteins, e.g., MnSOD[8] or succinyl-CoA:3-oxoacid CoA-transferase (SCOT),[9] has been reported. With the exception of cytochrome *c* all of the mitochondrial respiratory chain components have matrix faces. Peroxynitrite-induced inactivation of mitochondrial respiratory chain complexes I to IV has been reported by many groups (reviewed in Ghafourifar and Richter[10]). Although tyrosine nitration of cytochrome *c*, which is located in the low pH intermembrane environment, does not seem favorable in biology, $ONOO^-$ can dissociate cytochrome *c* from mitochondria.[11,12] How *S*-nitrosothiols or $ONOO^-$ are formed within mitochondria has not yet been fully elucidated. However, evidence suggests that the newly characterized mitochondrial NO synthase (mtNOS) might be the hitherto undetected source of NO and its congeners within mitochondria (Fig. 1).

Mitochondrial Nitric Oxide Synthase

In biology, NO is synthesized by NO synthase (NOS; EC 1.14.13.39) family members. NOS isozymes catalyze oxidation of the terminal guanidino nitrogen of L-arginine to NO and L-citrulline as the coproduct. So far, three distinct isoforms of NOS have been well characterized in mammalian tissues. Although none of these isozymes has a tissue-specific pattern of expression, they are referred to as endothelial NOS (eNOS), neuronal NOS (nNOS), and inducible NOS (iNOS). All presently characterized NOS isozymes are heme-containing proteins that are dimeric under native conditions with a monomer molecular mass of about 126–160 kDa. nNOS and eNOS are expressed constitutively, whereas iNOS is expressed when cells are stimulated by certain stimuli. The constitutive isoforms show a typical Ca^{2+}-calmodulin sensitivity, whereas the activity of iNOS, once expressed, is not altered by increased cytosolic Ca^{2+}.

Several laboratories have investigated the possible presence of a NOS within mitochondria. In 1997, we demonstrated for the first time the activity of a constitutively expressed Ca^{2+}-sensitive NOS within mitochondria (mtNOS).[13] We also demonstrated that mtNOS is associated with the mitochondrial inner membrane and regulates mitochondrial O_2 consumption and $\Delta\psi$. These results were later confirmed by other laboratories.[14,15] Since 1997, several laboratories have shown

[8] L. A. MacMillan-Crow, J. P. Crow, and J. A. Thompson, *Biochemistry* **37,** 1613 (1998).

[9] I. V. Turko, S. Marcondes, and F. Murad, *Am. J. Physiol. Heart. Circ. Physiol.* **281,** H2289 (2001).

[10] P. Ghafourifar and C. Richter, *in* "Mitochondrial Ubiquinone (Coenzyme Q10): Biochemical, Functional, Medical, and Therapeutical Aspects in Human Health and Disease" (M. Ebadi, J. Marwah, and R. Chopra, eds.), p. 437. Prominent Press, AZ, 2000.

[11] P. Ghafourifar, U. Schenk, S. D. Klein, and C. Richter, *J. Biol. Chem.* **274,** 31185 (1999).

[12] V. Borutaite, R. Morkuniene, and G. C. Brown, *Biochim. Biophys. Acta* **1453,** 41 (1999).

[13] P. Ghafourifar and C. Richter, *FEBS Lett.* **418,** 291 (1997).

[14] Z. Lacza, M. Puskar, J. P. Figueroa, J. Zhang, N. Rajapakse, and D. W. Busija, *Free Radic. Bio. Med.* **31,** 1609 (2001).

[15] A. J. Kanai, L. L. Pearce, P. R. Clemens, L. A. Brider, M. M. VanBibber, S. Y. Choi, W. C. de Groat, and J. Peterson, *Proc. Natl. Acad. Sci. U.S.A.* **98,** 14126 (2001).

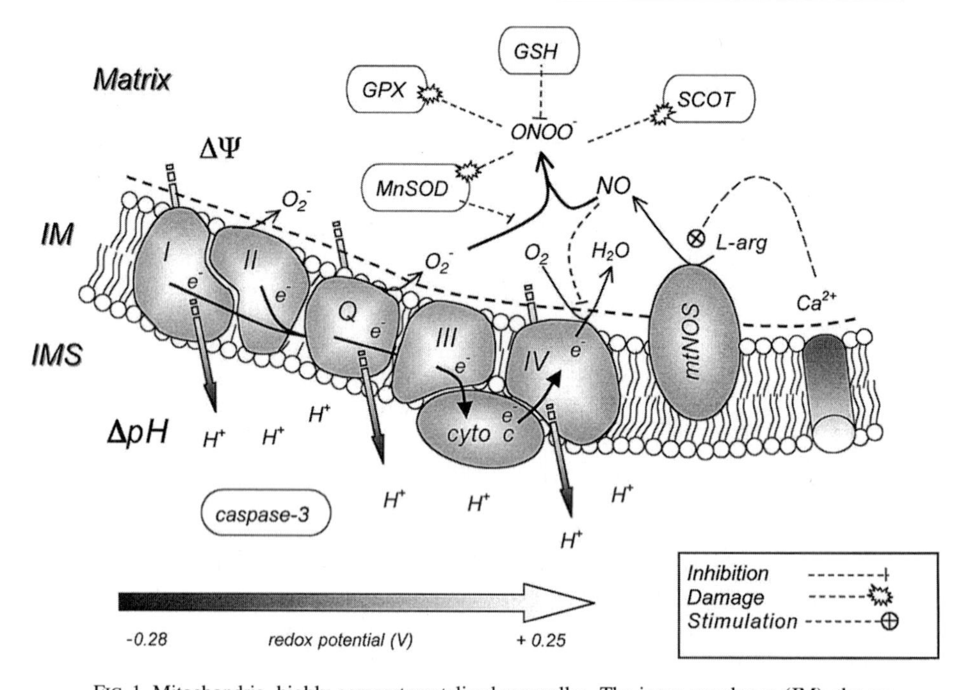

FIG. 1. Mitochondria, highly compartmentalized organelles. The inner membrane (IM), the matrix, and the intermembrane space (IMS) and their distinct composition, electrochemistry, and redox state are illustrated. IM: The respiratory chain complexes are embedded in this compartment. The chain consists of four complexes (I to IV), coenzyme Q (ubiquinone; Q) and cytochrome *c* (cyto c). The respiratory complexes are arranged functionally in an electrochemical hierarchy that provides a unique broad spectrum of redox potentials. Electrons flow down the chain to complex IV where they reduce O_2 to H_2O. Coupled to the electron flow, protons are pumped from the matrix into IMS. This proton extrusion establishes a transmembrane potential ($\Delta\psi$, negative inside) and an electrochemical gradient (ΔpH, alkaline inside) across IM. In response to $\Delta\psi$, mitochondria take up Ca^{2+}, which stimulates mtNOS activity. The produced NO inhibits the complex IV activity, which affects $\Delta\psi$, ΔpH, and mitochondrial Ca^{2+} homeostasis. The mitochondrial respiratory chain is one of the main cellular sources of O_2^-. Nitric oxide reacts readily with O_2^- to produce $ONOO^-$. Mitochondrial redox barriers, such as MnSOD, may affect the rate of $ONOO^-$ formation. Matrix: Key mitochondrial redox defense barriers, e.g., MnSOD, GSH, or glutathione peroxidase (GPX), are located within the matrix. Some matrix proteins, such as MnSOD and SCOT, are susceptible to $ONOO^-$-induced oxidative damage. IMS: Cytochrome *c* is the only respiratory chian member located in this compartment. Certain mitochondrial apoptogenic proteins, such as caspase 3, are also located in IMS.

interest in performing mtNOS research. Rat liver is the most commonly used source for isolating mitochondria. Mitochondria can be isolated by the conventional differential centrifugation as described.[16] However, the following are practical tips for specifically studying mtNOS in rat liver mitochondria.

[16] C. Richter, *Methods Enzymol.* **105,** 435 (1984).

Procedure

Fast the animals one night before the experiment, with water *ad libitum*. Decapitate and deplete the body of blood to limit the exposure of mitochondria to NO-reacting molecules, such as hemoglobin. Remove the liver quickly, place it in a dish on ice, and remove fat, ducts, connective tissues, and blood clots. Cut the liver into small pieces and wash several times in ice-cold buffer to remove remaining blood. Homogenize the washed pieces in a glass homogenizer with a Teflon pestle. It is highly recommended that the homogenizer, pestle, or centrifuge tubes not be washed with any detergent. Detergents released from the glass, Teflon, or centrifuge tubes can release membrane-associated proteins, such as mtNOS, from mitochondrial membranes. It is important to keep the temperature low during the homogenization, e.g., by placing the glass tube on ice. Do not overhomogenize because heat and the mechanical force produced during homogenization can strongly damage mtNOS.

During centrifugation, a red spot in the low-spin pellet may indicate contamination with blood. If there is a fat layer floating above the supernatant of the high-spin steps, remove it with a soft lint-free tissue. Discard the fluffy light brown layer above the high-spin pellet. This layer is not mitochondria. A good rat liver mitochondria preparation provides about 30–40 mg mitochondrial protein per rat liver. Higher protein contents may indicate contamination. There are several methods for assessing the purity of the preparation. We do this routinely by measuring the cytochrome *a* content spectrophotometrically at 605–630 nm using the extinction coefficient of 12 M^{-1} cm^{-1}.[17]

Determination of mtNOS Activity

NOS activity can be determined by different techniques. The rather unique structure and composition of mitochondria, however, limit the use of some of these techniques in mtNOS activity determination. Also, it must be noted that the activity of mtNOS, as well as that of many other mitochondrial enzymes, declines rapidly in isolated mitochondria. Thus, it is highly recommended that mtNOS-related measurements be performed only with freshly isolated mitochondria, generally within 4–6 hr of isolation.

Hemoglobin (Hb) Assay

In aqueous solutions, NO reacts rapidly with oxyhemoglobin (oxyHb) to produce methemoglobin (metHb). This reaction is stoichiometric and can be followed spectrophotometrically. With minor modifications, the standard Hb assay can be used as a reliable method for quantifying mtNOS activity. However, if the isolated

[17] R. S. Balaban, V. K. Mootha, and A. Arai, *Anal Biochem.* **237,** 274 (1996).

mitochondria are intact and not contaminated with cytosolic proteins, e.g., cytosolic forms of NOS, no metHb formation is observed. NO produced within mitochondria reacts with several mitochondrial NO traps such as thiols. This prevents intramitochondrially formed NO from reaching the extramitochondrial probe, oxyHb. Additionally, oxyHb cannot enter mitochondria because the mitochondrial inner membrane is impermeable to almost every molecule larger than 100–150 Da (oxyHb is 65 KDa). Thus, formation of metHb from oxyHb within intact freshly prepared mitochondria generally indicates artifacts, such as contamination with extramitochondrial proteins. mtNOS activity can be determined by the Hb assay if mitochondrial membranes are ruptured, i.e., in broken mitochondria (BM). Some investigators, however, have suggested that oxymyoglobin (18 kDa) might be used to detect mtNOS activity in intact mitochondria.[18]

Another confounding factor that limits the use of the Hb assay in mitochondria is O_2^-. Superoxide anion can potently react with NO, and mitochondria are one of the main cellular sources O_2^-. Thus, relatively high amounts of superoxide dismutase (SOD), e.g., $\geq$1 kU/ml, is highly recommended in this assay.

Preparation of Broken Mitochondria. Apply a hypoosmolar shock by adding 2–4 volumes of ice-chilled H_2O containing the protease inhibitors (10 μM each) leupeptin, pepstatin A, aprotinin, and phenylmethylsulfonyl fluoride (PMSF). Apply a mild sonification, e.g., 100–150 W, 50% duty cycle, 75 sec, to break the remaining intact mitochondria. To avoid reaction with oxygen, purge the water with N_2 for at least 15 min and apply N_2 over the suspension during the sonification. Osmolality can be readjusted by the addition of pH-adjusted concentrated buffer (e.g., a 10× buffer) to the BM suspension. Spin the suspension at 10,000*g* for 10 min. The supernatant is BM.[13]

Spectrophotometric Determination of mtNOS Activity. Mitochondria contain sufficient concentrations of the substrate and cofactors NOS requires, and Ca^{2+} per se seems sufficient to trigger the mtNOS activity in intact isolated mitochondria.[11,19] However, dilution or oxidation of some of these factors during the preparation of BM or submitochondrial particles (SMP) may require the addition of the following substrates or cofactors[13]:

1–10 μg/ml calmodulin: prepare a 1-mg/ml stock solution, make aliquots, and store at −80°. Avoid freeze-thawing.
1 m*M*-L-arginine: prepare a 100 m*M* stock solution and store at −20°.
0.2 m*M* NADPH: prepare a 20 m*M* stock solution daily.
5 μM FAD: prepare a 0.5 m*M* stock solution daily.
5 μM FMN: prepare a 0.5 m*M* stock solution daily.

[18] C. Giulivi, J. J. Poderoso, and A. Boveris, *J. Biol. Chem.* **273,** 11038 (1998).
[19] U. Bringold, P. Ghafourifar, and C. Richter, *Free Radic. Biol. Med.* **29,** 343 (2000).

10 μM tetrahydrobiopterin (BH_4): prepare a 2 mM stock solution in 10 mM HCl immediately prior to the experiment. It undergoes rapid autoxidation on dilution.

$\geq$1 kU/ml SOD: prepare a 100 kU/ml stock solution. Make aliquots and store at $-20°$. Avoid freeze-thawing.

Procedure

Add the 37° buffer (e.g., 0.1 M HEPES, pH 7.0) to a 37° thermostated cuvette and then add the mtNOS substrate(s) and 4 μM oxyHb. Mix and record the absorbance. Add 0.03 to 0.1 mg of BM or SMP/ml. More than 0.1 mg mitochondrial protein per milliliter can disturb the measurement due to too much light scattering. MetHb formation is shown by a continual increase in optical density. Several wavelengths can be used.[20] For example, at 401 nm the NO formation can be quantified using an $\varepsilon_{401(\text{metHb-oxyHb})}$ of 49 mM^{-1} cm^{-1}.

Citrulline Assay

Determination of radiolabeled L-citrulline produced from radiolabeled L-arginine is one of the widely used methods for NOS activity determination. This assay is preferred for intact mitochondria, as the Hb assay cannot be used. However, many of the buffers used traditionally to investigate mitochondrial functions contain high concentrations of Mg^{2+}, a well-known mitochondrial Ca^{2+} uptake blocker. The presence of Mg^{2+} decreases mtNOS activity drastically and can abolish its Ca^{2+} sensitivity. In the presence of 5 mM Mg^{2+}, mtNOS may even appear Ca^{2+} insensitive.[21]

Procedure

Incubate 0.1 to 3 mg mitochondrial protein with NOS substrates (as discussed earlier) supplemented with L-[^{3}H]arginine (30,000–50,000 cpm) at 37° in a total volume of 100 μl buffer. Terminate mtNOS activity by adding 1 ml of ice-chilled stop solution containing 2 mM EDTA, 1 mM unlabeled L-citrulline, and 20 mM sodium acetate buffer, pH 5.0. Basal mtNOS activity can be determined by supplementing mitochondria with 30,000–50,000 cpm L-[^{3}H]arginine without NOS substrates.[11,13]

Preparation of Cation-Exchange Columns. Prepare the exchange resin (Dowex 50W ×8, mesh size 200–400, H^+ form) as described.[22] Load the spin filters with

[20] M. Feelisch, D. Kubitzek, and J. Werringloer, *in* "Methods in Nitric Oxide Research" (M. Feelisch and J. S. Stamler, eds.), p. 455. Wiley, Chichester, 1996.

[21] S. French, C. Giulivi, and R. S. Balaban, *Am. J. Physiol. Heart Circ. Physiol.* **280,** H2863 (2001).

[22] B. Mayer, P. Klatt, E. R. Werner, and K. Schmidt, *Neuropharmacology* **33,** 1253 (1994).

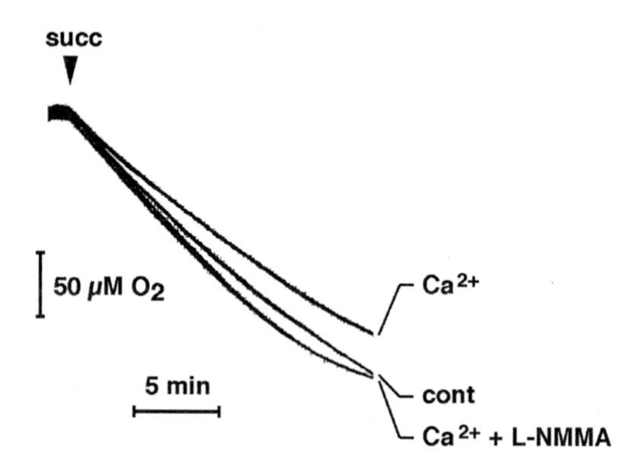

FIG. 2. mtNOS and mitochondrial oxygen consumption. Oxygen consumption of 1 mg/ml isolated rat liver mitochondria in 0.1 *M* HEPES, pH 7.1 (cont), was measured at room temperature with a tightly sealed Clarke-type electrode under continuous stirring. Respiration was supported by 0.4 m*M* succinate (succ). Ca^{2+} (100 μM) or L-NMMA (10 m*M*) was present in the buffer prior to the addition of mitochondria. Reproduced with permission from P. Ghafourifar, U. Schenk, S. D. Klein, and C. Richter, *J. Biol. Chem.* **274,** 31185 (1999).

0.6 to 1 ml of the prepared resin, spin at 5000*g* for 2 min, and wash with 500 μl H_2O. Load the columns with 500 μl of the radiolabeled mitochondrial samples and spin at 5000*g* for 2–5 min. Wash the resins twice with 200 μl H_2O. Transfer the effluent to scintillation vials, add 3 ml of scintillation fluid, and determine the radioactivity. Radioactive material passing through the column after mock incubation (no mitochondrial material added) is normally about 1% of the total radioactivity used in the experiment.

mtNOS and Mitochondrial Functions

mtNOS and Oxygen Consumption

mtNOS regulates mitochondrial oxygen consumption. In intact mitochondria, Ca^{2+} per se is sufficient to decrease respiration in an mtNOS-dependent manner (Fig. 2) given that mitochondria are energized, thus they can take up Ca^{2+}, and Ca^{2+} uptake is not blocked, e.g., by Mg^{2+}.

Procedure

Into a tightly sealed thermostated chamber equipped with an oxygen electrode, add the buffer, e.g., 0.1 *M* HEPES, the desired concentrations of Ca^{2+}, and 1 mg/ml of mitochondria or mitochondrial subfractions. If broken mitochondria are used, mtNOS substrates (as mentioned earlier) may also be required. After a 3- to 5-min incubation at room temperature with the mtNOS substrates or inhibitors, energize mitochondria by adding mitochondrial respiratory substrates. Relatively high concentrations of NOS inhibitors may be required to inhibit mtNOS activity. Most

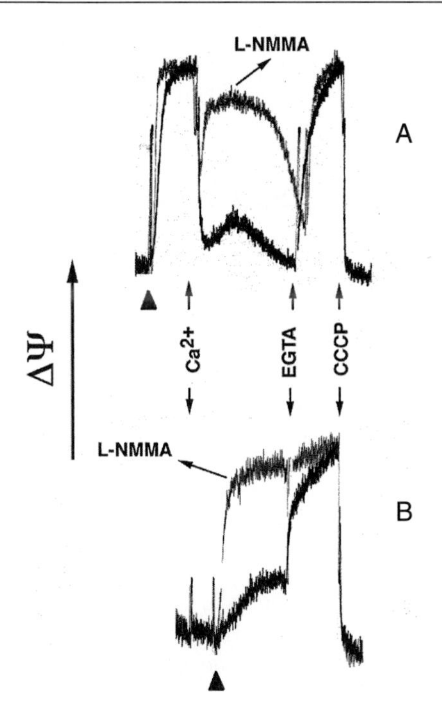

FIG. 3. mtNOS and $\Delta\psi$. Mitochondria were suspended in 0.1 *M* HEPES buffer, pH 7.4, containing 10 μM safranin. Electron transport was supported by 0.8 m*M* potassium succinate (▲). $\Delta\psi$ was measured by dual-wavelength spectrophotometry at 533–511 nm. The addition of EGTA (0.5 m*M*) or CCCP (1 μM) is indicated by arrows. Where indicated, 5 m*M* L-NMMA was present in the buffer. (A) Ca^{2+} (40 nmol/mg mitochondrial protein) was added where indicated. (B) Ca^{2+} (40 nmol/mg mitochondrial protein) was present in the buffer prior to the addition of succinate. Reproduced with permission from P. Ghafourifar and C. Richter, *J. Biol. Chem.* **380,** 1025 (1999).

of the commonly used NOS inhibitors are competitive L-arginine analogs, and the intramitochondrial L-arginine concentration is in the millimolar range.[23] Additionally, mitochondrial membranes are not very permeable to most of these inhibitors. If longer preincubations are needed, mitochondria can be preincubated with the NOS inhibitor on ice for up to 1 hr. Broken mitochondria or submitochondrial particles normally do not need extensive preincubation with NOS inhibitors.

mtNOS and Mitochondrial Membrane Potential

mtNOS regulates the mitochondrial transmembrane potential ($\Delta\psi$) (Fig. 3). The $\Delta\psi$ can be determined by several techniques, such as dual-wavelength spectroscopy.

[23] M. Dolinska and J. Albrecht, *Neurochem. Int.* **33,** 233 (1998).

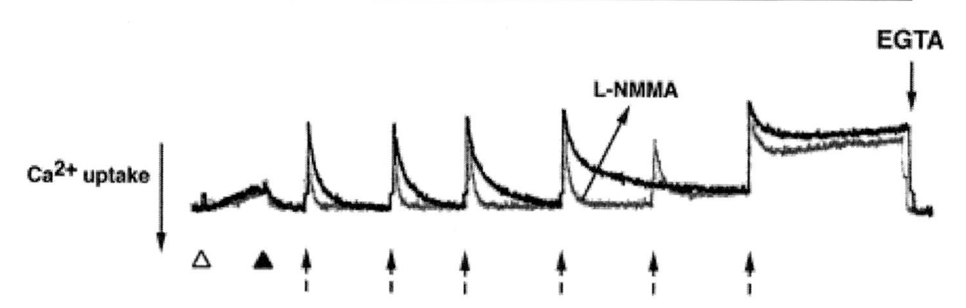

FIG. 4. mtNOS and mitochondrial Ca^{2+} homeostasis. Mitochondria were suspended in 0.1 *M* HEPES buffer, pH 7.4, containing 50 μM arsenazo III, and the respiratory chain was supported by 0.8 m*M* potassium succinate (▲) in the presence of 5 μM rotenone (Δ). Mitochondrial Ca^{2+} homeostasis was followed by dual-wavelength spectrophotometry at 685–675 nm. Repeated additions of 10 nmol Ca^{2+}/mg mitochondrial protein are shown by dashed arrows and 0.5 m*M* EGTA by the filled arrow. Where indicated, 5 m*M* L-NMMA was present in the buffer. Reproduced with permission from P. Ghafourifar and C. Richter, *J. Biol. Chem.* **380,** 1025 (1999).

Procedure

Incubate 1 mg/ml of mitochondria in buffer (e.g., 0.1 *M* HEPES, pH 7.35) in the presence of 10 μM safranin T. Energize mitochondria with mitochondrial respiratory substrates, such as 0.8 m*M* succinate, and record the optical density at 511–533 nm. Inhibition of mtNOS increases both the rate and the magnitude of $\Delta\psi$ formation, and its stimulation decreases it.

mtNOS and Mitochondrial Ca^{2+} Homeostasis

Until recently, the importance of mitochondria in cellular Ca^{2+} homeostasis was overlooked. However, recent findings indicate the crucial role of mitochondria in phasic Ca^{2+} homeostasis.[24] Mitochondria take up Ca^{2+} in response to their $\Delta\psi$ and release Ca^{2+} via distinct pathways.[25] Mitochondrial Ca^{2+} homeostasis can be followed by several techniques, such as dual-wavelength spectroscopy at 685–675 nm using Arsenazo III as a probe.

Procedure

Suspend 1 mg mitochondria/ml in 0.1 *M* HEPES buffer, pH 7.4, in the presence of 50 μM Arsenazo III and measure the optical density at 685–675 nm. Support respiration by adding mitochondrial respiratory substrates, such as 0.8 m*M* potassium succinate. mtNOS is stimulated upon mitochondrial Ca^{2+} uptake, and mtNOS inhibition can be achieved by preincubating mitochondria with a NOS inhibitor, as described earlier (Fig. 4).

[24] R. Rizzuto, P. Pinton, W. Carrington, F. S. Fay, K. E. Fogarty, L. M. Lifshitz, R. A. Tuft, and T. Pozzan, *Science* **280,** 1763 (1998).

[25] C. Richter, P. Ghafourifar, M. Schweizer, and R. Laffranchi, *Biochem. Soc. Trans.* **25,** 914 (1997).

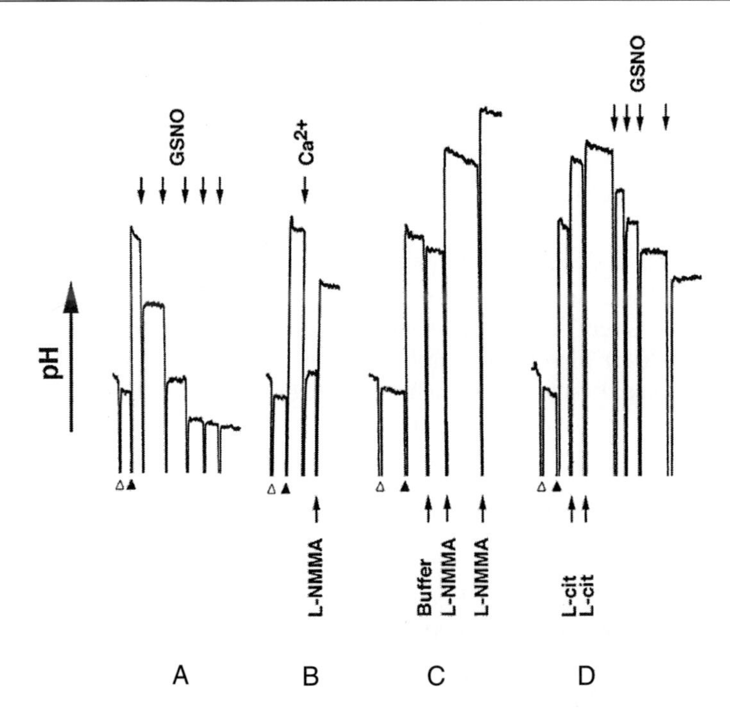

FIG. 5. mtNOS and mitochondrial matrix pH. Mitochondria were suspended in the incubation buffer as described in the text, and proton extrusion was supported by 0.8 m*M* potassium succinate (▲) in the presence of 5 μM rotenone (Δ). (A) NO was provided exogenously by the repeated addition of 0.5 m*M* nitrosoglutathione (GSNO); (B) mtNOS activity was stimulated by loading mitochondria with Ca^{2+} (40 nmol/mg mitochondrial protein) and was inhibited using L-NMMA (5 m*M*); (C) mtNOS basal activity was inhibited by 5 m*M* L-NMMA; and (D) mtNOS basal activity was inhibited by 5 m*M* L-citrulline (L-cit). Exogenous NO was provided by repeated additions of 0.5 m*M* GSNO. Reproduced with permission from P. Ghafourifar and C. Richter, *J. Biol. Chem.* **380,** 1025 (1999).

mtNOS and Mitochondrial Matrix pH

Mitochondrial respiratory complexes are functionally arranged in a redox potential (also called midpoint potential) hierarchy. Electrons enter the chain, flow to the downstream complexes, and reduce O_2 to water at the terminal respiratory complex, complex IV. Coupled to the electron flow, protons are pumped from the mitochondrial matrix into the mitochondrial intermembrane space. This establishes the ΔpH, a proton gradient across the coupling membrane that renders the membrane alkaline inside. Inhibition of the mitochondrial electron transport chain, e.g., by intramitochondrially formed NO, decreases ΔpH (Fig. 5). Mitochondrial matrix pH can be determined spectrofluorometrically using 2′,7′-bis(carboxyethyl)-5,6-carboxyfluorescein (BCECF) as a probe.[26]

[26] P. Ghafourifar and C. Richter, *J. Biol. Chem.* **380,** 1025 (1999).

Buffers

Loading buffer: 250 m*M* sucrose, 10 m*M* Tris–HCL, 0.1 m*M* EGTA–Tris, pH 7.4
Incubation buffer: 200 m*M* sucrose, 0.5 m*M* Tris–MOPS, pH 7.4

Procedure

Suspend 25 mg mitochondrial protein in 1 ml of the loading buffer containing 10 μg BCECF. Mix and incubate for 20 min at room temperature in the dark (because BCECF is very light sensitive) while shaking very gently. Spin the mitochondria at 10,000*g* for 10 min at 4°, wash the pellet with 1 ml ice-cold loading buffer, and recentrifuge. Resuspend the pellet in 1 ml of the ice-cold incubation buffer. Measure the matrix pH of 0.2 mg mitochondrial protein/ml in incubation buffer at room temperature with a spectrofluorometer with an excitation wavelength of 500 nm (4-mm slit) and an emission wavelength of 525 nm (5-mm slit).

Calibration

Incubate 0.2 mg of BCECF-loaded mitochondrial protein in 1 ml of calibration buffer containing 100 m*M* KCl, 30 μM EGTA–Tris, 2 μM rotenone, 0.8 μg/ml oligomycin, 50 n*M* carbonyl cyanide 3-chlorophenylhydrazone (CCCP), 0.25 μg/ml valinomycin, and 10 m*M* Tris–MOPS, pH ranging from 6.2 to 8.2. Intra- and extramitochondrial pH are equal under these conditions. Measure extramitochondrial pH with a glass electrode and the intramitochondrial fluorescence as explained earlier. Draw the working curve by plotting the fluorescence intensity against pH.

Acknowledgment

I greatly acknowledge the enormous intellectual freedom that Dr. Christoph Richter provided to me during the time I spent in his laboratory. The extremely valuable discussions with and the critical reading of the manuscript by Dr. Tammy Dugas are greatly appreciated.

Section IV

Peroxynitrite

[32] Peroxynitrite Formation from Biochemical and Cellular Fluxes of Nitric Oxide and Superoxide

By MARÍA NOEL ALVAREZ, MADIA TRUJILLO, and RAFAEL RADI

Introduction

Peroxynitrite anion ($ONOO^-$) is the product of the diffusion-controlled reaction ($k \sim 10^{10} M^{-1} s^{-1}$)[1] of the radical species nitric oxide (·NO) and superoxide ($O_2^{\cdot -}$) [Eq. (1)], and this species has been shown to contribute to oxidative signaling and damage.[2–6]

$$\cdot NO + O_2^{\cdot -} \rightarrow ONOO^- \tag{1}$$

The oxidative chemistry of peroxynitrite is complex and dictated by a series of reaction pathways, including direct one- and two-electron oxidations, carbon dioxide-dependent formation of carbonate ($CO_3^{\cdot -}$) and nitrogen dioxide ($\cdot NO_2$) radicals, and proton-catalyzed homolysis to hydroxyl radical (·OH) and $\cdot NO_2$.[1,6] Many of the experiments used to study the biological chemistry and actions of peroxynitrite have been performed using bolus additions[7–9] or infusions[10–12] of the oxidant. However, although useful for studying reaction kinetics, mechanisms, oxidation yields, and diffusion properties, this procedure has limitations when

[1] R. Radi, A. Denicola, B. Alvarez, G. Ferrer-Sueta, and H. Rubbo, *in* "Nitric Oxide" (L. Ignarro, ed.), p. 57. Academic Press, San Diego, 2000.

[2] J. Ara, S. Przedborski, A. B. Naini, V. Jackson-Lewis, R. R. Trifiletti, J. Horwitz, and H. Ischiropoulos, *Proc. Natl. Acad. Sci. U.S.A.* **95,** 7659 (1998).

[3] A. G. Estevez, J. P. Crow, J. B. Sampson, C. Reiter, Y. Zhuang, G. J. Richardson, M. M. Tarpey, L. Barbeito, and J. S. Beckman, *Science* **286,** 2498 (1999).

[4] C. R. White, T. A. Brock, L. Y. Chang, J. Crapo, P. Briscoe, D. Ku, W. A. Bradley, S. H. Gianturco, J. Gore, B. A. Freeman, and M. M. Tarpey, *Proc. Natl. Acad. Sci. U.S.A.* **91,** 1044 (1994).

[5] C. Brito, M. Naviliat, A. C. Tiscornia, F. Vuillier, G. Gualco, G. Dighiero, R. Radi, and A. M. Cayota, *J. Immunol.* **162,** 3356 (1999).

[6] R. Radi, G. Peluffo, M. N. Alvarez, M. Naviliat, and A. Cayota, *Free Radic. Biol. Med.* **30,** 463 (2001).

[7] R. Radi, J. S. Beckman, K. M. Bush, and B. A. Freeman, *J. Biol. Chem.* **266,** 4244 (1991).

[8] J. S. Beckman, H. Ischiropoulos, L. Zhu, M. van der Woerd, C. Smith, J. Chen, J. Harrison, J. C. Martin, and M. Tsai, *Arch. Biochem. Biophys.* **298,** 438 (1992).

[9] A. Denicola, J. M. Souza, and R. Radi, *Proc. Natl. Acad. Sci. U.S.A.* **95,** 3566 (1998).

[10] A. Denicola, H. Rubbo, D. Rodriguez, and R. Radi, *Arch. Biochem. Biophys.* **304,** 279 (1993).

[11] A. Trostchansky, C. Batthyany, H. Botti, R. Radi, A. Denicola, and H. Rubbo, *Arch. Biochem. Biophys.* **395,** 225 (2001).

[12] C. Batthyany, C. X. Santos, H. Botti, C. Cervenansky, R. Radi, O. Augusto, and H. Rubbo, *Arch. Biochem. Biophys.* **384,** 335 (2000).

0076-6879/02 $35.00

applied to biological processes. First, the free radical chemistry triggered by persistent and low steady-state concentrations (e.g., nanomolar) of peroxynitrite-derived radicals is sometimes different to that observed after the acute addition of milimolar amounts of the oxidant, which has a half-life of less than a second at pH 7.4 and 37°.[1] Second, the peroxynitrite precursors ·NO and $O_2^{\cdot-}$ participate in competing reactions, which are not always considered,[13–16] and therefore limit peroxynitrite yields and actions. Third, the ·NO to $O_2^{\cdot-}$ ratios may be distant to the 1 : 1 stoichiometry required for peroxynitrite formation, and the excess of one precursor radical may affect peroxynitrite-dependent oxidations.[14,16] Finally, the significantly distinct diffusion properties of ·NO and $O_2^{\cdot-}$[17,18] make peroxynitrite formation sites and yields in cells/tissues highly affected by compartmentalization.

The aims of this article are to (1) indicate methodologies for the controlled biochemical generation of peroxynitrite from fluxes of ·NO plus $O_2^{\cdot-}$ and (2) assess the biological formation of peroxynitrite from the simultaneous production of ·NO plus $O_2^{\cdot-}$ using activated macrophages as a cellular model.

Peroxynitrite from Biochemical Fluxes of Nitric Oxide and Superoxide

General Considerations

The simultaneous generation of ·NO plus $O_2^{\cdot-}$ can be accomplished by a series of biochemical or chemical systems, all of which have their advantages and disadvantages. Important issues to consider in selecting the most appropriate system are (a) the desired ·NO and $O_2^{\cdot-}$ fluxes and the time of exposure, (b) the influence of the various participating reactants on peroxynitrite formation rates and reaction pathways, and (3) the direct effects of the ·NO- and/or $O_2^{\cdot-}$-generating systems on the biotargets under study. The most commonly used ·NO-generating systems are ·NO donors, such as the diazeniumdiolates,[19] *S*-nitrosothiols,[20] and pure ·NO,[21] whereas for $O_2^{\cdot-}$ enzymatic formation from xanthine oxidase,[21] autooxidation of

[13] I. Fridovich, *J. Biol. Chem.* **272,** 18515 (1997).

[14] H. Rubbo, R. Radi, M. Trujillo, R. Telleri, B. Kalyanaraman, S. Barnes, M. Kirk, and B. A. Freeman, *J. Biol. Chem.* **269,** 26066 (1994).

[15] R. Radi, *Chem. Res. Toxicol.* **9,** 828 (1996).

[16] A. M. Miles, D. S. Bohle, P. A. Glassbrenner, B. Hansert, D. A. Wink, and M. B. Grisham, *J. Biol. Chem.* **271,** 40 (1996).

[17] A. Denicola, J. M. Souza, R. Radi, and E. Lissi, *Arch. Biochem. Biophys.* **328,** 208 (1996).

[18] R. E. Lynch and I. Fridovich, *J. Biol. Chem.* **253,** 4697 (1978).

[19] L. K. Keefer, R. W. Nims, K. M. Davies, and D. A. Wink, *Methods Enzymol.* **268,** 281 (1996).

[20] W. R. Mathews and S. W. Kerr, *J. Pharmacol. Exp. Ther.* **267,** 1529 (1993).

[21] L. Castro, M. N. Alvarez, and R. Radi, *Arch. Biochem. Biophys.* **333,** 179 (1996).

reduced compounds[22] and superoxide donors,[23] and addition of pure $O_2^{\cdot-}$ [24] have been utilized. Experiments using γ radiolysis for the simultaneous generation of ·NO and $O_2^{\cdot-}$ have been performed.[25] Valuable data can be obtained from this technology, which, however, is not readily available in most laboratories and only allows for a 1 : 1 ·NO to $O_2^{\cdot-}$ ratio.

Sources of ·NO

Diazeniumdiolates. Diazeniumdiolates (also known as NONOates or NOCs), containing a moiety of the type

$$>N-N(O^-)-N{=}O$$

have been used widely and successfully. They are thermally decomposed to two molecules of ·NO per molecule.[19,26] Nitric oxide fluxes from the different compounds depend on the initial concentration of the donor and their half-lives. For example, half-lives of propylamine-NONOate (NOC-7), spermine-NONOate, and DETA-NONOate (NOC-18) are 5 min, 39 min, and 20 hr, respectively, in phosphate buffer, pH 7.4 and 37°.[26] Decomposition rates are highly influenced by the composition of media, temperature, and pH and, thus, after initial selection of a diazeniumdiolate, direct determination of the *actual* ·NO fluxes in the reaction mixture must be determined by methods such as oxyhemoglobin oxidation or electrochemical detection.[21,27]

The unimolecular decomposition of the diazeniumdiolates follows first-order kinetics [Eq. (2)] and produces ·NO according to (Eq. 3):

$$[\text{NONOate}]_t = [\text{NONOate}]_o e^{-kt} \quad (2)$$

$$\frac{d[\cdot\text{NO}]}{dt} = 2k[\text{NONOate}]_t \quad (3)$$

where k is the first-order rate constant of decomposition related to the compound half-life ($t_{0.5}$), $t_{0.5} = \ln 2/k$. Thus, to obtain a constant ·NO flux throughout an experiment, the chosen NONOate should have a half-life $\geq$4 times longer than the experimental time so that less than 20% of the original compound is consumed.

S-Nitrosothiols. Low molecular weight *S*-nitrosothiols (RSNO), including *S*-nitrosocysteine (Cys-NO), *S*-nitrosoacetylpenicillamine (SNAP), and *S*-nitrosoglutathione (GSNO), have been used frequently as sources to produce ·NO.[16,20,28]

[22] H. P. Misra and I. Fridovich, *J. Biol. Chem.* **247,** 188 (1972).
[23] K. U. Ingold, T. Paul, M. J. Young, and L. Doiron, *J. Am. Chem. Soc.* **119,** 12364 (1997).
[24] J. S. Valentine, A. R. Miksztal, and D. T. Sawyer, *Methods Enzymol.* **105,** 71 (1984).
[25] S. Goldstein, G. Czapski, J. Lind, and G. Merenyi, *J. Biol. Chem.* **275,** 3031 (2000).
[26] A. L. Fitzhugh and L. K. Keefer, *Free Radic. Biol. Med.* **28,** 1463 (2000).
[27] M. Kelm, R. Dahmann, D. Wink, and M. Feelisch, *J. Biol. Chem.* **272,** 9922 (1997).
[28] F. Laszlo, B. J. Whittle, and S. Moncada, *Br. J. Pharmacol.* **115,** 498 (1995).

However, the production of ·NO from RSNO has been, in general, not well controlled, as various factors, including light, metals, and reductants, highly affect the decomposition. Thus, strict experimental conditions (even more than for NONOates!) are required for reproducible and constant ·NO fluxes, and although it has been assumed that, for example, Cys-NO is a short-lived RSNO, these compounds do not release ·NO spontaneously and the half-life is totally dependent on assay conditions.[29] Particularly important for the content of this article is that we[29] and others[30–32] have shown that $O_2^{\cdot-}$ promotes ·NO release from RSNO, with this becoming a critical aspect to consider when using RSNO in $O_2^{\cdot-}$-producing systems.[29] In addition, RSNO can also release nitroxyl anion (NO^-) or participate in transnitrosation reactions via their nitrosonium ion (NO^+)-like reactivity.[33]

Pure ·NO. Nitric oxide solutions with concentrations of up to 1.7–2.0 m*M* can be prepared conveniently using ·NO gas, which is dissolved in deoxygenated water or buffer as described elsewhere.[21] The ·NO stock solution can be used to load a gas-tight syringe that can deliver ·NO at controlled rates using a motor-driven pump as described previously.[21] An important limitation relies on the difficulty of reproducible and appropriate mixing of ·NO going into the reaction solution and the potential escape of ·NO to the gas phase. Thus, the reaction mixture must be under continuous stirring and, preferentially, under positive pressure in capped tubes/vessels having minimal headspace. An important advantage of this method is that there cannot be secondary reactions of the donors (i.e., diazeniumdiolates or RSNO) or their decomposition products (e.g., spermine, cystine) in the system under study.

Sources of $O_2^{\cdot-}$

Xanthine Oxidase. Xanthine oxidase (xanthine:oxygen oxidoreductase, EC 1.2.3.2) oxidizes a variety of substrates, including the purines hypoxanthine and xanthine, to uric acid and reduces molecular oxygen by one and two electrons to $O_2^{\cdot-}$ and hydrogen peroxide (H_2O_2), respectively.[34] The formation of $O_2^{\cdot-}$, known as "univalent flux," depends on a variety of factors, including substrate concentration, pH, and oxygen tension.[35] The univalent flux percentage is also dependent on the substrate molecule, with low turnover substrates providing the largest percentage of univalent flux.

[29] M. Trujillo, M. N. Alvarez, G. Peluffo, B. A. Freeman, and R. Radi, *J. Biol. Chem.* **273,** 7828 (1998).
[30] S. Aleryani, E. Milo, Y. Rose, and P. Kostka, *J. Biol. Chem.* **273,** 6041 (1998).
[31] D. Jourd'heuil, C. T. Mai, F. S. Laroux, D. A. Wink, and M. B. Grisham, *Biochem. Biophys. Res. Commun.* **246,** 525 (1998).
[32] E. Ford, M. N. Hughes, and P. Wardman, *J. Biol. Chem.* **4,** 2430 (2002).
[33] D. R. Arnelle and J. S. Stamler, *Arch. Biochem. Biophys.* **318,** 279 (1995).
[34] J. M. McCord and I. Fridovich, *J. Biol. Chem.* **243,** 5753 (1968).
[35] I. Fridovich, *J. Biol. Chem.* **245,** 4053 (1970).

Xanthine oxidase usually yields a constant flux of $O_2^{\cdot-}$ for 10–20 min after which fluxes decrease progressively due to (a) inactivation of the enzyme by reaction products, (b) purine substrate depletion, and/or (c) oxygen depletion. If longer incubation times are required, sequential additions of xanthine oxidase, oxygen, and substrate at different times may be performed to compensate for losses in $O_2^{\cdot-}$ formation rates. Commercial sources of xanthine oxidase typically contain iron traces, which are difficult to fully eliminate.[36,37] Thus, it is recommended that 100 μM DTPA be included in the buffer system to minimize Haber–Weiss chemistry. The use of desferrioxamine is not convenient, as desferrioxamine can inhibit peroxynitrite-dependent oxidations.[38] In addition, catalase (500–3000 U/ml) can be added to eliminate H_2O_2 and diminish enzyme autoinactivation and undesirable secondary reactions.

Xanthine oxidase activity is determined spectrophotometrically at 295 nm by following xanthine conversion to uric acid ($\varepsilon = 11$ mM^{-1}cm^{-1}) and $O_2^{\cdot-}$ formation by the reduction of cytochrome c^{3+} ($\varepsilon = 21$ mM^{-1}cm^{-1}).[21,29,39] Typically, 100–200 μM hypoxanthine is used as substrate ($K_m = 1$–3 μM[40]). However, because uric acid reacts with both peroxynitrite ($k \sim 150$–500 M^{-1}s^{-1}[41,42]) and peroxynitrite-derived radicals,[43,44] for incubations of >10 min, alternative substrates can be used, such as lumazine (50–200 μM[29,45]), pterin (50–200 μM[46]), and acetaldehyde (0.5–2 mM[21]), which yield less reactive products. In our hands, a reasonable upper limit of formation of $O_2^{\cdot-}$ by xanthine oxidase is of the order of ∼20 μM/min.

An important issue is whether ·NO and $ONOO^-$ inactivate xanthine oxidase. In turnover and under aerobic conditions, neither ·NO[47,48] nor peroxynitrite inhibit xanthine oxidase activity.[47] Also, RSNO, in general, do not affect xanthine

[36] R. V. Lloyd and R. P. Mason, *J. Biol. Chem.* **265,** 16733 (1990).

[37] B. E. Britigan, S. Pou, G. M. Rosen, D. M. Lilleg, and G. R. Buettner, *J. Biol. Chem.* **265,** 17533 (1990).

[38] A. Denicola, J. M. Souza, R. M. Gatti, O. Augusto, and R. Radi, *Free Radic. Biol. Med.* **19,** 11 (1995).

[39] R. C. Bray, *in* "The Enzymes" (P. G. Boyer, ed.). Academic Press, Orlando, FL, 1975.

[40] A. Hausladen and I. Fridovich, *Arch. Biochem. Biophys.* **306,** 415 (1993).

[41] C. X. Santos, E. I. Anjos, and O. Augusto, *Arch. Biochem. Biophys.* **372,** 285 (1999).

[42] G. L. Squadrito, R. Cueto, A. E. Splenser, A. Valavanidis, H. Zhang, R. M. Uppu, and W. A. Pryor, *Arch. Biochem. Biophys.* **376,** 333 (2000).

[43] T. Masuda, H. Shinohara, and M. Kondo, *J. Radiat. Res.* **16,** 153 (1975).

[44] M. G. Simic and S. V. Jovanovic, *J. Am. Chem. Soc.* **111,** 5778 (1989).

[45] D. Jourd'heuil, F. L. Jourd'heuil, P. S. Kutchukian, R. A. Musah, D. A. Wink, and M. B. Grisham, *J. Biol. Chem.* **276,** 28799 (2001).

[46] T. Sawa, T. Akaike, and H. Maeda, *J. Biol. Chem.* **275,** 32467 (2000).

[47] M. Houston, P. Chumley, R. Radi, H. Rubbo, and B. A. Freeman, *Arch. Biochem. Biophys.* **355,** 1 (1998).

[48] C. I. Lee, X. Liu, and J. L. Zweier, *J. Biol. Chem.* **275,** 9369 (2000).

oxidase activity, except Cys-NO, which may serve as a substrate and decrease $O_2^{\cdot-}$ formation rates.[29]

Autooxidation of Reduced Compounds and Superoxide Donors

Superoxide may be formed from the autooxidation of compounds such as flavins, quinones, phenols, and catechols.[22,49–51] These nonenzymatic reactions are usually poorly controlled, as they are sensitive to various factors such as light, metals, oxygen level, and pH and lead to the formation of secondary radicals that may themselves react avidly with ·NO. However, specific experiments can be designed using these redox-active compounds (e.g., at concentrations 10–200 μM) under rigorous and well-characterized reaction conditions to generate $O_2^{\cdot-}$ for extended periods of time. Because cytochrome *c* can be reduced directly by several of these compounds, $O_2^{\cdot-}$ detection must be confirmed by the addition of SOD.

An azo compound, the di(4-carboxybenzyl)hyponitrite known as SOTS-1 (superoxide thermal source-1),[23,52,53] has been developed as a "superoxide donor" in aqueous solutions. Superoxide formation occurs after thermal homolysis of the azo compound, which decomposes to yield electron-rich carbon-centered radicals oxidized by molecular oxygen to yield $O_2^{\cdot-}$ (e.g., 3.5 μM/min with 1 mM SOTS-1) and the corresponding carbocation.[23] The half-life of SOTS-1 is 80 min at pH 7.5 and 37° and yields 0.4 M equivalents of $O_2^{\cdot-}$. SOTS-1 is not available commercially and must be synthesized[23] and handled as indicated elsewhere.[52] This is a promising compound developed on a sound chemical basis. It must be tested further in more complex biological milieu, although the need for local synthesis limits its use.

Solutions of Superoxide

Preparation and quantitation of stable potassium superoxide (KO_2) solutions in dimethyl sulfoxide (DMSO) have been described elsewhere.[24] The quality and dryness of the DMSO are critical, as any water trace will promote $O_2^{\cdot-}$ dismutation and loss of concentration. Superoxide solutions have been used successfully for the formation of peroxynitrite after fast mixing with a pure ·NO solution using stopped-flow techniques.[54] However, they are not so easily applicable for generating $O_2^{\cdot-}$ fluxes, as mixing artifacts can arise when the DMSO-saturated $O_2^{\cdot-}$ solution (1–1.3 mM) is added slowly to the aqueous reaction milieu. Also, interference of excess DMSO in peroxynitrite-dependent oxidations must be considered.

[49] S. Pfeiffer, K. Schmidt, and B. Mayer, *J. Biol. Chem.* **275,** 6346 (2000).
[50] H. P. Misra and I. Fridovich, *J. Biol. Chem.* **247,** 3170 (1972).
[51] C. Szabo and A. L. Salzman, *Biochem. Biophys. Res. Commun.* **209,** 739 (1995).
[52] G. R. Hodges, J. Marwaha, T. Paul, and K. U. Ingold, *Chem. Res. Toxicol.* **13,** 1287 (2000).
[53] T. Paul, *Arch. Biochem. Biophys.* **382,** 253 (2000).
[54] C. D. Reiter, R. J. Teng, and J. S. Beckman, *J. Biol. Chem.* **275,** 32460 (2000).

Simultaneous Formation of ·NO and $O_2^{\cdot-}$ from SIN-1 Decomposition

SIN-1 (3-morpholinosydnonimine) can generate both ·NO and $O_2^{\cdot-}$ under aerobic conditions.[55] SIN-1 is first hydrolyzed to SIN-1A, which subsequently suffers a one-electron oxidation by molecular oxygen to yield $O_2^{\cdot-}$ and the cation radical of SIN-1A, SIN-$1^{\cdot+}$. The latter compound decomposes spontaneously to yield ·NO and SIN-1C. In simple reaction systems, SIN-1 produces similar yields of ·NO and $O_2^{\cdot-}$, as assessed by a method developed for the simultaneous detection of both radicals.[27] However, SIN-1 decomposition pathways vary in the presence of other oxidants, which attack SIN-1A[56,57] and result in the formation of ·NO (but not $O_2^{\cdot-}$!). In fact, oxidizing radicals arising from peroxynitrite reaction with targets may serve as SIN-1A oxidants and contribute to the generation of an excess of ·NO over $O_2^{\cdot-}$.

SIN-1 decomposition is highly pH dependent, as the hydrolysis step is base-catalyzed,[55] and therefore stock solutions should be prepared in dilute HCl (pH 5) and kept anaerobically at 0–4° until use. Typically, concentrations in the 0.1–1.0 m*M* range are used, with a half-life at pH 7.4, 37° and phosphate buffer on the order of 60 min.

Nitric Oxide and Superoxide Fluxes in Action: Peroxynitrite Detection

Various methods can be used to demonstrate and quantitiate peroxynitrite formation, including oxidation [e.g., dihydrodichlorofluorescein (DCFH) and dihydrorhodamine (DHR)] and nitration of probes (e.g., *p*-hydroxyphenylacetic acid) or endogenous biomolecules (e.g., tyrosine). Reviews covering peroxynitrite detection methods have been published elsewhere.[6,58]

Peroxynitrite Formation by Fluxes of Diazeniumdiolate-Derived ·NO and Xanthine Oxidase-Derived $O_2^{\cdot-}$. As seen in Fig. 1A, only the simultaneous formation of ·NO and $O_2^{\cdot-}$ (NOC + XO condition) leads to significant DHR oxidation. DHR oxidation yields were comparable with those obtained with pure peroxynitrite (~40%),[6,58] in agreement with a stoichiometric formation of peroxynitrite from its radical precursors. Similarly, DCFH and cytochrome c^{2+} were oxidized efficiently by fluxes of ·NO and $O_2^{\cdot-}$ (Fig. 1B). The inset in Fig. 1B shows the inhibitory effect of xanthine oxidase-derived uric acid on DCFH oxidation, which can be reverted either by using the alternative substrate lumazine (Lum) or by the addition of uricase[29] (e.g., 0.1 U/ml).

Excess of either ·NO or $O_2^{\cdot-}$ leads to a decrease of net DHR oxidation yields due to secondary reactions with DHR radical intermediates, as well as with

[55] H. Bohn and K. Schonafinger, *J. Cardiovasc. Pharmacol.* **14,** S6 (1989).

[56] R. J. Singh, N. Hogg, J. Joseph, E. Konorev, and B. Kalyanaraman, *Arch. Biochem. Biophys.* **361,** 331 (1999).

[57] Y. Wang, L. G. Rochelle, H. Kruszyna, R. Kruszyna, R. P. Smith, and D. E. Wilcox, *Toxicology* **88,** 165 (1994).

[58] M. Trujillo, M. Naviliat, M. N. Alvarez, G. Peluffo, and R. Radi, *Analysis* **28,** 518 (2000).

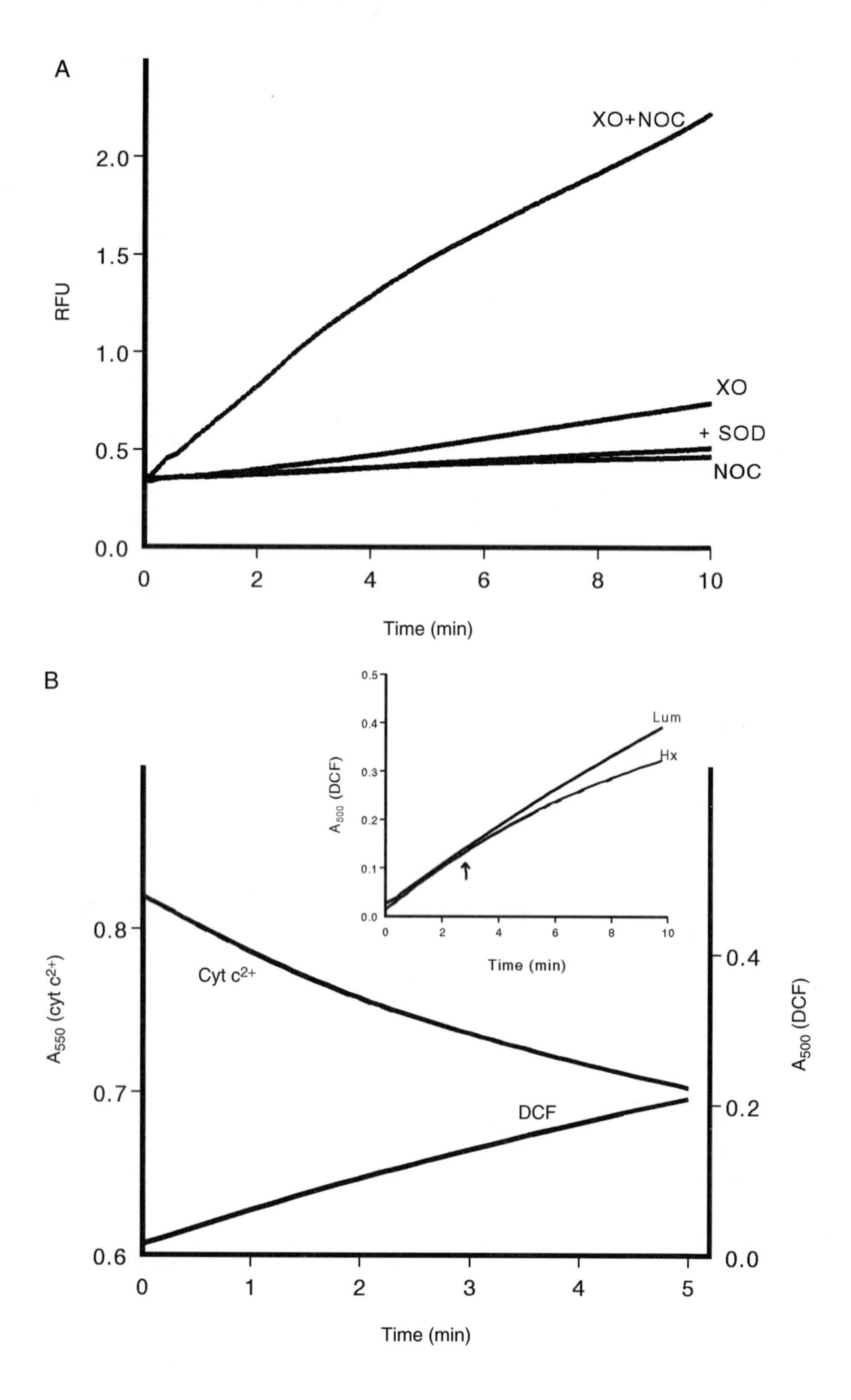

A
XO+NOC
XO
+ SOD
NOC
RFU
2.0
1.5
1.0
0.5
0.0
0
2
4
6
8
10
Time (min)
B
Lum
Hx
A_{500} (DCF)
0.5
0.4
0.3
0.2
0.1
0.0
0
2
4
6
8
10
Time (min)
Cyt c^{2+}
DCF
A_{550} (cyt c^{2+})
A_{500} (DCF)
0.8
0.7
0.6
0.4
0.2
0.0
0
1
2
3
4
5
Time (min)

peroxynitrite-derived $\cdot NO_2$.[16,59] Importantly, it has been determined that both $\cdot NO_2$ and $CO_3^{\cdot -}$ radicals can oxidize DHR and DCFH efficiently via one-electron oxidation to yield the corresponding probe radicals, which then dismutate to one molecule of oxidized (i.e., rhodamine and dichlorofluorescein) and one of parent compound (P. Wardman, personal communication). Therefore, the presence of CO_2 does not affect probe oxidation yields by $\cdot NO$ and $O_2^{\cdot -}$ fluxes.[58]

Peroxynitrite Formation from Superoxide-Mediated Decomposition of S-Nitrosothiols. Superoxide promotes the decomposition of *S*-nitrosoglutathione (GSNO) to $\cdot NO$ [Eq. (4)], which then reacts with a second molecule of $O_2^{\cdot -}$ to yield peroxynitrite[29]:

$$GSNO + O_2^{\cdot -} + H^+ \rightarrow GSH + \cdot NO + O_2 \quad (4)$$

Superoxide-mediated GSNO decomposition leads to the oxidation of DHR and cytochrome c^{2+} (Table I). As DHR and cytochrome c^{2+} oxidation yields by authentic peroxynitrite are around 40%,[6,58] it can be calculated that all $\cdot NO$ evolving from $O_2^{\cdot -}$-dependent GSNO decomposition is converted to peroxynitrite (Table I). Uric acid inhibits DHR oxidation by reacting with $\cdot NO_2$ and $\cdot OH$, while its reaction rate with peroxynitrite is not sufficient to inhibit cytochrome c^{2+} oxidation, which occurs mainly by a bimolecular reaction with peroxynitrite.[60] Methionine reacts directly with peroxynitrite[1] and inhibits both oxidations. For the DHR and reduced cytochrome *c* assay, initial rates should be measured to avoid confusing effects arising from either uric acid accumulation or interactions of $O_2^{\cdot -}$ with the oxidized form of cytochrome *c*.

Nitration by Fluxes of $\cdot NO$ and $O_2^{\cdot -}$

Peroxynitrite can nitrate tyrosine residues via free radical mechanisms involving the transient formation of tyrosyl and $\cdot NO_2$ radicals.[1,6,61,62] Nitration yields

[59] S. Goldstein, G. Czapski, J. Lind, and G. Merenyi, *Chem. Res. Toxicol.* **12,** 132 (1999).
[60] L. Thomson, M. Trujillo, R. Telleri, and R. Radi, *Arch. Biochem. Biophys.* **319,** 491 (1995).
[61] S. V. Lymar, Q. Jiang, and J. K. Hurst, *Biochemistry* **35,** 7855 (1996).
[62] H. Ischiropoulos, *Arch. Biochem. Biophys.* **356,** 1 (1998).

FIG. 1. Peroxynitrite formation from fluxes of $\cdot NO$ and $O_2^{\cdot -}$. (A) Nitric oxide and $O_2^{\cdot -}$ were formed at an equimolar flux of 0.05 μM/min from 50 μM NOC-18 [1-hydroxy-2-oxo-3,3-bis(2-aminoethyl)-1-triazene] and 0.04 mU/ml xanthine oxidase (XO) in the presence of 200 μM hypoxanthine (30% univalent flux). in 50 mM phosphate buffer, pH 7.4, 25°. DHR was 2 μM, and rhodamine formation was followed fluorimetrically (SLM Aminco fluorometer) at $\lambda_{ex} = 500$ nm and $\lambda_{em} = 525$ nm. SOD (0.05 μM) was added to the sample containing NOC-18 plus xanthine oxidase. (B) Conditions were as in (A), except that fluxes were 2.5 μM/min, with $\cdot NO$ produced by 50 μM NOC-7 (1-hydroxy-2-oxo-3-(*N*-methyl aminopropyl)-3-methyl-1-triazene). Cytochrome c^{2+} was 40 μM and DCFH was 100 μM ($\varepsilon_{500} = 59.5$ mM^{-1} cm^{-1} for the oxidized form, DCF). In the inset, lumazine was 200 μM and the arrow indicates the time after which uric acid accumulation starts to inhibit DCFH oxidation.

TABLE I
SUPEROXIDE-DEPENDENT *S*-NITROSOGLUTATHIONE DECOMPOSITION LEADS TO PEROXYNITRITE FORMATION[a]

Condition	$-d$[GSNO]/dt (μM/min)	d[RH]/dt (μM/min)	$-d$[Cyt c^{2+}]/dt (μM/min)
Control	0.5 ± 0.3	0	0.3 ± 0.2
Xanthine oxidase	2.4 ± 0.2	1.2 ± 0.1	1.3 ± 0.1
+ SOD (5 μM)	0.5 ± 0.2	0	0.6 ± 0.3
+ catalase (200 /ml)	2.1 ± 0.6	1.1 ± 0.1	1.1 ± 0.2
+ uric acid (100 μM)	1.8 ± 0.5	0	1.1 ± 0.1
+ methionine (1 mM)	2.2 ± 0.1	0	0.7 ± 0.1

[a] *S*-nitrosoglutathione (1 mM) was present in all the conditions. Then, xanthine oxidase (6 mU/ml) was incubated with hypoxanthine (150 μM) in air-equilibrated 100 mM potassium phosphate, pH 7.4, and 25°, either alone or in the presence of the compounds indicated. The univalent flux percentage was 25%, corresponding to 6 μM/min $O_2^{\cdot -}$. GSNO decomposition was followed at 336 nm ($\varepsilon = 0.87$ mM^{-1} cm^{-1}). DHR (100 μM) and cytochrome c^{2+} (50 μM) initial oxidation rates were followed at 500 ($\varepsilon = 78.8$ mM^{-1} cm^{-1} for rhodamine, RH) and 550 nm, respectively.

after the bolus addition of peroxynitrite can be 6–10 and 14–20% in phosphate buffer in the absence or presence of carbon dioxide, respectively.[1,62] However, peroxynitrite formed from fluxes of ·NO and $O_2^{\cdot -}$ may result in significantly lower nitration yields[25,49,52] than those obtained from the addition of peroxynitrite as a bolus. Indeed, under low steady-state concentrations of the radicals, which lead to tyrosine nitration such as $CO_3^{\cdot -}$, $\cdot NO_2$, and ·OH, a significant fraction of $\cdot NO_2$ reacts preferentially with excess tyrosine (e.g., >100 μM) instead of with the tyrosyl radical, which otherwise dimerizes to dityrosine.[25,61] Under low tyrosine concentration or high ·NO and $O_2^{\cdot -}$ fluxes, nitration yields approximate those of authentic peroxynitrite.[46,49,54]

Despite potentially low nitration yields, protein-tyrosine nitration has been shown after exposure to fluxes of ·NO plus $O_2^{\cdot -}$.[63–67] Oxidation of probes, however, is a more sensitive (although less specific) approach than nitration to measure

[63] M. Zou, C. Martin, and V. Ullrich, *Biol. Chem.* **378,** 707 (1997).

[64] A. Daiber, S. Herold, C. Schoneich, D. Namgaladze, J. A. Peterson, and V. Ullrich, *Eur. J. Biochem.* **267,** 6729 (2000).

[65] A. Gow, D. Duran, S. R. Thom, and H. Ischiropoulos, *Arch. Biochem. Biophys.* **333,** 42 (1996).

[66] J. M. Souza, E. Daikhin, M. Yudkoff, C. S. Raman, and H. Ischiropoulos, *Arch. Biochem. Biophys.* **371,** 169 (1999).

[67] Y. Kamisaki, K. Wada, K. Bian, B. Balabanli, K. Davis, E. Martin, F. Behbod, Y. C. Lee, and F. Murad, *Proc. Natl. Acad. Sci. U.S.A.* **95,** 11584 (1998).

peroxynitrite formation in biological systems, more closely resembling reaction yields obtained with pure peroxynitrite.

Cellular Fluxes of Nitric Oxide and Superoxide and Peroxynitrite Formation

General Considerations

Cells produce ·NO from constitutive or inducible isoforms of nitric oxide synthase (NOS).[68] In addition, various biological sources can contribute to $O_2^{\cdot -}$, including the NAD(P)H oxidases of immune and nonimmune cells, mitochondrial electron transport complexes, xanthine oxidase, redox cycling of drugs, and substrate or cofactor-depleted NOS.[13,69–72] In resting cells, where the steady-state concentrations of both ·NO and $O_2^{\cdot -}$ are low, activation or induction of NOS, as well as activation of NAD(P)H oxidases, can lead to significant fluxes of both radicals that could result in peroxynitrite formation. Various reports have indicated that cell-derived ·NO and $O_2^{\cdot -}$ can lead to peroxynitrite,[3,5,27,73–77] but direct and quantitative evidence is rather scarce. Because ·NO is readily diffusible through cell membranes, whereas $O_2^{\cdot -}$ is not, it is envisioned that peroxynitrite will be principally formed close to the site and in the same compartment of $O_2^{\cdot -}$ generation.

This section introduces a well-defined protocol for macrophage stimulation that leads to the simultaneous production of ·NO and $O_2^{\cdot -}$ and results in peroxynitrite with close to stoichiometric yields. This model can then be applied to explore the actions of cell-derived peroxynitrite on target biomolecules and cells.

Stimulation Protocol for Simultaneous Generation of ·NO and $O_2^{\cdot -}$

A near-confluent monolayer of J774.A1 macrophages is exposed to 200 U/ml interferon-γ (IFN-γ) plus 3 μg/ml lipopolysacharide (LPS), known inductors of iNOS. Significant levels of NOS protein and activity are obtained after a 3- to

[68] S. Moncada and A. Higgs, *N. Engl J. Med.* **329,** 2002 (1993).

[69] B. M. Babior, *Am. J. Med.* **109,** 33 (2000).

[70] B. M. Babior, *IUBMB Life* **50,** 267 (2000).

[71] C. R. White, V. Darley-Usmar, W. R. Berrington, M. McAdams, J. Z. Gore, J. A. Thompson, D. A. Parks, M. M. Tarpey, and B. A. Freeman, *Proc. Natl. Acad. Sci. U.S.A.* **93,** 8745 (1996).

[72] J. Vasquez-Vivar, B. Kalyanaraman, P. Martasek, N. Hogg, B. S. Masters, H. Karoui, P. Tordo, and K. A. Pritchard, Jr., *Proc. Natl. Acad. Sci. U.S.A.* **95,** 9220 (1998).

[73] H. Ischiropoulos, L. Zhu, and J. S. Beckman, *Arch. Biochem. Biophys.* **298,** 446 (1992).

[74] M. K. Shigenaga, H. H. Lee, B. C. Blount, S. Christen, E. T. Shigeno, H. Yip, and B. N. Ames, *Proc. Natl. Acad. Sci. U.S.A.* **94,** 3211 (1997).

[75] N. W. Kooy and J. A. Royall, *Arch. Biochem. Biophys.* **310,** 352 (1994).

[76] H. Possel, H. Noack, W. Augustin, G. Keilhoff, and G. Wolf, *FEBS Lett.* **416,** 175 (1997).

[77] A. G. Estevez, N. Spear, S. M. Manuel, R. Radi, C. E. Henderson, L. Barbeito, and J. S. Beckman, *J. Neurosci.* **18,** 923 (1998).

4-hr period, after which a constant rate of cell-derived ·NO is obtained in the range of 0.2–0.3 nmol/min/mg protein and for up to 15–16 hr. This stimulation does not lead to NADPH oxidase *activation*,[78,79] despite scattered reports in the literature that have suggested this to occur. Superoxide formation is triggered by the addition of particulate or nonparticulate stimuli that lead to assembly of the cytosolic and membrane components of NADPH oxidase, such as opsonized zymosan and phorbol myristate acetate (PMA, 2 μg/ml). The release of $O_2^{\cdot -}$ starts immediately after the activation of NADPH oxidase and lasts for 90–120 min at a rate of 0.3–0.4 nmol/min/mg. The simultaneous formation of ·NO and $O_2^{\cdot -}$ is obtained by exposure to IFN-γ and LPS (>4 hr), leading to iNOS induction followed by NADPH oxidase activation. Then, ·NO, $O_2^{\cdot -}$, and hence peroxynitrite are expected to be formed simultaneously for ~2 hr.

Formation of Peroxynitrite by Macrophage-Derived ·NO and $O_2^{\cdot -}$

The simultaneous formation of ·NO (IFN-γ/LPS condition) and $O_2^{\cdot -}$ (PMA condition) leads to a time-dependent oxidation of DCFH (Fig. 2). Oxidation is not observed when cells are stimulated to form ·NO or $O_2^{\cdot -}$ alone. Considering DCFH oxidation yields by peroxynitrite (~35%)[6] and the experimental rates obtained from Fig. 3 (0.13 nmol/min/mg), it is calculated that most ·NO and $O_2^{\cdot -}$ yielded peroxynitrite. If these peroxynitrite formation rates are extrapolated to intraphagosomal formation, it can be estimated to be ~50–100 μM/min, which represents a large and highly toxic flux of peroxynitrite and on line with previous calculations.[10,73]

This cell model and stimulation protocol is being utilized in our laboratory to study the action of macrophage-derived peroxynitrite on *Trypanosoma cruzi* infection processes and can be used further for a large variety of systems.

Formation of Peroxynitrite from Combined Biochemical and Cellular Sources of ·NO and $O_2^{\cdot -}$

As a way to further confirm that the independent production of ·NO and $O_2^{\cdot -}$ leads to peroxynitrite and that peroxynitrite in the macrophage model is mostly formed extracellularly, macrophages were stimulated to form ·NO or $O_2^{\cdot -}$ alone and supplemented with a biochemical source of either $O_2^{\cdot -}$ (i.e., xanthine oxidase) or ·NO (i.e., diazeniumdiolate), respectively, at similar fluxes than those cell derived.

As can be observed (Fig. 3), the combination of cells producing only one radical precursor (·NO or $O_2^{\cdot -}$) plus a donor system producing the second precursor radical result in maximal DCFH oxidation yields, indicating peroxynitrite formation.

[78] B. M. Babior, *Environ. Health Perspect.* **102**(Suppl. 10), 53 (1994).
[79] B. M. Babior, *Blood* **93,** 1464 (1999).

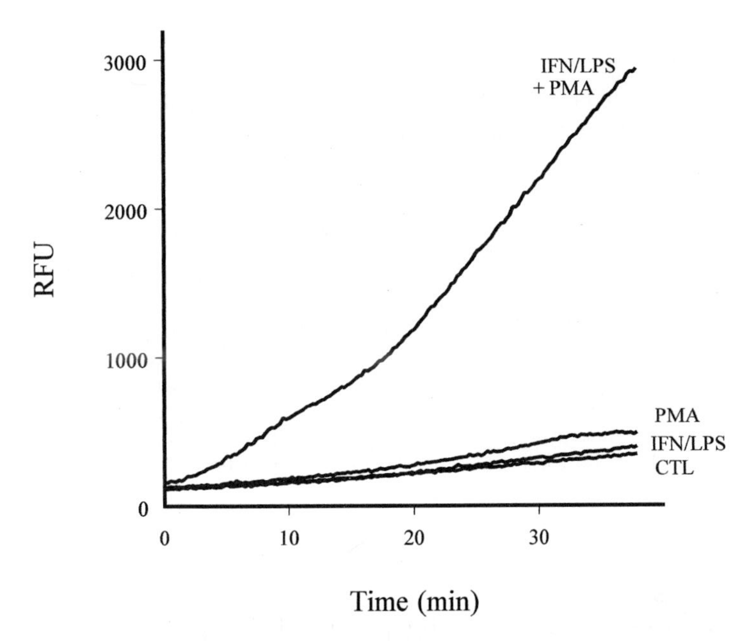

FIG. 2. Peroxynitrite formation from macrophage-derived ·NO and $O_2^{\cdot -}$ Macrophages were plated to confluence in 24-well plates, and the time course of DCFH oxidation was followed in a fluorescence plate reader at 37° (Fluostar, BMG Labtechnologies) with filters at $\lambda_{ex} = 485$ nm and $\lambda_{em} = 520$ nm. Incubations were performed in Dulbecco's-PBS with DCFH-diacetate (10 μM) under different stimulation conditions. Probe oxidation was followed immediately after NADPH oxidase activation.

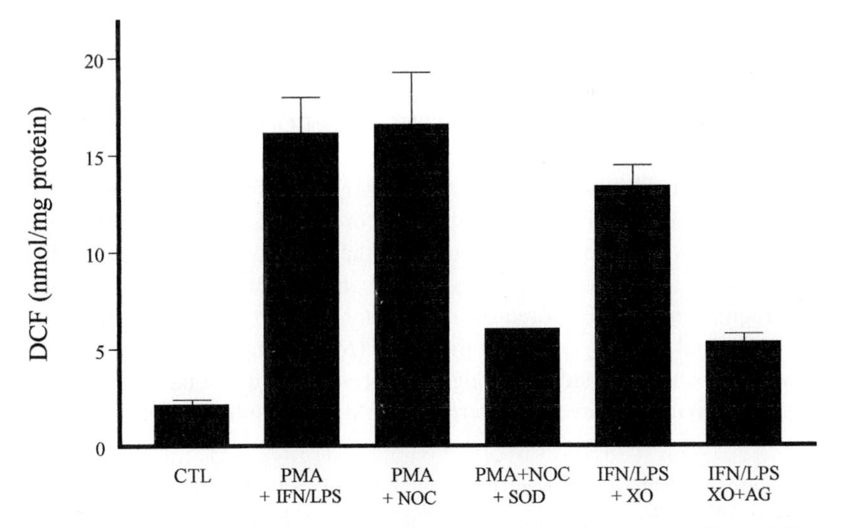

FIG. 3. Macrophage and chemical/biochemical-derived ·NO and $O_2^{\cdot -}$ can combine to yield peroxynitrite. Conditions as in Fig. 2. End point fluorescence measurements (Aminco-SLM) were obtained at 120 min with $\lambda_{ex} = 502$ nm and $\lambda_{em} = 523$ nm. Where indicated, xanthine oxidase (0.02 mU/ml; ~0.03 μM/min $O_2^{\cdot -}$) or NOC-18 (0.02 mM; 0.03 μM/min ·NO) was added.

The inhibitory effects of SOD and the iNOS inhibitor aminoguanidine further support that ·NO and $O_2^{\cdot -}$ fluxes lead to peroxynitrite in high yield.

Acknowledgments

We thank Dr. Gonzalo Peluffo for useful discussions and participation in parts of the experimental work presented herein. We are grateful to Dr. Kazumi Samamoto from Dojindo Laboratories (Kumamoto, Japan) for kindly supplying the NOCs. This work was supported by grants from the Howard Hughes Medical Institute (United States), Fogarty-National Institutes of Health (United States), International Centre for Genetic Engineering and Biotechnology (Italy), Swedish Agency for Research Cooperation (Sweden) to R.R., and Comisión Sectorial de Investigación Científica-Universidad de la República (Uruguay) to M.T. and M.N.A. M.N.A. is partially supported by a fellowship from the Programa de Desarrollo de Ciencias Básicas (Uruguay). R.R. is an International Research Scholar of the Howard Hughes Medical Institute.

[33] Evaluation of Antioxidant Activity Using Peroxynitrite as a Source of Radicals

By YURII V. GELETII and GILBERT G. A. BALAVOINE

Introduction

Reactive nitrogen and oxygen species (RNOS) are formed *in vivo* in side reactions of dioxygen reduction.[1–3] RNOS include free oxygen, $HO^{\cdot}$, $O_2^{\cdot -}$, and nitrogen, $NO_2^{\cdot}$, radicals as well as peroxides, H_2O_2, ROOH, and HOONO/$ONOO^-$ (peroxynitrite). Reacting with cell constituents, these species result in modifications of biological molecules and thus alter their function considerably.[1–3] Superoxide anion is a weak oxidant and therefore is unlikely to damage biological molecules. However, it can be a precursor of the hydroxyl radical, which is likely to be formed in the "superoxide anion-driven Fenton reaction."[1,2] The latter mechanism requires a presence of redox-active metal catalysts, which is an unresolved question in biological systems.[2] Nitric oxide (NO), being an important mediator of diverse biochemical and physiological processes,[4] is a free radical of very low oxidation activity. However, its overproduction may result in formation of more

[1] B. Halliwell and J. M. C. Gutteridge, "Free Radicals in Biology and Medicine," 2nd Ed. Clarendon Press, Oxford, 1989.

[2] B. Halliwell and J. M. C. Gutteridge, *Methods Enzymol.* **186,** 1 (1990).

[3] G. Cs-Szabo, *in* "Nitric Oxide" (L. Ignarro, ed.), p. 841. Academic Press, San Diego, 2000.

[4] S. Moncada and E. A. Higgs, "Endothelium, Nitric Oxide, and Atherosclerosis," p. 13. Futura Publishing Co., Armonk, NY, 1999.

0076-6879/02 $35.00

reactive nitrogen dioxide and peroxynitrite (PN). The latter is formed rapidly in recombination of NO and $O_2^{\cdot-}$ and is likely to be produced *in vivo*.[5,6,7] A stable anion, $ONOO^-$, once protonated under physiological conditions to give peroxynitrous acid, HOONO, produces very reactive "hydroxyl-radical like" species rapidly.[5,6,7] Despite current discussions on the nature of these radicals, the role of carbon dioxide, and transition metal ions in its reactions, PN is definitely a very toxic agent, resulting in tissue injuries.[3,5,6,7]

Different antioxidant systems protect cells from oxidative damage.[1,2] Enzymes remove superoxide anions (SOD) and peroxides (catalase, peroxidase) catalytically, whereas antioxidants are consumed stoichiometrically, scavenging active radicals and forming harmless radicals. Because antiradical activity depends on the nature of RNOS, a universal antioxidant assay does not exist, and the choice of chemical system to generate biologically relevant RNOS *in vitro* is a critical point. If a complex mixture of unknown content should be assayed, the question arises on the total concentration of antioxidants present. This article describes an inexpensive, fast, and simple approach to determine the antioxidant activity of different compounds, particularly their mixtures, such as natural plant extracts of unknown composition.

Antioxidant activity is usually characterized by the reaction rate constant of the given antioxidant with the given radical. Under physiological conditions, a radical flux and antioxidant concentrations are extremely low. Thus, a study of antioxidant activity requires techniques (1) to generate only specific radicals with low controlled rates and yields and (2) to monitor reaction progress at extremely low concentrations of all reagents. These tasks require the use of expensive and complicated equipment. However, good approximations of relative antioxidant activity can often be obtained using the method of competing reactions, which was developed to determine relative k values for methyl radical reactions with aromatic hydrocarbons.[8]

Radical Generation

The radicals and conditions used in evaluation of antioxidant activity *in vitro* should be preferably similar to those *in vivo* (aqueous media, neutral pH, ambient temperature).[1,2] Fenton's reagent ($H_2O_2 + Fe^{2+}$) is used frequently to mimic radical formation,[2] but biologically relevant iron concentrations are too small, $<10\ \mu M^2$, and comparable with possible iron contamination. Moreover, most antioxidants chelate Fe^{2+} or reduce Fe^{3+}, thus strongly interfering with the generation

[5] R. Radi, G. Peluffo, M. N. Alvarez, M. Naviliat, and A. Cayota, *Free Radic. Biol. Med.* **30,** 463 (2001).

[6] W. H. Koppenol, *Met. Ions Biol. Syst.* **36,** 597 (1999).

[7] B. Halliwell, K. Zhao, and M. Whiteman, *Free Radic. Res.* **31,** 651 (1999).

[8] M. Levy and M. Szwarc, *J. Am. Chem. Soc.* **77,** 1949 (1955).

of radicals.[1,2] At the same time, in most cases, PN decays with a rate independent of the presence of antioxidants and transition metal ions and produces radicals with a constant yield ~30%.[5,6,7] Fast bimolecular reactions of PN are rare and well documented.[5,6,7] Some transition metal ions (Co, Mn, Cu) may catalyze PN reactions (see Ref. 9 and references therein); however, chelators such as EDTA or DPTA completely inhibit such activity.[4,6] Hence, PN, for which synthesis, quantification, handling, and manipulation are rather simple, seems to be an excellent source of "biological" radicals.

Method of Competing Reactions and Monitoring Reaction Progress

The PN anion has a maximum at 302 nm, $\varepsilon = 1700\ \text{mol}^{-1}\ \text{liter cm}^{-1}$,[10] which overlaps with a much stronger absorbency of most antioxidants. In rare cases, the kinetics of PN decay can still be monitored with a stop-flow technique, but often it appears to be useless, as a rate-limiting step is a monomolecular PN decay to form radicals. Alternatively, reaction progress can be monitored using a method of competing reaction. If two antioxidants, a detecting molecule DH and a studied AH, are mixed and exposed to radicals R, formed from PN, DH and AH compete for R (see Eqs. (2) and (3)).

$$\text{PN} \rightarrow \text{R}^{\bullet}, \alpha k_o \tag{1}$$

$$\text{R}^{\bullet} + \text{DH} \rightarrow \text{D}^{\bullet} + \text{RH} -> \text{ products, } k_d \tag{2}$$

$$\text{R}^{\bullet} + \text{AH} \rightarrow \text{A}^{\bullet} + \text{RH} -> \text{ products, } k_a \tag{3}$$

where α is a radical yield, whereas k_o, k_d, and k_a are reaction rate constants. If AH scavenges a part of radicals, then it protects DH. At low conversions of antioxidants, this protection is described by Eq. (4):

$$\Delta[\text{DH}]_o/\Delta[\text{DH}]_a = 1 + (k_a/k_d)([\text{AH}]_o/[\text{DH}]_o) \tag{4}$$

where $\Delta[\text{DH}]_o$ and $\Delta[\text{DH}]_a$ are consumptions of DH in the absence and the presence of antioxidant AH by radicals formed from the same amount of PN. The ratio of rate constants k_a/k_d is a slope of a straight line of $\Delta[\text{DH}]_o/\Delta[\text{DH}]_a$ – $[\text{AH}]_o/[\text{DH}]_o$ plot. Equation (4) requires a precise control of the total amount of radicals generated in the system. Such a control can be achieved easily by keeping the amount of PN added constant. However, the effect of slight deviations in PN concentration can be avoided by using in Eq. (4) a reaction stoichiometry $S = \Delta[\text{DH}]/\Delta[\text{PN}]$ instead of $\Delta[\text{DH}]_o$ and $\Delta[\text{DH}]_a$ [Eq. (5)]:

$$S_o/S_a = 1 + (k_a/k_d)([\text{AH}]_o/[\text{DH}]_o) \tag{5}$$

[9] Yu. V. Geletii, A. J. Bailey, E. A. Boring, and C. L. Hill, *Chem. Commun.* 1484 (2001).
[10] D. S. Bohle, B. Hansert, S. C. Paulson, and B. D. Smith, *J. Am. Chem. Soc.* **116,** 7423 (1994).

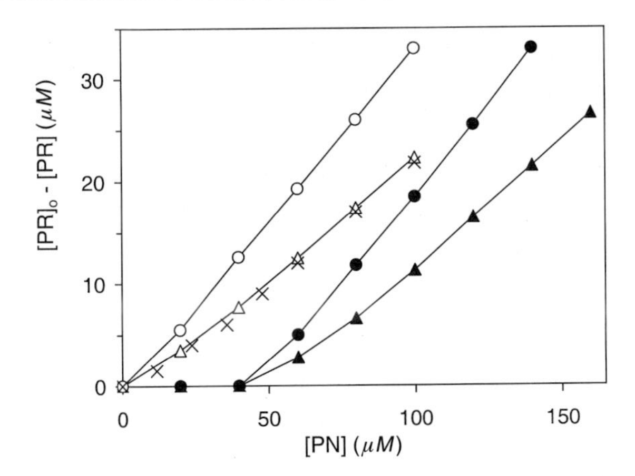

FIG. 1. Pyrogallol Red consumption in its reaction with PN in the absence (○) and presence of 75 μM Trolox (△), red wine concentrate (×, TAE = 24 μM), 10 μM ascorbic acid (●), and 75 μM of Trolox + 10 μM ascorbic acid (▲). pH 7, room temperature. $[PR]_o = 77\ \mu M$.

where S_o and S_a are the stoichiometries in the absence and the presence of antioxidant AH, respectively. Thus, an antioxidant activity of AH relative to DH, k_a/k_d, can be determined by monitoring the consumption of a detecting antioxidant (molecule) DH.

Detecting Molecule

A suitable detecting molecule should satisfy several requirements. It should (i) strongly absorb at $\lambda > 450$ nm to avoid overlapping with absorption at 300–400 nm of natural antioxidants; (ii) compete efficiently with known active antioxidants for radicals, but not react directly with PN; (iii) form products not absorbing in the same range as a starting compound; and (iv) be available commercially, reasonably inexpensive, and stable in solution.

After screening more than 30 dyes, three compounds, Pyrogallol Red (PR), gallocyanine, and carminic acid, are recommended for use.[11] These dyes are bleached by PN with a stoichiometry $\Delta[\text{dye}]/\Delta[\text{PN}] = 0.3$–$0.4$ in accordance with a radical yield from PN (Fig. 1, Table I). Their high extinction coefficients at $\lambda > 500$ nm (Table I) allow their use at concentrations <0.1 mM. The extent of their bleaching is lower in the presence of added active antioxidants and is concentration dependent according to Eq. (5).

Other detecting molecules were also suggested in the literature. Dihydrorhodamine 123 (DHR) in reaction with PN forms a fluorescent product.[12]

[11] G. G. A. Balavoine and Yu. V. Geletii, *Nitric Oxide* **3,** 40 (1999).

[12] N. W. Kooy, J. A. Royall, H. Ischiropoulos, and J. S. Beckman, *Free Radic. Biol. Med.* **16,** 149 (1994).

TABLE I
DYES RECOMMENDED FOR USE AS DETECTING MOLECULES, THEIR SPECTRAL PROPERTIES (ENTRIES 1 AND 2), REACTION STOICHIOMETRY (ENTRY 3), AND ANTIOXIDANT ACTIVITIES[a] OF INDIVIDUAL COMPOUNDS MEASURED WITH THESE DYES (ENTRIES 4–7)[b]

Entry	Pyrogallol Red	Carminic acid	Gallocyanine
1 λ_{max} (nm)	528	622	542
2 ε_{max} (mol^{-1} liter cm^{-1})	8.7×10^3	2.5×10^4	2.4×10^4
3 Δ[dye]:Δ[PN]	0.30 ± 0.06	0.40 ± 0.06	0.32 ± 0.04
4 4-Hydroxyphenylacetic acid	1	1	1
5 3,4-Dihydroxybenzoic acid	6.4	38	8.5
6 Uric acid	0.9	30	27
7 Trolox	3×10^2	5.5×10^2	—

[a] Activities are given relative to 4-hydroxyphenylacetic acid, as Trolox completely protects gallocyanine similarly to ascorbic acid; see text.
[b] Conditions: pH 7, ambient temperature.

A reaction stoichiometry $\Delta[\text{DHR}]/\Delta[\text{PN}] = 0.44$[13] is similar to numbers observed for our dyes and suggests the same radical reaction mechanism (Eqs. (1) to (3)). However, DHR is expensive and requires the use of more sophisticated equipment. Tyrosine was also used as a detector molecule.[14] The unreacted tyrosine and formed 3-nitrotyrosine were quantified using HPLC. Because tyrosine itself is a weak antioxidant, it is not a suitable detecting molecule for strong antioxidants.

Evaluation of Antioxidant Activity of Individual Compounds

Reagents

Peroxynitrite can be synthesized using different methods.[15] We use a flow reactor to mix hydrogen peroxide and nitrite, followed by quenching of PN by sodium hydroxide.[15] Obtained PN solutions are treated with MnO_2 to remove residual H_2O_2 and are then diluted immediately in 50 m*M* NaOH. Thus obtained ~20 m*M* PN stock solutions are kept frozen at -18° for a few weeks without significant decay. The desired concentration, ~5 m*M* is prepared daily by another dilution with 50 m*M* NaOH and kept in an ice bath. The PN concentration is determined spectrophotometrically, $\varepsilon_{302} = 1700$ mol^{-1} liter cm^{-1}.[10] Carbon dioxide catalyzes some PN reactions.[16] The use of fresh NaOH salt (not contaminated with $NaCO_3$)

[13] J. P. Crow, *Nitric Oxide* **1,** 145 (1997).
[14] A. S. Pannala, C. A. Rice-Evans, B. Hallywell, and S. Singh, *Biochem. Biophys. Res. Commun.* **232,** 164 (1997).
[15] W. H. Koppenol, R. Kissner, and J. S. Beckman, *Methods Enzymol.* **269,** 296 (1996).
[16] S. Lymar and J. K. Hurst, *Inorg. Chem.* **37,** 294 (1998).

in PN preparation and keeping its stock solutions in closed vials eliminate or minimize the effect of CO_2 catalysis.

Phosphate buffer (75 m*M*) is prepared from NaH_2PO_4 and Na_2HPO_4. If necessary, pH is adjusted to 7.0 ± 0.05 by HCl or NaOH. No special precautions are required against transition metal ion contamination; however, an addition of EDTA or DTPA ($\sim$5 μM) is advised.

Dyes, PR, gallocyanine, and carminic acid, are from Aldrich. Their stock solutions ($\sim$1 m*M*) are prepared in buffer and are kept in an ice bath during measurements. They can be stored at -18° for a few weeks. Commercial dyes, particularly PR, may contain some uncontrolled impurities with high antioxidant activity. These impurities are oxidized easily by air if solutions are kept for several hours at ambient temperature.

Because aqueous solutions of active antioxidants are often unstable, they should be prepared daily and stored in an ice bath. Some antioxidants may shift pH. In such cases pH should be adjusted to 7.0. Some antioxidants are difficult to solubilize in aqueous solutions. Therefore, they can be prepared in ethanol and then diluted by buffer. EtOH does not much affect the evaluation of activity if it is present up to 5% (v/v) in final solution.

General Procedure

Fill a test tube with 4 ml of 30–90 μM of a detecting molecule DH in phosphate buffer with or without antioxidant AH. Record a visible spectrum of this solution in a 1-cm cell; desired absorbency at λ_{max} should be in the range of 0.8–1.5. Place the content of the UV-VIS cell back to a test tube, and dry its wall if wet with a filter paper. Incline a tube and place 10 μm of PN stock solution ($\sim$5 m*M*) on a dry wall using a micropipette. Close the tube with a rubber stopper and shake it quickly. Record again the visible spectrum of this solution. Repeat this procedure; usually four to six portions of PN should be added, subsequently. The final pH after the addition of six to eight portions of PN should be in the 7.0–7.2 range. At least four to five different concentrations of AH should be used with each stock solution prepared separately.

Calculation of Antioxidant Activity

Plot consumption of the dye Δ[DH] versus Δ[PN] at a given concentration of AH. $\Delta[DH] = \Delta A/\varepsilon$, where ΔA and ε are the changes in optical density and extinction coefficient of dye used. A reaction stoichiometry is a slope of the straight line $S = \Delta[DH]/\Delta[PN]$ (Fig. 1). Impurities in the dye used may result in a weak deviation from a straight line for the first portion of PN. In such cases the first point should be ignored. Plot S_o/S_a for different concentrations of antioxidant against $[AH]_o/[DH]_o$. The antioxidant activity k_a/k_d of AH relative to DH is a slope of the straight line (Fig. 2). Antioxidant activity should preferably be compared with

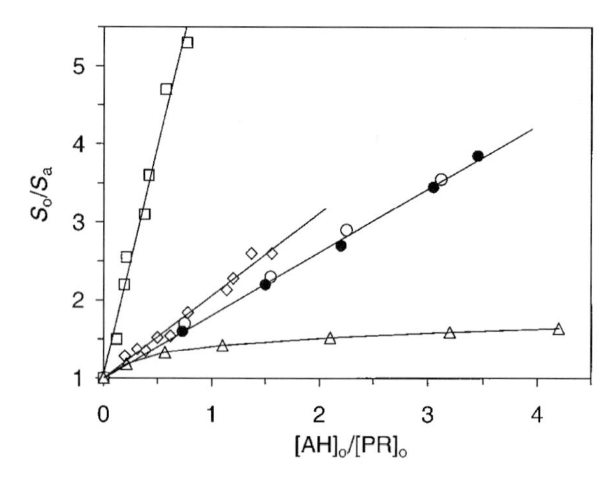

FIG. 2. Dependence of S_o/S_a on $[AH]_o/[PR]_o$ in the presence of Trolox (○, ●), L-glutathione (□), red wine concentrate (◇), and thymol (△). pH 7, room temperature. $[PR]_o = 77\ \mu M$ ($40\ \mu M$ for ○).

the activity of a well-known standard antioxidant. Trolox, a water-soluble analog of vitamin E, can be recommended as such a standard. The activities of other compounds relative to Trolox, k_a/k_T, can be calculated as shown in Eq. (6):

$$k_a/k_T = (k_a/k_d)/(k_T/k_d) \tag{6}$$

where k_T/k_d is the experimental value for the activity of Trolox measured with a given DH. The activities of several antioxidants measured with PR are given in Table II.

Total Concentration of Antioxidants in a Mixture of Unknown Composition

Many natural antioxidants are present as a mixture of unknown composition in plant extracts, juice, wine, tea, etc. An analysis of such mixtures is a very complex task. If a concentration of antioxidants is not determined, their average activity cannot be evaluated. Antioxidants are good reductants and therefore can be titrated routinely by an oxidant. We have found that $ABTS^{+\cdot}$ [2,2′-azinobis (3-ethylbenzothiazoline-6-sulfonic acid)] is a very selective reagent oxidizing active antioxidants in the presence of other reducing agents.[11,17] Based on this reaction, a simple procedure has been developed to determine the total concentration of antioxidants in their mixtures.

[17] Yu. V. Geletii, G. G. A. Balavoine, O. N. Efimov, and V. S. Kulikova, *Russ. J. Bioorg. Chem.*, in press (2002).

TABLE II
ANTIOXIDANT ACTIVITY RELATIVE TO TROLOX OF DIFFERENT COMPOUNDS DETERMINED WITH PYROGALLOL RED, THEIR REACTION STOICHIOMETRIES WITH $ABTS^{+\cdot}$, AND THEIR SPECIFIC ANTIOXIDANT ACTIVITIES (SSA)

Antioxidant, AH	Activity	Stoichiometry, $\Delta[AH]/\Delta[ABTS^{+\cdot}]$	TAE/[AH]	SAA[a]
Trolox	1	2.0 ± 0.15	1.0	1
Quercetin	11	4.8 ± 0.7	2.4	4.6
Gallic acid	7.7	4.0 ± 0.2	2.0	3.8
Caffeic acid	0.35	4.2 ± 0.2	2.1	0.17
Catechin	0.7	6.3 ± 0.5	3.1	0.22
Thymol	0.2	1.7 ± 0.2	0.9	0.2
Uric acid	0.003	2.0 ± 0.2	1.0	0.003
L-Glutathione	9.4	1.1 ± 0.1	0.5	19
Ascorbic acid	See text	2.0 ± 0.15	1.0	—
Red wines	—	—	—	1.5–2.3[b]

[a] SAA, Antioxidant activity/TAE, where TAE is Trolox antioxidant equivalent (see text).
[b] See Ref. 17 for detail.

Reagents

ABTS is available from Sigma and can be used without additional purification. $ABTS^{+\cdot}$ is synthesized easily by oxidation of ABTS with solid PbO_2. Dissolve 50 mg of ABTS in 50 ml of phosphate buffer solution (75 m*M*, pH 7.0) and then under agitation add 150–220 mg of PbO_2 (7- to 10-fold excess over ABTS). The solution immediately becomes blue-green. After 2–3 min of agitation, thoroughly remove all powder from the solution using a fine filter. Reaction time may vary depending on the surface area of PbO_2. ABTS conversion to $ABTS^{+\cdot}$ should be ~20–30%. If it is higher, $ABTS^{+\cdot}$ slowly decays and in several hours its concentration reaches the value corresponding to 20–30% conversion. $ABTS^{+\cdot}$ can be quantified by measuring its absorbency at 660 nm, $\varepsilon = 1.2 \times 10^4$ mol^{-1} liter cm^{-1}.[18] Its concentration should be checked before and after use. $ABTS^{+\cdot}$ stock solutions (~0.5 m*M*) are rather stable at ambient temperature. The $ABTS^{+\cdot}$/ABTS mixture irreversibly precipitates at $<10°$.

General Procedure

Make a solution of an antioxidant or their mixture (usually 0.025–0.1 m*M*) in phosphate buffer (75 m*M*, pH 7.0). Place 2 ml of it into a test tube or a small vial

[18] L. G. Forni, V. O. Mora-Arellano, J. E. Parker, and R. L. Wilson, *J. Chem. Soc. Perkin Trans.* **2,** 1 (1986).

and, under vigorous stirring, titrate it dropwise with the $ABTS^{+\cdot}$ stock solution. On addition, $ABTS^{+\cdot}$ is reduced quickly by antioxidants to colorless ABTS. A relatively stable blue-green color appears after complete consumption of antioxidants. If color does not disappear after 10–20 sec, this is the titration end point. A time interval of ~20 sec between drops should be kept near the end point. A total volume of $ABTS^{+\cdot}$ stock solution added should be preferably in the 0.2- to 0.6-ml range. Repeat this procedure two to three times and calculate a mean value of $ABTS^{+\cdot}$ equivalents. Then repeat the procedure with another antioxidant concentration. Complications occur in rare cases when antioxidants on oxidation produce intense color, often rose or pink. In such cases, the end point is a change of color to dark violet or other darker colors. Another complication occurs in the titration of weak antioxidants, which react with $ABTS^{+\cdot}$ slower ($\tau_{1/2} \sim 20$–60 sec), resulting in a more difficult determination of the end point. In such cases, an increase in the time interval between drops from the usual ~20 to 30–40 sec near the end point is recommended. Titration in a UV-VIS cell with a simultaneous spectra recording is described in Ref. 17.

This method is closely related to the "$ABTS^{+\cdot}$ decolorization assay."[19,20] However, in our protocol, there is always a slight excess of antioxidants over $ABTS^{+\cdot}$ and the reaction time is much shorter compared with the 1.5 min at 30° in the assay.[19] Thus, in our case, a slow reduction of $ABTS^{+\cdot}$ by weak antioxidants is excluded.

Interpretation of Titration Data

The concentration of antioxidants can be expressed in $ABTS^{+\cdot}$ equivalent units. Because $ABTS^{+\cdot}$ is a 1-electron oxidant, two of its molecules (equivalents) are consumed to oxidize typical antioxidant such as ascorbic acid or Trolox (Table II). Because Trolox is used as a standard antioxidant, the total concentration of antioxidants in their mixtures can be expressed in Trolox antioxidant equivalents (TAE). Thus, TAE of a given solution of antioxidants is the total amount of consumed $ABTS^{+\cdot}$($\Delta[ABTS^{+\cdot}]$, in mole/liter) divided by 2, $TAE = (1/2)\Delta[ABTS^{+\cdot}]$. For individual antioxidants, their TAE is often equal to their concentrations. However, some antioxidants may scavenge more (or less) than two radicals per molecule, and hence more (or less) than two $ABTS^{+\cdot}$ molecules can be consumed per one antioxidant (Table II). For that reason, TAE can differ from antioxidant concentration. For example, TAE/[AH] = 1.0 for Trolox itself, ascorbic acid, and uric acids, whereas glutathione, gallic acid, quercetin, and catechin have TAE/[AH] equal to 0.5, 2.0, 2.4, and 3.1, respectively (Table II). TAE of a mixture of antioxidants AH_i is a sum of their TAE_i, $TAE = \Sigma TAE_i$ and may also be different than a simple sum of the concentrations of individual compounds.

[19] C. A. Rice-Evans and N. Miller, *Methods Enzymol.* **234,** 279 (1994).

[20] C. Romay, C. Pascual, and E. M. Lissi, *Braz. J. Med. Biol. Res.* **29,** 175 (1996).

Antioxidant Activity of a Mixture of Antioxidants

Reagents, general procedure, and antioxidant preparation are the same as described earlier.

Calculation of Antioxidant Activity

The procedure is the same as that described for individual antioxidants, but TAE should be used instead of concentrations. In this case, S_o/S_a should be plotted against $[TAE]_o/[DH]_o$. The average antioxidant activity k_a/k_d of a mixture of antioxidants is a slope of the straight line (Fig. 2, line $\diamond$). Similar to the case of individual antioxidants, the activity for a mixture can be expressed relative to Trolox as $k_a/k_T = (k_a/k_d)/(k_T/k_d)$.

Thus, measured activity can be called specific antioxidant activity (SAA). It is the mean activity of all antioxidants present normalized per their TAE [Eq. (7)]:

$$\mathrm{SSA} = (\Sigma[\mathrm{AH_i}] \times k_i/k_d)/(\mathrm{TAE}) \tag{7}$$

where (k_i/k_d) is an activity of AH_i. SAA can differ from mean antioxidant activity, $k/k_d = (\Sigma[AH_i] \times k_i/k_d)/(\Sigma[AH_i])$; examples are presented in Table II.

What Is Measured by This Assay?

The simplest model, Eqs. (1)–(3) and (5), was used to treat data obtained by the method of competing reactions. For that reason, relative antioxidant activity is expressed as a ratio of reaction rate constants, k_a/k_d. As a result, in the literature, this ratio is often interpreted erroneously as a real ratio of rate constants. However, a much more complex reaction mechanism than Eqs. (1)–(3) should be considered for competing reactions if free radicals are involved. If a radical is formed, it always begets another radical in reactions with nonradical compounds. If one of the antioxidants is oxidized to a radical in the mixture of two antioxidants, DH and AH, Eq. (8) always takes place in the case of competing reactions:

$$\mathrm{D^{\bullet}} + \mathrm{AH} \rightleftharpoons \mathrm{DH} + \mathrm{A^{\bullet}} \tag{8}$$

An extended kinetic model should also include the termination reactions for all radicals involved (recombination of $D^{\bullet}$ and $A^{\bullet}$ radicals as well as their cross reaction). This kinetic model cannot be resolved in the form of a simple equation. A shift of Eq. (8) to one side certainly results in the preferential formation of one radical ($D^{\bullet}$ or $A^{\bullet}$), which in turn may alter a major termination reaction and finally may cause faster consumption of one antioxidant at the expense of a second one. Equation (8) also demonstrates an ability of antioxidants to regenerate one another. Therefore, antioxidants can be classified not only in the order of their rates toward radicals [(Eqs. (2) and (3)], but also according to their ability to regenerate one another through Eq. (8). The second hierarchy is known as the "pecking order" of antioxidants.[21] If Eq. (8) is shifted to the right, the antioxidant DH lies higher than AH in the pecking order. The pecking order significantly affects the order of antioxidant depletions in their mixtures. For example, DH may react faster with radicals, but due to its regeneration through Eq. (8), AH can be depleted faster.

Commonly, Eq. (8) and radical termination reactions are not considered, and the simple Eq. (4) [or Eq. (5)] is used to treat experimental data. Indeed, data plotted as S_o/S_a-/[AH]$_o$/[DH]$_o$ often give a straight line, but in fact its slope depends in a complex manner on the rate constants k_a, k_d, k_8, and k_{-8}, as well as on the rate constants of termination reactions. The ability of antioxidants to regenerate other compounds is linked directly to their protective effect against damage caused by radicals. Indeed, a mechanism of antioxidant protection includes competing reactions: a free radical either reacts with its target and damages it or reacts with an antioxidant. If a radical formed primarily from a target molecule is reduced by an antioxidant, then a damaged molecule is regenerated. A detecting molecule can be considered as a model compound for cell constituents. Thus, antioxidant activity measured by this method is related directly to the ability of a given antioxidant to protect cell constituents. Moreover, because a mixture of antioxidants always operates *in vivo,* it would be very important to determine the activity of a studied compound relative to one of a well-known (standard) antioxidant using a method of competing reactions. Trolox is used as a standard, but not directly as a detecting molecule. Therefore, two questions arise: (i) How correctly are detecting molecules chosen? (ii) Do measured activities depend on the nature of these molecules? Table I shows that such dependence does exist in some cases. Despite a similarity in structure of dyes used, that the gallocyanine/Trolox couple is completely different from the PR/Trolox and (carminic acid)/Trolox couples, whereas the latter two have similar relative activities. Trolox in combination with gallocyanine exhibits a very particular behavior; the same as that of ascorbic acid (see later). For this reason, gallocyanine is not as convenient to use as PR. On oxidation, carminic acid produces a product with a spectrum that partially overlaps with the starting acid. Therefore, carminic acid is also less convenient for use than PR. The activities and structures of PR and Trolox are very close, therefore k_8 is likely to be close to 1 for the PR/Trolox couple, thus minimizing an interference of this equilibrium. Thus, PR, as a typical polyphenol, seems to be the most suitable molecule to evaluate the activity of polyphenols. Certainly, if all three dyes are used to determine the activity of a given antioxidant and they give the same activity, then a determined value is likely to be close to the ratio of the reaction rate constants k_a/k_T.

Deviations

An ability of antioxidants to regenerate one another through Eq. (8) is often seen as a deviation from the straight line of the S_o/S_a-[AH]$_o$/[DH]$_o$ plot with thymol

[21] G. R. Buettner, *Arch. Biochem. Biophys.* **300,** 535 (1993).

as an example (Fig. 2, curve Δ). An increase in the [thymol]/[PR] ratio results in faster formation of thymyl radicals, which are reduced by PR. Consequently, the S_o/S_a-[thymol]$_o$/[PR]$_o$ curve reaches a plateau. An opposite case is the (ascorbic acid)/PR couple. Radicals formed from PR are entirely reduced by ascorbic acid and PR is not consumed until all ascorbic acid is depleted (Fig. 1). Thus, deviations of the S_o/S_a-[AH]$_o$/[DH]$_o$ plot from linearity may provide useful information on the "pecking order" of antioxidants. At the same time, perfect linearity does not ensure that Eq. (8) is not involved. If deviations are observed, the activity can be estimated from the initial slope of the S_o/S_a-[AH]$_o$/[DH]$_o$ curve.

A another complexity may derive from a fast, direct bimolecular reaction of an antioxidant with PN. Assuming that an antioxidant AH also reacts directly with PN with a second-order rate constant k'_a [Eq. (9)]:

$$\text{AH} + \text{PN} \rightarrow \text{products}, \ k'_a \tag{9}$$

and considering also Eqs.(1)–(3) and (9), one can obtain Eq. (10) instead of Eq. (5):

$$S_o/S_a = (1 + (k'_a/k_o)[\text{AH}]_o)\{1 + (k_a/k_d)([\text{AH}_o/[\text{DH}]_o)\} \tag{10}$$

Equation (10) predicts a quadratic (not linear) dependence of S_o/S_a on $[\text{AH}]_o$. In the simplest case when AH activity derives exclusively from a direct scavenging of PN [in Eq. (9)], Eq. (10) simplifies to Eq. (11):

$$S_o/S_a = (1 + (k'_a/k_o)[\text{AH}]_o) \tag{11}$$

S_o/S_a depends again linearly on $[\text{AH}]_o$, but not on the $[\text{AH}]_o/[\text{DH}]_o$ ratio as in Eq. (5). Hence, antioxidant activity should be evaluated from a minimum of two different $[\text{DH}]_o$, whereas both Eqs. (5) and (11) should be considered to fit experimental data.

An incorrect use of Eq. (5) may result in an erroneous conclusion. For example, polyphenols, such as flavonoids, do not react with PN,[22] but they were reported to scavenge PN faster than $10^6\ M^{-1}\ s^{-1}$.[23,24] These extremely high numbers were calculated using Eq. (5) instead of Eq. (11) using $k_a = 2 \times 10^6\ M^{-1}\ s^{-1}$ for ebselen as the reference antioxidant. However, in fact, this rate constant is k'_a for the direct reaction of ebselen with PN.[25]

Antioxidant Activity of Different Compounds

The method described has been used to screen the activity of hundreds of different compounds (examples are given in Refs. 11 and 17, and in Table II).

[22] S. Tibi and W. H. Koppenol, *Helv. Chim. Acta* **83,** 2412 (2000).

[23] G. R. M. H. Haenen, J. B. G. Paquay, R. E. M. Korthouwer, and A. Bast, *Biochem. Biophys. Res. Commun.* **236,** 591 (1997).

[24] J. B. G Paquay, G. R. M. H. Haenen, R. E. M. Korthouwer, and A. Bast, *J. Agric. Food Chem.* **45,** 3357 (1997).

[25] H. Masumoto, R. Kissner, W. H. Koppenol, and H. Sies, *FEBS Lett.* **398,** 179 (1996).

Polyhydroxyphenols were the most active, particularly gallic acid derivatives and quercetin. Monophenols were less active, and often deviations from Eq. (5) similar to those in the case of thymol were observed.

Thiols react quickly with PN in a second-order reaction with an apparent pH-dependent rate constant of $1–6 \times 10^3 M^{-1}s^{-1}$ (pH 7.4, 37°).[5,6,26–28] L-Cysteine partly traps PN directly through Eq. (9) at >0.025 mM. This results in a very complex interpretation of experimental data. The same reaction of L-glutathione is slower compared to that of cysteine [$1.35 \times 10^3 M^{-1}s^{-1}$ and $4.5–5.9 \times 10^3 M^{-1}s^{-1}$ (pH 7.4, 37°), accordingly[26–28]]. Therefore, high activity of L-glutathione at its low concentrations can be definitely attributed to radical scavenging (Fig. 2, Table II).

Ethanol, methanol, isopropanol, and mannitol are well-known and widely used scavengers of hydroxyl radicals. However, at their low concentrations (0.1 M), they did not show any activity with the PR assay.

A combination of "ABTS" and PR assays allows determination of SAA of different plant extracts (from tea, olives, etc.), wine, juice concentrates, and other beverages. Their activity is similar to that of polyhydroxyphenols. For example, the total content of antioxidants in red wine is in the range of 8–20 mM, with SAA being 1.5–2.3 times higher than that of Trolox (Table II).

Ascorbic Acid as an Antioxidant

Ascorbic acid is often considered a distinguished antioxidant, particularly due to its ability to regenerate ("repair") other antioxidants through Eq. (8). For example, ascorbic acid reduces α-tocopheroxyl (see Refs. 1, 29–31, and references therein), flavonoid aryloxyl,[32] uracyl,[33] and DHR radicals,[34] but does not reduce the final products of antioxidant oxidation. A similar process occurs when PR is mixed with ascorbic acid. PR consumption is entirely inhibited by ascorbic acid, approximately 2.5 ± 0.5 mol of PN per mol of ascorbic acid is required to start PR bleaching (Fig. 1). This stoichiometry corresponds to ~40% radical yield from PN close to that known in the literature.[4,6] It is important to note that if some

[26] R. Radi, J. S Beckman, K. M. Bush, and B. A. Freeman, *J. Biol. Chem.* **266,** 4244 (1991).
[27] W. H. Koppenol, J. J. Moreno, W. A. Pryor, H. Ischiripoulos, and J. S. Beckman, *Chem. Res. Toxicol.* **5,** 834 (1992).
[28] C. Quijano, B. Alvarez, R. M. Gatti, O. Augusto, and R. Radi, *Biochem. J.* **322,** 167 (1997).
[29] B. Frei, L. England, and B. N. Ames, *Proc. Natl. Acad. Sci. U.S.A.* **86,** 6377 (1989).
[30] R. H. Bisby and A. W. Parker, *Arch. Biochem. Biophys.* **317,** 170 (1995).
[31] K. Mukai, M. Nishimura, and S. Kikuchi, *J. Biol. Chem.* **266,** 274 (1991).
[32] W. Bors, C. Michel, and S. Schikora, *Free Radic. Biol. Med.* **19,** 45 (1995).
[33] M. G. Simic and S. V. Jovanovic, *J. Am. Chem. Soc.* **111,** 5778 (1989).
[34] M. Kirsch and H. De Groot, *J. Biol. Chem.* **275,** 16702 (2000).

antioxidant and ascorbic acid are added together, ascorbic acid is consumed in the same manner as in the absence of this additional antioxidant (Fig. 1). After consumption of ascorbic acid, PR is bleached by PN with the same stoichiometry as if only this antioxidant were added. The amount of PN needed to deplete ascorbic acid (the "lag" concentration, Fig. 1, curves ▲ and ●) increases with the growth of the latter one. A plot of this "lag" against ascorbic acid concentration is a calibration curve for its determination in mixtures with other antioxidants. So far, after screening of hundreds of compounds using PR as detecting molecules, no antioxidant with the same behavior as that of ascorbic acid was found. Thus, this approach allows determination of ascorbic acid in the presence of other antioxidants.[17]

Acknowledgments

This work was supported by "Pierre Fabre Santé" (France), "Institut de Recherche Pierre Fabre" (France), and INTAS (Grant 99-209).

[34] Peroxynitrite Reactions with Heme and Heme-Thiolate (P450) Proteins

By ANDREAS DAIBER and VOLKER ULLRICH

Introduction

Peroxynitrite (PN) is often considered a potent oxidant, but in exact terms it is in most cases the subsequent charge compensation by a proton, an electrophile or a Lewis acid, which predisposes the O–O-bond to homolytic cleavage. In the case of protons ($pK_a = 6.8$), OH and NO_2 radicals are formed, but in biological systems, carbonyl groups (e.g., CO_2) or metal ions (e.g., Fe^{3+} in proteins) effectively compete for the PN anion, as they form more stable intermediates, which then oxidize in a secondary reaction (see Scheme 1).

In the case of transition metals, the reaction results in even more stable oxo or hydroxy complexes and the NO_2 radical (see Scheme 2).

Such systems can oxidize or nitrate organic matter because they can provide two oxidizing equivalents and a low activation energy. The thermodynamically favorable isomerization to nitrate will always be competing, but seems to be restricted kinetically. Phenols are preferred substrates, as they are oxidized easily by high-valent metal-oxo species and the subsequent addition of the NO_2 radical does not require any activation energy (Scheme 3).

0076-6879/02 $35.00

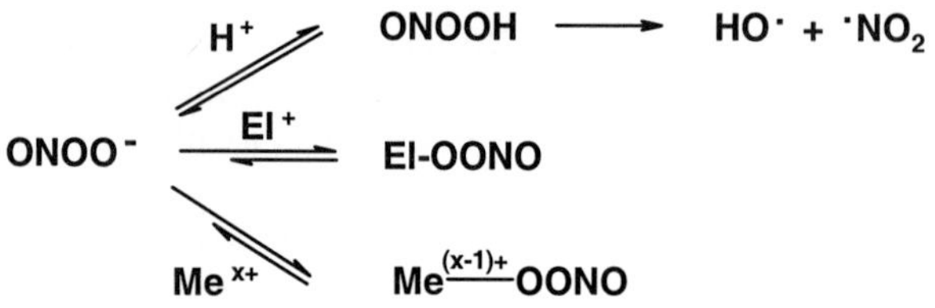

$$ONOO^- \xrightarrow{H^+} ONOOH \longrightarrow HO^\cdot + {}^\cdot NO_2$$

$$ONOO^- \xrightleftharpoons{El^+} El\text{-}OONO$$

$$ONOO^- \xrightarrow{Me^{x+}} Me \overset{(x-1)+}{—} OONO$$

SCHEME 1

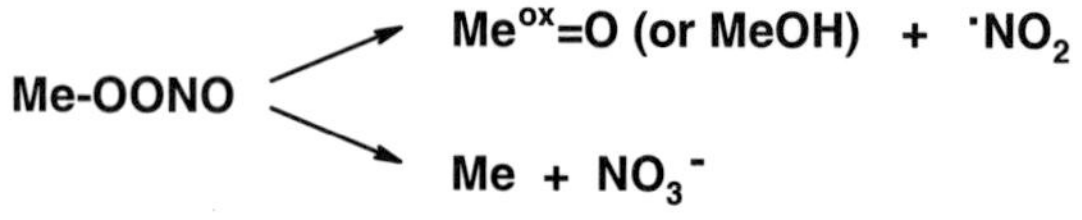

$$Me\text{-}OONO \longrightarrow Me^{ox}{=}O \text{ (or MeOH)} + {}^\cdot NO_2$$

$$Me\text{-}OONO \longrightarrow Me + NO_3^-$$

SCHEME 2

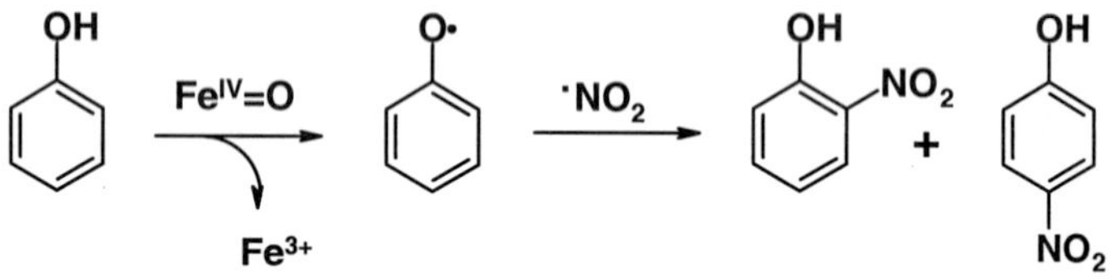

SCHEME 3

The nitration of tyrosine residues in heme and heme-thiolate proteins follows this mechanism[1–3] and leads to 3-nitrotyrosine (3-NT), as the para position is occupied. Our group has reported a very rapid Tyr nitration and inactivation of prostacyclin synthase, a P450 protein that has led us deeper into the mechanism of PN reactions with heme and heme-thiolate proteins, in particular.[4,5] In essence, we could show that the postulated mechanism for tyrosine nitration is correct, but that the sterical access to the tyrosine is a main determinant. Furthermore, the methods involved deserve a critical evaluation and are also discussed.

Materials

PN is synthesized from nitric oxide and superoxide according to the method of Kissner and Koppenol.[6] Bovine aortic microsomes containing prostacyclin

[1] R. Floris, S. R. Piersma, G. Yang, P. Jones, and R. Wever, *Eur. J. Biochem.* **215,** 767 (1993).

[2] M. Mehl, A. Daiber, S. Herold, H. Shoun, and V. Ullrich, *Nitric Oxide* **3,** 142 (1999).

[3] A. Daiber, S. Herold, C. Schöneich, D. Namgaladze, J. A. Peterson, and V. Ullrich, *Eur. J. Biochem.* **267,** 6729 (2000).

[4] M. H. Zou, C. Martin, and V. Ullrich, *Biol. Chem.* **378,** 707 (1997).

[5] M. H. Zou, A. Daiber, H. Shoun, J. A. Peterson, and V. Ullrich, *Arch. Biochem. Biophys.* **376,** 149 (2000).

[6] R. Kissner, T. Nauser, P. Bugnon, P. G. Lye, and W. H. Koppenol, *Chem. Res. Toxicol.* **10,** 1285 (1997).

synthase (PGIS) are prepared as published previously,[7] and PGIS is purified from these microsomes as described in Ref. 8. Bacterial monooxygenase-3 ($P450_{BM-3}$) wild-type and F87Y mutant were a kind gift of J. A. Peterson.[9,10] Cytochrome P450 NO-reductase ($P450_{NOR}$) was kindly donated by N. Takaya and H. Shoun,[11] P450 campher-5-monooxygenase ($P450_{CAM}$) was provided by C. Jung,[12] and chloroperoxidase (CPO) was provided by L. Hager.[13] Pronase from *Streptomyces griseus* was obtained from Fluka (Steinheim, Germany). 3-Morpholinosydnonimine hydrochloride (Sin-1) and (Z)-1-[*N*-[3-aminopropyl]-*N*-[4-(3-aminopropylammonio) butyl]-amino]diazen-1-ium-1,2-diolate (Spermine NONOate) are from Calbiochem (La Jolla, CA). Xanthine oxidase grade III (XO) and Cu,Zn-superoxide dismutase (SOD) are from Sigma (Steinheim, Germany).

Detection and Quantitation of Nitrotyrosine in Proteins

Most attempts to identify tyrosine nitrated proteins were based on the use of antibodies first introduced by the Beckman group.[14] Mono- and polyclonal antibodies are available commercially and both should be used, as their specificity depends on a rather narrow concentration range and varies among the different nitrated proteins, as will be shown later. Controls in the presence of 3-NT (1–10 m*M*) should also be done, as well as pretreatment of the protein sample with dithionite to reduce the nitro group to the amine. Although we have not obtained false positives in our experiments, direct proof for 3-NT by HPLC is required or direct MALDI MS spectra showing the nitrated peptide. For total hydrolysis of the protein, the usual acidic chemical procedure can give false positives if small amounts of nitrite are present. Pronase digestion is recommended, but sometimes may take up to 7 days to complete, as in the case of PGIS (unpublished observations). Peaks colocalized with 3-NT in HPLC chromatograms have also been reported.[15] With regard to enzymatic digestion with trypsin and other specific proteases, we have indications that with PGIS the nitration or other similar modifications of the protein may disturb specificity (unpublished results). The following sections describe the conditions used in our laboratory.

[7] V. Ullrich, L. Castle, and P. Weber, *Biochem. Pharmacol.* **30,** 2033 (1981).

[8] D. L. DeWitt and W. L. Smith, *J. Biol. Chem.* **258,** 3285 (1983).

[9] S. S. Boddupalli, B. C. Pramanik, C. A. Slaughter, R. W. Estabrook, and J. A. Peterson, *Arch. Biochem. Biophys.* **292,** 20 (1992).

[10] S. Graham-Lorence, G. Truan, J. A. Peterson, J. R. Falck, S. Wei, C. Helvig, and J. H. Capdevila, *J. Biol. Chem.* **272,** 1127 (1997).

[11] K. Nakahara, T. Tanimoto, K. Hatano, K. Usuda, and H. Shoun, *J. Biol. Chem.* **268,** 8350 (1993).

[12] C. Jung, G. Hui Bon Hoa, K. L. Schröder, M. Simon, and J. P. Doucet, *Biochemistry* **31,** 12855 (1992).

[13] D. R. Morris and L. P. Hager, *J. Biol. Chem.* **241,** 1763 (1966).

[14] Y. Z. Ye, M. Strong, Z. Q. Huang, and J. S. Beckman, *Methods Enzymol.* **269,** 201 (1996).

[15] H. Kaur, L. Lyras, P. Jenner, and B. Halliwell, *J. Neurochem.* **70,** 2220 (1998).

Detection of Nitrated Proteins by Western Blot Analysis with 3-Nitrotyrosine Antibodies

Direct detection of nitrated proteins even in a complex mixture of different proteins (cell and tissue homogenates) can be achieved using Western blot analysis with mono- or polyclonal 3-nitrotyrosine antibodies (NT-Ab) developed by Ye *et al.*[14] Proteins are nitrated by PN, Sin-1, or XO/Spermine NONOate in 0.1 *M* potassium phosphate buffer, pH 7.4, at room temperature and separated on Novex 8–12% acrylamide gels from Invitrogen (Carlsbad, CA). Each sample (2.5–20 μl) in Laemmli buffer is loaded onto the gel. The electrolyte consists of 25 m*M* Tris, 192 m*M* glycin, and 5 m*M* SDS at a current of 35 mA for 1 hr. Separation is followed by a standard semidry Western blot procedure on a nitrocellulose membrane (normally 0.8–0.9 mA/cm^2 for 100–120 min).[16] The quality of the transfer is controlled by Ponceau S staining on the membrane. Incubation with the antibodies follows the standard washing procedure.[17] A mouse monoclonal 3-NT antibody from Upstate Biotechnology (Hamburg, Germany) is used at a dilution of 1 : 1000 with a secondary antibody (GAM-POX, 1 : 7500), which is obtained from Pierce (Rockford). In addition a rabbit polyclonal 3-NT antibody from Upstate Biotechnology is used at a dilution of 1 : 1000 with a secondary antibody (GAR-POX, 1 : 3000) from Pierce. Finally, the blot is developed using a Super Signal ECL kit from Pierce. To compare monoclonal and polyclonal stainings, the membranes are stripped after monoclonal staining and then treated with the polyclonal antibody. PGIS samples are stained first with the PGIS antibody and then stripped one or even two times before applying the NT antibody.

PGIS was found to be the only nitrated protein in microsomal fractions of bovine coronary arteries after treatment with submicromolar concentrations of PN or low steady-state concentrations of PN generated by Sin-1 or XO/spermine NONOate (see Fig. 1A, and Refs. 4 and 18). PGIS was also 3-NT positive in immunohistochemically stained tissue samples after treatment with LPS (unpublished observations) or from atherosclerotic lesions.[19] Because not enough purified PGIS was available, we studied other P450 proteins as models for PGIS nitration and inactivation. The F87Y variant of P450$_{BM-3}$ with a tyrosine at the active site turned out to be a valid model, as this protein was nitrated at submicromolar concentrations of PN (see Fig. 1B) and low fluxes of cogenerated PN by Sin-1 and XO/spermine NONOate (see Fig. 1C).[3] Also, other proteins, such as P450$_{CAM}$, could be nitrated under similar conditions and detected by NT-Ab Western blot analysis.[17]

Overall, NT-Abs recognized nitrated proteins with high selectivity and sensitivity, but for some nitrated proteins, these antibodies showed a higher affinity than

[16] M. H. Zou, M. Jendral, and V. Ullrich, *Br. J. Pharmacol.* **126,** 1283 (1999).

[17] A. Daiber, C. Schöneich, P. Schmidt, C. Jung, and V. Ullrich, *J. Inorg. Biochem.* **81,** 213 (2000).

[18] M. H. Zou and V. Ullrich, *FEBS Lett.* **382,** 101 (1996).

[19] M. H. Zou, M. Leist, and V. Ullrich, *Am. J. Pathol.* **154,** 1359 (1999).

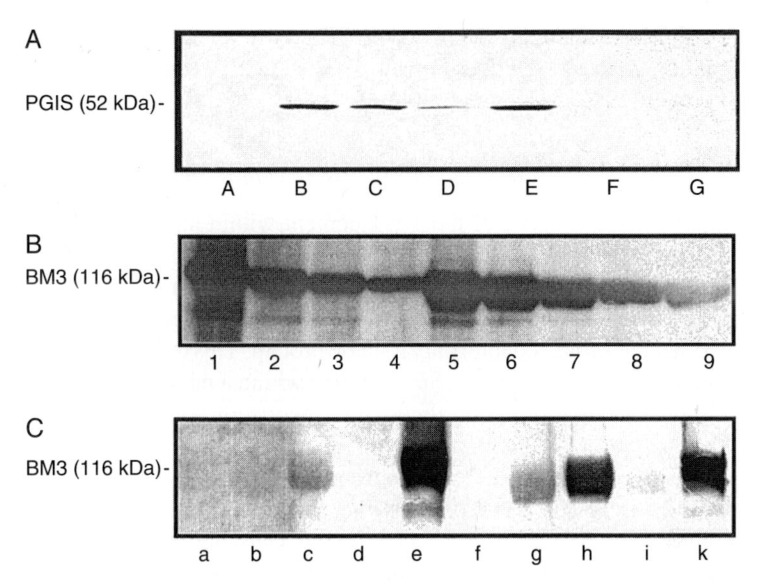

FIG. 1. (A) Anti-NT Western blot of purified PGIS samples. (A) Untreated PGIS (2.5 μM) and treated with (B) PN (1 μM), (C) tetranitromethane (1 mM), (D) PN (1 μM) in the presence of the PGIS inhibitor U46619 (1 mM), (E) spermine-NONOate (100 μM) and XO/hypoxanthine (5 mU/ml/1 mM), (F) spermine-NONOate alone (100 μM), and (G) XO/hypoxanthine alone (5 mU/ml/1 mM). All experiments were performed at pH 7.4 and 37°. Incubation times were 60 min for all samples. (B) Anti-NT Western blot of $P450_{BM-3}$ wild-type and F87Y variant after treatment with different concentrations of PN. (9) untreated $P450_{BM-3}$ F87Y variant (5 μM) and treated with (8) 0.5, (7) 5, (6) 50, and (5) 100 μM PN. $P450_{BM-3}$ wild-type treated with (4) 5, (3) 50, (2) 100, and (1) 500 μM PN. All experiments were performed at pH 7.4 and 25°. Incubation times were 5 min for all samples. (C) Anti-NT Western blot of $P450_{BM-3}$ wild-type and F87Y variant after treatment with different concentrations of Sin-1, spermine NONOate, and/or XO/hypoxanthine. $P450_{BM-3}$ wild-type (2 μM) treated with (a) XO (2.8 mU/ml), (b) spermine NONOate (100 μM), (c) both, (d) both in the presence of Cu,ZnSOD (200 U/ml), and (e) Sin-1 (100 μM). $P450_{BM-3}$ F87Y variant (2 μM) treated with (f) XO (2.8 mU/ml), (g) spermine NONOate (100 μM), (h) both, (i) both in the presence of Cu,Zn-SOD (200 U/ml), and (k) Sin-1 (100 μM). All experiments were performed at pH 7.4 and 25° in 100 mM potassium phosphate buffer, which contained protease inhibitor cocktail for general use (Sigma) and hypoxanthine (1 mM). Incubation times were 120 min for all samples [A. Daiber, S. Herold, C. Schöneich, D. Namgaladze, J. A. Peterson, and V. Ullrich, *Eur. J. Biochem.* **267,** 6729 (2000) with permission of Blackwell Science Ltd.].

for others.[20] Moreover, when proteins with large differences in molecular weight were investigated on the same blot, the blotting conditions could not be optimized for all proteins. This means that some proteins (e.g., PGIS) were not yet transfered completely to the membrane, whereas others (e.g., Mn-SOD or subunits of Hb)

[20] E. T. Morgan, V. Ullrich, A. Daiber, P. Schmidt, N. Takaya, H. Shoun, J. C. McGiff, A. Oyekan, C. J. Hanke, W. B. Campbell, C. S. Park, J. S. Kang, H. G. Yi, Y. N. Cha, D. Mansuy, and J. L. Boucher, *Drug Metabol. Dispos.* **29,** 1366 (2001).

were already blotted through the membrane. Therefore, Western blot analysis only yields qualitative results but does not allow a quantitative comparison of different nitrated proteins.

Quantitation and Localization of 3-Nitrotyrosine within Proteins

For quantitative analysis of the 3-NT content within nitrated proteins, enzymatic proteolytic methods should be used. Pronase, a mixture of several proteases, has turned out to be highly efficient and allows direct quantitation of 3-NT in nitrated proteins.[20] To assign the detected 3-NT levels to a specific protein, the investigated sample should contain only a single protein. This requires high purity, or one must ensure that only one protein is nitrated within a mixture, e.g., by Western blot analysis. Nitrated proteins are digested by incubation with pronase (2 mg/ml), $CaCl_2$ (1 m*M*) and 5 vv% acetonitrile at pH 7.4 and 37° for 8 hr. Then the samples are injected onto an HPLC system from Jasco (PU-980 pump, two UV-975 detectors and a LG-980-02 low-pressure mixer) with a C_{18} Nucleosil (125 × 4.6) 100-3 column from Macherey and Nagel (Düren, Germany). A mobile phase gradient is used: 0–7 min, 0 vv% B; 7–9 min, 0–100 vv% B; 9–15 min, 100 vv% B; 15–22 min 100-0 vv% B (A: 50 m*M* potassium phosphate buffer, pH 6, with 5 vv% acetonitrile; B: 80 vv% acetonitrile in water). The flow is 0.8 ml/min, and 3-NT has a typical retention time of 6 min. The products are detected at 270 and 365 nm; and tyrosine, phenylalanine, tryptophan, and 3-NT are identified by internal and external standards. 3-NT could also be quantified at 428 nm after a postcolumn pH change of the mobile phase to 9.[3,17] The quality of the digest could be monitored at 270 nm by following the peak area of tyrosine and tryptophan. Some proteins require longer digestion times (e.g., PGIS), probably because of their poor solubility and tertiary structure (e.g., tight folding).

A variety of different proteins were investigated by this digest method, which proved to be reproducible (see Ref. 20; unpublished data). Figure 2A shows chromatograms recorded for nitrated $P450_{CAM}$ samples at 365 nm. The 3-NT peak increased with higher concentrations of PN.

To retrieve the localization of nitrated tyrosine residues within a protein, specific proteolysis to peptides is required (e.g., bromocyan degradation or trypsin, Lys C, Glu N digest), followed by MS (desirably MS/MS) analysis. Nitrated proteins are denatured by the addition of 20 vv% acetonitrile and by keeping the sample for 10 min at 95°. Then the pH is adjusted to 8.0, $CaCl_2$ (1 m*M*) and trypsin (1 : 10 based on the concentration of the nitrated proteins) are added, and the solution is incubated at 37° for 10–24 hr. Samples are injected onto a Jasco HPLC system (see earlier discussion) with a column [C_{18} Nucleosil (250 × 4.6) 100-5] from Macherey and Nagel.[3,17] A mobile phase gradient is used: 0–75 vv% B in 0–40 min (A: 10 vv% acetonitrile and 0.1 vv% TFA in water; B: 80 vv% acetonitrile and 0.08 vv% TFA in water). The flow is 1 ml/min; 3-NT-positive peptides are detected at 365 or 428 nm. The quality of the digest could be monitored

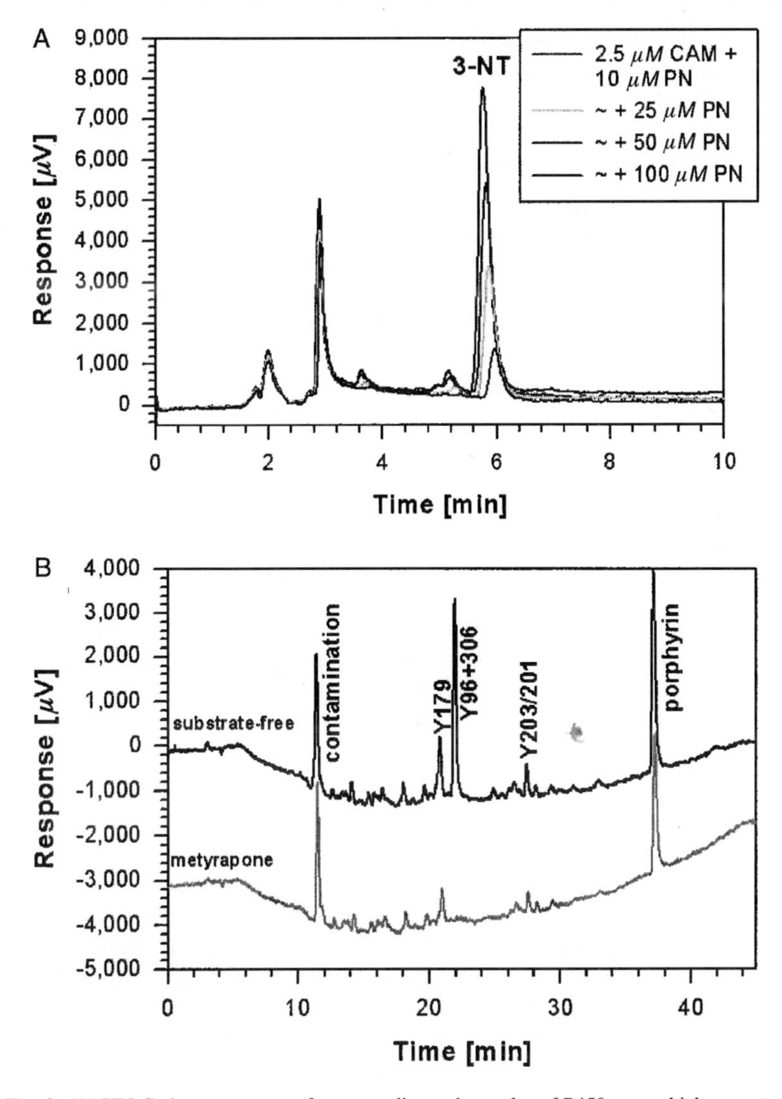

FIG. 2. (A) HPLC chromatograms of pronase-digested samples of $P450_{CAM}$, which were treated with different concentrations of PN. Conditions were as described in the corresponding section in the text. (B) HPLC chromatograms of trypsin-digested samples of $P450_{CAM}$, which were treated with PN in the presence and absence of the active site blocker metyrapone (50 μM). All NT-positive peptide peaks were isolated and analyzed by LC-MS. For each peak, the nitrated tyrosine residues and other identified compounds are indicated [A. Daiber, C. Schöneich, P. Schmidt, C. Jung, and V. Ullrich, *J. Inorg. Biochem.* **81,** 213 (2000)].

at 270 nm by following the absorption of Phe-, Tyr-, or Trp-containing peptides. 3-NT-positive peptide peaks are collected and undergo MS analysis using the method of Viner *et al.*[21]

The localization of nitrated tyrosine residues could be determined for PN-treated $P450_{BM-3}$ and $P450_{CAM}$.[3,17] For both proteins, most of the 3-NT was localized within the heme pocket, suggesting an autocatalytic nitration mechanism. Figure 2B shows the chromatograms recorded for trypsin digests of PN-treated $P450_{CAM}$ in the presence and absence of metyrapone, which blocks the active site of the protein.[22] Nitration of the metyrapone-blocked protein was decreased significantly compared to the substrate-free sample. Not only the signal of the active-site located Y_{96} decreased, but also the signals of all the other, even surface located, tyrosines (see Fig. 2B). This clearly indicates the catalytic role of the heme-thiolate iron in the overall nitration mechanism.

Catalytic Effects of Metal Complexes and Metallo-Proteins on the Reactivity of Peroxynitrite

Metal-Catalyzed Nitration of Phenolic Compounds by Peroxynitrite

The nitration of phenolic compounds by PN has been investigated extensively,[23,24] and nitration of tyrosine residues in proteins has been proposed by Crow and Beckman[25] as a footprint for PN *in vivo*. The nitration of phenolic compounds is catalyzed by CO_2[26,27] and metal complexes.[28] P450 proteins turned out to be highly efficient nitration catalysts when they were present in reaction mixtures of PN and phenol. CPO catalyzed this nitration reaction at much lower concentrations than microperoxidase (MP-11), followed by hemin and iron(III)EDTA (see Fig. 3A). CPO increased the yield of 2- and 4-nitrophenol about 10-fold at a concentration of 1 μM, whereas MP-11 had to be employed at 10 μM to show similar effects. Hemin and iron(III)EDTA could not compete with this catalytic efficiency, even when employed at 25 and 600 μM, respectively. The nitration products were measured by HPLC as described.[5,17,24] In several studies, the catalysis of PN-mediated nitration of added phenol by heme and heme-thiolate proteins was investigated and allowed us to conclude that P450 proteins, especially

[21] R. I. Viner, D. A. Ferrington, T. D. Williams, D. J. Bigelow, and C. Schoneich, *Biochem. J.* **340,** 657 (1999).

[22] T. L. Poulos and A. J. Howard, *Biochemistry* **26,** 8165 (1987).

[23] M. S. Ramezanian, S. Padmaja, and W. H. Koppenol, *Chem. Res. Toxicol.* **9,** 232 (1996).

[24] A. Daiber, M. Mehl, and V. Ullrich, *Nitric Oxide* **2,** 259 (1998).

[25] J. P. Crow and J. S. Beckman, *Adv. Pharmacol.* **34,** 17 (1995).

[26] A. Denicola, B. A. Freeman, M. Trujillo, and R. Radi, *Arch. Biochem. Biophys.* **333,** 49 (1996).

[27] A. Gow, D. Duran, S. R. Thom, and H. Ischiropoulos, *Arch. Biochem. Biophys.* **333,** 42 (1996).

[28] J. T. Groves, *Curr. Opin. Chem. Biol.* **3,** 226 (1999).

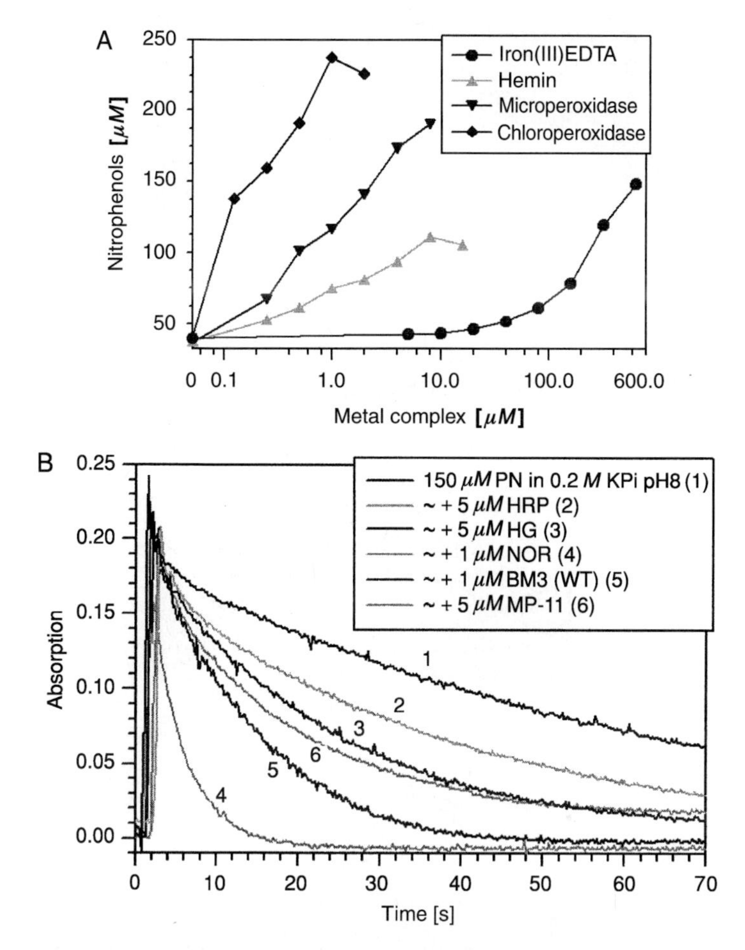

FIG. 3. (A) Yield of 2- and 4-nitrophenol from reactions of phenol (5 mM) with PN (800 μM) in dependence of the concentration of different metal complexes or metalloproteins. All reactions were performed in 0.1 M potassium phosphate buffer at pH 7 and 25°. (B) Kinetic traces of the decomposition of PN (150 μM) in the presence of different metal complexes or metalloproteins. All reactions were performed in 0.1 M potassium phosphate buffer at pH 8 and 10°. The decomposition of PN was followed on an Aminco DW-2 dual-wavelength photometer at 302 nm vs 350 nm, and PN was injected with a syringe through a septum into a cuvette, which was equipped with a magnetic stirrer [M. H. Zou, A. Daiber, H. Shoun, J. A. Peterson, and V. Ullrich, *Arch. Biochem. Biophys.* **376,** 149 (2000)].

CPO and $P450_{NOR}$, but also $P450_{BM\text{-}3}$ and $P450_{CAM}$ were much more efficient in catalyzing the nitration of phenol compared to horseradish peroxidase (HRP), catalase, cytochrome *c*, or met-Hb.[5] These results support earlier conclusions that heme-thiolate proteins can also catalyze the nitration of their own tyrosine residues more efficiently than other heme and nonheme proteins.

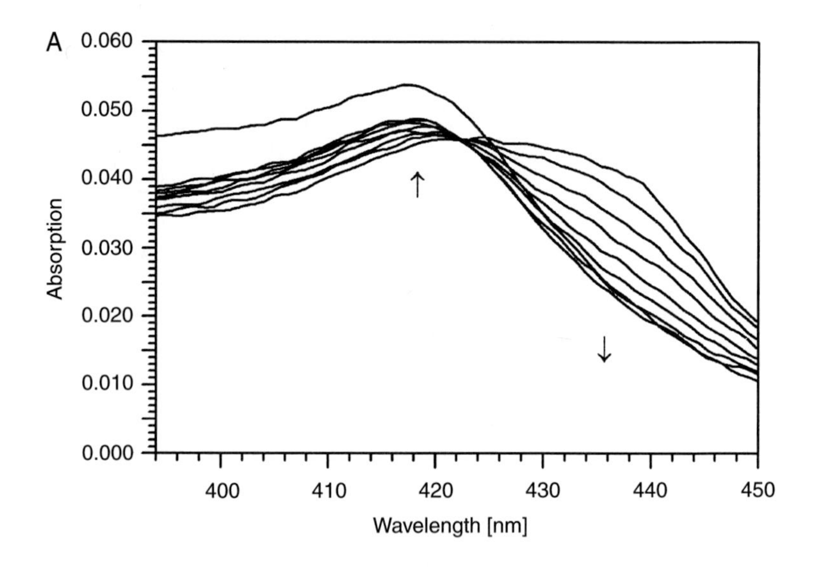

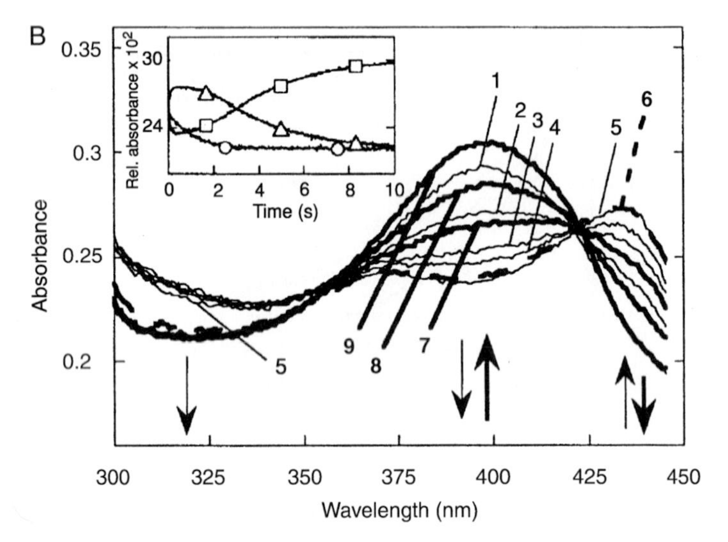

FIG. 4. Stopped-flow spectra of P450 compound II species, which were formed in the reactions with PN. (A) Reaction of $P450_{NOR}$ (1 μM) and PN (200 μM) in 0.1 M potassium phosphate buffer at pH 6.7 and 12°. The Soret band of the ferric enzyme absorbs at 417 nm (↑) and the compound II species at 436 nm (↓). Spectra were collected every 80 ms [M. Mehl, A Daiber, S. Herold, H. Shoun, and V. Ullrich, *Nitric Oxide* **3,** 142 (1999)]. (B) Rapid-scan UV/VIS spectra of the reaction of 2 μM CPO with 30 μM PN in 0.05 M phosphate buffer at pH 6.8, 12°. Scans 1–5 represent the formation of CPO compound II, whereas scans 6–9 (bold) show its decay back to native CPO after consumption of PN. Scans are shown after the following times: 1, 5 ms; 2, 16 ms; 3, 32 ms; 4, 48 ms; 5, 128 ms; 6, 1.41 sec; 7, 3.65 sec; 8, 6.89 sec; 9, and 27.01 sec. (Inset) Kinetic traces are shown at 302 (PN, ○), 397 (Soret band, □), and 434 (ferryl, △) nm [A. Daiber, S. Herold, C. Schöneich, D. Namgaladze, J. A. Peterson, and V. Ullrich, *Eur. J. Biochem.* **267,** 6729 (2000) with permission of Blackwell Science Ltd.].

Mechanistic Considerations

Reactions in the system heme protein/PN are complex, depending on the heme protein and the concentration of PN. In biological systems, the relative concentrations of $^{\bullet}NO$ and $O_2^{\bullet -}$, as a source of PN, and the level of CO_2 also have to be taken into account. The latter does not seem to influence heme-thiolate catalysis, as these reactions are fast, as seen from Fig. 3B. Such traces reflect the turnover for PN and probably are determined by the reaction of the compound II intermediate (ferryl state), with PN leading to $ONOO^{\bullet}$, which then decomposes to O_2 and $^{\bullet}NO$, which together with $^{\bullet}NO_2$ forms N_2O_3 as the source of nitrite.[29] As reported previously for heme proteins[1] and iron porphyrin complexes,[28] compound II could be detected by stopped-flow spectroscopy (Fig. 4) in the case of $P450_{BM\text{-}3}$ WT, CPO, and $P450_{NOR}$.[2,3] It was interesting that the $P450_{BM\text{-}3}$ F87Y mutant did not form this complex in agreement with the fact that the introduced tyrosine locates close to the iron and could interrupt the catalytic cycle. P450 proteins, which are nitrated only weakly by PN, like $P450_{NOR}$, on addition of phenol can increase the nitration of phenol catalytically.[2,5,17] No hydroxylation of phenol is observed as seen with PN alone, which again indicates primary formation of the ferryl species with heme proteins.[2,17] Phenoxy radicals must be formed, as evidenced by their dimerization products.[17] By HRP, these biphenols are formed in high yields in agreement with its preferred 1e oxidations. Thus, tyrosine nitration by PN and heme proteins starts with the formation of a ferryl species concomitant with the NO_2 radical. It then depends on whether an adjacent tyrosine is oxidized by the Fe^{IV} or whether it reacts with an added phenol or even with a second PN molecule.

PGIS seems to be special in having a tyrosine closely attached, probably even as a tyrosinate residue. This would allow an almost concerted attack of the ferryl form and would explain the high affinity for PN. It remains to be seen whether other heme proteins also use nitrations by PN as a means of regulatory enzyme activity. So far they are only models. Attention should always be paid to the physiological levels of PN and to the possibility that nitrite can also be a source of nitration in acidic aerobic solutions or in the presence of peroxidases and H_2O_2.[30] At last, there are indications that nitrotyrosine in proteins can be somehow reduced or that these nitrated proteins are degraded rapidly, which also may add to the difficulties involved.[31] Concerning PGIS, ongoing experiments have confirmed its high resistance towards proteolysis and the 3-NT residue could be identified (to be published).

Acknowledgments

This work was supported by the Deutsche Forschungsgemeinschaft. We also thank Drs. M.-H. Zou, S. Herold, C. Schöneich, and J. S. Beckman for their contributions.

[29] S. Goldstein, G. Czapski, J. Lind, and G. Merenyi, *Chem. Res. Toxicol.* **12,** 132 (1999).

[30] B. Halliwell, *FEBS Lett.* **411,** 157 (1997).

[31] Y. Kamisaki, K. Wada, K. Bian, B. Balabanli, K. Davis, E. Martin, F. Behbod, Y.-C. Lee, and F. Murad, *Proc. Natl. Acad. Sci. U.S.A.* **95,** 11584 (1998).

[35] Gas Chromatography/Mass Spectrometry Assay for 3-Nitrotyrosine

By MICHAEL BALAZY

Introduction

Measurements of 3-nitrotyrosine levels in biological samples are being used to evaluate exposure to reactive nitrogen species *in vivo*. Because protein tyrosine nitration has been detected in many pathophysiological conditions,[1–3] it is thought that potent nitrating compounds are generated during disease processes. A product of nitric oxide (NO) and superoxide, the peroxynitrite ($ONOO^-$) has been shown originally to cause tyrosine nitration.[4] Recent work also suggests that other nitrating molecules, such as nitryl chloride and nitrogen dioxide, may be generated *in vivo* from the oxidation of nitrite by peroxidases.[5,6] Measurement of 3-nitrotyrosine in animal studies, as well as studies involving human subjects, can lead to a better understanding of the basic biochemical principles involved in free radical damage mediated by reactive nitrogen species. Furthermore, pharmacological control of the synthesis of 3-nitrotyrosine-labeled proteins might lead to alleviation of many of the symptoms associated with such damage. One of the factors that inhibit progress in this area is the difficulty of quantifying 3-nitrotyrosine in proteins, as sensitive methods are required to detect small quantities of 3-nitrotyrosine.

Methods currently available for the measurement of 3-nitrotyrosine based on competitive binding with immunoglobulins raised against the 3-nitrotyrosine- or 3-nitrotyrosine-containing peptides and on HPLC suffer from a lack of sensitivity and/or specificity. Various immunoassay techniques, including enzyme-linked immunoassay, have been reported to measure 3-nitrotyrosine-containing proteins.[7,8] These assays represent very sensitive means by which one can detect these particular products. Although the antibodies are useful to localize 3-nitrotyrosine in tissues, it is difficult to quantify 3-nitrotyrosine using these techniques.[9,10] Several

[1] K. A. Hanafy, J. S. Krumenacker, and F. Murad, *Med. Sci. Monit.* **7,** 801 (2001).
[2] S. A. Greenacre and H. Ischiropoulos, *Free Radic. Res.* **34,** 541 (2001).
[3] C. Oldreive and C. Rice-Evans, *Free Radic. Res.* **35,** 215 (2001).
[4] H. Ischiropoulos, L. Zhu, J. Chen, M. Tsai, J. C. Martin, C. D. Smith, and J. S. Beckman, *Arch. Biochem. Biophys.* **298,** 431 (1992).
[5] J. P. Eiserich, M. Hristova, C. E. Cross, and A. D. Jones, *et al., Nature* **391,** 393 (1998).
[6] L. Gebicka, *Acta Biochim. Pol.* **46,** 919 (1999).
[7] J. Khan, D. M. Brennan, N. Bradley, and B. Gao, *et al., Biochem. J.* **330,** 795 (1998).
[8] Y. Z. Ye, M. Strong, Z. Q. Huang, and J. S. Beckman, *Methods Enzymol.* **269,** 201 (1996).
[9] A. van der Vliet, J. P. Eiserich, H. Kaur, and C. E. Cross, *et al., Methods Enzymol.* **269,** 175 (1996).
[10] J. P. Crow and H. Ischiropoulos, *Methods Enzymol.* **269,** 185.

METHODS IN ENZYMOLOGY, VOL. 359
0076-6879/02 $35.00

HPLC assays have been reported that can separate and detect 3-nitrotyrosine with ultraviolet or electrochemical detectors,[10,11] and detection limits in the range of several micrograms per milligram of protein have been reported.[10] Reduction of 3-nitrotyrosine with dithionite generates 3-aminotyrosine, which can be detected using fluorescence or electrochemical detectors, and enhances the sensitivity 10- to 50-fold relative to the ultraviolet detection.[10,12] A major disadvantage of the assays based on HPLC is the lack of specificity because artifacts frequently interfere with 3-nitrotyrosine peaks.[13]

Mass spectrometry is, in principle, an ideal method for detection and quantification of 3-nitrotyrosine because of its sensitivity, specificity, and selectivity.[14] Methods that have been developed mostly rely on converting 3-nitrotyrosine into a more volatile molecule by a series of reactions called derivatization. In this process the carboxyl, hydroxy, and amino groups are converted into an ester, ether, and amide. The final derivative should produce a sharp chromatographic peak and a useful ion for quantification. In an ideal case, a 3-nitrotyrosine derivative that generates a single ion without other fragments will provide the highest sensitivity because it will be converted quantitatively into a detectable ion. This task is difficult to achieve because known 3-nitrotyrosine derivatives produce significant fragmentation largely because they employ several reactions with different chemicals.[14–17] While heptafluorobutyryl and pentafluoropropionyl derivatives of amino acids have been synthesized[18] and used for 3-nitrotyrosine quantification,[17,19] it appears that noncharacteristic ions such as $C_2F_5COO^-$, $[C_2F_5COO_2H - F]^-$, and $C_2F_4COO^-$ are more abundant than ions specific for the amino acid structure.[18] In addition, the electron capture mass spectra of these derivatives depend strongly on the amount of sample entering the ion source and, therefore, show considerable variation across the width of the gas chromatographic peak.[18] This could lead to problems in quantitative analysis when working with pentafluoropropionate derivatives.[18] Use of these reagents poses an additional complication because the 3-nitrotyrosine hydroxyl group needs to be further protected with another reagent.

[11] K. Hensley, M. L. Maidt, Q. N. Pye, C. A. Stewart, M. Wack, T. Tabatabaie, and R. A. Floyd, *Anal. Biochem.* **251,** 187 (1997).

[12] M. K. Shigenaga, H. H. Lee, B. C. Blount, S. Christen, E. T. Shigeno, H. Yip, and B. N. Ames, *Proc. Natl. Acad. Sci. U.S.A.* **94,** 3211 (1997).

[13] H. Kaur, L. Lyras, P. Jenner, and B. Halliwell, *J. Neurochem.* **70,** 2220 (1998).

[14] M. T. Frost, B. Halliwell, and K. P. Moore, *Biochem. J.* **345**(Pt 3), 453 (2000).

[15] E. Schwedhelm, D. Tsikas, F. M. Gutzki, and J. C. Frolich, *Anal. Biochem.* **276,** 195 (1999).

[16] S. Pennathur, V. Jackson-Lewis, S. Przedborski, and J. W. Heinecke, *J. Biol. Chem.* **274,** 34621 (1999).

[17] J. R. Crowley, K. Yarasheski, C. Leeuwenburgh, J. Turk, and J. W. Heinecker, *Anal. Biochem.* **259,** 127 (1998).

[18] G. K. C. Low and A. M. Duffield, *Biomed. Mass Spectrom.* **11,** 223 (1984).

[19] C. Leeuwenburgh, M. M. Hardy, S. L. Hazen, P. Wagner, S. Oh-ishi, U. P. Steinbrecher, and J. W. Heinecke, *J. Biol. Chem.* **272,** 1433 (1997).

Pentafluorobenzyl (PFB) esters are unique in that they can generate abundant carboxylate anions in negative ion chemical ionization mass spectrometry for compounds such as acidic lipids.[20] We found that 3-nitrotyrosine can be converted into a novel PFB derivative in a single step, and this derivative produces a single ion suitable for a sensitive and quantitative GC/MS assay.[21]

Chemicals

Pentafluorobenzyl bromide (α-bromo-2,3,4,5,6-pentafluorotoluene), diisopropylethylamine, nitronium borofluorate (NO_2BF_4), and 3-nitro-L-tyrosine are purchased from Aldrich Chemical Co. (Milwaukee, WI)

L-4-Hydroxyphenyl-[$^{13}C_6$]alanine ([$^{13}C_6$]tyrosine, isotopic abundance >99%; [^{12}C]tyrosine <0.01%) is from Isotec Inc. (Miamisburg, OH)

[3,5-$^3H_2$]tyrosine (specific activity 59.7 Ci/mmol) is from New England Nuclear (Cambridge, MA)

Peroxynitrite has been synthesized in a quenched-flow reactor as described previously[22]

All solvents are of chromatographic purity from Fisher Scientific.

Preparation of ^{13}C-Labeled 3-Nitrotyrosine as an Internal Standard for Mass Spectrometry

One hundred micrograms of [$^{13}C_6$]tyrosine and 10 μCi of ^3H-labeled tyrosine (as separate solutions in water/ethanol, 98 : 2) are placed in a dry glass tube, and the solvent is evaporated under vacuum to dryness. The residue is dissolved in 100 μl of a 1 *M* solution of NO_2BF_4 in 0.1 *M* HCl with vigorous vortexing. The NO_2BF_4 solution should be prepared shortly before use. The tube is then capped with a Teflon cap and placed in a Pierce laboratory heater, with temperature adjusted to 70°, for 60 min with occasional shaking. After the reaction mixture is cooled, aliquots (20 μl) are injected into a C18 column (150 × 4.6 mm, Beckman, Fullerton, CA) installed in a HP1050 HPLC system (Hewlett-Packard, Palo-Alto, CA). Samples are eluted with a gradient of acetonitrile in water (0 to 100% in 20 min) at a flow rate of 1 ml/min. The reaction products can be detected by a radioactivity monitor (Packard, Meriden, CT) with an Ecolite (ICN, Costa Mesa, CA) scintillation cocktail and/or a UV detector (205 nm). The effluent from the column is collected into glass tubes as 1-ml fractions using a Gilson FC203B fraction collector. The radiolabeld products in these fractions are identified by scintillation counting of 50-μl aliquots. [$^{13}C_6$]-3-Nitrotyrosine is collected in a single fraction eluting at 5.5–6.5 min. The final stock solution is obtained in 5 ml

[20] M. Balazy, *J. Biol. Chem.* **266,** 23561 (1991).
[21] H. Jiang and M. Balazy, *Nitric Oxide* **2,** 350 (1998).
[22] M. Balazy, *Pol. J. Pharmacol.* **46,** 593 (1994).

of 0.1 *M* HCI. The concentration of this stock solution is finally determined via scintillation counting of radioactivity in a 100-μl aliquot. Finally, a solution having a concentration of 0.1 ng [$^{13}C_6$]-3-nitrotyrosine/ml is prepared.

Our attempts to synthesize [$^{13}C_6$]-3-nitrotyrosine using tetranitromethane[23] have not been successful. We experienced low yields and difficulty removing a major closely eluting contaminant by HPLC chromatography. Treatment of tyrosine with NO_2BF_4 gives typical yields of 3-nitrotyrosine in the range of 80–90%. A minor product, 3-nitrohydroxyphenyllactic acid (10–20%), is well separated from 3-nitrotyrosine (Fig. 1). Thus the reaction of NO_2BF_4 with tyrosine occurs via electrophilic aromatic nitration, whereas substitution of the amino group by a hydroxyl, possibly via an unstable diazonium salt intermediate, is a minor process. Because nitration occurs at carbons labeled with tritium, this reaction also generates a tritium-labeled hydroborotetrafluoric acid by-product, which elutes as an early peak of radioactivity (Fig. 1). In our experience, use of the radioactive tracer, [3,5-$^3H_2$]tyrosine, is essential to monitor the progress of the reaction, as well as in establishing the final concentrations of [$^{13}C_6$]-3-nitrotyrosine solutions. It is also important to note that because one tritium was removed and replaced by an NO_2 group, the specific activity of the final product decreases by 50% relative to the specific activity of the original radiolabeled tyrosine. This observation should be taken into account when calculating the final concentration of the [$^{13}C_6$]-3-nitrotyrosine stock solution. Analysis by GC/MS of the product eluting at 5.5–6.5 min confirmed the structure of pure ^{13}C-labeled 3-nitrotyrosine (Fig. 1).

Preparation of Pentafluorobenzyl Derivative of 3-Nitrotyrosine

The PFB derivative of 3-nitrotyrosine is prepared by the addition of PFB bromide (40 μl) and diisopropylethylamine (40 μl) followed by vigorous mixing and heating at 70° for 40 min. Both reagents for derivatization are prepared as 10% solutions in acetonitrile. Derivatization is terminated by the evaporation of reagents under a gentle stream of nitrogen. The residue is then dissolved in 50 μl of methanol and is subjected to purification by HPLC using an acetonitrile gradient in water (50 to 100% in 20 min; C18 column, 150 $\times$ 4.6 mm, Beckman). The PFB derivative of 3-nitrotyrosine is typically collected at 13–14 min. After evaporation of the solvent under vacuum, samples are dissolved in 50 μl of *n*-decane and analyzed by GC/MS (Fig. 2).

The reaction of PFB bromide with 3-nitrotyrosine in the presence of catalytic amounts of diisopropylethylamine quantitatively generates a new derivative (Fig. 3). Treatment of this derivative with a siliconizing compound, BSTFA [*N,O*–bis(trimethylsilyl)trifluoroacetamide], does not alter its mass spectrum, indicating that, in addition to the carboxyl, both the hydroxyl and the amino group also reacted with PFB bromide and, therefore, could not be converted into a TMS

[23] M. Sokolovsky, J. F. Riordan, and B. L. Vallee, *Biochemistry* **5,** 3582 (1966).

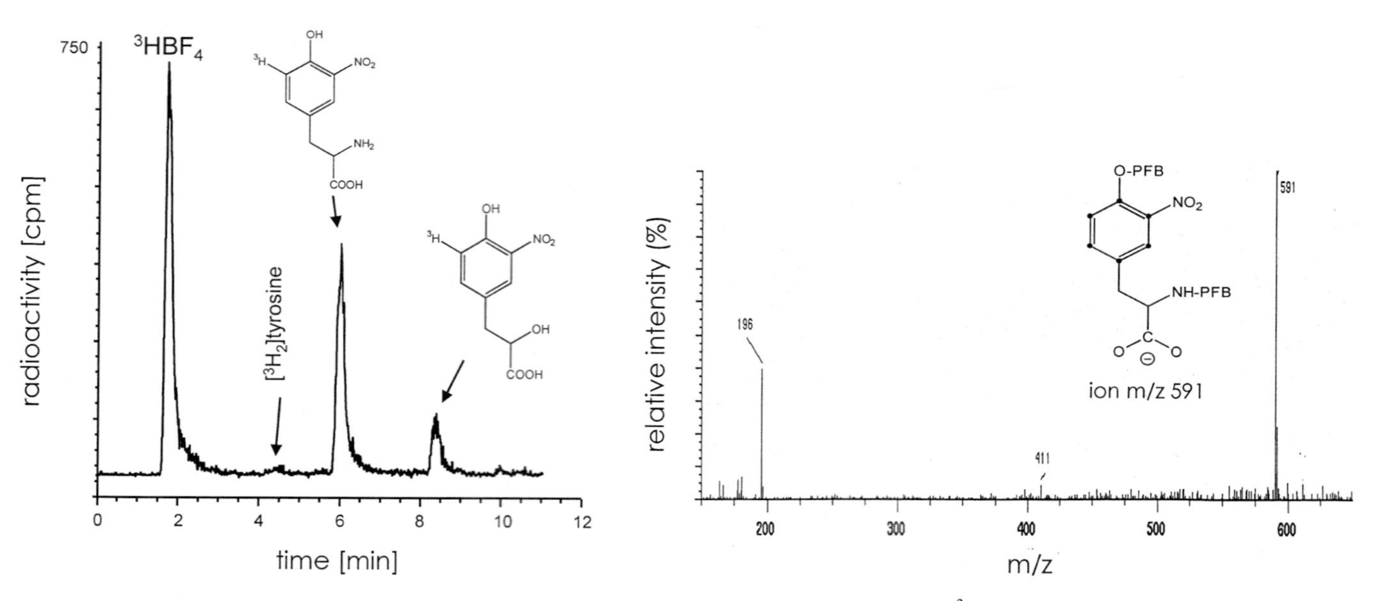

FIG. 1. (Left) Radiochromatogram obtained following the reaction of tyrosine (100 μg plus 10 μCi of [3,5-$^3H_2$]tyrosine) with 100 μl of NO_2BF_4 (1 *M*) in 0.1 *N* HCl at 70° for 60 min. An aliquot (10 μl) was analyzed on a C18 HPLC column, which was eluted with a gradient of acetonitrile in water (0 to 100%; rate 5%/min). Radioactivity was detected using an on-line radioactivity detector. (Right) Mass spectrum of [$^{13}C_6$]-3-nitrotyrosine as a tri-PFB derivative obtained by GC/MS analysis using methane negative ion chemical ionization (electron capture). Dots in the benzene ring denote specific labeling with carbon 13. PFB is $CH_2C_6F_5$.

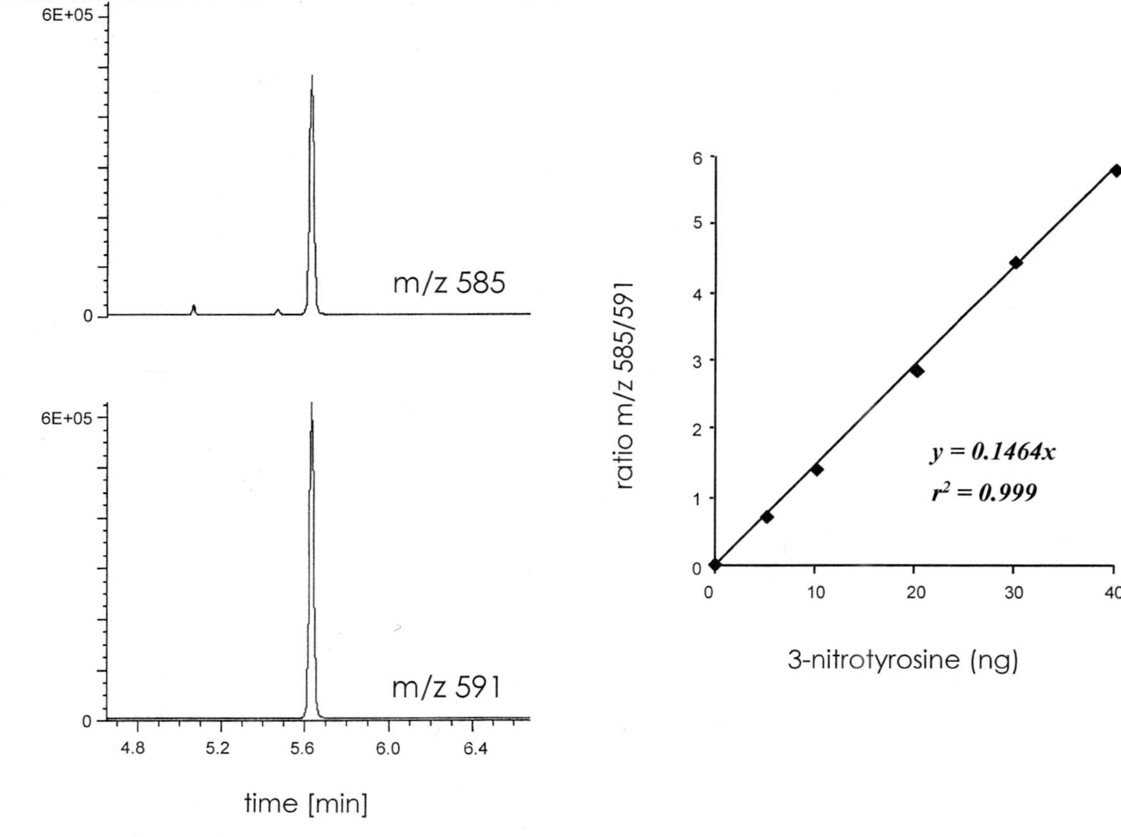

FIG. 2. (Left) GC/MS identification of total 3-nitrotyrosine levels in human platelets (1.5 mg protein) exposed to a single dose of peroxynitrite (final concentration 150 μM). Chromatograms recorded abundances of ions at *m/z* 585 and 591, which corresponded to endogenous 3-nitrotyrosine and ^{13}C-labeled 3-nitrotyrosine (internal standard, 5 ng), respectively. In this experiment, the amount of 3-nitrotyrosine was 5.3 ng/mg of protein. (Right) Standard curve for 3-nitrotyrosine. Solutions containing authentic standards of 3-nitrotyrosine were prepared and analyzed by selected ion monitoring.

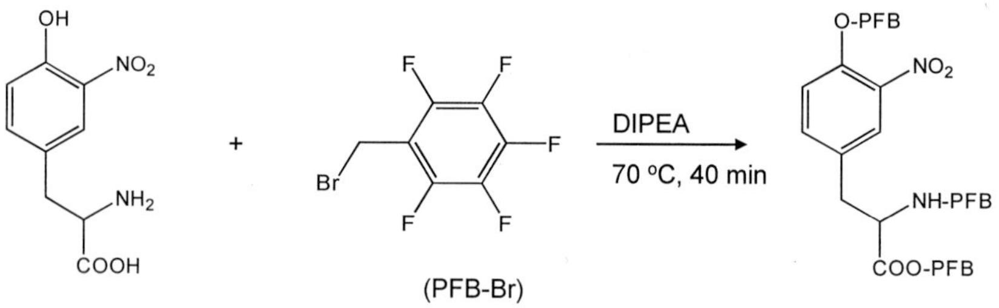

FIG. 3. Formation of the tri-PFB derivative of 3-nitrotyrosine.

(trimethylsilyl) derivative. A hydroxyl vicinal to a nitro group in 3-nitrotyrosine has a pK_a of approximately 7.5, which makes it sufficiently acidic to react quantitatively with PFB bromide yielding a PFB ether. This is an uncommon reaction, as primary and secondary alcohols usually do not react with PFB bromide. The heptafluoro derivative of 3-nitrotyrosine requires additional derivatization of the hydroxyl with TBDMS prior to GC/MS analysis.[14]

GC/MS Analysis of 3-Nitrotyrosine PFB Ester

Analysis can be performed on a suitable GC/MS system equipped with a chemical ionization source and negative ion optics. We have used a HP5989A mass spectrometer interfaced to a HP 5890 gas chromatograph (Hewlett-Packard). Samples (1 μl) are injected with an autoinjector into a capillary chromatographic column (DB-1 fused silica, 10 m, 0.25 mm i.d., 0.25 μm film thickness, J&W Scientific, Rancho Cordova, CA) via a splitless injector having a temperature of 250°. The oven temperature is programmed from 150 to 300° at 30° /min during the analysis. Helium is used as the carrier gas with a linear velocity of 4 m/sec. Electron-capture ionization is carried out with methane as a moderating gas delivered to produce a pressure in the ion source of 2.8 Torr. The other mass spectrometer parameters are as follows: ion source temperature, 200°; electron energy, 220 eV; and transfer line temperature, 250°. Selected ion monitoring is used to record ion abundances at *m/z* 585 and 591, which correspond to the PFB derivatives of 3-nitrotyrosine and [$^{13}C_6$]-3-nitrotyrosine (internal standard), respectively. Areas under chromatographic peaks defined for ions at *m/z* 585 and 591 are obtained from GC/MS integration data (Chemstation software). A standard curve is prepared by mixing 3-nitrotyrosine (0 to 40 ng) and [$^{13}C_6$]-3-nitrotyrosine (5 ng) followed by derivatization, purification, and GC/MS analyses as described earlier. The amount of 3-nitrotyrosine in the sample is calculated from the regression line that shows good linearity ($r^2 = 0.999$) when plotted as a function of 3-nitrotyrosine concentration, and the ratio of the peak areas is generated by ions *m/z* 585 and 591 (Fig. 2). A detection limit of 10 pg for the 3-nitrotyrosine PFB derivative has been achieved with a signal-to-noise ratio greater than 3.

The tri-PFB derivative of 3-nitrotyrosine shows a sharp and symmetrical chromatographic peak and a simple mass spectrum (Figs. 1 and 2). A prominent ion at *m/z* 591 corresponds to a carboxylate anion of $[^{13}C_6]$-3-nitrotyrosine and originates from the electron-capture-mediated loss of the PFB radical typically observed in these esters. Formation of such an anion is a favorable process because of its high efficiency and intrinsic stability of the carboxylate anion. While many amino acids can be potentially converted to PFB derivatives, we have not observed negative ions from amino acids other than 3-nitrotyrosine following purification of the hydrolysate from platelets by HPLC.

Determination of 3-Nitrotyrosine in Human Platelets

Platelets lose their function when exposed to peroxynitrite,[24,25] thus we compared the 3-nitrotyrosine content following exposure of platelet suspension to various nitrating compounds and hormones. Platelets are prepared by centrifugation as described[20,21] and are suspended in 10 ml of HEPES buffer containing glucose (5.56 m*M*) and sodium bicarbonate (11.9 m*M*). The final concentration of the platelet suspension is 6.8×10^8 cells/ml using a Coulter T 540 cell sorter (Coulter Electronics, Hialeah, FL). The protein content of the platelet suspension is determined using the Bradford assay (Bio-Rad). The average protein content is 1.5 mg/ml. One milliliter of the platelet suspension is used for 3-nitrotyrosine analysis following treatment of the platelets with test compounds or controls. The peroxynitrite solution is added to a stirred platelet suspension using a 10-μl syringe. NO_2 gas (1 μl) is taken with a gas-tight syringe and is added immediately to the stirred platelet suspension. NO_2Cl is generated from nitrite and sodium hypochlorite.[21] Platelet samples are stirred for an additional 3–5 min and are then mixed with 5 ml of ice-cold acetone, vortexed, and kept for 1 hr at $-20°$.

Isolation of 3-Nitrotyrosine

The protein precipitate is isolated by low-speed centrifugation and hydrolyzed with 6 *M* HCl (200 μl) supplemented with 1% phenol (v/w) at 110° for 14–18 hr. Prior to hydrolysis, 5 ng of $[^{13}C_6]$-3-nitrotyrosine is added to the protein precipitate. After cooling, the samples are extracted twice with 200 μl of ethyl acetate to remove lipids. The water phase is dried, dissolved in 10% trichloroacetic acid, and filtered through C18 Sep-Pak cartridges (Waters), which are preconditioned with 5 ml methanol and equilibrated with 5 ml water. 3-Nitrotyrosine is eluted quantitatively in 2 ml of 25% methanol/water. The samples are then concentrated and purified by HPLC, derivatized, and analyzed by GC/MS. Control experiments do not show

[24] M. A. Moro, V. M. Darley-Usmar, D. A. Goodwin, N. G. Read, R. Zamora-Pino, M. Feelisch, M. W. Radomski, and S. Moncada, *Proc. Natl. Acad. Sci. U.S.A.* **91,** 6702 (1994).

[25] C. Boulos, H. Jiang, and M. Balazy, *J. Pharmacol. Exp. Ther.* **293,** 222 (2000).

3-nitrotyrosine artifacts. A protocol using alkaline protein hydrolysis that avoids nitration artifacts has been reported.[14]

Detection of 3-Nitrotyrosine in Human Platelets

Intact, washed human platelets contain 1.4 ± 0.6 ng of 3-nitrotyrosine per milligram of platelet proteins ($n = 6$).[21] Exposure of platelets to peroxynitrite (10 to 300 μM) produces a dose-dependent increase in platelet 3-nitrotyrosine (4.4- to 535-fold). Decomposed peroxynitrite is without effect (3.4 ± 1.4 vs 1.4 ± 0.6 ng 3-nitrotyrosine/mg protein). NO_2 (final concentration, 43 μM) generates 214 ± 95 ng of 3-nitrotyrosine per milligram of platelet proteins. Treatment of platelets with NaOCl (50 μM) and nitrite (50 μM) generates 4.4 ± 0.3 ng of 3-nitrotyrosine per milligram of protein. Incubations with nitroglycerin (100 μM), nitric oxide (50 μM), nitrite (50 μM), and nitrite (50 μM) plus hydrogen peroxide (50 μM) have not increased 3-nitrotyrosine levels. Similarly, platelet agonists (thrombin, ADP, platelet activating factor, arachidonic acid) have no effect on 3-nitrotyrosine levels in platelet proteins.[21]

Concluding Remarks

Despite complex mechanisms of antioxidant defense, platelets readily undergo nitration by peroxynitrite, which appears to be the most efficient among the compounds tested. In concentrations that are likely to be found *in vivo,* peroxynitrite caused significant nitration of platelet proteins, which is known to cause loss of platelet function.[25] It also appears that protein nitration is unlikely to be induced by hormonal stimulation. Rather, extracellular peroxynitrite or NO_2 from other cells or tissues could cause platelet protein nitration. Our methodology for measurements of 3-nitrotyrosine has two useful features. One is the clean method for preparation of the internal standard, ^{13}C-labeled nitrotyrosine. The use of radiolabeled tracer allows preparing its solutions with a precisely determined concentration. The other is a simple one-step preparation of a tri-PFB derivative of 3-nitrotyrosine, which has good chromatographic and mass spectrometric properties. The disadvantage of this method is the rather lengthy purification protocol prior to GC/MS analysis. We are developing this assay further so that the tri-PFB derivative will be analyzed by LC/MS (liquid chromatography/mass spectrometry). More recent methods for 3-nitrotyrosine have utilized LC/MS for direct analysis with detection limits of about 1 ng/ml.[26,27] More advanced techniques of mass spectrometry, such as matrix-assisted laser desorption ionization time-of-flight mass spectrometry, have been

[26] J. S. Althaus, K. R. Schmidt, S. T. Fountain, M. T. Tseng, R. T. Carroll, P. Galatsis, and E. D. Hall, *Free Radic. Biol. Med.* **29,** 1085 (2000).

[27] D. Yi, B. A. Ingelse, M. W. Duncan, and G. A. Smythe, *J. Am. Soc. Mass Spectrom.* **11,** 578 (2000).

used for the sequencing of proteins containing nitrotyrosine to identify those tyrosine residues that have greater susceptibility for nitration.[28,29] Future developments in this area are anticipated to focus on methodologies for the rapid identification and sequencing of nitrated proteins and their metabolism. In addition, methods for 3-nitrotyrosine metabolites found in urine (3-nitrohydroxyphenyllactic acid and 3-nitrohydroxyphenylacetic acid) need to be developed. Our method has the potential of being adapted for measurement of these metabolites.

Acknowledgments

This research was supported by grants from the American Heart Association, N.Y. State Affiliate (9850104), and the National Institutes of Health (GM 62453).

[28] A. Sarver, N. K. Scheffler, M. D. Shetlar, and B. W. Gibson, *J. Am. Soc. Mass Spectrom.* **12,** 439 (2001).

[29] A. Gaur, P. Kowalski, and M. Balazy, "Proceedings of the 49th ASMS Conference on Mass Spectrometry and Allied Topics," Chicago, IL, May 27–31, 2001.

[36] Quantitation and Localization of Tyrosine Nitration in Proteins

By PATRICK S.-Y. WONG and ALBERT VAN DER VLIET

The formation of reactive oxidized metabolites of nitric oxide ($NO^{\bullet}$) such as peroxynitrite ($ONOO^-$)[1,2] and nitrogen dioxide ($NO_2^{\bullet}$)[1,2] is generally believed to play a contributing role in the pathology of inflammatory diseases. In contrast to their precursor nitric oxide ($NO^{\bullet}$), these reactive nitrogen species (RNS) can covalently modify many classes of biomolecules, thereby potentially inducing functional changes. Important modifications in proteins include the oxidation of amino acid residues such as cysteine, methionine, tryptohan, and tyrosine,[3–6] the formation of protein carbonyls,[4] and protein fragmentation or cross-linking. More

[1] H. Ischiropoulos, *Arch. Biochem. Biophys.* **356,** 1 (1998).

[2] J. S. Beckman, *Chem. Res. Toxicol.* **9,** 836 (1996).

[3] B. Alvarez, G. Ferrer-Sueta, B. A. Freeman, and R. Radi, *J. Biol. Chem.* **274,** 842 (1999).

[4] M. Tien, B. S. Berlett, R. L. Levine, P. B. Chock, and E. R. Stadtman, *Proc. Natl. Acad. Sci. U.S.A.* **96,** 7809 (1999).

[5] W. Wu, Y. Chen, and S. L. Hazen, *J. Biol. Chem.* **274,** 25933 (1999).

[6] J. P. Eiserich, M. Hristova, C. E. Cross, A. D. Jones, B. A. Freeman, B. Halliwell, and A. van der Vliet, *Nature* **391,** 393 (1998).

0076-6879/02 $35.00

characteristic protein modifications by RNS include nitros(yl)ation and nitration products, which have often been implicated in protein dysfunction or cellular changes specifically associated with nitrosative stress. Nitration of protein tyrosine residues has been the subject of many studies as a biological marker of RNS formation and has indeed been observed in a wide range of pathological conditions. Based on such observations, the formation of RNS is believed to be an integral component of disease pathology. In addition to representing a biomarker for endogenous RNS formation, tyrosine nitration has also been implicated in RNS-mediated functional changes through the inactivation of several enzymes systems that contain critical and susceptible tyrosine residues.

Several decades ago, the nitrating reagent tetranitromethane (TNM) was used to identify the presence of protein tyrosine residues essential for enzyme structure or function; their selective nitration by TNM was often accompanied by enzymatic changes. Such observations have been rekindled by the more recent discovery of RNS formation in biological systems and of endogenous protein tyrosine nitration and have fostered the idea that endogenously produced RNS could affect biological systems through tyrosine nitration. Nevertheless, despite many studies in which nitration of a critical tyrosine was held responsible for enzyme inactivation by RNS,[7–15] a direct cause–effect relationship has rarely been established. In most cases, overall tyrosine nitration has been correlated with enzymatic or structural changes, but this is in itself inadequate proof for a direct causal relationship, as tyrosine nitration is generally accompanied by other forms of protein oxidation. Table I lists several studies that were designed to identify specific sites of nitration in individual proteins following exposure to RNS and the contribution of such nitration to functional or structural changes. However, the quantitation of nitration in such studies is often inaccurate and is usually based on immunological and spectroscopical methods that lack specificity. A perhaps even bigger confounding

[7] T. Shimokawa, R. J. Kulmacz, D. L. DeWitt, and W. L. Smith, *J. Biol. Chem.* **265,** 20073 (1990).

[8] E. S. Roberts, H. Lin, J. R. Crowley, J. L. Vuletich, Y. Osawa, and P. F. Hollenberg, *Chem. Res. Toxicol.* **11,** 1067 (1998).

[9] G. R. Janig, R. Kraft, J. Blanck, O. Ristau, H. Rabe, and K. Ruckpaul, *Biochim. Biophys. Acta* **916,** 512 (1987).

[10] B. S. Berlett, R. L. Levine, and E. R. Stadtman, *Proc. Natl. Acad. Sci. U.S.A.* **95,** 2784 (1998).

[11] J. A. Beckingham, N. G. Housden, N. M. Muir, S. P. Bottomley, and M. G. Gore, *Biochem. J.* **353,** 395 (2001).

[12] L. A. MacMillan-Crow, J. P. Crow, and J. A. Thompson, *Biochemistry* **37,** 1613 (1998).

[13] B. Blanchard-Fillion, J. M. Souza, T. Friel, G. C. T. Jiang, K. Vrana, V. Sharov, L. Barron, C. Schoneich, C. Quijano, B. Alvarez, R. Radi, S. Przedborski, G. S. Fernando, J. Horwitz, and H. Ischiropoulos, *J. Biol. Chem.* **276,** 46017 (2001).

[14] S. Mierzwa and S. K. Chan, *Biochem. J.* **246,** 37 (1987).

[15] O. Guittet, P. Decottignies, L. Serani, Y. Henry, P. Le Marechal, O. Laprevote, and M. Lepoivre, *Biochemistry* **39,** 4640 (2000).

TABLE I
ENZYME SYSTEMS IN WHICH NITRATION OF TYROSINE RESIDUES HAS BEEN IMPLICATED TO AFFECT ACTIVITY AND/OR FUNCTION

Enzyme	Nitrotyrosine analysis	Tyrosine role	Reference
COX-1	Protein digestion, HPLC/UV analysis, and amino acid sequencing	Tyr radical in active site	7
Cytochrome P450	GC/MS and HPLC of intact protein, sequencing of tryptic fragments	Unknown, electron transfer?	8, 9
Glutamine synthetase	HPLC and UV analysis of tryptic peptide fragments	Enzyme (in)activation	10
Glutathione-*S*-transferase	HPLC separation of tryptic fragments, MALDI/MS, and HPLC/EC analysis	GSH binding and activation	24
Protein L	HPLC/UV analysis of intact protein	Binding of protein to IgG	11
Mn-SOD	HPLC/UV, amino acid sequencing, and MS/MS analysis of peptides	Catalytic activity?	12, 35
Tyrosine hydroxylase	Western blot, protein digestion, and HPLC/UV analysis	Unknown	13
Cytochrome *c*	Protein digestion, HPLC/UV, and MS analysis of peptides	Catalytic activity?	36
Fe-SOD	HPLC/ESI-MS	Catalytic activity?	37
α_1-Antitrypsin	Amino acid sequencing of peptide fragments	Catalytic domain	14
Ribonucleotide reductase	HPLC/UV analysis, protein digestion, and amino acid sequencing	Catalytic activity	15

factor in such studies is the fact that establishing the degree to which nitration of a tyrosine residue changes its role in enzyme catalytic activity or in protein structural interactions is exceedingly difficult. Although tyrosine nitration is associated with physical changes in electronegativity, pK_a, and/or steric factors, the contribution of such changes to alterations in protein structure or function is difficult to establish.

This article describes approaches to more directly address the question of whether tyrosine nitration can be quantitatively linked to RNS-induced changes in protein structure or function. We will discuss methodology to elucidate the localization of tyrosine nitration within a protein and to quantitate nitration of specific tyrosine residues. Such studies will be needed to increase our understanding of the potential biological significance of tyrosine nitration.

Chemical Nitration of Purified Proteins

Studies that address the functional significance of tyrosine nitration often involve the use of purified proteins (recombinant enzymes, commercially available proteins, or proteins purified from cultured cells or tissue homogenates by standard procedures). Proteins can be nitrated *in vitro* by a variety of mechanisms, including exposure to peroxynitrite,[1,2] tetranitromethane,[16] or myeloperoxidase and hydrogen peroxide with nitrite as a substrate.[17] Because the RNS responsible for endogenous nitration are not always identified conclusively, it is not clear what system most adequately reflects *in vivo* nitration mechanisms. For this reason, it is advisable to use multiple nitrating systems that may have varying target specificity based on differences in steric factors and hydrophobicity. Appropriate methods to generate RNS *in vitro* can be found in references listed in Table I.

Following chemical nitration, contaminating RNS metabolities should be removed by extensive dialysis, microfiltration, affinity chromatography, protein precipitation with acetonitrile, or HPLC separation. In such procedures, care should be taken to avoid conditions that induce artificial nitration (especially during protein precipitation procedures), such as acidification, metal contamination, and autoxidation. Following removal of RNS by-products and concentration of the proteins to dryness in a Speed-Vac or by lyophilization, proteins can be digested using a variety of proteolytic enzymes, such as trypsin, chymotrypsin, and endoproteinase Lys C. The protease of choice may depend on the size of the protein of interest and the number and relative location of tyrosine residues, and multiple proteolytic steps may, in some cases, be required to separate all tyrosine residues. Sequencing-grade trypsin, which cleaves at the carboxy-terminal side of basic amino acid residues (lysine and arginine), is used most commonly and yields positively charged peptide fragments, which are optimal for mass spectroscopy analysis.[18] Proteins are unfolded by incubation with urea and dithiothreitol (DTT) at 50°, subsequently carboxymethylated with iodoacetimide to block thiol groups, and then digested enzymatically with the protease of interest for 24 hr at 37°.

Peptide Analysis

The obtained mixture of peptide fragments can be analyzed directly by mass spectroscopy as described by Sarver *et al.*[19] to obtain a qualitative view on which peptides may contain nitrated tyrosines. More rigorous analysis of the resulting

[16] M. Sokolovsky, J. F. Riordan, and B. L. Vallee, *Biochemistry* **5,** 3582 (1966).

[17] A. van der Vliet, J. P. Eiserich, B. Halliwell, and C. E. Cross, *J. Biol. Chem.* **272,** 7617 (1997).

[18] K. L. Stone and K. R. Williams, *in* "Practical Guide to Protein and Peptide Purification for Microsequencing" (P. Matsudaira, ed.), p. 43. Academic Press, San Diego, 1993.

[19] A. Sarver, N. K. Scheffler, M. D. Shetlar, and B. W. Gibson, *J. Am. Soc. Mass. Spectr.* **12,** 439 (2001).

peptide fragments can be conducted by HPLC separation and collected individually for analysis by mass spectroscopy or nitrotyrosine (NO_2-Tyr) analysis as described later. The nature of the peptides can then be determined by a variety of peptide sequencing or by MALDI or ESI MS/MS procedures. Comprehensive reviews of these methods are contained in Refs. 20–25.

Identification of peptide fragments can be accomplished by predicted fragmentation based on amino acid sequences available in the literature or catalogued in a variety of computerized databases such as SWISS-PROT and TrEMBL (us.expasy.org/). Based on these amino acid sequences and predicted enzymatic digest fragments, the calculated molecular mass and HPLC elution profile for these peptides can be obtained using a publically available computer program such as ProteinProspector (prospector.ucsf.edu/) or PeptideMass (us.expasy.org/tools/peptide-mass.html). Information obtained from these sources can be used in combination with MALDI-TOF mass spectroscopy or amino acid sequencing results for additional peptide identification.

Quantitation of Tyrosine Nitration in Peptide Fragments

Although ESI or MALDI mass spectrometry is extremely helpful in identifying and localizing RNS-induced protein modifications, they are less suitable for obtaining reliable, usually quantitative data, and additional procedures are required to quantitate the degree of tyrosine nitration within a certain protein or peptide. In order to determine relative amounts of NO_2-Tyr in individual peptide fragments, they can be separated by HPLC on a C18 reverse phase column, using linear gradient elution with increasing acetonitrile concentration and 1% TFA. Peptides containing aromatic amino acid can be detected by monitoring UV absorbance (280 nm), and NO_2-Tyr-containing peptides are usually detected at 360 nm. Tyrosine-containing peptides can be identified more specifically by monitoring its intrinsic fluorescence (excitation: 284 nm; emission: 410 nm).[23] Comparison of HLPC profiles of unmodified and nitrated glutathione *S*-transferase-μ (GST-μ)[24] shows that nitrated peptides coelute with their unmodified counterparts and are thus collected simultaneously (Fig. 1). MALDI-TOF analysis of several tyrosine-containing peptide fractions indeed shows the presence of expected

[20] J. R. Chapman, "Protein and Peptide Analysis by Mass Spectrometry." Humana Press, Totowa, NJ, 1996.

[21] B. M. Dunn and M. W. Pennington, "Peptide Analysis Protocols." Humana Press, Totowa, NJ, 1994.

[22] P. T. Matsudaira, "A Practical Guide to Protein and Peptide Purification for Microsequencing." Academic Press, San Diego, 1993.

[23] T. Covey, *in* "Protein and Peptide Analysis by Mass Spectometry" (J. R. Chapman, ed.), Vol. 61, p. 83. Humana Press, Totowa, NJ, 1996.

[24] P. S. Y. Wong, J. P. Eiserich, S. Reddy, C. L. Lopez, C. E. Cross, and A. van der Vliet, *Arch. Biochem. Biophys.* **394,** 216 (2001).

[25] H. Ohshima, I. Clean, L. Chazotte, B. Pignatelli, and H. F. Mower, *Nitric Oxide* **3,** 132 (1999).

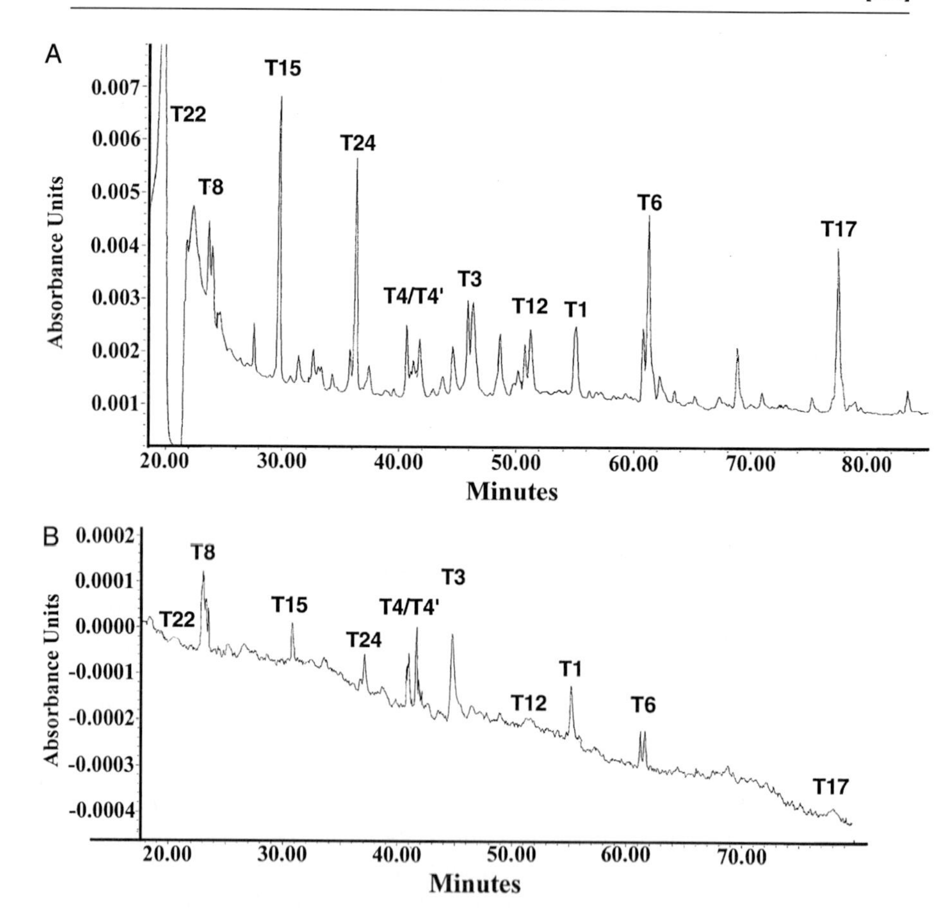

FIG. 1. HPLC separation and UV detection of tryptic fragments of mGST-μ. Untreated and $ONOO^-$ treated (100 μM) mouse mu GST (mGST-μ; 32 μg/100 μl; ~14 μM) was digested with trypsin, and the resulting peptides were separated by HPLC and detected using photodiode array detection (PDA; 240–400 nm). Chromatograms represent UV absorbance at 280 nm of $ONOO^-$ treated GST (A) and UV absorbance at 360 nm of $ONOO^-$ treated GST (B). Modified from Ref. 24.

peptide mass ions, as well as some ions with increased mass, reflecting nitrated peptides. As illustrated in Fig. 2, the collected peptide fraction containing the active site of GST-μ shows mass ions corresponding to the unmodified peptide (1381 Da), as well as oxidized and nitrated peptides (peptides with +16 or +45 Da mass, reflecting addition of an O atom or an NO_2 group, respectively). Analysis of this mixture allows determination of the relative extent of nitration within this peptide by comparing the relative amount of nitrated peptide or (more precisely) the nitrotyrosine/tyrosine ratio. Analysis of the NO_2-Tyr content of these peptides can

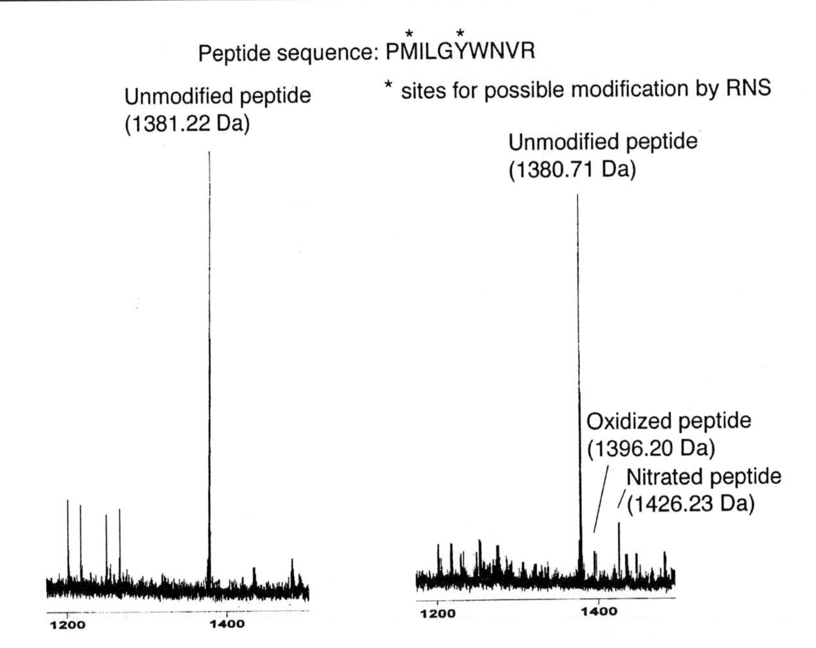

FIG. 2. MALDI-TOF MS spectra of isolated tyrosine containing tryptic fragments of mGST-μ. Mass spectra are shown for the peptide fragment that contains the active site tyrosine residue in mGST-μ. (Left) Peptide from unmodified mGST-μ. (Right) Similar peptide from $ONOO^-$ treated mGST-μ revealing the presence of peptides with a mass increase of +16 and +45.

be achieved by hydrolysis and direct UV detection or, more ideally, by amino acid derivatization and HPLC electrochemical detection or LC- or GC-MS detection, as described by various references.[25–28]

X-ray diffraction analysis has elucidated the three-dimensional structure of many proteins, and such data can be used to visualize the location and intensity of tyrosine nitration within proteins. Crystal structures can be obtained from a variety of databases, such as the protein database (PDB; www.pdblite.org), which gives molecular coordinates for protein amino acids. This information can then be used to produce a map of tyrosine nitration using a molecular visualization program such as Ras-Mol (www.umass.edu/microbio/rasmol/index.html). This allows visualization of nitrated tyrosine residues, and from their location relative to certain adjacent or vicinal amino acid side chains, some predictions can be made regarding

[26] M. T. Frost, B. Halliwell, and K. P. Moore, *Biochem. J.* **345,** 453 (2000).

[27] M. K. Shigenaga, *Methods Enzymol.* **301,** 27 (1999).

[28] A. van der Vliet, J. P. Eiserich, H. Kaur, C. E. Cross, and B. Halliwell, *Methods Enzymol.* **269,** 175 (1996).

the potential consequences of such nitration, e.g., through altered hydrogen bonding, electrostatic interactions, or steric effects.

Identification of Tyrosine Nitration in Murine Glutathione *S*-Transferase M1

We have employed these strategies (which are summarized schematically in Scheme 1) to analyze RNS-induced tyrosine nitration in purified glutathione *S*-transferase M-1 (a μ class isozyme of the GST family).[24] Purified mouse mu GST (mGST-μ; 32 μg/100 μl; ~12.8 μM) is exposed to various RNS, and the

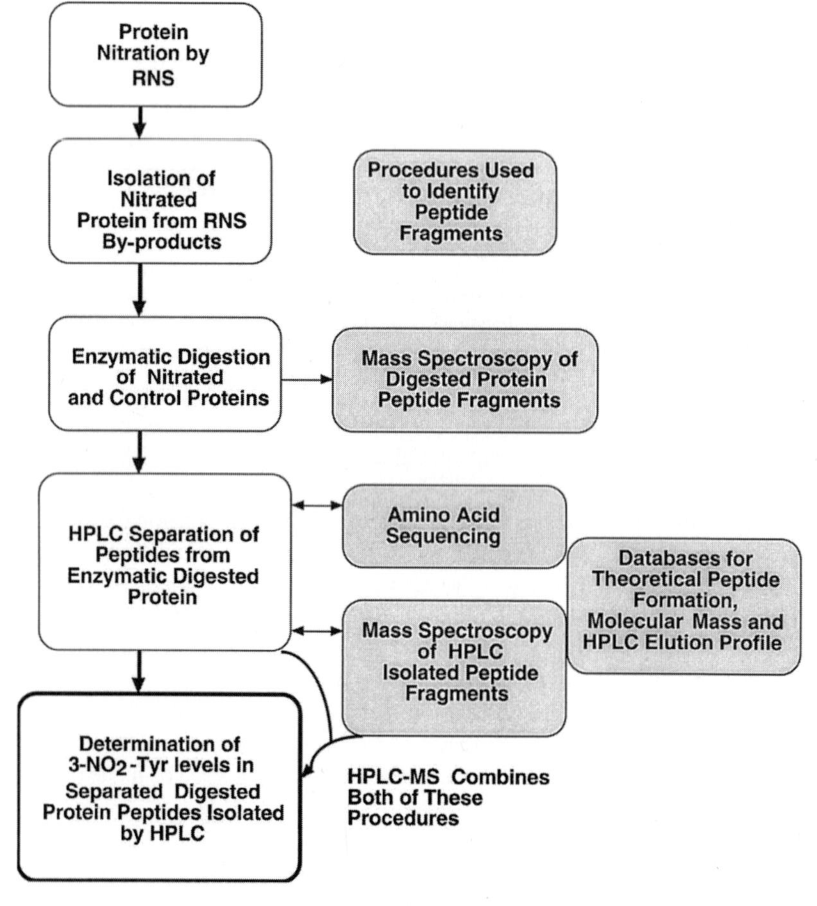

SCHEME 1. Flow chart summarizing the various procedures involved in the analysis of protein nitration.

mGST-μ is purified by microfiltration on Centricon filters, dehydrated, and incubated in 25 μl of 8 *M* urea/0.4 *M* NH_4HCO_3 and 5 μl of 45 m*M* dithiotheitol (DTT) at 50° for 15 min. The protein is carboxymethylated with 5 μl of 100 m*M* iodoacetamide for 15 min at 25°, which is terminated by the addition of 60 μl of water, and is then digested by the addition of 1.25 μg of sequence grade trypsin (Sigma) for 24 hr at 37°. The resulting tryptic fragments are separated by HPLC on a Vydac 214TP54 C18 column (25 cm × 4.6 mm internal diameter, 30 nm pore size)[29] and eluted using a linear gradient of 0–45% acetonitrile in water (1% TFA) over 90 min, 45–60% acetontrile in water (1% TFA) for 20 min, 60–0% acetonitrile in water (1% TFA) for 5 min, and 1% TFA in water for 5 min at a flow rate of 1 ml/min. Figure 1 illustrates a typical chromatogram of tyrosine (280 nm; Fig. 1A) and NO_2-Tyr-containing (360 nm; Fig. 1B) peptides obtained from mGST-μ.

MALDI-TOF analysis of the active site containing a peptide of either unmodified or RNS-treated GST illustrates the presence of unmodified, oxidized, and nitrated peptides (the latter characterized by mass increases of +16 or +45). The +16 mass increase is most likely due to oxidation of the Met residue in this peptide, whereas the +45 mass increase reflects addition of an NO_2 moiety, presumably within the Tyr residue. To determine the extent of tyrosine nitration within selected peptides, collected peptide fractions (containing both unmodified and nitrated peptides) are hydrolyzed in 1 ml of 6 *M* HCl at 110°, and the resulting amino acids are derivatized for HPLC-EC detection, as described previously.[27] The results of such HPLC analysis of each tyrosine-containing peptide are illustrated schematically in Fig. 3, which is a Ras-Mol representation of GST-μ in which the tyrosine residues are color labeled, based on the degree of nitration by RNS. As shown, only several tyrosine residues are targeted by RNS, and the extent of nitration was found to range from 5 to 20% under the conditions used. Measurement of GST activity after RNS treatment under similar conditions revealed nearly complete enzyme inactivation; hence tyrosine nitration (which only affected a relatively small fraction of even essential tyrosine residues) is unlikely to be the only cause of inactivation, and other amino acid modifications (e.g., cysteine or methionine) must have contributed as well. Overall, such quantitative analysis of tyrosine nitration is important in determining its potential contribution to changes in enzyme function, and these results with GST demonstrate that tyrosine nitration may not always be the major factor.

Further Considerations

The procedures described earlier allow for qualitative and quantitative assessment of RNS-induced NO_2-Tyr formation in proteins, but are still not sufficient

[29] A. Alin, B. Mannervik, and H. Jornvall, *Eur. J. Biochem.* **156,** 343 (1986).

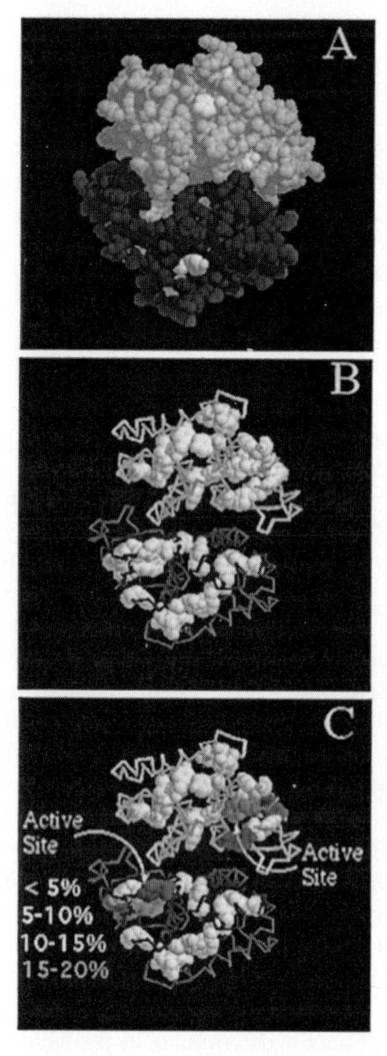

FIG. 3. Ras-Mol representation of NO_2-Tyr levels in GST exposed to $ONOO^-$ or MPO/GO/NO_2^- system; adapted from data contained in Ref. 24. (A) The two identical subunits in the structure of mGST-μ. Nonshaded molecules represent tyrosine residues located within the protein. (B) The space-filling representation of tyrosine residues on the amino acid backbone chain. (C) Levels of NO_2-Tyr formation in mu GST (32 μg ~1.36 nmol) exposed to 10 nmol $ONOO^-$. Increased shading indicates increased percentage of NO_2-Tyr formation for the represented tyrosine.

in conclusively answering the question to what degree tyrosine nitration affects protein structure or function. First, some methodological limitations warrant caution with interpretation of the results. For instance, oxidative protein modifications other than tyrosine nitration cause additional changes in peptide mass, which may hinder identification and quantitation of specific individual modifications. Also, UV detection at 360 nm is normally used to detect nitrotyrosine-containing peptides, but protein modifications other than nitration (e.g., tryptophan oxidation, other nitration or nitrosation products) may contribute to absorbance changes at 360 nm. The focus on nitrotyrosine often ignores the role of other amino acid modifications (cysteine, methionine, etc.), which may be more extensive and functionally more important.[30,31] Chemical and/or biochemical reversal of such oxidative effects of RNS using thiol-reducing agents such as DTT[32] or reversal of methionine oxidation by methionine sulfoxide reductase[33] may be used to dissect various oxidative modifications and their relative contributions to protein functional changes.

Another issue that is not addressed in such investigations is whether and to what degree nitration actually affects the role of a specific tyrosine residue within a protein. Such effects are implied from the changes in physical and chemical properties of nitrotysine compared to tyrosine (steric effects, effects on hydrogen bonding, ionization/proton exchange, redox potential, etc.). Given the diverse roles of tyrosine residues in enzyme catalysis (see Table I), the actual degree of enzyme changes on nitration of such a residue may be variable, depending on the protein of interest. Even if a specific tyrosine residue is essential for catalytic activity, based on site-directed mutagenesis, this does not provide conclusive proof that nitration of such a residue will completely abolish its function. Conversely, studies of nitration in proteins in which specific tyrosines have been mutated may help reveal selective tyrosine targets for nitration by RNS,[34] but this does not answer the question of whether such nitration affects protein function directly.

As illustrated in Table I, several studies have indicated that tyrosine nitration affects only one specific residue,[35] although most studies have indicated the nitration of several tyrosine residues. Several studies to factors that may influence tyrosine nitration have indicated both steric and electronic factors, and nitration occurs primarily in surface-exposed tyrosine residues, although not necessarily all

[30] D. M. Kuhn and T. J. Geddes, *J. Biol. Chem.* **274,** 29726 (1999).

[31] D. M. Kuhn, C. W. Aretha, and T. J. Geddes, *J. Neurosci.* **19,** 10289 (1999).

[32] R. Singh, G. V. Lamoureux, W. J. Lees, and G. M. Whitesides, *Methods Enzymol.* **251,** 167 (1995).

[33] V. S. Sharov and C. Schoneich, *Free Radic. Biol. Med.* **29,** 986 (2000).

[34] J. M. Souza, E. Daikhin, M. Yudkoff, C. S. Raman, and H. Ischiropoulos, *Arch. Biochem. Biophys.* **371,** 169 (1999).

[35] F. Yamakura, H. Taka, T. Fujimura, and K. Murayama, *J. Biol. Chem.* **273,** 14085 (1998).

surface-exposed residues are equally susceptible.[36] Some selectivity of tyrosine nitration may also depend on the nitrating mechanism, especially if it involves enzymatic pathways. Nevertheless, there is still very indirect and inconclusive evidence in support of a critical role of tyrosine nitration in the regulation of protein or enzyme function *in vivo,* and additional studies that identify specific nitration targets more quantitatively will be needed to address this question further.

[36] A. M. Cassina, R. Hodara, J. M. Souza, L. Thomson, L. Castro, H. Ischiropoulos, B. A. Freeman, and R. Radi, *J. Biol. Chem.* **275,** 21409 (2000).

Section V

Nitric Oxide Synthases

[37] Neuronal Nitric Oxide Synthases in Brain and Extraneural Tissues

By MARÍA CECILIA CARRERAS, MARIANA MELANI, NATALIA RIOBÓ, DANIELA P. CONVERSO, EMILIA M. GATTO, and JUAN JOSÉ PODEROSO

Neuronal nitric oxide synthase (nNOSα or NOS I) was the first NOS isoform characterized and purified in mammalian tissues.[1,2] Since then, nNOS has been found in specific neurons of the central and peripheral nervous system in widespread distribution. nNOS is likely not only to play an important role in physiologic neuronal functions, such as neurotransmitter release, neural development, regeneration, synaptic plasticity, and regulation of gene expression, but also in a variety of neurological disorders in which excessive production of nitric oxide (NO) leads to neural injury.[3] In addition, splice isoforms such as 136-kDa nNOSβ and 125-kDa nNOSγ exist.[4] In the past few years, nNOS isoforms have been reported in nonneural tissues such as rat and human skeletal muscle,[5] heart,[6] and blood circulating cells such as neutrophils.[7] Skeletal and cardiac muscle express an alternatively spliced nNOS isoform, a 160-kDa nNOSμ, which colocalizes with α_1-syntrophin and dystrophin at the inner membrane of sarcolemma, near the mitochondria (Table I).[8]

The subcellular localization of nNOS protein varies greatly among the cell types studied. In this sense, and depending on the study, the particulate enzyme represents between 30 and 60% of the total neuronal nNOS protein. Neuronal NOS is a cytosolic enzyme anchored to postsynaptic PDZ domains and is expressed in neurons with different densities, including the neocortex, hippocampus, and brain stem.

In 1995, Bates *et al.*[9] presented immunocytochemical evidence of the presence of nitric oxide synthase in nonsynaptic brain mitochondria, showing a high

[1] B. Mayer, M. John, B. Heinzel, E. R. Werner, H. Watcher, G. Schultz, and E. Böhme, *FEBS Lett.* **288,** 187 (1991).

[2] D. S. Bredt and S. H. Snyder, *Proc. Natl. Acad. Sci. U.S.A.* **87,** 682 (1991).

[3] U. Förstermann, J. P. Boissel, and H. Kleinert, *FASEB J.* **12,** 773 (1998).

[4] M. J. L. Elliason, S. Blackshaw, M. J. Schell, and S. Snyder, *Proc. Natl. Acad. Sci. U.S.A.* **94,** 3396 (1997).

[5] J. S. Stamler and G. Meissner, *Physiol. Rev.* **81,** 209 (2001).

[6] A. J. Kanai, L. L. Pearce, P. R. Clemens, L. A. Birder, M. M. VanBibber, S. Y. Choi, W. C. De Groat, and J. Peterson, *Proc. Natl. Acad. Sci. U.S.A.* **98,** 14126 (2001).

[7] T. Wallerath, I. Gata, W. E. Aulitzky, J. S. Pollack, H. Kleinert, and U. Förstermann, *Thromb. Haemost.* **77,** 163 (1997).

[8] L. Kobzik, M. B. Reid, D. S. Bredt, and J. S. Stamler, *Nature* **372,** 546 (1994).

[9] T. E. Bates, A. Loesch, G. Burnstock, and J. B. Clark, *Biochem. Biophys. Res. Commun.* **281,** 40 (1996).

METHODS IN ENZYMOLOGY, VOL. 359
0076-6879/02 $35.00

TABLE I
CHARACTERISTICS OF nNOS VARIANTS IN DIFFERENT TISSUES AND SPECIES

Type	M_r	Tissue	Subcellular localization	Activity (pmol [^{3}H]L-Cit/ min mg protein)	Function
nNOSα	157,000	Rat brain	Postsynaptic PDZ–domain interaction	150 ± 16	Regulation of synapsis
nNOSβ	135,000	Mice brain Human smooth muscle	Brain extracts Membrane homogenates	0.2–7% of nNOS[a]	Unknown
nNOSγ	125,000	Mice brain	Brain extracts	ND	Unknown
nNOSμ	160,000	Rat skeletal muscle	Sarcolemma	2–25[b]	Regulation of resting potential and contractility
nNOS$_{144}$	144,000	Rat brain	Synaptosomes	ND	Developmental modulation[c]
nNOS	155,000	Human neutrophils	Cytosol	23 ± 8	Production of peroxynitrite
Brain mtNOS	144,000	Rat brain	Inner membrane	24 ± 2	Developmental modulation
Heart mtNOS	146,000	Rat heart	Inner membrane	24 ± 6	Modulation of energy levels
Muscle mtNOS	159,000	Rat gastrocnemius muscle	ND	19 ± 5	Mediator in thyroid hormone effects

[a] From P. L. Huang, T. M. Dawson, D. S. Bredt, S. H. Snyder, and M. C. Fishman, *Cell* **75,** 1273 (1993).
[b] From J. S. Stamler and G. Meissner, *Physiol. Rev.* **81,** 209 (2001).
[c] From P. Ogilvie, K. Schilling, M. L. Billingsley, and H. H. H. W. Schmidt, *FASEB J.* **9,** 799 (1995).

percentage of mitochondria immunolabeled for eNOS at the inner membrane. Since then, the presence of different NOS isoforms in isolated mitochondria of different tissues has been reported. In 1998, Tatoyan *et al.*[10] purified and characterized liver iNOS-like mitochondrial NOS. We detected mitochondrial variants related to nNOS in purified organelles from brain, skeletal muscle, and heart, which appear to be slightly different in Western blot studies (Fig. 1). Kanai *et al.*[6] also added strong evidence for the presence of functional neuronal NOS in individual cardiac mitochondria by direct porphyrinic microsensor measurements. These results confirm the existence of different mtNOS and exclude possible contamination of mitochondrial preparations by cellular isoforms. However, some cross-reactivity of the antibodies against different NOS isoforms could be observed.

[10] A. Tatoyan and C. Giulivi, *J. Biol. Chem.* **273,** 11044 (1998).

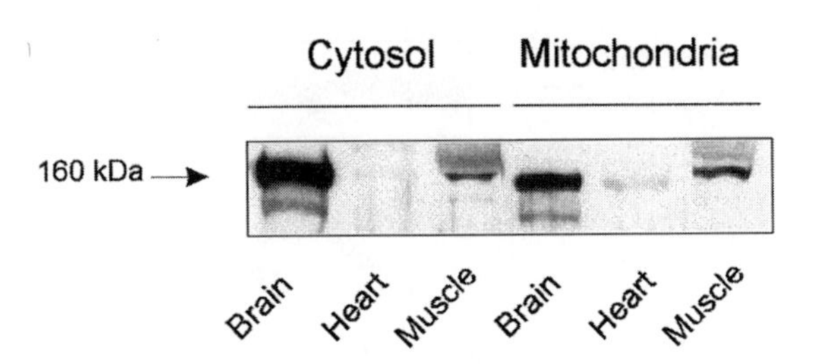

FIG. 1. Western blot of nNOS proteins in cytosol and mitochondria from different rat tissues. Proteins were separated on a 7.5% polyacrylamide gel and detected immunologically with an antibody directed to the 1409–1429 C-terminal domain.

In brain, mtNOS exhibits a V_{max} about 10 times lower than nNOS from whole brain homogenates [14 ± 3 vs 180 ± 10 pmol L-[^{3}H]citrulline/min mg protein, respectively].[11] Both mtNOS (144 kDa) and classical nNOSα (157 kDa) are detected in adult brain synaptosomes with antibodies against the C-terminal reductase domain of nNOS (Fig. 2). This 144-kDa protein is not recognized by antibodies against the N-terminal (1–181) region of NOS I or against the other isoforms but, eventually, it could react with antibodies detecting longer epitopes in the N-terminal domain, as mentioned by Boveris *et al.* in [30] of this volume.

Isolation and Purification of nNOS Rat Brain Mitochondria and Synaptosomes

Adult Wistar rats are sacrificed by decapitation, and the brains are removed rapidly and homogenized in 320 m*M* sucrose, 20 m*M* HEPES, pH 7.2, 1 m*M* EDTA, 1 m*M* dithiothreitol, 10 μg/ml leupeptin, 2 μg/ml aprotinin, and 10 μg/ml phenylmethylsulfonyl fluoride. Mitochondria are purified in a Percoll gradient according to the procedure of Giulivi *et al.*[12] in MSHE buffer (0.23 *M* mannitol, 0.07 *M* sucrose, 0.5 m*M* EGTA, 2 m*M* HEPES, pH 7.4). Synaptosomes are also isolated after Percoll gradient centrifugation at a lower density and processed like mitochondria. The cytosolic fraction is obtained after 100,000*g* centrifugation of brain homogenates. Purified mitochondria are stored frozen at −70°. To avoid contamination with the prominent cytosolic nNOS, experiments with mtNOS require a careful purification process. It is convenient to test the purity and enrichment of the mitochondrial fractions by measuring specific enzymatic activities of other subcellular fractions, such as lysosomal acid phosphatase.

[11] N. A. Riobó, M. Melani, and J. J. Poderoso, *Nitric Oxide* **4,** 185 (2000).
[12] C. Giulivi, J. J. Poderoso, and A. Boveris, *J. Biol. Chem.* **273,** 11038 (1998).

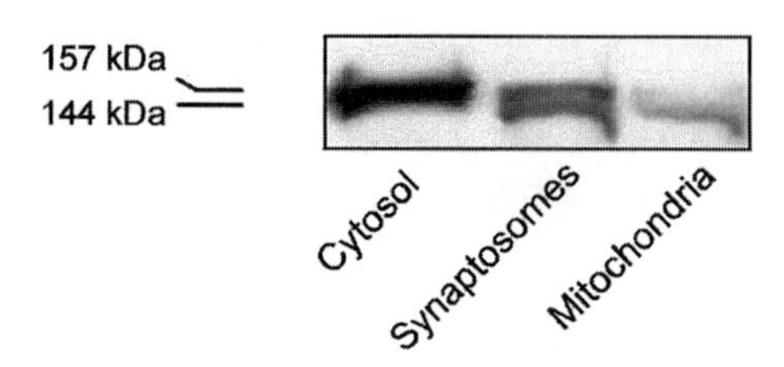

FIG. 2. Western blot of major nNOS proteins in brain fractions; proteins were separated on a 6% polyacrylamide gel.

mtNOS Activity (nNOS Type)

In brain and in other tissue, activity from nNOS-like mtNOS is determined by the conversion of L-[^{3}H]arginine to L-[^{3}H]citrulline in frozen-thawed mitochondria. As purified mitochondria were stored at −70° before use, we tested whether the frozen-thawed process modifies mtNOS activity. Nonfrozen mitochondria showed the same activity as mitochondria frozen and thawed once, but mitochondria frozen three times exhibited very little activity.

The reaction medium consists of 0.1 μM L-[^{3}H]arginine, 120 μM L-arginine, 0.1 mM NADPH, 0.3 mM $CaCl_2$, 0.1 μM calmodulin, 10 μM BH_4, 1 μM FAD, 1 μM FMN, and 2 mM L-valine in 50 mM potassium phosphate buffer (pH 7.4) and 0.1 mg of mitochondrial protein (100 μl final volume). After a 10-min incubation at 37°, the reaction is stopped by the addition of 300 μl of an ice-cold stop solution containing 20 mM HEPES and 2 mM EDTA at pH 5.5. Then, 600 μl of a cationic exchanger resin (Dowex 50 W-X8 100-200, H form) is added to remove the unconverted L-[^{3}H]arginine from the medium and is centrifuged for 30 sec. Four hundred microliters of the supernatant is mixed with 1 ml of scintillation liquid, and the L-[^{3}H]citrulline formed is detected with a scintillation counter. mtNOS specific activity is determined by subtracting the radiolabel obtained in the presence of an NOS inhibitor (1.2 mM N^G-methyl-L-arginine) from the ones obtained in the free inhibitor reaction medium and is expressed in picomoles per milligram protein per minute.

The activity of neuronal-like mtNOS variants was also measured in other tissues, such as rat skeletal muscle and heart and human neutrophils, and compared with data in the referenced literature (Table I).

Other Assay Conditions

Brain mtNOS activity was highly dependent on the presence of Ca-calmodulin, NADPH, and BH4 in the reaction medium, but less sensitive to flavins, as has already been shown for nNOSα (Fig. 3A).

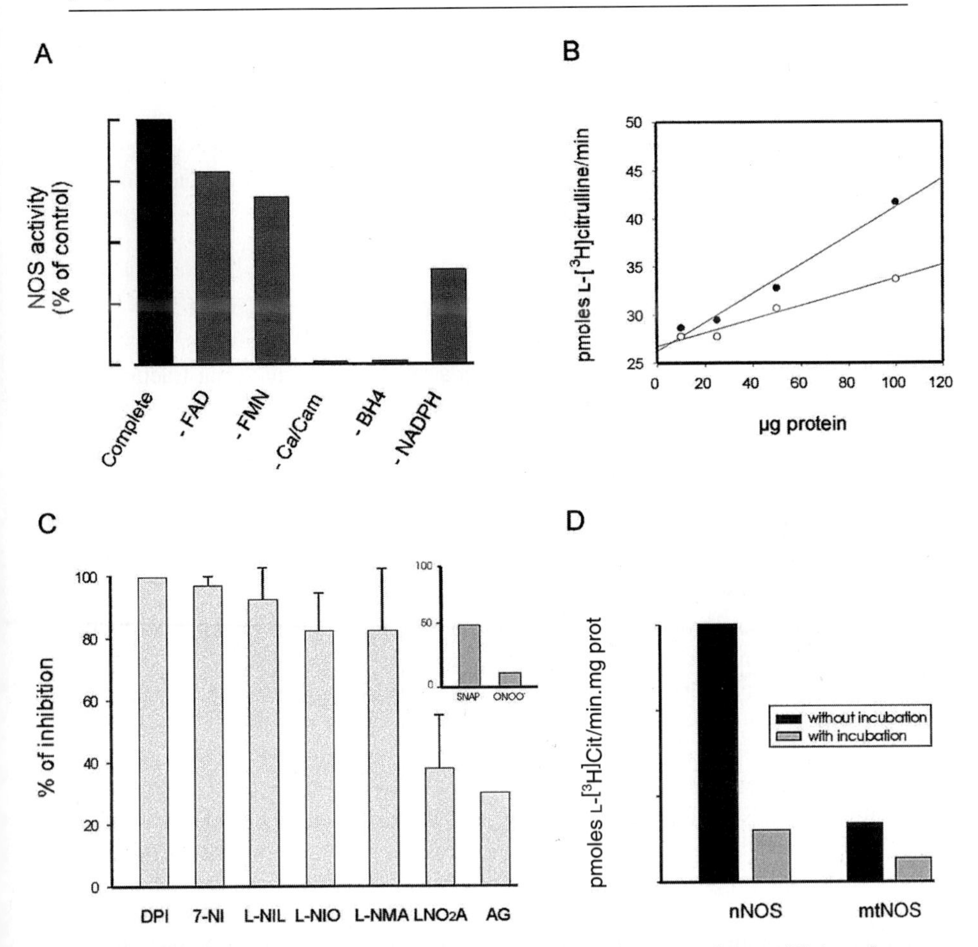

FIG. 3. Characterization of brain mtNOS activity. (A) Dependence on cofactors; (B) dependence on protein concentration; (C) effects of NOS inhibitors and of SNAP and peroxynitrite (inset); and (D) activity of brain nNOS variants after a 30-min preincubation with all reactants, without L-Arg, at 25°.

The brain mtNOS activity was linearly dependent on protein concentration of up to 100–150 μg mitochondrial protein (Fig. 3B).

The more potent inhibitors of brain mtNOS were N^G-methyl-L-arginine (L-NMMA), *N*-iminoethyl-L-ornithine (L-NIO), N^6-(1-iminoethyl)-L-lysine (L-NIL), 7-nitroindazole, and diphenyleniodonium (DPI) (80–95% inhibition). N^G-Nitro-L-arginine and aminoguanidine had significant inhibitory effects (50% inhibition) (Fig. 3C). All inhibitors were tested at concentrations 10 times greater than the substrate.

Incubation in the presence of a NO donor (1.2 m*M* SNAP) also decreased mtNOS activity (50%), whereas 200 μM peroxynitrite had no effect on mtNOS activity in contrast to nNOSα (Fig. 1C, inset).

Preincubation of purified mitochondria for 30 min at 25° in the presence of all the required cofactors, except for L-arginine, followed by incubation for 10 minutes at 37° with the addition of the substrate resulted in a 60% decrease of the activity (Fig. 3D).

Recommendations for mtNOS Assay

1. Measurements should be done using a purified mitochondrial fraction enriched in mitochondrial markers, such as succinate-cytochrome *c* reductase. The mitochondrial fraction should contain no more than 5–15% of the lysosomal activity, such as from acid phosphatase. The purification method can be that proposed by either Giulivi *et al.*[12] or Lacza *et al.*[13]

2. Determination of mtNOS activity by citrulline assay should be done in purified mitochondria frozen and thawed *only one time;* repeating thc process in the same sample will lead to an almost complete loss of activity.

3. Preincubation of the samples with cofactors before substrate addition should be avoided because it decreases mtNOS activity.

4. All mitochondrial NOS require calcium and calmodulin supplementation to be active.

5. In some tissues such as brain, activation of mitochondrial arginase II may metabolize the substrate producing radioactive [^{3}H]urea rather than [^{3}H]citrulline, resulting in, misleading results. In this case, the use of arginase inhibitor valine (2–50 m*M*) will be necessary, depending on the arginase II content in mitochondria.

6. Considering that mtNOS activity is about 25% of nNOSα and to avoid lack of specifity in measurements of radioactive counting, it is imperative to substract the activity not inhibited by NOS inhibitors such as L-NMMA or, alternatively, by flavin inhibitor DPI or by calcium substraction with EDTA or EGTA.

7. Specific mtNOS activity can be enriched by 10- to 20-fold by obtaining submitochondrial particles and then measuring the activity in the insoluble membrane fraction after treatment with CHAPS.

8. It is preferable to use fresh mitochondria, not older than 1 week, after isolation either to measure activity or to perform Western blots. Depending on biological conditions, mtNOS undergoes degradation in a relatively fast fashion. Mitochondria contain different enzymes such as calpains and hemoxigenase II that attack mtNOS.

[13] Z. Lacza, M. Puskar, J. P. Figueroa, J. Zhang, N. Rajapakse, and D. W. Busija, *Free Radic. Biol. Med.* **31,** 1609 (2001).

Functions of nNOS Variants

Cytosolic nonmitochondrial nNOS has many functions in the regulation of synapsis. Two of the most relevant functions of NO related to memory and learning are long-term potentiation (LTP) and long-term depression (LTD). LTP occurs principally at the hippocampus and represents the enhancement of synaptic plasticity secondary to repeated stimulation over time (tetanic condition), which occurs when the stimulation of an excitatory synapse (NMDA receptors) causes postsynaptic depolarization and a concomitant release of neurotransmitters from the presynaptic terminal (glutamate). Moreover, NO participates in axonal remodeling during development by interacting with proteins present in axonal terminals during axonal growth and synaptogenesis and finally modulates the release of various neurotransmitters, such as dopamine and acethylcholine.[14]

The diversity of nNOS proteins and mRNA in the different tissues has been considered "a major characteristic of nNOS gene expression."[15] It has also been reported that nNOS gene expression might be tightly regulated at transcriptional and splicing levels. Several nNOS mRNA, which arise from alternative splicing, have been identified,[16] and specific expression patterns have also been observed during development. Residual nNOS catalytic activity of knockout mice (carrying a deletion of exon 2) appeared to be reciprocal to the classical nNOS activity.[17] Interestingly, brain mtNOS expression and activity proved to be reciprocal to those of nNOSα. In agreement, it was reported previously that, during development, nNOS activity and expression were low or undetectable in mouse and rat brain and that nNOS or its variants were sharply induced in periods coincident with synapotogenesis. Furthermore, Ogilvie *et al.*[18] detected an nNOS immunoreactive protein, M_r 144,000, in developing brain. We found high expression of mtNOS (M_r 144,000) in brain mitochondria from newborn rats, whereas nNOSα was almost undetectable. The opposite expression pattern was found in brain mitochondria from adult animals.

The existence of an mtNOS generating nitric oxide in brain mitochondria is consistent with the modulation of some critical mitochondrial functions. Mitochondria play a principal role in the life and death of neurons through the process of oxidative phosphorylation, the release of proapoptotic factors, and the production of oxidant species. In the brain, NO effects include inhibition of respiration, de-energization of synaptic neurons, and the stimulation of Ca^{2+} release from the mitochondrial matrix. As for mtNOS variants, two effects are considered: (1) mitochondrial production of NO is likely the main factor in brain H_2O_2

[14] A. Law, S. Gauthier, and R. Quirion, *Brain Res. Rev.* **35,** 73 (2001).

[15] D. Saur, H. Paehge, V. Schusdziarra, and H.-D. Allescher, *Gastroenterology* **118,** 849 (2000).

[16] M. A. Lee, L. Cai, N. Hubner, Y. A. Lee, and K. Lindpaintner, *J. Clin. Invest.* **100,** 1507 (1997).

[17] P. L. Huang, T. M. Dawson, D. S. Bredt, S. H. Snyder, and M. C. Fishman, *Cell* **75,** 1273 (1993).

[18] P. Ogilvie, K. Schilling, M. L. Billingsley, and H. H. H. W. Schmidt, *FASEB J.* **9,** 799 (1995).

steady-state concentration and, (2) it is important in the developmental set-up of the brain redox state. This agrees recent reports that stated that NO and H_2O_2 participate in the development and maturation of the nervous system.

In addition, as suggested previously, neuronal mtNOS variants driving NO to the small volume matrical compartment may be an excellent regulator of oxidative phosphorylation. We and others have reported decreased O_2 uptake by the inhibition of cytochrome oxidase at 30–50 n*M* NO.[6,19] Moreover, the modulatory effects of NOS and NO may be part of complex regulatory feedback. In this way, hypoxia seemed to up regulate brain NOS activity and expression in mice, although changes were attributed to an eNOS-like 140-kDa mtNOS variant.[13] In other tissues, such as skeletal muscle, mtNOS or classic nNOS and NO may decrease muscle contractility and mitochondrial ATP synthesis. In agreement, in *mdx* mice with dystrophin gene deficiency, cardiac muscle lacks eNOS but it overexpresses heart neuronal mtNOS, leading to a reversible NO inhibition of respiration and contractility.[6] In addition, other mtNOS such as the liver iNOS-like one are modulated positively or negatively in the course of hypothyroidism and hyperthyroidism.[20] These studies suggest a potential role of the mtNOS variants in physiology and pathology.

Neurodegeneratives Disorders and Nitric Oxide

The neuropathological features of neurodegenerative diseases, such as Parkinson's disease (PD), Huntington's disease, and Alzheimer's disease (AD), involve the degeneration of circumscribed groups of neurons that may be connected functionally or neuroanatomically. Although genetic and epigenetic risk factors have been implicated, the intrinsic pathophysiological mechanism that induces this selective neuronal loss remains under discussion, and a role of nitric oxide (NO) and NOS in neurodegeneratives disease has been suggested. NO appears to be neurotoxic or neuroprotective, depending on the structures and functions involved and the respective increased or decreased NO production rate.[21]

Nitric oxide neurotoxicity may occur directly or indirectly via peroxynitrite ($ONOO^-$), a potent species produced by the reaction of superoxide anion and nitric oxide, which, in turn, induces lipid peroxidation and secondary functional alterations in proteins and DNA. nNOS may be uncoupled (by deficit of cofactors such as tetrahydrobiopterin or substrate L-arginine), leading to concomitant generation of superoxide anion (O_2^-), thus increasing $ONOO^-$ production. An additional source of O_2^- and peroxynitrite comes from reactions that account for the utilization of NO in mitochondria. The formation of peroxynitrite is responsible for

[19] J. A. Boveris and J. J. Poderoso, *in* "Nitric Oxide: Biology and Pathobiology" (L. J. Ignarro, ed.). Academic Press, San Diego, 2000.

[20] M. C. Carreras, J. G. Peralta, D. P. Converso, P. V. Finocchietto, I. Rebagliati, A. A. Zaninovich, and J. J. Poderoso, *Am. J. Physiol. Heart Circul. Physiol.* **281,** H2282 (2001).

[21] A. W. Deckel, *J. Neurosci. Res.* **64,** 99 (2001).

some alterations, such as the inhibition of mitochondrial complex I,[22] which was also observed in PD. Additionally, NO neurotoxicity could be induced via other reactions, such as nitrosylation of a variety of proteins, e.g., protein kinase C and glyceraldehyde-3-phosphate dehydrogenase.[21]

Are Changes in Extraneural nNOS Representative of Brain nNOS Activity?

An important problem in the study of nNOS in neurodegenerative diseases in humans is to access the enzyme in tissues other than those of the nervous system. The presence of nNOS, cytosolic, and mitochondrial types, in extraneural tissues allows analysis of the modulation of the enzyme in pathology. In the same way, mitochondrial abnormalities related to NO have been identified in the central nervous system and in circulating cells (platelets and lymphocytes) from patients with PD and AD. For instance, β-amyloid$_{40-42}$, the relevant constituent of senile plaque, is detectable in polymorphonuclear cells.[21] These cells are able to generate free radicals and to modulate NO production by neuronal nitric oxide synthase via protein kinase C.

Increased NO Production and Neurotoxicity: Parkinson's Disease

We reported enhanced NO production of neutrophils from patients with Parkinson's disease stimulated with PMA. The rate of NO production by PMN from PD patients *de novo* or receiving L-DOPA alone resulted in about a 42% increase as compared to controls. The increased NO production rate of PMN depended on an overexpression of neutrophil nNOS.[23] Moreover, overexpression of the nNOS gene was found in the brain of patients with PD[24] and also a 5′ flanking region polymorphism of the nNOS gene.[25] Although actual rates of NO release in the substantia nigra in PD are unknown, some clinical and experimental evidence favors increased NO production in this condition, such as nitrotyrosine detection in the central core of Lewy bodies, which constitutes the pathologic hallmark of PD.[26] The formation of peroxynitrite could be related to the impairment of mitochondrial complex I activity.

[22] N. A. Riobó, E. Clemente, M. Melani, A. Boveris, E. Cadenas, S. Moncada, and J. J. Poderoso, *Biochem. J.* **359,** 139 (2001).

[23] E. M. Gatto, N. A. Riobo, M. C. Carreras, A. Cherñavski, A. Rubio, M. L. Satz, and J. J. Poderoso, *Nitric Oxide* **4,** 534 (2000).

[24] D. S. Eve, A. P. Nisbet, A. E. Kingsbury, E. L. Hewson, S. E. Daniel, A. J. Lees, C. D. Marsden, and D. S. Foster, *Brain Res. Mol. Brain Res.* **63,** 62 (1998).

[25] H. S. Lo, E. L. Hogan, and B. W. Soong, *J. Neurol.* **194,** 11 (2002).

[26] S. Przedborski, V. J. Lewi, R. Yokohama, T. Shibata, V. L. Dawson, and E. Dawson, *Proc. Natl. Acad. Sci. U.S.A.* **93,** 4565 (1996).

Low nNOS, Decreased Function, and Neurodegeneration

However, some neurodegenerative disorders may be associated with reduced nNOS activity and expression as reported for Alzheimer and Huntington's diseases.

Alzheimer disease is the most common type of dementia and has been creating profound economic and social impacts as the aging population continues to rise. Although the presence of β-amyloid$_{40-42}$ plaques and neurofibrillary tangles constitute the pathological hallmarks of AD, a putative link between NO and AD has been suggested. In this illness, different authors have reported a decrease in nNOS but, in contrast, expression of other isoforms such as that of microglial iNOS could be increased.[21]

NO Production by Polymorphonuclear Cells from Patients with AD

We measured nitric oxide and hydrogen peroxide production by activated polymorphonuclear cells from patients with Alzheimer's disease. PMN were isolated by dextran sedimentation followed by Ficoll–Hypaque centrifugation and hypotonic lysis of contaminant erythrocytes at room temperature. NO production was measured by the spectrophotometric assay of oxymyoglobin oxidation and H_2O_2 by the buffer as described previously[27] in the presence of 0.1 μg/ml PMA or 1 μM N-formyl-methionyl-leucyl-phenylalanine (fMLP). In the absence of inflammatory mediators, neutrophil activity should be attributed to nNOS activity. The study included 10 untreated patients, fulfilling NINCDS-ADRDA criteria for AD, and 17 age and sex-matched healthy controls (mean age 57 years) without extrapyramidal disorders, cognitive impairment, and/or familial history of AD. The following exclusion criteria were applied to all groups: intake of antioxidants, dopamine agonists, or aspirin and the presence or previous history of severe systemic disease.

The NO production rate of PMN activated with PMA was similar in AD and in control groups. In contrast, a significant difference (-30%) was detected after fMLP activation ($p = 0.013$). The H_2O_2 production rate, reflecting the activity of NADPH oxidase, was not different in AD and controls (Table II). These data suggest that, in AD, circulating neutrophils are not "primed" as proposed for brain microglial cells. In addition, the results may represent variations in the predominant nNOS activity. In this way, the decreased response to fMLP agrees with alterations in defective G protein-dependent signal transduction mechanisms and with the reported decreased expression and/or activity of nNOS in AD. In addition, an inverse correlation between β-amyloid concentrations and nNOS levels has been proposed in AD.[28]

[27] M. C. Carreras, J. J. Poderoso, E. Cadenas, and A. Boveris, *Methods Enzymol.* **269,** 65 (1996).
[28] L. Gargiulo, M. Bermejo, and A. Liras, *Rev. Neurol.* **30,** 301 (2000).

TABLE II
NO AND H_2O_2 PRODUCTION BY STIMULATED HUMAN NEUTROPHILS FROM PATIENTS WITH ALZHEIMER DISEASE[a]

	Controls	Alzheimer
+1 μg/ml PMA		
NO production (nmol/min/10^6 cells)	0.58 ± 0.03	0.53 ± 0.02
H_2O_2 production (nmol/min/10^6 cells)	1.12 ± 0.18	0.97 ± 0.09
+1 μM fMLP		
NO production (nmol/min/10^6 cells)	0.20 ± 0.02	0.14 ± 0.01[b]
H_2O_2 production (nmol/min/10^6 cells)	0.59 ± 0.05	0.48 ± 0.09

[a] Data are mean $\pm$ SEM from 10 samples of each group.
[b] $p < 0.05$ by Student t test.

Increases in iNOS in microglia could be related to a loss of nNOS; an elegant study showed that nNOS could have a tonic inhibitory effect on the expression of iNOS in brain.[29]

Concluding Remarks

Neuronal NOS constitutes a broad spectrum, which comprises the classic 157-kDa protein and other isoforms, including mitochondrial variants in brain and in extraneural tissues. The role of cytosolic or mtNOS could be different and rather specific in different tissues. Changes in expression of the different 144-kDa nNOS variants could correlate with developmental changes as in brain plasticity and maturation. The recently found genetic control of neuronal-type mtNOS expression by the *mdx* gene in mice emphasizes the NO-dependent modulation of muscle energy and function. However, increased or decreased nNOS activity is associated with major neurodegenerative diseases, which can be expressed by parallel variations in nonneural tissues. Further studies are required to ascertain whether changes in mtNOS activity are related to neurological or to other human illnesses.

Acknowledgment

This work was supported by research grants from the University of Buenos Aires (TM47), National Ministry of Health, and the Fundación Perez Companc, Buenos Aires, Argentina.

[29] H. Togashi, M. Sasaki, E. Frohman, E. Taira, R. R. Ratan, T. M. Dawson, and V. L. Dawson, *Proc. Natl. Acad. Sci. U.S.A.* **94,** 2676 (1997).

[38] Visualization and Distribution of Neuronal Nitric Oxide Synthase-Containing Neurons

By MEI XU, YEE KONG NG, and PETER T.-H. WONG

Introduction

A great deal of our understanding of nitric oxide (NO) functioning as a unique biological messenger has derived from studies of its biosynthetic enzyme nitric oxide synthase (NOS).[1,2] Three distinct isoforms of NOS have been cloned: neuronal NOS (nNOS), inducible NOS, and endothelial NOS. nNOS was first purified from rat and porcine cerebellum.[3,4] Its primary amino acid sequence is broadly divided into a reductase domain at the COOH terminus, an oxidative domain at the NH_2 terminus, and several cofactor recognition domains, including a nicotinamide adenine dinucleotide phosphate (NADPH)-binding site.[5] The enzyme is found constitutively in discrete groups of neurons.[6] To delineate the distribution of nNOS-containing neurons, it is imperative that there are sensitive and reliable detection techniques for the visualization of nNOS.

This article describes methods that have been successfully applied specifically for nNOS. With appropriate modifications, these techniques can be used for the visualizations of other isoforms of NOS.

Visualization of nNOS-Containing Neurons

Immunolocalization

The abundance of nNOS in the central nervous system facilitates the isolation and purification of the enzyme and permits the production of monoclonal or polyclonal antibodies for immunohistochemical study. Hence immunohistochemistry is a widely used technique for the identification of nNOS-containing neurons. Antibody raised against the specific amino acid sequence of nNOS is employed to bind the antigen in the tissues under investigation. To detect the bound antibody, enzyme- or fluorochrome-conjugated Fab fragment is commonly employed. They are used to conjugate a secondary antibody that binds the primary anti-nNOS

[1] L. J. Ignarro, M. E. Gold, G. M. Buga, R. E. Byrns, K. S. Wood, G. Chaudhuri, and G. Frank, *Circ. Res.* **64,** 315 (1989).

[2] U. Förstermann and N. J. Dun, *Methods Enzymol.* **268,** 510 (1996).

[3] D. S. Bredt, P. M. Hwang, and S. H. Snyder, *Nature* **347,** 768 (1990).

[4] B. Mayer, M. John, and E. Böhme, *FEBS Lett.* **277,** 215 (1990).

[5] D. S. Bredt, P. M. Hwang, C. E. Glatt, C. Lowenstein, R. R. Reed, and S. H. Snyder, *Nature* **351,** 714 (1991).

[6] U. Förstermann, J. P. Boissel, and H. Kleinert, *FASEB J.* **12,** 773 (1998).

0076-6879/02 $35.00

antibody to achieve optimal results. Peroxidase and alkaline phosphatase are the most common enzyme conjugates, which can be visualized with the respective chromogenic substrates, whereas the fluorescein conjugate can be visualized directly with a fluorescence microscope. The methodology used in our laboratory is as follows.

Light Microscopic: Immunoperoxidase

Rats are anesthetized and perfused with Ringer's solution before being fixed with 4% paraformaldehyde in 0.1 *M* phosphate-buffered saline (PBS) at pH 7.4. The brain and spinal cord are obtained and postfixed in the same fixative for 2 hr at 4°. The tissues are then cryoprotected in 20% sucrose in PBS at 4° overnight. Tissue blocks are cut at 40 μm thickness and collected on gelatin-coated slides. Frozen sections are washed with 0.1 *M* PBS containing 0.1% Triton X-100 (PBS-TX) for 15 min. Normal serum acting as a blocking agent is applied on the sections for 1 hr. Sections are then incubated overnight with the anti-nNOS (primary) antibody at a dilution of 1 : 100 (Transduction Laboratories) at room temperature (25°). After incubation, they are washed in PBS-TX for 15 min, and peroxidase-conjugated secondary antibodies against specific host animals (1 : 200) are applied to the sections for 1 hr. Sections are futher incubated for 1 hr in avidin–biotin peroxidase complex (ABC) solution (Vector Laboratories; final dilution 1 : 100, prepared not more than 30 min before use). They are then rinsed three times for 5 min in PBS before being developed in the substrate 3,3′-diaminobenzidine tetrahydrochloride (DAB, Sigma). Sections are dehydrated and mounted (Fig. 1). Control sections

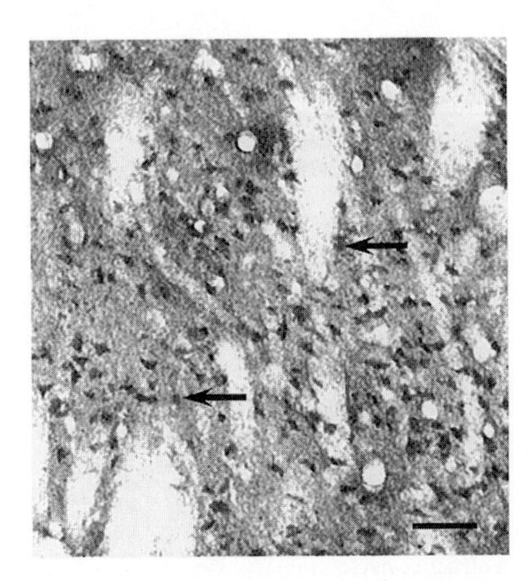

FIG. 1. nNOS immunoreactivity visualized by the immunoperoxidase method. A large number of intensely stained neurons are observed (arrows) around the injection site in the striatum after kainic acid injection. Scale bar: 50 μm.

are obtained by omitting the primary antibody or by using an irrelevant antibody to evaluate the nonspecific peroxidase staining.

Light Microscopic: Immunofluorescence

After incubation with the anti-nNOS antibody, frozen sections are washed in 0.1 *M* PBS-TX. They are placed in either Oregon Green-conjugated anti-mouse IgG (1 : 100) (Molecular Probes) or CY3-conjugated anti-rabbit IgG (1 : 100) (Molecular Probes) for 1 hr. Finally, they are washed three times with 0.1 *M* PBS, picked up on gelatin-coated slides, and mounted using the SlowFade antifade kit (Molecular Probes). The sections are examined under a fluorescence microscope (Leitz, Aristoplan, Germany) with appropriate filters (blue 13, excitation 436–437 nm; N2.1 green, excitation 515–560 nm) (Fig. 2a).

Electron Microscopic: Immunoperoxidase

Tissue blocks are cut transversely at 100-μm thickness using an Oxford vibratome. Free-floating sections are processed for nNOS immunostaining as described earlier. Areas of interest are trimmed into 1-mm^3 blocks under the dissecting

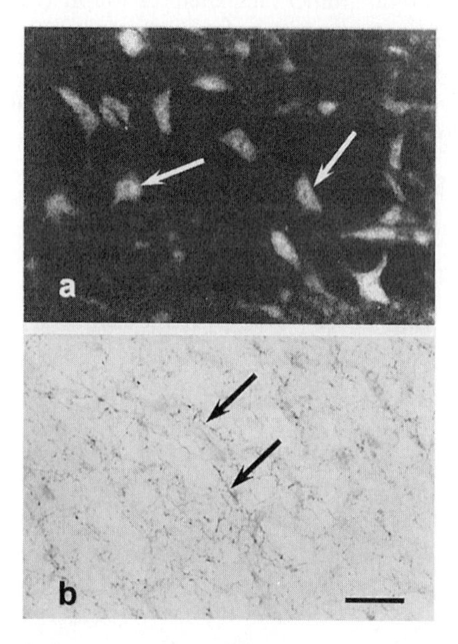

FIG. 2. (a) nNOS immunoreactive neurons are observed in the red nucleus (arrows), as visualized by the immunofluorescence method. (b) Similar section as (a) but stained by the NADPH-diaphorase histochemical method. Positively stained neurons are not observed. Only a few stained nerve fibers are noted (arrows). Scale bar: 20 μm.

microscope. The trimmed sections are fixed in 1% osmium tetroxide containing 1.5% potassium ferrocyanide for 1 hr. After osmication, they are washed briefly in deionized water and dehydrated in an ascending series of ethanol (5%–absolute) by sequential 5-min incubations. The sections are finally infiltrated and embedded in Araldite. Light golden-colored ultrathin sections are obtained by cutting the blocks with a Reichert Jung Ultracut E ultramicrotome. They are then picked up onto 100 meshed copper grids coated with formvar film, stained with lead citrate for 5 min, and viewed in a transmission electron microscope (Fig. 3).

NADPH-Diaphorase Histochemistry

The molecular structure of nNOS reveals highly conserved consensus sequences for binding of NADPH, and purified rat cerebellar nNOS has been shown to contain catalytic NADPH-diaphorase activity.[5] Based on the molecular and biochemical identification of nNOS and its associated NADPH-diaphorase activity,

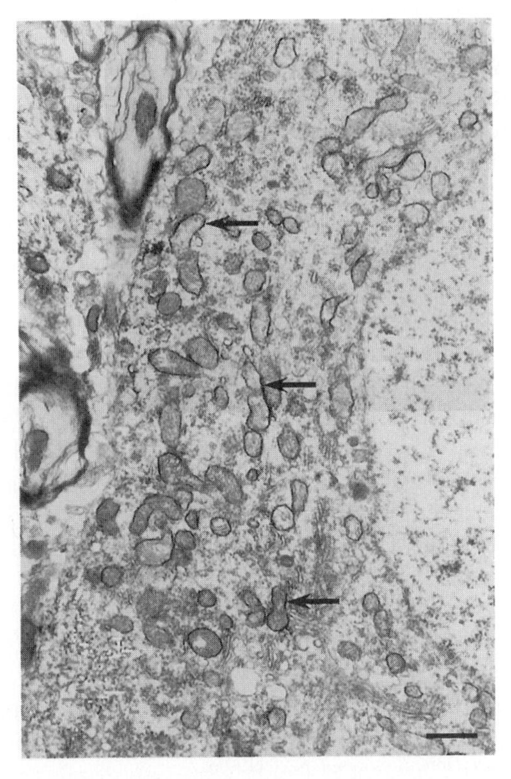

FIG. 3. Electron micrograph showing a nNOS immunoreactive neuron. Arrows indicate immunoreactive product found on the mitochondrial membrane. Scale bar: 1 μm.

histochemical NADPH-diaphorase staining has been known as a marker for nNOS activity in neurons.[7] This histochemical staining relies on the substrates NADPH and an electron acceptor, nitroblue tetrazolium (NBT), as NADPH-diaphorase can reduce NBT to an insoluble blue-colored formazan product.[8] Our laboratory uses a modified method according to Yamamoto *et al.*[9]

Briefly, animal perfusion is done by 4% paraformaldehyde. Tissue preparation and sectioning are as described for immunohistochemistry. Sections on gelatin-coated slides are washed in 0.05 *M* Tris buffer (TB) at pH 7.6 for 10 min. They are then incubated in 0.05 *M* TB containing 0.5% TX, 0.02% nitroblue tetrazolium chloride, 0.1% β-NADPH, and 0.2% calcium chloride for 1 hr in the dark (Sigma). The reaction is stopped by immersion of sections in TB. Subsequently, the sections are washed in three changes of distilled water, dried, and then dehydrated (Fig. 2b). Control sections are performed with the omission of the enzyme substrate β-NADPH from the incubating medium.

In Situ Hybridization

Spatial distribution of neurons expressing nNOS mRNA can be examined by *in situ* hybridization. For many years, radiolabeled probes provided a highly sensitive method for the detection of low-level mRNA.[10] Currently, nonradioisotopic *in situ* hybridization has become a real alternative to radioactive methods.[11] Among numerous nonradioactive methods, digoxigenin (DIG, Boehringer Mannheim, Germany)-labeled complementary DNA (cDNA) probes have been used successfully. DIG is a steroid hapten, found exclusively in digitalis plants. nNOS cDNA probes are generated by RT-PCR using a 23-mer forwarding primer, 5′-GAATACCAGCCTGATCCATGGAA-3′, and a 26-mer reverse primer, 5′-TCCTCCAGGAGGGTGTCCACCGCATG-3′.[12] To incorporate DIG into the cDNA probes, another round of PCR is run using DIG-11-dUTP to replace some of the dTTP in the dNTP mixture (70 μM DIG-11-dUTP, 130 μM dTTP, 200 μM dATP, 200 μM dGTP, 200 μM dCTP). Cryostat sections are postfixed in 4% paraformaldehyde for 4 hr, followed by washing in PBS for 15 min. They are washed further in 0.1 *M* glycine for 5 min, subjected to digestion in 1 μg/ml proteinase K at 37° for 20 min, and followed by a series of washings in 0.1 *M* PBS and 2× standard saline citrate (SSC). Prehybridization in a solution containing 50%

[7] D. S. Bredt, C. E. Glatt, P. M. Hwang, M. Fotuhi, T. M. Dawson, and S. H. Synder, *Neuron* **7,** 615 (1991).

[8] E. Thomas and A. G. E. Pearse, *Acta Neuropathol.* **3,** 238 (1964).

[9] R. Yamamoto, D. S. Bredt, S. H. Snyder, and R. A. Stone, *Neuroscience* **54,** 189 (1993).

[10] D. Sassoon and N. Rosenthal, *Methods Enzymol.* **225,** 384 (1993).

[11] D. G. Wilkinson, *Curr. Opin. Biotechnol.* **6,** 20 (1995).

[12] P. W. Shaul, A. J. North, S. H. Snyder, and R. A. Star, *Am. J. Respir. Cell Mol. Biol.* **13,** 167 (1995).

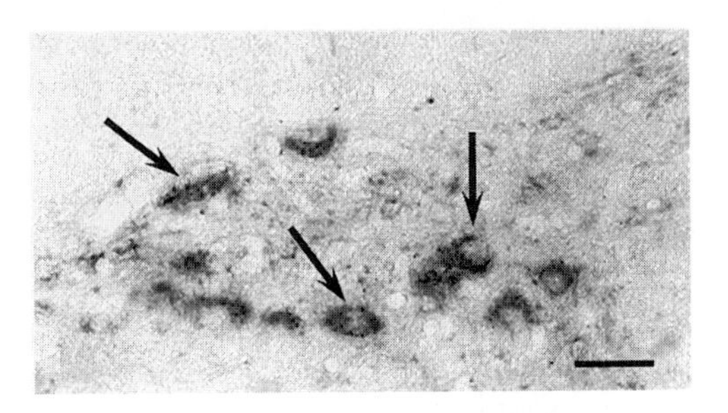

FIG. 4. nNOS mRNA observed in neurons of the nucleus dorsalis (arrows), as visualized by the *in situ* hybridization method. The rat was subjected to lower thoracic spinal cord hemisection. Scale bar: 30 μm.

formamide, 5× SSC, and 40 μg/ml salmon sperm DNA is performed at 55° for 2 hr. The sections are incubated overnight in hybridization buffer containing the 0.5 μg/ml DIG-labeled probe at 55° and are then washed consecutively in 2× SSC and in 0.1× SSC for 1 hr each before being reacted with alkaline phosphatase-coupled anti-DIG antibody for 2 hr. For visualization, the color is developed using NBT and 5-bromo-4-chloro-3-indolyl phosphate as substrates. The sections are dehydrated and mounted with Permount (Fig. 4).

Real-Time Detection of NO

The most important advantage of a real-time technique is that it allows the imaging of endogenous NO released in living cells. In the presence of hydrogen peroxide, NO forms peroxynitrite, which in turn generates photons when it reacts with luminol.[13] Luminol chemiluminescence has been used to visualize NO release induced by electrical stimulation in guinea pig myenteric plexus and rabbit hypogastric nerve trunk.[14]

Alternatively, visualization of NO formation can be made by using selective fluorescence indicators, diaminofluoresceins (DAFs), which are innovatively designed and synthesized for this purpose.[15] Membrane-permeable DAF-2 DA (diacetate) was first used for visualizing NO in rat hippocampal slices.[15] It penetrates through the cell membrane and is hydrolyzed to yield DAF-2. DAF-2 has a relatively low membrane permeability and therefore remains in the cell. It reacts with NO in the presence of dioxygen to form fluorescent DAF-2 T (the triazole form). However, DAF-2 T is pH sensitive and therefore should not be used when

[13] R. Radi, T. P. Cosgrove, J. S. Beckman, and B. A. Freeman, *Biochem. J.* **290,** 51 (1993).

[14] N. P. Wiklund, S. Cellek, A. M. Leone, H. H. Iversen, L. E. Gustafsson, L. Brundin, V. W. Furst, A. Flock, and S. Moncada, *J. Neurosci. Res.* **47,** 224 (1997).

[15] H. Kojima, N. Nakatsubo, K. Kikuchi, S. Kawahara, Y. Kirino, H. Nagoshi, Y. Hirata, and T. Nagano, *Anal. Chem.* **70,** 2446 (1998).

pH is unstable. Another improved indicator, diaminorhodamine 4-methyl iodide (DAR-4M AM), has been developed,[16] which shows no pH dependency above pH 4.0 and provides reliable NO visualization with a detection limit of 7 n*M*.

Distribution of nNOS-Containing Neurons in the Nervous System

In the Central Nervous System

Immunohistochemical studies delineate that nNOS immunoreactive neurons in rats are widely distributed at differing density in all areas and across all layers of the cerebral cortex.[17,18] Dense areas include cuingular area 1, piriform cortex, frontal motor area, and in the medial visual association area, whereas primary sensory areas contain a medium density of nNOS-containing neurons.[18] In the diencephalon, nNOS immunoreactive neurons are largely localized in both the hypothalamus and the thalamus.[17] In the hypothalamus, nNOS-positive neurons are typically found in the perivascular-neurosecretory systems and mamillary bodies.[17,19] Comparing two common rodent species, the number of nNOS-positive neurons and their intensity of staining are much higher in rats than in mice in most nuclei of the hypothalamus.

Particularly, dense clusters of nNOS-positive neurons are detected in the paraventricular and supraoptic nuclei in rats in contrast to the scarcity of those in mice.[19] In the rat hippocampus, only a small proportion of neurons express nNOS. These neurons are mainly localized to the pyramidal layer of the subiculum, stratum radiatum of Ammon's horn, and subgranular zone of the dentate gyrus.[20] Neurons expressing nNOS in the mesencephalon are found in nuclei of the central gray, peripeduncular nucleus, red nucleus, substantia nigra pars lateralis, geniculate nucleus, and superior and inferior colliculi (Figs. 2a, 2b, and 3).[17,21] In the pons, nNOS-containing neurons are seen principally in the pedunculopontine and laterodorsal tegmental nuclei, the ventral tegmental nucleus, the reticulotegmental pontine nucleus, the parabrachial nucleus, and locus ceruleus.[17,22] nNOS-containing neurons are also detected in the principal sensory trigeminal nucleus, trapezoid body, dorsal and median raphe nuclei, pontine reticular nuclei, prepositus

[16] H. Kojima, M. Hirotani, N. Nakatsubo, K. Kikuchi, Y. Urano, T. Higuchi, Y. Hirata, and T. Nagano, *Anal. Chem.* **73,** 1967 (2001).

[17] J. Rodrigo, V. Riveros-Moreno, M. L. Bentura, L. O. Uttenthal, E. A. Higgs, A. P. Fernandez, J. M. Polak, S. Moncada, and R. Martinez-Murillo, *J. Comp. Neurol.* **378,** 522 (1997).

[18] H. J. Bidmon, J. Wu, A. Godecke, A. Schleicher, B. Mayer, and K. Zilles, *Neuroscience* **81,** 321 (1997).

[19] Y. K. Ng, Y. D. Xue, and P. T. H. Wong, *Nitric Oxide* **3,** 383 (1999).

[20] J. G. Valtschanoff, R. J. Weinberg, V. N. Kharazia, M. Nakane, and H. H. Schmidt, *J. Comp. Neurol.* **331,** 111 (1993).

[21] M. Xu, Y. K. Ng, and S. K. Leong, *Brain Res.* **88,** 23 (1998).

[22] N. J. Dun, S. L. Dun, and U. Förstermann, *Neuroscience* **59,** 429 (1994).

hypoglossal nucleus, medial and spinal vestibular nuclei, dorsal cochlear nucleus, reticular nucleus, nucleus of the solitary tract, gracile and cuneate nuclei, the dorsal motor nucleus of the vagus, nucleus ambiguus, and spinal trigeminal nucleus.[17,22] In the cerebellum, nNOS immunoreactivity is detected virtually in all basket and granule cells.[17]

In the rat spinal cord, intense nNOS-positive neurons occur in laminae I–IV and X throughout the entire rostrocaudal extent.[23,24] Weakly stained neurons are scattered in laminae V and VI.[24] However, in the nucleus dorsalis of lamina VII, positive neurons are scarcely seen.[21] In the ventral horn, particularly in the medial portion of laminae VIII and IX, rare nNOS-positive cells are noted.[24] As for the intermediolateral cell column of the thoracic, lumbar, and sacral spinal cord, neurons are heavily stained. The cells are organized as discrete clusters with dendrites projecting rostrocaudally between clusters.[23,24] nNOS immunoreactive neurons concentrated in the aforementioned regions of the spinal cord are also demonstrated in mouse, cat, squirrel monkey, and human.[25]

In the Peripheral Nervous System

For sensory ganglia, nNOS-containing neurons are present predominately in the thoracic and lumbar, but a few are noted in cervical and sacral dorsal root ganglion.[25,26] In nodose ganglia, a moderate number of intensely positive neurons are also noted, whereas in the trigeminal ganglia, positive neurons are few.[26]

nNOS is present in sympathetic and parasympathetic neurons of peripheral tissues and in various autonomic ganglia.[27] nNOS-containing neurons occur in the myenteric plexus throughout the gastrointestinal tract and in the submucous plexus of the stomach, colon, and rectum.[28,29] Topographic patterns show the increasing density of nNOS-positive neurons down the digestive tube longitudinal axis from the stomach to the distal colon.[29] In addition, nNOS-containing neurons are abundant in myenteric ganglia in comparison with those in submucous ganglia in mammals, including humans.[30] They are also seen in intrapancreatic ganglia of all these species.[30]

[23] N. J. Dun, S. L. Dun, U. Förstermann, and L. F. Tseng, *Neurosci. Lett.* **7,** 217 (1992).

[24] S. Saito, G. J. Kidd, B. D. Trapp, T. M. Dawson, D. S. Bredt, D. A. Wilson, R. J. Traystman, S. H. Snyder, and D. F. Hanley, *Neuroscience* **59,** 447 (1994).

[25] G. Terenghi, V. Riveros-Moreno, L. D. Hudson, N. B. Ibrahim, and J. M. Polak, *J. Neurol. Sci.* **118,** 34 (1993).

[26] P. Alm, B. Uvelius, J. Ekstrom, B. Holmqvist, B. Larsson, and K. E. Andersson, *Histochem. J.* **27,** 819 (1995).

[27] S. Ceccatelli, J. M. Lundberg, X. Zhang, K. Aman, and T. Hokfelt, *Brain Res.* **656,** 381 (1994).

[28] T. M. Dawson, D. S. Bredt, M. Fotuhi, P. M. Hwang, and S. H. Snyder, *Proc. Natl. Acad. Sci. U.S.A.* **88,** 7797 (1991).

[29] J. B. Furness, Z. S. Li, H. M. Young, and U. Förstermann, *Cell Tissue Res.* **277,** 139 (1994).

[30] E. Ekblad, P. Alm, and F. Sundler, *Neuroscience* **63,** 233 (1994).

nNOS-containing neurons are abundant in major pelvic ganglia, composed of both sympathetic and parasympathetic neurons, which innervate the genitourinary system in rats.[26] These ganglia display regional differences with respect to the density of nNOS-positive neurons, which may relate to the distinct topographical innervation of pelvic organs, such as the urinary bladder and penis, by these ganglia.[26] Furthermore, nNOS-positive neurons are also documented in intramural ganglion cells of the urinary bladder with predominance in the bladder base of guinea pigs.[31]

Differences Revealed by nNOS Immunohistochemical and NADPH-Diaphorase Staining

As an alternative to immunohistochemistry, NADPH-diaphorase histochemistry has become a widely used tool for localizing nNOS in the nervous system since the colocalization of NADPH-diaphorase and nNOS was described.[28,32] The NADPH-diaphorase is found to be identical to nNOS in many neurons of the central nervous system, including those in the cerebral cortex, striatum, hippocampus, laterodorsal tegmental nucleus, and pedunculopontine nucleus.[28,32] In the peripheral nervous system, an identical presence of NADPH-diaphorase and nNOS is reported in the trigeminal, nodose, dorsal root, major pelvic, otic ganglia,[26] the ganglion neurons of the myenteric plexus,[28] pancreas, and urinary bladder.[31] However, some inconsistencies have been demonstrated in rats between NADPH-diaphorase histochemistry and nNOS immunohistochemistry. NADPH-diaphorase staining revealed more extensive localizations in peripheral ocular tissues, the ciliary ganglion, and the retina when compared to immunohistochemical staining.[9] Conversely, in red and arcuate nuclei, nNOS-immunopositive neurons are observed, whereas NADPH-diaphorase histochemistry revealed only some stained nerve fibers (Figs. 2a and 2b).[19,33]

nNOS accounts for only a small fraction of NADPH-diaphorase activity in unfixed tissue homogenates.[32] Fixation by paraformaldehyde abolishes a high proportion of enzymatic NADPH-diaphorase activity, whereas nNOS immunoreactivity remains largely intact. Thus, the coexistence of NADPH-diaphorase with nNOS immunostaining may depend on the inactivation of most non-nNOS NADPH-diaphorase activity during the fixation procedure.[34] This may explain the inconsistencies in the results obtained by these two staining methods.

[31] Y. Zhou, C. K. Tan, and E. A. Ling, *J. Anat.* **190,** 135 (1997).

[32] B. T. Hope, G. J. Michael, K. M. Knigge, and S. R. Vincent, *Proc. Natl. Acad. Sci. U.S.A.* **88,** 2811 (1991).

[33] M. Xu, Y.-K. Ng, and S.-K. Leong, *Nitric Oxide* **4,** 483 (2000).

[34] T. Matsumoto, M. Nakane, J. S. Pollock, J. E. Kuk, and U. Förstermann, *Neurosci. Lett.* **155,** 61 (1993).

[39] Quantitative Measurement of Endothelial Constitutive Nitric Oxide Synthase

By JANET MEURER, ERIC BLASKO, ANN ORME, and KATALIN KAUSER

Introduction

Due to the important roles of endothelial constitutive nitric oxide synthase (ecNOS) and its enzymatic reaction product, nitric oxide (NO), extensive research has been performed on this enzyme. As a result, there has been an active effort to develop reliable and quantitative methods for measuring human ecNOS protein in cells, for use in studying ecNOS regulation.

Historically, much of the research on the ecNOS protein has employed enzyme activity measurements. When studying the mechanism of intracellular ecNOS protein regulation, however, these measurements have limitations due to the fact that enzyme activity is not a direct measurement of ecNOS protein. One technique available to directly study ecNOS protein expression in cells is Western blot analysis. This methodology, however, is relatively insensitive and does not permit a precise quantitation of ecNOS per cell. An ecNOS ELISA, conversely, is a convenient, quantitative method available for measuring ecNOS protein levels in cultured cells and is currently the most sensitive method for investigating ecNOS protein expression and regulation.

We have previously reported the development of a capture ELISA, which is specific for the detection of the ecNOS isoform.[1] The assay allows for a rapid, quantitative measurement of ecNOS from as few as 6000 human endothelial cells cultured in 96-well microtiter plates and has been used to investigate differences in ecNOS expression in various human cells. This article describes the further refinement of the ecNOS ELISA and the method to quantitate ecNOS from various human endothelial cells. In addition, we describe the use of the ELISA to study the relative levels of ecNOS in bovine, rat, and murine endothelial cells, as well as in murine tissue samples. Finally, data are presented demonstrating that the ELISA is a convenient assay that can be used to quantitatively measure changes in ecNOS protein in human endothelial cells following treatment with chemical and biological agents.

[1] S. Aberle, T. A. Young, P. Medberry, J. Parkinson, G. M. Rubanyi, and K. Kauser, *Nitric Oxide* **1,** 226 (1997).

0076-6879/02 $35.00

Measurement of ecNOS Protein from Cells and Tissues by Sandwich ELISA

Reagents

All reagents are from Sigma (St. Louis, MO) unless otherwise indicated.

Antibodies against ecNOS. For generation of a polyclonal antibody to ecNOS, a peptide from amino acid residues 599 to 614 of human ecNOS, PYNSSPR PEQHKSYK-C is synthesized.[2] This sequence is identical to the corresponding sequence found in bovine ecNOS. Polyclonal antibodies generated against this region of ecNOS have proved to be powerful capture antibodies in the ecNOS ELISA. The peptide is coupled to keyhole limpet hemocyanin (KLH), purified, and used to immunize rabbits (BAbCo, Richmond, CA). Bleeds are collected over a 4-month period. ELISA and Western blot analyses are performed by standard procedures to characterize the antisera to human ecNOS and ecNOS peptide, respectively. Immunopositive rabbit antiserum is run over a protein G Sepharose column (Amersham Pharmacia Biotech, Piscataway, NJ) to isolate the IgG fraction. In some cases, affinity-purified anti-ecNOS peptide antibodies are isolated using a column containing the ecNOS peptide coupled to Sepharose 4B (Amersham Pharmacia Biotech, Piscataway, NJ) and employing standard antibody purification procedures.

Mouse monoclonal antibodies to ecNOS are from BD Biosciences (San Diego, CA). Two ecNOS-specific monoclonal antibodies are tested in the ecNOS ELISA. Both antibodies are made against a 20.4-kDa amino acid protein fragment corresponding to amino acids 1030 to 1209 of human ecNOS. Although both monoclonal antibodies have been raised against the same region of ecNOS, we have found that only one monoclonal antibody can successfully detect human ecNOS in the ELISA. This antibody can also detect ecNOS from various mammalian species, including bovine, rat, and mouse.

Cell Culture for ecNOS Measurement. Human aortic endothelial cells (HAEC) and human umbilical vein endothelial cells (HUVEC) are from Clonetics Corporation (San Diego, CA). Human embryonic kidney cells (HEK293), transfected with human ecNOS, are developed in house as described previously.[1] Primary cultures of bovine aortic endothelial cells (BAEC) are isolated as described by Ryan and colleagues.[3] Rat aortic endothelial cells (RAEC) are obtained from Vec Technologies (Rensselaer, NY). EOMA, a mouse hemangioendothelioma cell line,[4] was a gift from Dr. P. Lelkes (University of Wisconsin Medical School).

HAEC and HUVEC are cultured in endothelial growth medium (EGM) from Clonetics Corporation, which is supplemented with 10% fetal bovine serum (FBS; Hyclone Laboratories, Inc., Logan, UT). BAEC are grown in Ryan's Red medium

[2] L. Buscoi and T. Michel, *J. Biol. Chem.* **268,** 8410 (1993).

[3] J. W. Ryan, A. Chung, L. C. Martin, and U. S. Ryan, *Tissue Cell* **10,** 535 (1978).

[4] J. Obeso, J. Weber, and R. Auerbach, *Lab. Invest.* **63,** 259 (1990).

containing M199 medium from Life Technologies (Rockville, MD) supplemented with 100 μM thymidine, 2 mM L-glutamine, 5% newborn calf serum, and 5% FBS.[3] RAEC are cultured in MCDB-131 medium (Vec Technologies, Rensselaer, NY) in flasks coated with 0.2% gelatin. EOMA cells are grown in Dulbecco's modified Eagle's medium (DMEM) (Life Technologies) supplemented with 10% FBS and 2 mM L-glutamine.

Antibody Labeling with Europium

For use as a reagent to quantitate ecNOS, we previously described the utilization of the ecNOS monoclonal antibody in combination with alkaline phosphatase-conjugated affinity-purified rabbit anti-mouse IgG antibody to detect captured ecNOS in the ELISA.[1] Although this procedure produced sufficient signal-to-background in the ecNOS ELISA, we have found that direct labeling of the anti-ecNOS monoclonal antibody with the lanthanide, europium (Eu), produces equivalent results and shortens the overall ELISA procedure.

To label the antibody with Eu, 2 mg of antibody (carrier free, without sodium azide) is dialyzed against 4 liters of buffer containing 50 mM $NaHCO_3$ (pH 8.5) and 150 mM NaCl for 3 hr at 4°. Following a buffer change, dialysis is continued for an additional 3 hr at 4°. One milligram of dialyzed antibody is then added per microfuge tube containing 150 nmol DELFIA Eu-N1-DTA powder (PerkinElmer Life Sciences, Boston, MA). The microfuge tubes are protected from light and mixed, end over end, at room temperature overnight. Eu-labeled antibodies are isolated by purification over a PD-10 desalting column (Amersham Pharmacia Biotech, Piscataway, NJ) as described by the manufacturer. Following purification, the protein concentration is determined by measuring absorbance at A_{280} in a spectrophotometer. The amount of Eu labeling per antibody is estimated relative to a standard solution (1 nM) of Eu (PerkinElmer Life Sciences) by measuring time-resolved fluorescence emission at 615 nm in a Wallac 1420 VICTOR^2 multilabel counter (PerkinElmer Life Sciences). Eu-labeled antibody can be stored at 4° for over 1 month or at $-20°$ for over 6 months.

Cell Plating and Cell Lysis

For human endothelial cell incubations in microtiter plates, HAEC are cultured to passage 5 or 6, and HUVEC are cultured to passage 3 or 4, harvested by gentle trypsinization, and seeded at a cell density of 2×10^4 cells/well in a 96-well flat-bottom tissue culture plate using 100 μl/well Clonetics EGM supplemented with 10% FBS. Cells are incubated at 37° in 5% CO_2 for 24 hr. Thereafter, media are removed by aspiration, and the cells are incubated for an additional 24 hr in 100 μl/well Clonetics EGM containing 0.1% gelatin, 10 mM HEPES (pH 7.4), and 50 μg/ml gentamicin (Life Technologies, Rockville, MD), with or without chemical or biological reagents [i.e., tumor necrosis factor (TNF)α, simvastatin]. Media are removed, plates are blotted dry, and cells are lysed with 80 μl/well of 50 mM Tris (pH 7.4) containing 100 mM NaCl, 0.5% NP-40, 1 mM EDTA, and protease inhibitors (400 μM pefabloc SC, 10 μg/ml antipain, 10 μg/ml leupeptin, 25 μg/ml aprotinin, and 2 μg/ml pepstatin).

Endothelial cells from nonhuman species are cultured in 96-well plates as follows: BAEC are plated at 3.5×10^4 cells per well in Ryan's Red medium (see earlier discussion). RAEC are plated at 3.5×10^4 cells per well in "MCDB-131 Complete," a proprietary formula from Vec Technologies (Rensselaer, NY). EOMA cells are plated at 3.5×10^4 cells per well in DMEM supplemented with 10% FBS and 2 mM L-glutamine. All cell types are incubated at 37° in 5% CO_2 for 24 hr. The following day, media are removed by aspiration, and each of the cell types is incubated for an additional 24 hr in 100 μl/well Clonetics EGM containing 0.1% gelatin, 10 mM HEPES (pH 7.4), and 50 μg/ml gentamicin. Following this incubation, media are removed, plates are blotted dry, and cells are lysed as described earlier for human endothelial cells.

Enzyme-Linked Immunosorbent Assay (ELISA) Procedure

ELISA Using Cell Lysates. Cell lysates from human, bovine, rat, and mouse endothelial cells have been analyzed successfully using the ecNOS ELISA. One hundred microliters per well of 15 μg/ml rabbit polyclonal anti-ecNOS peptide antibody, diluted in 50 mM carbonate buffer (pH 9.5), is adsorbed to EIA/RIAplates (Corning Costar, Corning, NY) for 18 to 48 hr at 4°. The antibody solution is aspirated, and wells are blocked with 150 μl/well of either 0.5% I-Block (Tropix, Bedford, MA) in phosphate-buffered saline (PBS) or 4% bovine serum albumin (BSA) in PBS for 1 hr at 37° (plates can also be incubated overnight at 4° for this step). Just prior to use, the blocking solution is aspirated and the plate is washed twice with PBS containing 0.05% Tween-20. Sixty microliters of cell lysate (or purified ecNOS standard ranging from 0.5 to 50 ng/ml) is added to the wells and incubated for 1–2 hr at room temperature. The wells are washed three times with PBS containing 0.05% Tween-20 and blotted dry. One hundred microliters per well of Eu-labeled mouse monoclonal anti-ecNOS detection antibody, diluted to approximately 125 ng/ml in DEFIA assay buffer (PerkinElmer Life Sciences), is added to the plates and incubated for 1 hr at room temperature (the exact amount of detection antibody may need adjustment based on the amount of Eu labeling). The wells are washed three times with PBS containing 0.05% Tween-20 and blotted dry. One hundred microliters per well of DEFIA enhancement solution (PerkinElmer Life Sciences) is added and incubated for 1–2 hr at room temperature or overnight at 4°. Time-resolved fluorescence is measured by monitoring emission at 615 nm in a Wallac 1420 VICTOR^2 multilabel counter (PerkinElmer Life Sciences). The fluorescence from wells incubated with the sample buffer (in parallel with the samples and ecNOS standard) is subtracted as background from the sample and standard fluorescence values. For ecNOS quantitation, fluorescence values from the purified ecNOS standard are fitted to a linear equation, which is compared with sample values.

ELISA Using Tissue Samples. Tissue homogenates from mouse lung and heart have also been analyzed successfully using the ecNOS ELISA. Tissue samples are excised, frozen at −80°, and diced using razor blades prior to homogenization in buffer containing 25 m*M* Tris (pH 7.8), 10% glycerol, 0.2% NP-40, and protease inhibitors (protease inhibitor complete mini tablet, Roche Molecular Biochemicals, Indianapolis, IN). Cofactors for enzyme activity are also added: 4 μM each of FAD, FMA, and BH_4; 3 m*M* dithiothreitol (DTT), 3 m*M* $CaCl_2$, and 0.125 μM calmodulin (Roche Molecular Biochemicals). The tissue is homogenized with a Kinematics polytron probe. After centrifugation (5 min at full speed in an Eppendorf microcentrifuge), the supernatant is transferred to a clean tube, and the pellet is extracted a second time with 75 μl of homogenization buffer using a ground glass Duall homogenizer. Following a second centrifugation (see earlier discussion), the two supernatants are combined and centrifuged a third time. The resulting supernatant is isolated and used for analysis by ELISA. Utilizing this procedure, the final concentration for each tissue sample is approximately 100 mg/ml when beginning with tissue wet weights of 160 mg for both lung and heart. For analysis by ELISA, 60 μl of tissue homogenate supernatant is added to the coated, blocked ELISA plates and incubated for 1–2 hr at room temperature, as described earlier for the cell lysates. The rest of the ELISA procedure is identical to that described earlier for the cell lysates.

Linearity and Sensitivity of the ecNOS ELISA

Linearity

The ecNOS capture ELISA was analyzed for linearity using 0.15 to 1000 ng/ml of purified ecNOS protein or using control lysate from ecNOS-transfected HEK293 cells[1] and testing a range (2 to 80 μg/ml) of lysate concentrations. Using purified ecNOS, the assay shows good linearity up to 300 ng/ml of purified protein (correlation coefficient, $r = 0.995$). The detection limit is approximately 0.3 ng/ml ecNOS (Fig. 1A), which is over 50 times greater than the detection of ecNOS using standard Western blot analysis. In comparison, using HEK293 cell lysates expressing recombinant ecNOS, linearity is seen up to 10 μg/ml of total lysate protein (correlation coefficient, $r = 0.984$), which is equivalent to 32 ng/ml ecNOS. The limit of detection of ecNOS in this lysate is approximately 4–5 ng/ml ecNOS (Fig. 1B). The decrease in both the linearity of detection and in the detection limit of ecNOS is likely due to interference from components of the cell lysate.

Specificity

The three nitric oxide synthase protein isoforms, ecNOS, brain NOS (bNOS), and inducible NOS (iNOS), are 50 to 60% homologous to one another. The ecNOS

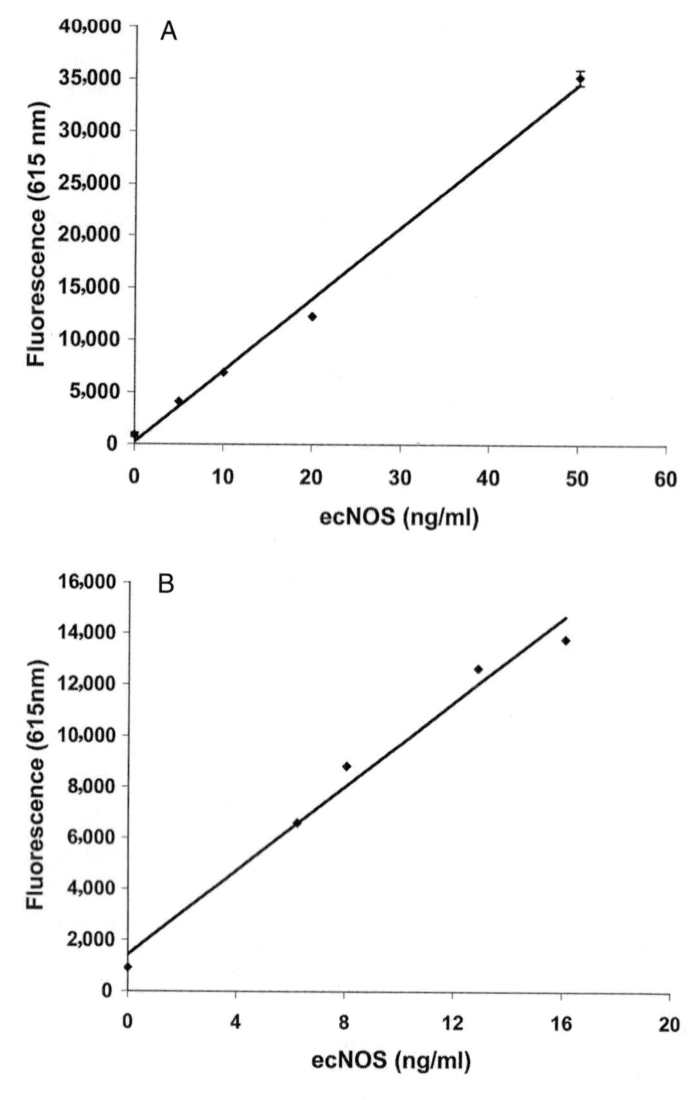

FIG. 1. Determination of the sensitivity of the ecNOS capture ELISA and linearity of detection. (A) Purified ecNOS protein (0.15 to 50 ng/ml) and (B) lysate from ecNOS-transfected HEK293 cells (1.8 to 5 μg/ml lysate) were tested in the capture ELISA. For this experiment, the background fluorescence units are not subtracted from the sample fluorescence at 615 nm. The average and standard deviation of three replicates are shown.

ELISA was designed specifically to use antibodies that were raised against epitopes of ecNOS that were weakly or not homologous to the other two nitric oxide synthase isoforms, bNOS and iNOS. The selectivity of the ecNOS ELISA was demonstrated previously by analyzing ecNOS, bNOS, and iNOS in the assay.[1] Purified ecNOS and lysate from ecNOS-transfected HEK293 cells were tested with samples of purified iNOS and a lysate from bNOS-transfected Chinese hamster ovary (CHO) cells, as well as with lysates from the untransfected parental cell lines, HEK293 and CHO. Strong signals were obtained with the positive controls: 25 ng/ml of purified ecNOS and 10 μg/ml of the ecNOS-HEK lysate. However, 10 μg/ml of purified iNOS, 40 μg/ml of bNOS-CHO lysate, and 40 μg/ml of the parental lysates (from HEK and CHO cells) could not be detected by the assay. This demonstrated that there was no cross-reactivity of the ecNOS monoclonal detection antibody with the other two NOS isoforms in the ecNOS ELISA.

Detection of ecNOS in Various Cells and Tissues

Detection and Quantitation of ecNOS in Cultured Human Endothelial Cells

ecNOS levels in cultured HAEC and HUVEC from several donors were measured and compared, as reported previously.[1] HAEC and HUVEC are seeded at 6.25×10^3 to 5×10^4 cells per well and cultured for 24 hr. Cells are lysed in 50 μl lysis buffer per well, and the total lysates are transferred directly into the antibody-coated and blocked ELISA plates (note that 1.25×10^4 cells per well lysed in 50 μl correspond to 2.5×10^5 cells/ml). Data indicate that ecNOS can be detected in as few as 6000 human endothelial cells per well.

To quantitate the amount of ecNOS in these cells, more concentrated lysates of HAEC (passages 5 to 6) and HUVEC (passages 3 to 4) from different donors were serially diluted and compared with purified ecNOS (Fig. 2).[1] The detection of ecNOS in these cells was linear up to 2.5×10^5 cells/ml ($r \geq 0.99$). More concentrated cell lysates resulted in some loss of linearity. A quantitation from these data yielded 16.8, 17.1, and 21.0 ng ecNOS per 10^6 HAEC for donors 1, 2, and 3, respectively, and 10.8, 10.9, and 9.4 ng ecNOS per 10^6 HUVEC for donors 4, 5, and 6, respectively. These results indicate relatively little heterogeneity between donors, but a trend for less ecNOS protein expression in HUVEC relative to HAEC. An averaging of data calculates approximately 8.3×10^4 molecules ecNOS per HAEC and 4.7×10^4 molecules ecNOS per HUVEC.

Measurement of ecNOS Levels in Endothelial Cells from Various Mammalian Species

As stated earlier the sequence of the peptide used to raise the ecNOS capture antibody is identical to the corresponding sequence in bovine ecNOS. Similarly, the protein fragment used to raise the ecNOS detection antibody is 94% homologous

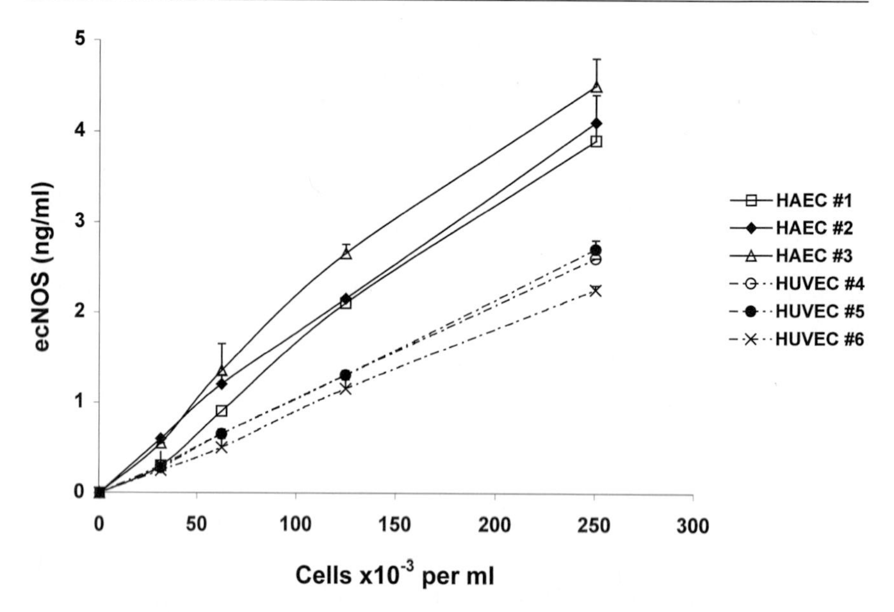

FIG. 2. Quantitation of ecNOS expression per cell for HAEC and HUVEC. Concentrated cell lysates from three donors of HAEC (#1, #2, #3) (passages 5 to 6) and HUVEC (#4, #5, #6) (passages 3 to 4), respectively, were titrated and compared with the purified ecNOS standard in the capture ELISA. The ecNOS concentration per sample is calculated from the ecNOS standard curve by linear regression analysis. The average and standard deviation of four replicates are shown. Reproduced with permission from S. Aberle, T. A. Young, P. Medlberry, J. Parkinson, G. M. Rubanyi, and K. Kauser, *Nitric Oxide* **1**, 226 (1997).

to the corresponding region in bovine ecNOS. For both of these antigens, the corresponding mouse and rat homologies are not known, as the amino acid sequences are not available for rat and mouse ecNOS. Given this information, the detection of ecNOS from cultured bovine and rat endothelial cells (BAEC and RAEC) and murine EOMA cells was tested and compared with HAEC (Fig. 3). Each cell type was plated as outlined earlier and lysed with 80 μl per well lysis buffer. The lysates were assayed for total protein content using the BCA protein assay (Pierce, Rockford, IL), with BSA as standard, according to the manufacturer's instructions. The lysate concentrations were between 100 and 300 μg/ml total protein for all cells tested. These lysates were diluted further with lysis buffer to final total protein concentrations ranging from 8 to 100 μg/ml and transferred directly to the antibody-coated and blocked ELISA plate. As seen in Fig. 3, the assay detects ecNOS in all of the species of endothelial cell lysates tested, with a linear response up to 100 μg/ml total protein. Although it is not known whether the antibodies used in the ELISA bind with equal affinity to human, bovine, rat and murine ecNOS, endothelial cells from all three nonhuman species appear to

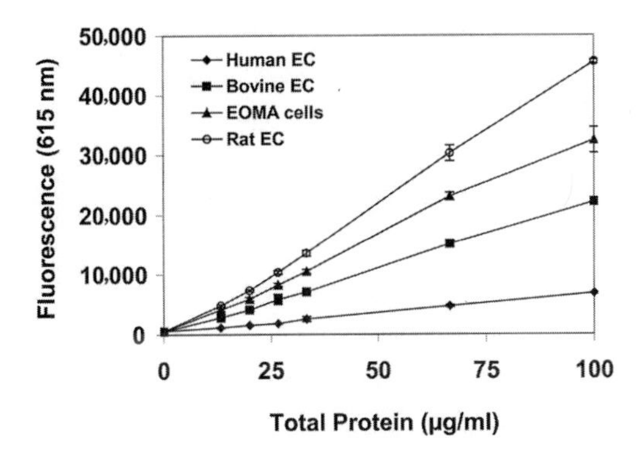

FIG. 3. Detection of ecNOS from human, bovine, mouse, and rat endothelial cells (EC). Human EC, bovine EC, mouse EOMA cells, and rat EC were seeded at 2 to 3.5×10^4 cells per well in a 96-well microtiter plate and cultured for 24 hr. The cells were lysed in 80 μl lysis buffer, and total protein was measured. Lysates were diluted in lysis buffer to final protein concentrations ranging from 8 to 100 μg/ml and added to the ELISA plate in a final volume of 60 μl. Diluted lysates were tested by the ecNOS capture ELISA. The average and standard deviation of three replicates are shown.

express significantly more ecNOS compared to human endothelial cells. In relation to HAEC, endothelial cells from rat express approximately 5.4-fold more ecNOS, the mouse EOMA cells express 4-fold more, and BAEC express 2.8-fold more.

Measurement of ecNOS Levels in Lung and Heart Tissue Homogenates from ecNOS-Deficient and Wild-Type Mice

The ELISA has been used successfully to measure ecNOS levels in homogenates from lungs and hearts excised from C57B1/J6 and ecNOS-deficient (ecNOS-KO) mice (Jackson Laboratories, Bar Harbor, ME) (Fig. 4). Following tissue isolation and homogenization, 60 μl of homogenate supernatant was transferred to the antibody-coated and blocked ELISA plate. Purified human ecNOS was used as a standard for this study. Note that as the standard for the ELISA is human ecNOS, the values given are relative rather than absolute. Data indicate that wild-type lungs contain 5.2 ± 0.96 ng ecNOS/mg tissue, whereas ecNOS-KO lungs contain no detectable amounts of ecNOS ($>0.09 \pm 0.01$ ng ecNOS/mg tissue). Wild-type heart values are slightly higher at 7.0 ± 0.6 ng ecNOS/mg tissue, whereas ecNOS-KO hearts showed levels less than 0.07 ± 0.03 ng ecNOS/mg tissue. Although ecNOS-KO mice do not contain any ecNOS protein, they were shown to have normal amounts of bNOS and iNOS (data not shown).

Regulation of ecNOS Protein in Cells

The major strength of the ecNOS capture ELISA is that it facilitates the investigation of ecNOS regulation greatly. The sensitivity of the ELISA enables one to

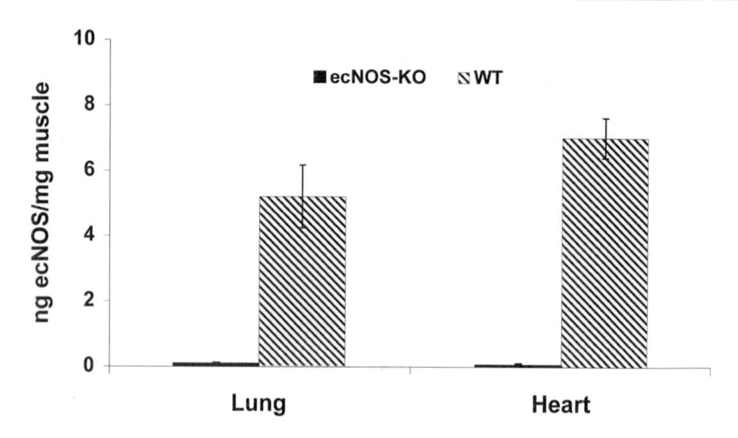

FIG. 4. Detection of ecNOS from control wild-type C57B1/7C (WT) and ecNOS-deficient (ecNOS-KO) mouse tissues. Liver and heart tissues were isolated from WT and ecNOS-KO mice and homogenized in tissue homogenization buffer. Sixty microliters of homogenate supernatant per well was tested by the ecNOS capture ELISA. Fluorescence values were compared with the ecNOS ELISA standard curve, and the ecNOS concentration per sample was calculated from the standard by linear regression analysis. The average and standard deviation of three replicates are shown.

culture and treat both human and nonhuman endothelial cells in 96-well microtiter plates, even at subconfluent cell densities. The microtiter incubations allow several parameters to be tested concurrently with multiple replicates. The precise quantitation of human ecNOS enzyme expression offers an advantage for investigating the regulation of ecNOS and its basal expression.

For the regulation studies presented here, we found that culturing the control and treated human endothelial cells in microtiter plates and testing the lysates by ELISA yielded very consistent and reproducible results. ecNOS expression in the untreated endothelial cells throughout the plate measured a 10% coefficient of variance (CV), demonstrating low well-to-well variability. A 15% CV for ecNOS levels was detected between plates ($n = 4$). Results from the 96-well microtiter plate incubations of HAEC demonstrated changes in ecNOS protein in response to agents similar to data obtained using Western blot analysis. Thus, the ELISA can substitute for the other more tedious and time-consuming methods used to detect changes in ecNOS protein expression.

Measurement of Changes in ecNOS Protein Expression in Endothelial Cells Treated with TNFα, Phorbol Ester, or Simvastatin

To investigate ecNOS downregulation, HAEC are incubated with a concentration range of 10 to 0.1 ng/ml TNFα for 24 hr. Consistent with reports from

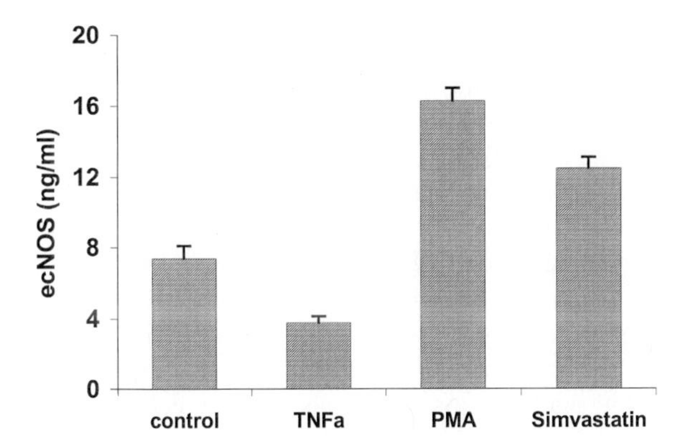

FIG. 5. Measurement of changes in ecNOS protein levels in HAEC treated with TNFα, PMA, or simvastatin. HAEC were seeded at 2×10^4 cells per well in a 96-well microtiter plate for 24 hr and then treated for an additional 24 hr with 3 ng/ml TNFα, 50 n*M* PMA, or 2 μM simvastatin. Cells were lysed in 80 μl lysis buffer. Sixty microliters of lysate was transferred to the ELISA plate and tested in the ecNOS capture ELISA. Fluorescence values were compared with the ecNOS ELISA standard curve, and the ecNOS concentration per sample was calculated from the eNOS standard by linear regression analysis. Data expressed have been normalized for total cellular protein. The average and standard deviation of three replicates are shown.

other investigators,[5–8] ecNOS expression is reduced significantly by treatment with TNFα (50% decrease) over the 24-hr period. The maximum decrease in ecNOS is seen at 3–10 ng/ml TNFα (Fig. 5). The samples exhibit no loss in cell viability during the course of the experiment. In order to study ecNOS upregulation, HAEC are incubated with a concentration range of 100 to 0.01 n*M* phorbol myristate acetate (PMA) or 10 to 0.04 μM of simvastatin for 24 hr. Both PMA and simvastatin dose dependently increase the level of ecNOS expression in HAEC by up to 2.2- and 1.7-fold, respectively, which is consistent with previous findings in the literature.[9,10] Maximum increases in ecNOS protein were observed at 50 n*M* PMA and 2 μM simvastatin, respectively (Fig. 5). As was seen in the study using TNFα, HAEC exhibited no loss in cell viability during the course of treatment and the ecNOS ELISA results were consistent and reproducible.

[5] K. L. MacNaul and N. I. Hutchinson, *Biochem. Biophys. Res. Commun.* **196,** 1330 (1993).

[6] N. Marczin, A. Antonov, A. Papetropoulos, D. Munn, R. Virmani, F. Kolodgie, R. Gerrity, and J. Catravas, *Arterioscler. Thromb. Vasc. Biol.* **16,** 1095 (1996).

[7] F. Mohamed, J. C. Monge, A. Gordon, P. Cernacek, D. Blais, and D. J. Stewart, *Arterioscler. Thromb. Vasc. Biol.* **15,** 52 (1995).

[8] M. Yoshizumi, M. A. Perrella, J. C. Burnett, Jr., and M. E. Lee, *Circ. Res.* **73,** 205 (1993).

[9] U. Laufs, V. LaFata, J. Plutzky, and J.-K. Liao, *Circulation* **97,** 1129 (1998).

[10] Y. Ohara, H. S. Sayegh, J. J. Yamin, and D. G. Harrison, *Hypertension* **25,** 415 (1995).

Additional Remarks

This article describes the quantitation of ecNOS protein from cultured human endothelial cells, as well as the relative measurement of ecNOS from several nonhuman endothelial cells and murine tissues, using a specific ecNOS ELISA. The ELISA procedure described here is simple and relatively inexpensive. R&D Systems (Minneapolis, MN) introduced the only commercially available human ecNOS sandwich ELISA. A comparison of the commercial ecNOS ELISA with the one described in this article demonstrated that subnanomolar levels of ecNOS protein, as expressed in human endothelial cells and murine tissues, can also be detected readily using the kit. We have not determined whether the commercially available assay can also detect ecNOS from bovine and rat.

The ELISA described here has also been adapted successfully to capture active ecNOS enzyme, using the polyclonal anti-ecNOS capture antibody, and to measure ecNOS activity on the enzyme bound to the ELISA plate. This has allowed us to quantitate ecNOS protein, as well as measure ecNOS enzyme activity from the same sample. The key to such a combination is extraction of the ecNOS enzyme, from cells or tissue homogenates, with the cofactors needed for full enzymatic activity (FMN, FAD, BH_4, DTT, calmodulin, and calcium), which presumably keeps ecNOS in its active state. An additional advantage of this modified procedure is that ecNOS enzyme activity can be determined without the interference from other NOS isoforms such as bNOS or iNOS. The details of this method are described elsewhere (manuscript in preparation).

Acknowledgments

We are pleased to acknowledge the assistance of Rhonda Humm, Zhong Liu, and Linda Cashion for the production of stable bNOS and ecNOS cell lines, as well as the contribution of Eileen Paulo for culturing the different endothelial cells. Robert Mintzer confirmed the specificity of the antibodies by Western blot analysis.

[40] Quantitation of Expressed Message for Inducible Nitric Oxide Synthase

By RICHARD R. ALMON and DEBRA C. DUBOIS

Introduction

Inducible nitric oxide synthase (NOS2) (EC 1.14.13.39) is one of three different nitric oxide synthases, the other two being neuronal nitric oxide synthase (NOS1) and endothelium nitric oxide synthase (NOS3).[1] All three forms of NOS catalyze the reaction

$$\text{L-arginine} + \text{NADPH} + O_2 = \text{citrulline} + \text{nitric oxide} + \text{NADP}^+$$

Both NOS1 and NOS3 are constitutive, whereas NOS2 is inducible and plays a complex role in inflammatory responses. NOS2 is expressed in macrophages, as well as other cell types, such as liver, retina, bone, lung, epithelial cells, microglia in the central nervous system, and in a variety of tumor cells. NO is involved in not only immune/inflammatory responses, but also in both vasodilation and neurotransmission. Because NOS1 is expressed in nonneuronal tissues, such as skeletal muscle, and NOS3 is expressed in vascular endothelium, it is not uncommon for all three genes to be expressed at the same time. For example, all three forms of NOS are involved following muscle tendon damage.[2] Because of the possibility of coexpression of the three forms of NOS, a major aspect necessary to the measurement of NOS2 is to separate its expression from the expression of both NOS1 and NOS3.

Available Approaches to NOS2 mRNA Quantitation

A number of experimental approaches to measure mRNA expression have been developed and applied to NOS2 (e.g., see Refs. 3–9). These include Northern

[1] K. S. Christophersen and D. S. Bredt, *J. Clin. Invest.* **100,** 2424 (1997).

[2] J. H. Lin, M. X. Wang, A. Wei, W. Zhu, A. D. Diwan, and G. A. Murrell, *J. Orthopaed. Res.* **19,** 136 (2001).

[3] J. A. Hewett, S. J. Hewett, S. Winkler, and S. E. Pfeiffer, *J. Neurosci. Res.* **56,** 189 (1999).

[4] M. Naassila, F. Roux, F. Beauge, and M. Daoust, *Eur. J. Pharmacol.* **313,** 273 (1996).

[5] D. Xia, A. Sanders, M. Shah, A. Bickerstaff, and C. Orosz, *Transplantation* **72,** 907 (2001).

[6] H. Guhring, I. Tegeder, J. Lotsch, A. Pahl, U. Werner, P. W. Reeh, K. Rehse, K. Brune, and G. Geisslinger, *Inflamm. Res.* **50,** 83 (2001).

[7] Y. Kohmura, T. Kirikae, F. Kirikae, M. Nakano, and I. Sato, *Int. J. Immunopharmacol.* **22,** 765 (2000).

[8] Y. Vodovotz, A. G. Geiser, L. Chesler, J. J. Letterio, A. Campbell, M. S. Lucia, M. B. Sporn, and A. B. Roberts, *J. Exp. Med.* **183,** 2337 (1996).

[9] A. R. Amin, M. Attur, P. Vyas, J. Leszczynska-Piziak, D. Levartovsky, J. Rediske, R. M. Clancey, K. A. Vora, and S. B. Abramson, *J. Inflamm.* **47,** 190 (1995).

0076-6879/02 $35.00

hybridization, conventional reverse transcription-polymerase chain reaction (Rt-PCR), and, more recently, fluorescence-based real-time Rt-PCR. Most of these assays have applied a semiquantitative approach, which allows experimental samples to be expressed as an "*x*-fold change" relative to some control sample measured within the same experiment. Absolute quantitation can be achieved by including cRNA standards in these techniques. The use of *in vitro*-transcribed cRNA standards of known concentrations allows molar quantitation and provides a basis for comparision of experimental outcomes between experiments (see later). However, with any assay employed for the measurement of mRNA expression, appropriate selection of probe and/or primers is necessary in order to ensure specificity. Because nature tends to be repetitious in the use of structure, one cannot randomly select a sequence from the coding region of a gene for use as probe and/or primers. For example, many enzymes use the same cofactor. The binding site for a particular cofactor most likely will be the same regardless of the enzyme. A similar problem is presented by multigene families. It is therefore necessary to determine exactly what message is to be measured and find a region of the message that is unique. The best way to identify specific regions is to submit the gene sequence to BLASTn, which is part of the NCBI group of interlocking databases.

The NOS2 Sequence

Choosing an appropriately selective sequence for use in the measurement of expression of inducible nitric synthase is a relatively difficult problem. The first problem involves selecting a sequence that distinguishes nitric oxide synthase 2 (NOS2) from NOS1 and NOS3. This a significant problem because the active site is highly conserved and quite similar among the three NOS forms.[10] In addition, all three forms of NOS contain distinct coding domains for heme, FNM, FAD, and NADP. In the mouse, NOS2 is calcium independent, but NOS1 and NOS3 are regulated by calcium/calmodulin, providing a possible area of distinction. In contrast, in humans, all three NOS forms are regulated by calcium/calmodulin, eliminating this difference. Also to be considered in selecting a sequence are different forms of the NOS2 message. In the mouse, several different splice variants have been reported. In the human, three distinct NOS2 genes all coded on chromosome 17 have been described. Small but significant sequence variations exist among the three human genes, NOS2A, NOS2B, and NOS2C. It is therefore necessary to decide in advance if one wishes results to reflect these distinctions.

[10] T. O. Fischmann, A. Hruza, X. D. Niu, J. D. Fossetta, C. A. Lunn, E. Dolphin, A. J. Prongay, P. Reichert, D. J. Lundell, S. K. Narula, and P. C. Weber, *Nature Struct. Biol.* **6,** 233 (1999).

Use of cRNA Standards for Quantitative Analysis of Gene Expression

We have employed *in vitro*-transcribed cRNA standards for molar quantitation in both Northern analysis and Rt-PCR. cRNA standards can be produced by inserting an appropriate sequence into a plasmid containing at least one viral promotor suitable for *in vitro* transcription as described originally by Krieg and Melton.[11] We routinely use a plasmid such as pGEM3Z (Promega), which has an expanded multicloning region and two promoters (T7 and SP6). In this way it is possible to construct sense cRNA standards and, if desired the antisense riboprobe. The important consideration is in selecting an "appropriate sequence" to use as a cRNA standard. The requirement for these standards is that they be specific for the expressed message being measured and of a reasonable size such that full-length cRNA transcripts can be obtained in reasonable quantity for validation and quantification. Initially obtaining full-length transcripts was a problem if the sequence was too long (i.e., greater than approximately 800 bp), and obtaining sufficient material required the pooling of multiple reactions. However, commercially available products such as the MegaScript line of transcription kits produced by Ambion (Austin TX) are capable of producing both longer products and more substantial yields. Following transcription, we digest template DNA with RNA-free DNase, phenol–chloroform extract, and subject the cRNA to three rounds of ethanol precipitation in the presence of 2.5 *M* ammonium acetate to ensure complete removal of free nucleotides.[11] We routinely assess the concentration of *in vitro*-transcribed cRNA standards by triplicate determinations of optical density at 260 nm. Such samples typically have a 260/280 ratio ranging from 1.8 to 2.0. Purity and integrity are assessed further by electrophoresis in 5% acrylamide/8 *M* urea gels. We have established that cRNA prepared by these procedures under RNase-free conditions can be stored for many years at -80° without loss of integrity.

In our original 1993 paper on molar quantification of expressed message by Northern hybridization,[12] we used such cRNAs as reference standards both to correct for yield of total RNA and to quantify the expression of mRNA for the enzyme tyrosine aminotransferase. The basis of this method was to construct concentration curves of known amounts of cRNA standards on each formaldehyde gel, which could be used to quantify the expressed message signal. We have used this approach to quantify expression of a number of messages in several different tissues.[13,14]

[11] P. A. Krieg and D. A. Melton, *in* "Recombinant DNA" (R. Wu, ed.), Vol. 155, p. 397. Academic Press, San Diego, 1987.

[12] D. C. DuBois, R. R. Almon, and W. J. Jusko, *Anal. Biochem.* **210,** 140 (1993).

[13] Y. N. Sun, D. C. DuBois, R. R. Almon, and W. J. Jusko, *J. Pharmacokinet. Biopharma.* **26,** 289 (1998).

[14] Y. N. Sun, L. I. McKay, D. C. DuBois, W. J. Jusko, and R. R. Almon, *J. Pharmacol. Exp. Therap.* **288,** 720 (1999).

An example of this approach is presented in Fig. 1. Because experimental signals are expressed in moles of message, this approach allows for the integration of data across multiple hybridization matrices in the same or different experiments. Although this procedure for quantitative Northerns is highly reproducible and can be used for quantifying the expression of NOS2, it does have several limitations. The first is that it is a very time-consuming method, which limits the number of samples that can be analyzed in a reasonable amount of time. The second is sensitivity. Even when a riboprobe is used, the lower limit of measurement is in the range of 10^{-15} mol of message per gram of tissue. One ramification of this limit is that a relatively large amount of tissue is necessary to obtain the 60 μg of total RNA needed to analyze a single sample in triplicate.

We incorporated the use of cRNA standards for our more recently published quantitative Rt-PCR for iNOS.[15] We chose to use Rt-PCR for the measurement of iNOS because of the increased sensitivity of this method over Northern analysis. This method was developed specifically for mouse NOS2 because of the difficulty in obtaining significant amounts of tissue expressing this gene from mouse. However, the approach should be applicable for NOS2 from any species once appropriate primers and cRNA standards are identified. This technique involves using an *in vitro*-generated cRNA as an internal standard, which is present in both the reverse transcription and PCR reactions, as described originally by Wang *et al.*[16] The internal standard is a deletion construct of the wild-type gene, which differs in size from the wild-type product by no more than 10%. The essential aspect of this deletion standard is that the deletion is internal to the sequence so that the same primers are used for the standard and the tissue mRNA. The largest variable in Rt-PCR is primer annealing efficiency. Primer annealing efficiency varies widely from primer to primer, and with amplification the distortion expands geometrically. In our approach, a known amount of the cRNA internal standard is reverse transcribed and amplified simultaneously with the sample total RNA in the same tube. The term "internal standard" is therefore used for this approach, not because the standard is endogenous to the tissue sample, but because it is subjected to Rt-PCR in the same tube as the unknown. Because the primer binding sites are identical for the internal standard and the wild-type message, the amplification efficiencies of both templates are the same.

We have used cRNA internal standards in murine cell extracts for the measurement of NOS2 by noncompetitive Rt-PCR.[15] For these studies, we used phenol/chloroform/ether extraction to release nucleic acids from the cells and directly quantify iNOS mRNA from the extracts without further isolation of RNA. This approach both reduces the amount of work involved and eliminates the problem of variability in RNA recovery. In these experiments, we quantified NOS2

[15] B. Han, D. C. DuBois, K. M. K. Boje, S. J. Free, and R. R. Almon, *Nitric Oxide Biol. Chem.* **3,** 281 (1999).

[16] A. Wang, M. Doyle, and D. Mark, *Proc. Natl. Acad. Sci. U.S.A.* **86,** 9717 (1989).

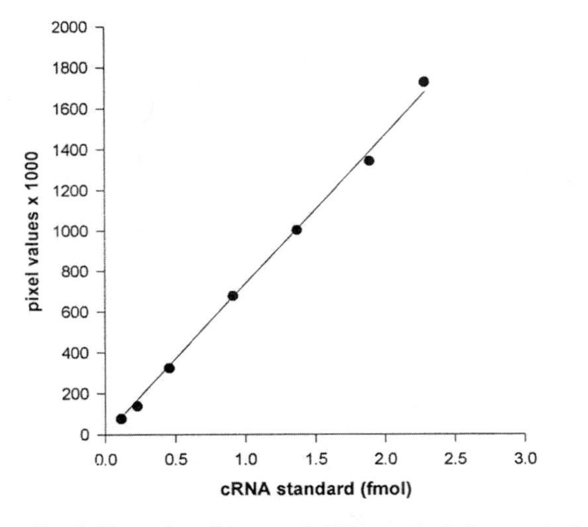

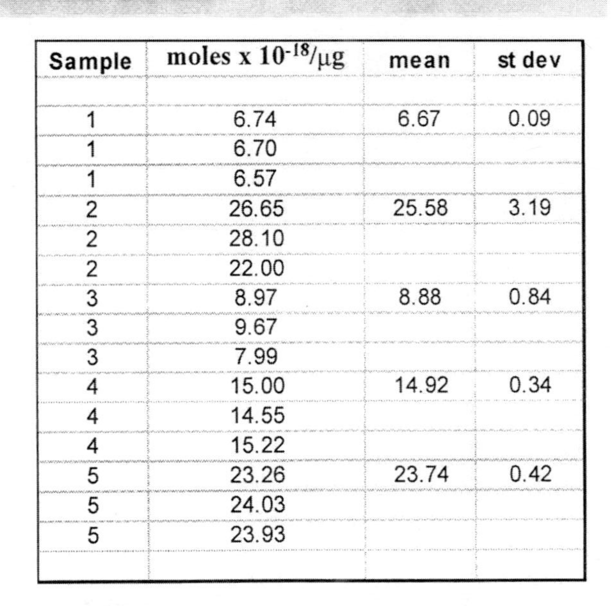

Sample	moles x 10^{-18}/μg	mean	st dev
1	6.74	6.67	0.09
1	6.70		
1	6.57		
2	26.65	25.58	3.19
2	28.10		
2	22.00		
3	8.97	8.88	0.84
3	9.67		
3	7.99		
4	15.00	14.92	0.34
4	14.55		
4	15.22		
5	23.26	23.74	0.42
5	24.03		
5	23.93		

FIG. 1. Illustration of the use of cRNA standards for quantitation in Northern hybridizations. (Top) A phosphorimage of a representative Northern blot. The first 15 lanes (left to right) include five different total RNA samples run in triplicate. The remaining 7 lanes represent a series of known concentrations of a cRNA standard. (Bottom left) Linear regression analysis of signal intensities of the standards (pixel values) versus concentration. The r^2 value is 0.999. (Right) Estimations of moles of mRNA per microgram total RNA in the five experimental samples.

concentration using cell number as a reference base. However, the use of internal standards can also be applied to Rt-PCR using isolated total RNA, assuming that an adequate reference base for message expression is established.

In our experiments, a known amount of cRNA standard is added to the cells at the beginning of extraction and is subjected to both reverse transcription and PCR reactions along with the cell RNA. The two PCR products are then size separated on high-resolution agarose gels (1.5% NuSeive Agarose, FMC Corp., in 1× TAE), and PCR products are visualized using Southern hybridization with a radiolabeled probe. Reactions are run at relatively low concentrations of template and for no more than 25 PCR cycles in order to maintain linearity in the PCR reaction and avoid heteroduplex formation. Because of these constraints, more sensitive Southern hybridization must be used for signal visualization in place of less work-intensive ethidium bromide staining. It should be pointed out that the sequence selected for probe synthesis must be present in both the expressed message and the deletion standard (i.e., should not encompass the deleted region) so that it will hybridize equally with both PCR products. The concentration of expressed mRNA in the experimental sample is calculated as:

$$C_s = A_i/A_s \times C_i$$

where C_s is moles of message in the experimental sample; C_i is moles of cRNA standard added to the assay; A_i is intensity of internal standard PCR product; and A_s is intensity of endogenous mRNA PCR product. Figure 2 presents an analysis of one such experiment.

Fluorescence-based real-time RT-PCR (kinetic RT-PCR) is a relatively new technique for the quantification of steady-state mRNA levels. The advantages of this technique include extremely high sensitivity (equivalent to conventional RT-PCR but much greater than Northern analysis), extremely high reproducibility (greater than either technique), and high throughput capability. The disadvantage of this technique is that it requires a relatively expensive spectrofluorimetric thermal cycler.

Kinetic RT-PCR is based on the ability of the instrumentation to detect a fluorescently labeled probe. Fluorescence values are recorded during every cycle and are proportional to the amount of product present at that cycle. Data are calculated as the Ct (threshold cycle), which represents the initial cycle where the fluorescent signal appears significantly above background and is directly proportional to the starting template. Because measurements are made during the exponential phase of amplification, they are not biased by reaction components becoming limited, as occurs with measurements made in the plateau phase. No post-PCR manipulations are required using this approach: all steps from RT to data output are fully automated, minimizing both experimental error and contamination, which both increases reproducibility and leads to very high throughput capability.

Like both Northern hybridizations and conventional Rt-PCR, kinetic Rt-PCR may be semiquantitative or quantitative. Absolute quantitation can be obtained

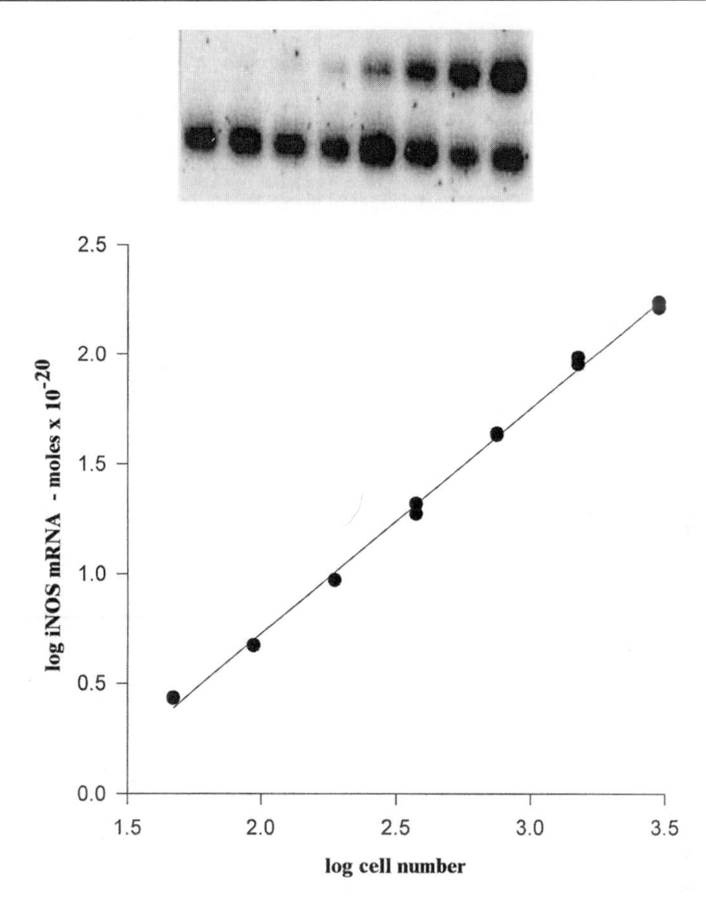

FIG. 2. Representative analysis of iNOS expression in lipopolysaccharide (LPS) + interferon (IFN)-γ-induced murine macrophage (J774.2) cells by Rt-PCR using cRNA standards. (Top) A phosphorimage analysis of a Southern hybridization of PCR signals. Lane 1 (left), control cells (6000 cells/tube); lanes 2–8, increasing amounts of induced cells (94–6000 cells/tube). (Bottom) An estimation of iNOS mRNA (moles/cell) as a function of cell number.

by including standard curves consisting of known amounts of *in vitro*-transcribed cRNA standards in an approach identical to that used in our quantitative Northern analysis. This approach has been used to examine differential expression of a variety of genes.[17] Although not yet applied to NOS2 expression, real-time Rt-PCR utilizing cRNA standards for molar quantitation is a promising tool for the study of NOS2 induction.

[17] S. Bustin, *J. Mol. Endocrinol.* **25,** 169 (2000).

Summary and Conclusions

Methodology for the analysis of expressed message has evolved rather rapidly since the early 1990s and will undoubtedly continue to evolve with new technology. cRNA standards are a useful tool, which allow for absolute rather than relative quantification of an expressed message. Selecting an appropriate sequence(s) for analysis has become even more important as techniques have evolved from Northern hybridization where possible multiple hybridization bands can be observed to real-time Rt-PCR approaches where only a signal is recorded. NOS2 presents a particular problem in this regard because of the high conservation of sequence with NOS1 and NOS3, as well as the existence of both multiple genes and splice variants, which may be of experimental importance.

[41] Inducible Nitric Oxide Synthase Knockout Mouse and Low-Density Lipoprotein Oxidation

By XI-LIN NIU, YI CHEN, YUTAKA SHOYAMA, KAZUO ISHIWATA, RURIKO OBAMA, and HIROE NAKAZAWA

Introduction

Oxidatively modified low-density lipoprotein (ox-LDL) is present in atherosclerotic lesions and appears to be fundamental in the early formation and progression of atherosclerotic lesions.[1] Nitric oxide (NO) has been implicated in various aspects of the atherogenic process and has been shown to possess both anti- and proatherogenic properties. Among the three isoforms of NO synthase (NOS), eNOS has been shown to protect the vessel wall from atherosclerosis. iNOS, which is usually not detectable in undiseased vascular tissue, has been shown to be present in the atherosclerotic lesions of both humans and animals.[2] However, the direct influence of iNOS on atherosclerosis was not examined until our recent study.[3] We showed that iNOS induction does not affect the size of atheroma, but rather decreases extracellular collagen content. Low collagen content in the atheroma plaque suggests that NO from iNOS may decrease plaque stability.[3] These results

[1] D. Steinberg, S. Parthasarathy, T. E. Carew, J. C. Khoo, and J. L. Witztum, *N. Engl. J. Med.* **320,** 915 (1989).

[2] L. D. Buttery, D. R. Springall, A. H. Chester, T. J. Evans, E. N. Standfield, D. V. Parums, M. H. Yacoub, and J. M. Polak, *Lab. Invest.* **75,** 77 (1996).

[3] X. L. Niu, X.Y. Yang, K. Hoshiai, K. Tanaka, S. Sawamura, Y. Koga, and H. Nakazawa, *Circulation* **103,** 1115 (2001).

0076-6879/02 $35.00

were obtained through experiments using iNOS knockout (iNOS$^{-/-}$) and wild-type (iNOS$^{+/+}$) mice fed a high cholesterol diet. Consistent with previous reports, our study demonstrated that macrophages are the major cells that express iNOS. Thus, after our study, the role of iNOS on *in vivo* atherosclerosis was evident, but its effect on macrophage-mediated LDL oxidation *in vitro* remains to be established.

Monocyte/macrophages are the most important cells that contribute to foam cell formation in atherosclerosis, and several studies show that macrophages promote LDL oxidation. However, several reports demonstrate antioxidant activities.[4] A plausible explanation for these opposing functions may be the marked dependence of *in vitro* LDL oxidation on individual experimental conditions. The degree of LDL oxidation is influenced by many cellular and extracellular factors. Particularly, it depends on the type of medium used for supporting LDL oxidation and the concentration of the metal, cysteine, tyrosine, or phenol used to supplement the medium. Furthermore, the different cell lines of macrophages have been shown to exhibit completely different abilities to promote or suppress LDL oxidation even in the same medium, such as in RPMI medium.[4] This article focuses on how to measure macrophage-mediated LDL oxidation using Ham's F-10 medium as a standard medium. It then shows how to elucidate the effect of iNOS on LDL oxidation.

Interferon (IFN)-γ is the most commonly used cytokine for inducing iNOS. However, when used to induce iNOS for observing the influence of NO on macrophage-mediated LDL oxidation, data must be interpreted with caution, as IFN-γ may have diametrical effects on LDL oxidation: a prooxidative one through priming macrophages to produce reactive oxygen intermediate and an antioxidative effect through NO generated from induced iNOS. To differentiate the individual roles in this complicated system, we compared LDL oxidation by macrophages from iNOS$^{-/-}$ and iNOS$^{+/+}$ mice.

Materials

Butylated hydroxytoluene (BHT), thiobarbituric acid (TBA), and Dulbecco's modified Eagle's medium (DMEM) are from Sigma Chemical Co. (St. Louis, MO). Ham's F-10 is from GIBCO/BRL. Recombinant murine IFN-γ is from Genzyme Company. N^{G}-Monomethyl-L-arginine (L-NMMA) is from Alexis Corporation.

Animals

Female C57Bl/6J mice and iNOS knockout mice with a cross background of 129 SvEv/C57 Bl/6J (8 to 12 weeks old) are from Clea Japan, Inc. and Merck & Co., Inc., respectively. These mice are maintained under pathogen-free conditions and are given standard mouse chow and water. All experiments comply with Tokai University guidelines for animal experiments.

[4] D. M. van Reyk, W. Jessup, and R. T. Dean, *Arterioscler. Thromb. Vasc. Biol.* **19,** 1119 (1999).

Isolation of Mouse Peritoneal Macrophages

Resident macrophages are obtained by peritoneal lavage with 5 ml ice-cold DMEM. The cells are cultured in DMEM medium supplemented with 100 U/ml penicillin, 100 μg/ml streptomycin, 10% heat-inactivated (56° for 30 min) fetal calf serum (FCS), and 5% CO_2 in air at 37°. After overnight culture, nonadherent cells are removed by washing three times with fresh DMEM medium. Macrophages cultured on 12-well plastic culture plates at 2×10^6 cells/well are used for the measurement of LDL oxidation.

Isolation of LDL

LDL is isolated from fresh human plasma (drawn from healthy normal-lipidemic donors and anticoagulated with 1 mg/ml of EDTA) by means of a single discontinuous density gradient ultracentrifugation procedure.[5] LDL is washed at $d = 1.063$ g/ml to remove traces of contaminating albumin. Following isolation, LDL is dialyzed against 10 m*M* phosphate-buffered saline (PBS) (pH 7.4) containing 0.3 m*M* EDTA and 100 U/ml penicillin G, filter sterilized, and stored at 4° until use. Before the experiments, LDL is applied to a Pharmacia Sephadex G-25M PD-10 column to remove EDTA and other low molecular weight contaminants and is eluted with 10 m*M* PBS (pH 7.4). The protein concentration of LDL solution is determined using bovine serum albumin as a standard protein.

Methods to Measure Macrophage-Mediated LDL Oxidation

Macrophages cultured in 12-well plates are washed three times with serum-free Ham's F-10 medium and incubated in 1 ml of serum-free Ham's F-10 medium containing 100 μg/ml LDL. After 24 hr of incubation, the medium is centrifuged immediately at 200*g* for 10 min to remove detached cells. The supernatant is collected and used for measurement of LDL oxidation after the addition of 20 μM BHT and 2 m*M* EDTA (final concentrations) to prevent any further oxidation. Oxidative modification of LDL is assayed by measuring the formation of thiobarbituric acid-reactive substances (TBARS)[6] and by the relative electrophoretic mobility (REM) of the modified LDL.[7]

The principle for the TBARS assay is based on TBA reacting with the LDL oxidation product, malondialdehyde (MDA), to produce TBA-reactive substances

[5] B. H. Chung, J. P. Segrest, M. J. Ray, J. D. Brunzell, J. E. Hokanson, R. M. Krauss, K. Beaudrie, and J. T. Cone, *Methods Enzymol.* **128,** 181 (1986).

[6] R. P. Patel and M. Darley-Usmar, *Methods Enzymol.* **269,** 375 (1996).

[7] X. L. Niu, K. Ichimori, X. Yang, Y. Hirota, K. Hoshiai, M. Li, and H. Nakazawa, *Free Radic. Res.* **33,** 305 (2000).

(TBARS). TBARS can be measured at 532 nm with reference to a standard MDA solution prepared by acid hydrolysis of 1,1,3,3-tetraethooxypropane. Although this is not a specific method for lipid oxidation, it is a quick and simple method used commonly for determing LDL oxidation. Briefly, 400 μl of medium containing 40 μl LDL protein is mixed with 1 ml 25% (w/v) trichloracetic acid and 1 ml of 0.67% TBA and heated to 100° for 45 min. Samples are centrifuged for 5 min, and absorbance is measured at 532 nm.

The oxidation of LDL is characterized by an increase in negative charge, which results in an increase in electrophoretic mobility on agarose gels. This increase can be expressed relative to the mobility of native LDL and is therefore termed the REM. Briefly, at the end of incubation, 50 μl of medium containing 5 μg of LDL is directly examined on a 1% agarose gel using barbital buffer (pH 8.6) at 90 V for 45 min. The gel is fixed in 5% acetic acid in 70% ethyl alcohol and stained with Sudan black B. The REM is calculated as the ratio of migration distance of modified LDL to that of native LDL.

iNOS$^{-/-}$ and iNOS$^{+/+}$ Macrophages Promote LDL Oxidation to the Same Extent in Ham's F-10 Medium

In experiments of macrophage-mediated LDL oxidation, researchers have used various types of cell culture medium and buffers, such as PBS, Hanks' balanced salt solution (HBSS), minimal essential medium (MEM), RPMI, and Ham's F-10 medium. Some researchers have even added transition metals, copper-chelated ceruloplasmin, or EDTA to the medium to obtain conditions suitable to support LDL oxidation. As experimental conditions have been known to determine the extent of macrophage-mediated LDL oxidation from enhancement to even inhibition, it is essential to use standardized conditions to evaluate the role of macrophages on LDL oxidation. For example, macrophages can inhibit LDL oxidation if HBSS or RPMI is used as the supporting medium without addition of metals, presumably by sequestering the metals in the medium and by removing the hydroxyperoxides.[4] Although the mechanism underlying cell-mediated LDL oxidation is not completely understood, it is commonly accepted that metal supplementation is definitely required. The so-called cell-mediated LDL oxidation, in fact, means that cells, including macrophages, are able to promote LDL oxidation induced by metals present in the medium. In our experiment, we used Ham's F-10 medium as the supporting medium in iNOS$^{-/-}$ macrophage-mediated LDL oxidation. This medium was already supplemented with 0.01 and 3 μM of Cu and Fe, respectively, by the manufacturer. We found that iNOS$^{-/-}$ macrophages were able to promote LDL oxidation in Ham's F-10 medium without supplementation of any additional metals. TBARS formation from LDL amounted to 17.3 nmol TBARS/mg protein after 24 hr incubation in the Ham's F-10 medium alone. In the presence of macrophages, it reached 31.5 nmol TBARS/mg protein for iNOS$^{-/-}$ macrophages and 30.2 nmol TBARS/mg protein for iNOS$^{+/+}$ macrophages, indicating that there is no significant difference in LDL-oxidizing ability between iNOS$^{+/+}$ and iNOS$^{-/-}$ macrophages in Ham's F-10 medium without supplementation of additional metals.

Macrophage-Mediated LDL Oxidation in iNOS$^{-/-}$ and iNOS$^{+/+}$ Mice Stimulated with IFN-γ

We and others have shown that iNOS was induced in atherosclerotic lesions, mainly in macrophages and T lymphocytes.[2,3] IFN-γ, one of the cytokines produced by T lymphocytes, was also shown to be colocalized with macrophages and lymphocytes in atherosclerotic lesions,[8] indicating that immune or autoimmune mechanisms are involved in the development of atherosclerosis. Regarding the effect of IFN-γ on LDL oxidation, the published results are controversial; IFN-γ was shown to prime macrophages[9] and neutrophils[10] to produce reactive oxygen intermediates, which are known initiators of LDL oxidation. However, IFN-γ was also shown to inhibit murine macrophage-mediated LDL oxidation by inducing iNOS expression and NO production in the cells.[11] Thus, a comparison of LDL oxidation mediated by iNOS$^{-/-}$ and iNOS$^{+/+}$ macrophages can distinguish the effect of NO from that of IFN-γ.

iNOS$^{+/+}$ and iNOS$^{-/-}$ macrophages cultured in 12-well plates are first preincubated with 200 units/ml IFN-γ in DMEM medium containing 10% heat-inactivated FCS for 24 hr. Control cells are prepared by incubation in the same medium containing only 10% FCS. Cell-free control incubations are set up in parallel. After the 24-hr incubation, media are collected and centrifuged at 200g for 10 min to remove detached cells. Nitrite in the supernatant is measured to check NO production using the Griess reagent as previously described.[12] The ability of these pretreated cells to oxidize LDL is examined during the subsequent 24 hr as described in the method section using Ham's F-10 medium as the supporting medium. The experiment is also performed in the presence or absence of 100 μM L-NMMA in the wells. Results show that pretreatment with IFN-γ for 24 hr induced iNOS expression in iNOS$^{+/+}$ macrophages, but not in iNOS$^{-/-}$ macrophages, as evidenced by immunostaining (data not shown). Nitrite production in untreated control cells is very low for both iNOS$^{-/-}$ and iNOS$^{+/+}$ macrophages

[8] G. K. Hansson, J. Holm, and L. Jonasson, *Am. J. Pathol.* **135,** 169 (1989).
[9] H. W. Murray, G. L. Spitalny, and C. F. Nathan, *J. Immunol.* **134,** 1619 (1985).
[10] S. D. Tennenberg, D. E. Fey, and M. J. Lieser, *J. Leukocyte Biol.* **53,** 301 (1993).
[11] W. Jessup, *Curr. Opin. Lipidol.* **7,** 274 (1996).
[12] B. B. Aggarwal and K. Mehta, *Methods Enzymol.* **269,** 166 (1996).

TABLE I
EFFECTS OF IFN-γ ON MACROPHAGE NITRITE PRODUCTION[a]

Group	Treatment	Nitrite production (nmol/well/24 hr)
iNOS$^{+/+}$ macrophages		
Control	None	0.07 ± 0.05
IFN-γ treated	None	3.81 ± 0.13[b]
	L-NMMA	1.34 ± 0.09[c]
INOS$^{-/-}$ macrophages		
Control	None	0.08 ± 0.06
IFN-γ treated	None	0.07 ± 0.08
	L-NMMA	0.08 ± 0.08

[a] Macrophages were preincubated in DMEM containing 10% FCS in the presence or absence of IFN-γ for 24 hr. The cells were then incubated in 1 ml of serum-free Ham's F-10 medium containing 100 μg/ml LDL with or without L-NMMA for 24 hr. Nitrite production was determined. Data represent mean $\pm$ SD ($n = 8$).

[b] $P < 0.01$ compared to control macrophages.

[c] $P < 0.01$ compared to INF-γ-treated macrophages. Reprinted with permission from X. L. Niu, *Nitric Oxide* **4,** 363 (2000).

(less than 0.1 nmol/well/24 hr). After a 24-hr treatment with IFN-γ, nitrite production by iNOS$^{+/+}$ macrophages is increased to 3.17 ± 0.29. Nitrite production remains below 0.1 nmol/well/24 hr in iNOS$^{-/-}$ macrophages after treatment with IFN-γ (data not shown). Table I shows nitrite production during the LDL oxidation experiment in the subsequent 24 hr. IFN-γ treated iNOS$^{+/+}$ macrophages release NO continuously in the subsequent 24 hr after washout of IFN-γ. In the presence of the NOS inhibitor, L-NMMA (100 μM), nitrite production is inhibited by 64.5% for IFN-γ-treated iNOS$^{+/+}$ macrophages. L-NMMA has no effect on nitrite production in iNOS$^{-/-}$ macrophages. As shown in Fig. 1, in the presence of IFN-γ, TBARS formation from LDL is suppressed in iNOS$^{+/+}$ macrophages, whereas it is increased in iNOS$^{-/-}$ macrophages. The suppressive effect of IFN-γ in iNOS$^{+/+}$ macrophages is abolished in the presence of L-NMMA, and TBARS formation is even increased to a level above that of untreated iNOS$^{+/+}$ macrophages. The REM of the modified LDL shows a similar pattern of change to that seen in TBARS formation in response to IFN-γ. IFN-γ decreases REM in iNOS$^{+/+}$ macrophages and increases REM in iNOS$^{-/-}$ macrophages by 14.7 and 31.1%, respectively. In the presence of L-NMMA, the IFN-γ induced decrease of REM in iNOS$^{+/+}$ macrophages is abolished and REM is even increased by 16.8%, but the IFN-γ-induced increase of REM in iNOS$^{-/-}$ macrophages is not affected. L-NMMA has no effect on LDL oxidation in cell-free cultures (14.91 ± 2.58 without L-NMMA vs 16.23 ± 1.91 nmol TBARS/mg protein with L-NMMA; $p > 0.05$, $n = 8$).

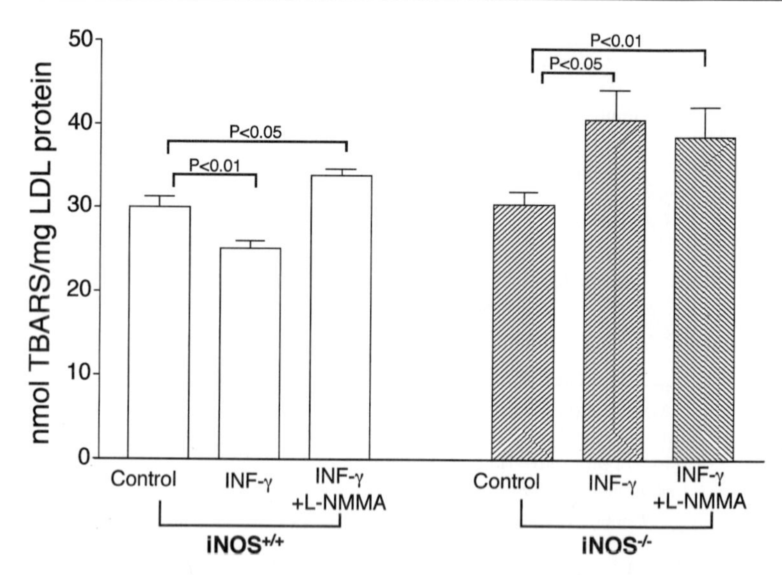

FIG. 1. Effect of IFN-γ on TBARS formation. Macrophages were preincubated in DMEM containing 10% FCS in the presence or absence of IFN-γ for 24 hr. Cells were then incubated in 1 ml of serum-free Ham's F-10 medium containing 100 μg/ml LDL with or without L-NMMA for 24 hr. Data represent mean $\pm$ SD ($n = 8$). Reprinted with permission from X. L. Niu, *Nitric Oxide* **4,** 363 (2000).

Summary

We have presented experimental procedures that examine macrophage-mediated LDL oxidation using Ham's F-10 medium. By comparing iNOS$^{-/-}$ and iNOS$^{+/+}$ macrophages, an antioxidant effect for NO and a prooxidant effect for IFN-γ were demonstrated. The methods outlined here should allow for the investigation on the mechanism of *in vitro* LDL oxidation and how the macrophage-mediated LDL oxidation process is affected by various factors, one of which was the effect of iNOS induction by IFN-γ.

[42] Nitric Oxide and Resolution of Inflammation

By SONSOLES HORTELANO, MIRIAM ZEINI, and LISARDO BOSCÁ

Introduction

Innate immunity plays an important role in host defense by providing reactive and signaling molecules that account for the rapid reaction of the immune system against pathogens and for the initiation of the adaptive immunity.[1–3] Therefore, the innate immune response is intended to control the early stages of infection. Activation of cells involved in innate immunity encompasses a previously unexpected repertoire of receptors, Toll-like receptors (TLRs), that allow distinguishing between different molecules present in pathogenic virus, bacteria, and fungi.[4] In the research into TLRs, which are key mediators in the activation of macrophages and dendritic cells, several molecules have been identified, allowing the recognition of different motifs in the pathogens.[3,5,6] One biologically relevant feature of TLRs is the capacity for a reduced number of receptors to recognize a large number of chemically unrelated motifs present in pathogens and ranging from bacterial DNA to flagellin, a component of bacterial flagella and recognized as a virulence factor in many organisms including flies, plants, and mammals.[7,8] In this vein, activation of the host immune system by gram-negative bacteria can be reproduced *in vitro* by the incubation of monocyte/macrophages with lipopolysaccharide (LPS) and proinflammatory cytokines. In turn, macrophages participate actively in the onset of inflammation and immune system activation by releasing cytokines [tumor necrosis factor (TNF)-α and interleukin (IL)-1β] that amplify the initial inflammatory stimulation, bioactive lipids [e.g. prostaglandins (PGs) and leukotrienes], and reactive oxygen (ROI) and nitrogen intermediates (RNI) that exert cytotoxic effects against pathogens and tumor cells.[9–11] As result of this activation sequence, macrophages express enzymes involved in inflammation,

[1] R. Medzhitov and C. J. Janeway, *N. Engl. J. Med.* **343,** 338 (2000).
[2] R. Medzhitov and C. J. Janeway, *Immunol. Rev.* **173,** 89 (2000).
[3] J. A. Hoffmann, F. C. Kafatos, C. A. Janeway, and R. A. Ezekowitz, *Science* **284,** 1313 (1999).
[4] E. B. Kopp and R. Medzhitov, *Curr. Opin. Immunol.* **11,** 13 (1999).
[5] A. Aderem and R. J. Ulevitch, *Nature* **406,** 782 (2000).
[6] H. D. Brightbill and R. L. Modlin, *Immunology* **101,** 1 (2000).
[7] F. Hayashi, K. D. Smith, A. Ozinsky, T. R. Hawn, E. C. Yi, D. R. Goodlett, J. K. Eng, S. Akira, D. M. Underhill, and A. Aderem, *Nature* **410,** 1099 (2001).
[8] L. Gomez-Gomez and T. Boller, *Mol. Cell* **5,** 1003 (2000).
[9] J. E. Albina and J. S. Reichner, *Cancer Metast. Rev.* **17,** 39 (1998).
[10] L. Boscá and S. Hortelano, *Cell Signal.* **11,** 239 (1999).
[11] J. MacMicking, Q. W. Xie, and C. Nathan, *Annu. Rev. Immunol.* **15,** 323 (1997).

0076-6879/02 $35.00

such as NOS-2 and COX-2 [enzymes responsible for the high-output synthesis of nitric oxide (NO) and PGs, respectively] and matrix metalloproteinases that allow tissue remodeling.[12]

Resolution of inflammation is accomplished by the presence of anti-inflammatory cytokines (e.g., IL-4, IL-10, and IL-13) and by the negative regulation of the activation process exerted by some of the effector molecules released by activated macrophages, particularly ROI, RNI, and cyclopentenone prostaglandins.[12] At the end of the inflammatory response, cells that have participated in the process are removed by apoptosis.[9,12,13] Induction of apoptosis in activated macrophages has been recognized as a physiological and altruistic mechanism that helps reduce inflammatory stress and avoid the establishment of persistent infection by intracellular pathogens.[14,15] The contribution of NO to trigger apoptosis in macrophages has been well established.[10,12] The high-output synthesis of NO due to the expression of NOS-2 releases mitochondrial mediators that initiate caspase activation, leading to characteristic programmed cell death.[16–18] Moreover, under certain conditions, the coordinate release of ROI and NO at the end of the inflammatory process leads to the formation of peroxynitrite, a more potent and efficient inducer of apoptosis than NO.[12]

Apoptosis occurs through a series of sequential biochemical changes that define a precise phenotype.[19] NO-dependent induction of apoptosis in macrophages involves changes in the mitochondrial inner membrane potential ($\Delta\Psi_m$) compatible with the generation of mitochondrial-dependent apoptosis mediators.[12,16,17] NO promotes a rapid increase of $\Delta\Psi_m$ in macrophages, followed by an apoptotic volume decrease (AVD), which has been recognized as an early event occurring in almost all apoptotic cells.[20] This is accompanied by release of cytochrome *c*, activation of caspases, and DNA fragmentation (Fig. 1).

The protocol that follows is intended for evaluating the sequential events that contribute to apoptosis in peritoneal macrophages and macrophage cell lines using flow cytometry techniques.

[12] S. Hortelano, A. Castrillo, A. M. Alvarez, and L. Boscá, *J. Immunol.* **165,** 6525 (2000).
[13] J. Savill, *J. Leukocyte Biol.* **61,** 375 (1997).
[14] C. B. Thompson, *Science* **267,** 1456 (1995).
[15] G. P. Anderson, *Trends. Pharmacol. Sci.* **17,** 438 (1996).
[16] S. Hortelano, B. Dallaporta, N. Zamzami, T. Hirsch, S. A. Susin, I. Marzo, L. Boscá, and G. Kroemer, *FEBS Lett.* **410,** 373 (1997).
[17] S. Hortelano, A. M. Alvarez, and L. Boscá, *FASEB J.* **13,** 2311 (1999).
[18] G. Kroemer, N. Zamzami, and S. A. Susin, *Immunol. Today* **18,** 44 (1997).
[19] G. Kroemer, L. Boscá, N. Zamzami, S. Hortelano, and A. C. Martinez, "The Immunological Methods Manual" (R. Lefkovits, ed.), Vol. 14.2, p. 1111. Academic Press, 1997.
[20] E. Maeno, Y. Ishizaki, T. Kanaseki, A. Hazama, and Y. Okada, *Proc. Natl. Acad. Sci. U.S.A.* **97,** 9487 (2000).

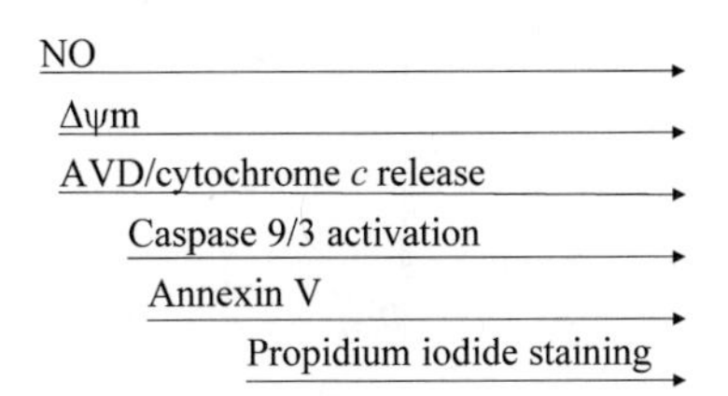

FIG. 1. Time line of apoptotic events. A conventional sequence of changes in cell parameters involved in the development of apoptotic cell death.

Materials, Methods, and Assays

Isolation of Elicited Peritoneal Macrophages and Culture Conditions of Macrophage Cell Lines

Four days prior to use, 10-week-old Balb/c mice maintained free of pathogens are injected intraperitoneally with 1 ml of sterile thioglycollate broth.[21] Resident peritoneal macrophages are obtained as follows: light ether-anesthetized elicited mice are killed by cervical dislocation and injected intraperitoneally with 5 ml of sterile RPMI 1640 medium at 37°. After 10 min of gentle distribution of the medium in the peritoneal cavity, the ascitic fluid is aspirated carefully, avoiding hemorrhage. The peritoneal cell suspension (use at least four animals) is kept at 4° to prevent adhesion of the macrophages to the plastic. After centrifugation for 10 min at 200*g* and 4°, the cell pellet is resuspended once with 45 ml of ice-cold phosphate-buffered saline (PBS) and centrifuged again. Cells are counted and seeded at $1 \times 10^6/cm^2$ in RPMI 1640 medium supplemented with 10% of heat-inactivated fetal calf serum (FCS) and antibiotics (50 μg/ml of penicillin, streptomycin, and gentamicin). After incubation for 1 hr at 37° in a 5% CO_2 atmosphere, nonadherent cells are removed by extensive washing with PBS at 37°. Cells can be activated 24 hr after isolation and maintained with RPMI 1640 medium containing 0.5 m*M* arginine and 10% of heat-inactivated FCS.

Murine macrophage cell lines (e.g., RAW 264.7, ATCC#TIB-71; or J774, ATCC#HB-197) are cultured at a density of $6-8 \times 10^4$ cells/cm^2 in RPMI 1640 medium supplemented with 2 m*M* glutamine and 10% of FCS. After 2 days in culture to reach near confluence, the cell layers are washed with PBS, and the culture medium is replaced by fresh culture medium as described for macrophages.

Assessment of Apoptosis in Macrophages

Induction of apoptosis in LPS and interferon (IFN)-γ-activated cells (200 ng/ml of *Escherichia coli* LPS and 10 units/ml of recombinant mIFN-γ, respectively) is

[21] F. Terenzi, M. J. Diaz-Guerra, M. Casado, S. Hortelano, S. Leoni, and L. Bosca, *J. Biol. Chem.* **270,** 6017 (1995).

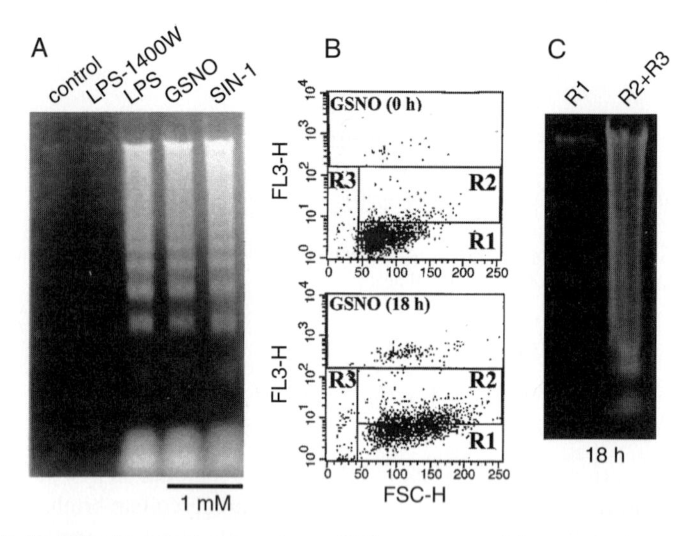

FIG. 2. Analysis of apoptotic macrophages. DNA was extracted from macrophages treated for 18 hr with 200 ng/ml of LPS or the indicated concentration of NO donors and analyzed in agarose gels (A). Alternatively, the cells were labeled with PI and analyzed in a flow cytometer. Nonapoptotic cells were identified in the R1 region, whereas apoptotic cells were recovered in the R2 and R3 regions (B). To confirm the nature of viable and apoptotic cells, the macrophages were sorted, equal amounts of cells (4×10^5) were homogenized, and the degraded DNA present in the cytosol was analyzed in an agarose gel [S. Hortelano, A. López-Collazo, and L. Boscá, *Br. J. Pharmacol.* **126,** 1139 (1999)] (C). 1400W was used at 100 μM.

very dependent on the synthesis of NO,[12] as selective NOS-2 inhibitors, such as *N*-(3-aminomethylbenzyl)acetamidine (1400W), abolish this effect. Alternatively, apoptosis can be induced after treatment of macrophages with NO donors that release NO in the range of the concentrations that prevail in activated cells (e.g., 5–10 m*M* of DETA-NO) (Fig. 2A). To evaluate the extent of apoptosis by flow cytometry, cells are resuspended in ice-cold PBS by pipetting the medium onto the cell layer. The cell viability under these conditions is higher than 90% (Trypan Blue exclusion criteria). If attachment of the macrophages to the plastic is strong, treatment of the cell layer with a minimal amount of trypsin can be employed prior to the ice-cold PBS treatment. Alternatively, a selection of plastic brands should be performed to ensure a secure resuspension of the cells (Costar, Cambridge, MA, and TPP, Trasadingen, Switzerland, plastics are suitable for these purposes). Propidium iodide (PI) staining is performed after incubation of the cells for 5 min with 0.005% PI in the dark.[12,22] Cells are carefully resuspended and analyzed in a FACScan cytometer (Becton Dickinson, San Jose, CA) equipped with a 25-mW

[22] S. Hortelano, A. López-Collazo, and L. Boscá, *Br. J. Pharmacol.* **126,** 1139 (1999).

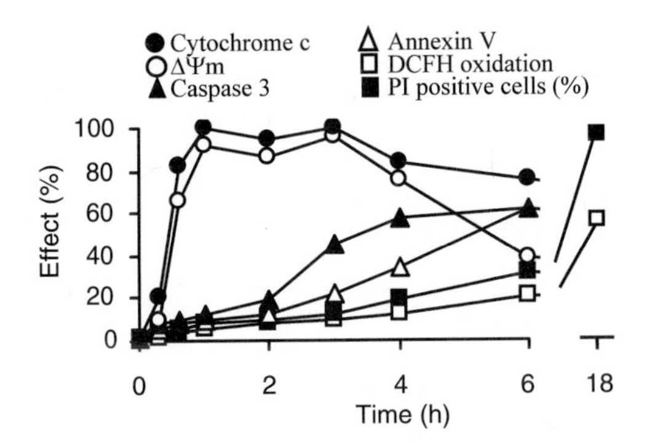

FIG. 3. Time course of changes in apoptotic parameters using flow cytometry techniques. Macrophages were challenged with 1 mM GSNO and labeled with CMXRos to evaluate the mitochondrial membrane potential ($\Delta\Psi$m); Oregon green-labeled antidenaturated cytochrome *c* mAb[17] to detect the presence of cytochrome *c* in the cytosol; PhyPhyLux to measure caspase 3 activity; annexin V-FITC for the exposure of phosphatidylserine in the plasma membrane; DCFH (10 min of incubation at each point) to measure the synthesis of ROI (mainly peroxynitrite); and PI to detect the percentage of apoptotic cells.

argon laser. Quantification of the percentage of apoptotic cells is achieved using a dot plot of the forward scatter against propidium iodide fluorescence. Apoptosis is expressed as a percentage of cells in the R2+R3 regions of the dot plot distribution (Fig. 2B). To confirm the apoptotic nature of the cells gated in these regions, cell sorting was performed and extranuclear DNA was analyzed by electrophoresis in agarose gels[22] (Fig. 2C).

Flow Cytometric Analysis of $\Delta\Psi_m$

Cells are incubated for 15 min at 37° in the presence of any of the following potential-sensitive probes: chloromethyl X-rosamine (CMXRos), 3,3′-dihexyloxacarbocyanine iodide [$DiOC_6(3)$], rhodamine 123 (Rh 123), or tetrachlorotetraethylbenzimidazolylcarbocyanine iodide (JC-1), all at 40 nM,[17] followed by activation with proinflammatory cytokines and other stimuli and analysis in a FACScan flow cytometer (Fig. 3). Fluorescence intensity in the presence of 10 μM of the uncoupling agent *m*-chlorophenylhydrazone carbonylcyanide (*m*-ClCCP) can be considered as the maximal loss of fluorochrome and allows comparison among different experimental conditions. An additional control by confocal microscopy is recomended to ensure that the changes in fluorescence occur mainly in the mitochondria.[17]

Measurement of Cell Volume

Analysis of cell volume is performed by flow cytometry based on the reference to the size of latex beads of known volume.[20] The cell size may also be evaluated

TABLE I
FLOW CYTOMETRY MEASUREMENTS

Probe	λ	Parameter
PI	$\lambda_{ex} = 488$ nm; $\lambda_{em} = 575$–630 nm	Nuclear staining
CMXRos	$\lambda_{ex} = 488$ nm; $\lambda_{em} = 590$–610 nm	$\Delta\Psi m$
$DiOC_6(3)$	$\lambda_{ex} = 484$ nm; $\lambda_{em} = 510$–520 nm	$\Delta\Psi m$
Rh123	$\lambda_{ex} = 488$ nm; $\lambda_{em} = 529$ nm	$\Delta\Psi m$
JC-1	$\lambda_{ex} = 488$ nm; $\lambda_{em} = 529$ nm	$\Delta\Psi m$
PhiPhiLux	$\lambda_{ex} = 488$ nm; $\lambda_{em} = 529$–605 nm	$\Delta\Psi m$
DCFH	$\lambda_{ex} = 488$ nm; $\lambda_{em} = 515$–525 nm	Caspase 3 activity
HE	$\lambda_{ex} = 488$ nm; $\lambda_{em} = 530$ nm	$ONOO^-$, H_2O_2
	$\lambda_{ex} = 488$ nm; $\lambda_{em} = 585$–620 nm	O_2^-

electronically by confocal microscopy in an MRC 1024 microscope (Bio-Rad) using commercial software.

Measurement of Caspase Activation

The activation of caspase 3 can be measured in intact stimulated cells after labeling with PhiPhiLux-G1D2 (a derivative of the GDEVDGI peptide that yields a fluorescent moiety after caspase cleavage; OncoImmunin, Gaithersburg, MD). In addition to this noninvasive technique, DEVDase (corresponding mainly to caspases 3 and 7) and caspase 9 activities can be determined in cell lysates using *N*-acetyl-DEVD- and *N*-acetyl-LEHD-labeled peptides (e.g., with coumarin or fluorescein from Calbiochem or Becton Dickinson) that yield a fluorogenic peptide after specific caspase cleavage. Under these conditions, the caspase assay is linear over a 30-min reaction period.

Measurement of ROI and RNI Synthesis

Activated cells are incubated for short periods of time (10–15 min) with 10 μM of 2′,7′-dichlorofluorescein diacetate (DCFH) to measure the synthesis of peroxynitrite and, to a minor extent, H_2O_2. Superoxide anion synthesis can be detected after loading the cells with 10 μM hydroethidine (HE), and fluorescence corresponding to the oxidized probe is followed by analysis in a flow cytometer.[12,23,24] Simultaneous incubation of the cells with the probes and with exogenously added peroxynitrite, H_2O_2, and superoxide should be performed to evaluate the sensitivity of these methods in the particular conditions of the cells used. Treatment of cells

[23] H. Ischiropoulos, A. Gow, S. R. Thom, N. W. Kooy, J. A. Royall, and J. P. Crow, *Methods Enzymol.* **301,** 367 (1999).

[24] G. Kroemer, B. Dallaporta, and M. Resche-Rigon, *Annu. Rev. Physiol.* **60,** 619 (1998).

with 50 μM *t*-butyl hydroperoxide (t-BH) may be used as a positive control of ROI release.

Multiple Measurement of Apoptotic Parameters

Taking advantage of the availability of different detection channels on most of the available FACS, a series of simultaneous measurements of apoptotic parameters can be accomplished. A list of the compatibility is shown in Table I.

Acknowledgments

This work was supported by Grants PM98-0120 and 2FD97-1432 from Comisión Interministerial de Ciencia y Tecnología (Spain).

Author Index

Numbers in parentheses are footnote reference numbers and indicate that an author's work is referred to although the name is not cited in the text.

A

B

C

D

E

F

G

H

K

L

M

N

O

P

Q

R

S

T

U

V

X

Y

Z

Subject Index

A

B

C

D

E

N

O

P

R

S

T

ISBN 0-12-182262-1
90038
9 780121 822620